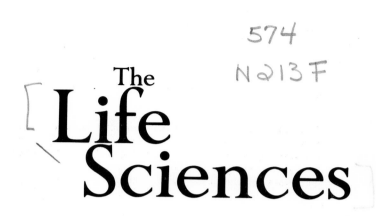

The Life Sciences

Recent Progress and Application to Human Affairs
The World of Biological Research
Requirements for the Future

Committee on Research in the Life Sciences
of the Committee on Science and Public Policy
National Academy of Sciences

NATIONAL ACADEMY OF SCIENCES Washington, D.C. 1970

ISBN 0-309-01770-X

Available from
Printing and Publishing Office
National Academy of Sciences
2101 Constitution Avenue
Washington, D.C. 20418

Library of Congress Catalog Card Number 71-606918

Printed in the United States of America

September 9, 1970

Dear Dr. Handler:

I take special pleasure in transmitting this report of the Committee on Research in the Life Sciences, since that committee undertook the study reported here under your chairmanship, prior to your election as President of the Academy. On behalf of the Committee on Science and Public Policy, I should like to express our profound appreciation for the prodigious work invested by you and your committee and its panels in the preparation of this report, as well as in the preparation of the book *Biology and the Future of Man,* recently published by the Oxford University Press.

The Committee on Science and Public Policy recognizes that many of the problems of support now facing the life sciences and described in the following report are common to all the natural and social sciences at the present time. In large part, our present difficulties stem from the fact that in a period of increasingly tight budgets, the research activities of each of the mission-oriented agencies are being restricted by their increasingly severe operational responsibilities. Thus, in the health field, as the report so graphically indicates, the $2 billion being spent on research is gradually being displaced by the $60 billion being spent on health care, even though in fact the savings in research expenditures are insufficient to make significant improvement in the delivery or quality of health care, and will probably result in increased costs of health care and unwise investments in the future. Savings in basic research are not resulting in any real transfer of resources to applications or to health care, but on the contrary are resulting in idle resources and unused highly trained talent.

The report is a lively and fascinating description of the accomplishments and future potential of the life sciences. It documents vividly the degree to which past progress and future developments in the control and prevention of disease are dependent on basic knowledge of life processes. There are few other areas of science in which the link between basic science and applications is closer. The committee is to be congratulated also on the excellent statistical characterization of the life science research enterprise. This pioneering effort will be a fruitful data base for future policy studies not only in the life sciences but also, by extrapolation, in other areas of science as well. The report is particularly valuable in documenting the interdependence of the various parts of the life sciences, and the strong links between the life sciences and the physical sciences, particularly in the use of physical instrumentation and in the pervasiveness of biochemical concepts and techniques throughout all areas of research in the life sciences, even at the levels of greater complexity such as population biology and ecology.

This report is commended to all those concerned with the future of American science, education, medicine, agriculture, and indeed of the biosphere.

Sincerely yours,

HARVEY BROOKS, *Chairman*
Committee on Science and Public Policy

COMMITTEE ON RESEARCH IN THE LIFE SCIENCES

PREFACE

In 1963, the Committee on Science and Public Policy of the National Academy of Sciences embarked upon a series of surveys of the status of various scientific disciplines. Each survey has attempted to summarize the most recent accomplishments of the discipline at its frontiers, the extent to which the findings of the disciplines have been translated into societal benefit in recent times, the nature and magnitude of research endeavors, and the requirements to assure that future research efforts would be vigorous and commensurate with perceived national needs. To date, reports published in this series have summarized the findings of surveys in ground-based astronomy, solid earth geophysics, chemistry, physics, mathematics, and the social and behavioral sciences.

For several years an equivalent study of the biological sciences was deferred. Whereas the physical sciences are usefully divided along conventional lines, no equivalently justifiable division of the life sciences seemed rational, and the entirety of the life sciences appeared to be so broad as to escape the grasp of any survey committee. Nonetheless, in 1966 the need for such a study seemed so compelling that the Committee on Research in the Life Sciences, which is responsible for the present report, was appointed by the Committee on Science and Public Policy to undertake the task.

The work of the Committee was soon organized into two major, essentially independent efforts. In attempting to appraise the "state of the art," the Committee agreed that the classical subdisciplines of biology are no longer sufficiently instructive or suitable as approaches to current understanding and appreciation of the phenomena of life. Thus, instead of organizing the study according to conventional categories such as zoology, botany, and microbiology, the Committee appointed panels charged, respectively, with review of molecular biology, biochemistry, cellular biology, the biology of development, the functions of tissues and organs, the structure of living forms, the nervous system, the biology of behavior, ecology, heredity and evolution, the diversity of life, and the origin of life. This classification of approaches to understanding of the living world will be apparent in Chapter One of this report and in subsequent analyses of the nature and magnitude of the research effort in biology. Moreover, the Survey Committee believes that this organization of biological understanding is appropriate to the organization of formal biological instruction.

A separate panel was asked to address itself to the utilization of the digital computer in the life sciences, because of its growing and unique role. Another panel was concerned with education in the life sciences, both for future investigators and teachers and for citizens generally. The results of these studies will be found in Chapters Seven and Six, respectively. An additional set of panels collaborated in evaluation of the contributions of biological understanding to agricultural practice, to medical practice, to management of renewable resources, to industrial technology, and to the problems of environmental health. These deliberations are summarized in Chapter Two.

To place these matters in perspective, an independent panel was asked to address itself to "Biology and the Future of Man." The results of these efforts, edited by the undersigned, were gathered in a single volume entitled *Biology and the Future of Man,* published by the Oxford University Press in May 1970. Chapters One and Two of this report represent an abbreviated digest of that volume; Chapter Nine, entitled "Biology and the Future of Man," is reproduced in its entirety from that work.

In the second phase of our study, a pair of questionnaires, designed by the executive board of the Survey Committee, was distributed to 25,964 life scientists, of whom 23,967 qualified as "investigators" by our criteria, and to 2,277 individuals who had been identified as chairmen of academic departments in the life sciences in American universities. These questionnaires will be found in Appendixes A and B, as will an analysis of the validity of the biologically concerned universe represented in the responses. It is the belief of the Survey Committee that the results obtained by this questionnaire procedure are adequate to describe quantitatively the research and education effort in the fundamental biological sciences, including those normal to the "preclinical" medical curriculum as these existed in 1967–1968.

Regrettably, the returns of questionnaires both from individual clinical investigators and from the chairmen of clinical departments represented much smaller fractions of those two communities than did the returns from investigators in the fundamental biological sciences or the related department chairmen. There is no reason to consider that that return reflected any specific bias, and it undoubtedly constitutes an adequate sample, but it rests on a smaller sampling of the total population so engaged (40 percent) than does that in the fundamental sciences (64 percent).

The data encompassed in these questionnaire returns were transferred to magnetic tape and analyzed by appropriate computer programs. With some refinements, these analyses are summarized in Chapters Three, Four, and Five, undoubtedly the most comprehensive description of the world of biological sciences yet available.

Withal, it must be recognized that our questionnaires and their responses reflect the situation in the last year of the post-World War II growth of federal support of research and research training. Had the Survey Committee had a vision of the subsequent abrupt alteration in the rate of federal funding, the questionnaires would surely have been designed somewhat differently. In any case, the data presented totally fail to reflect the impact of subsequent changes in federal philosophy and consequently in funding of the research and education effort generally or in the life sciences in particular. However, the impact of these changes was well known to our Survey Committee and our panels as this report was in preparation, particularly the chapter entitled "Conclusions and Recommendations." It was in the light of this collective experience as well as the understanding gained from analysis of the questionnaire returns that these conclusions and recommendations were constructed, although we lacked an adequate, comprehensive data base with which to support some of our recommendations, which must, necessarily, rest on largely anecdotal evidence and personal experience.

As the report will reveal, the task of its preparation was formidable. Such success as we may have encountered we owe to the generous support of the National Institutes of Health, the National Science Foundation, and the Smithsonian Institution, which underwrote our major expenses and whose staffs were cooperative in all regards; to our panelists and Committee, who gave of their time and effort without stint; to the Committee on Science and Public Policy, our sponsors, reviewers, and constructive critics; to numerous biological scientists who undertook specific writing assignments; to Milton Levine and Herbert H. Rosenberg of the National Science Foundation and the National Institutes of Health, respectively, and Roland

Bonato of the George Washington University for their generous assistance in the design and analysis of our questionnaires; to U. H. Leach, Jr., Herbert Soldz, Seymour Jablon, and A. Hiram Simon of the Academy staff for invaluable assistance with data processing and preliminary analyses, which saved us from disaster; to the publications staff of the Academy; to Robert Green, Executive Officer of the Committee on Science and Public Policy, and his secretary, Mary Van Demark, for assistance in many ways large and small; to the crew of young men and women who transferred the questionnaire data into forms suitable for transfer to the computer tapes; to Donna Teplitz, Brenda Hendon, and particularly to Gail Clark, who patiently managed our office and faultlessly prepared manuscripts and tables, and to Marilyn Swann and Saundra Greene, who aided me in all ways; and finally to Herbert Pahl, executive director of this study in its early phases, and Laura Greene, who succeeded him and managed the entire questionnaire effort, meticulously supervising each detail thereof as well as the final preparation of the tables and figures in this report and all aspects of the publication process. All warrant our gratitude and deep appreciation.

We conclude much as we began. Four years ago we *believed* public support of the research endeavor in the life sciences to be among the greatest investment bargains available to the American people. Today we *know* that to be true. Accordingly, we are pleased to offer this report to all those concerned: to responsible administrators of the executive branch of the federal government, to the Congress, to our colleagues in science, to academic administrators, to foundation executives, to students, and to practitioners of fundamental or applied life sciences and all their counterparts outside our own national borders.

PHILIP HANDLER, *Chairman*
Committee on Research in the Life Sciences

Woods Hole, Massachusetts
August 1970

CONTENTS

MAJOR
CONCLUSIONS
AND
RECOMMENDATIONS

PROLOGUE

For several centuries, research in the life sciences has constituted one of the great human adventures. While developing an independent style and value system, biologists have utilized the growing understanding of the physical universe to illuminate man's dim past, establish his kinship with all living creatures, and enable comprehension of the nature of life itself. This knowledge and understanding underlie some of the great advances that characterize our civilization: prolific agricultural productivity, a longer span of enjoyable and productive human life, and the potential to ensure the quality of the environment. A brief glimpse of this great saga is presented in subsequent chapters.

We are confident that research in the life sciences can surely contribute tomorrow at least as much to human welfare as it has in the past. The living scene continues to present numerous fascinating and perplexing mysteries. If, indeed, "the proper study of man is man," what remains to be investigated certainly exceeds in scope, in experimental difficulty, and in potential benefit to society all that has been learned throughout recorded history.

1

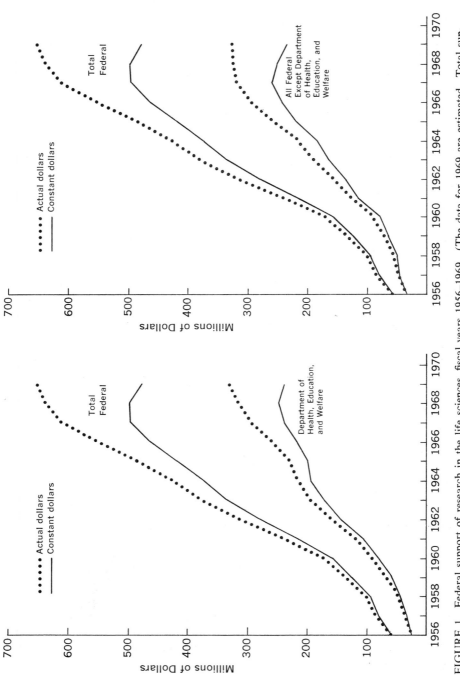

FIGURE 1 Federal support of research in the life sciences, fiscal years 1956–1969. (The data for 1969 are estimated. Total support and that from the Department of Health, Education, and Welfare are shown in absolute dollars and are also shown normalized to constant dollars, with the purchasing power of the 1956 dollar taken as the base. The data are those reported by federal agencies as "basic research," but the definition of "basic," as contrasted with "applied," research, of necessity, is neither rigorous nor con-

The community of life scientists cannot guarantee that future research will alleviate, much less eradicate, any specific one of the long list of diseases to which plants, animals, and man are still subject; nor can it guarantee unlimited food production for mankind or conservation of natural resources. But it is clear that if we fail to prosecute such research—if we fail to follow up the promising beginnings that have been made—then surely there can be no such cures—no new modes of disease prevention; no new foods; no new species of plants or animals; no new disease-resistant strains of otherwise susceptible crops; no new approaches to the mental disorders that inflict so much grief and suffering; no new means to cope with the ravages that our technology imposes upon the quality of urban and rural life—nor even adequate early recognition of the manner in which the changing environment may threaten humanity. We must continue because research in the life sciences is a truly noble endeavor, both the performance and results of which enrich and illumine our civilization while paving the way for a healthier, wealthier society in which free citizens may strive to realize their fullest human potential.

The research tradition of the United States is a gift of our European heritage. In our country, research in the life sciences developed sporadically in the eighteenth and nineteenth centuries and came to full flower after World War II. It was firmly lodged in the universities by the land-grant college (Morrill) acts of 1860 and 1892, by the reform of medical schools early in the twentieth century, and by the support of the great philanthropic foundations. In-house government biological research capability began within the Public Health Service and the Department of Agriculture in the last century and developed sporadically. Federal biological laboratories burgeoned after World War II when the Public Health Service and the Department of Agriculture constructed and equipped some of the world's finest laboratories. This advance was followed shortly by the creation of a multiplicity of biological laboratories within the Department of Defense and among the contract centers of the Atomic Energy Commission, as well as by the establishment of special laboratories within the National Aeronautics and Space Administration and the Department of the Interior. Private foundations pioneered in developing orderly procedures for the award of research grants, providing precedent for the nationally competitive research project grants and contracts programs of a variety of federal agencies, a system that, when it was adequately funded, brought American science to a peak heretofore unrivaled in history. Figures 1 and 2 illustrate the extent of federal support of research in the life sciences, 1956–1969.

This period coincided with the introduction of sophisticated physical and chemical approaches to the understanding of life processes. A young discipline, biochemistry, flourished, and America early achieved world leader-

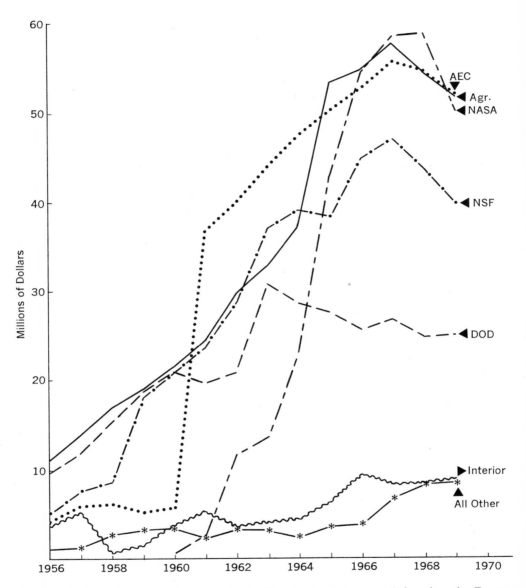

FIGURE 2   Federal support of research in the life sciences, by agency (other than the Department of Health, Education, and Welfare), in constant dollars, fiscal years 1956–1969. (The data for 1969 are estimated. Support is normalized to constant dollars, with 1956 as the base. As in Figure 1, the data refer to "basic" research.)

ship in this field. In turn, biochemical thinking and procedures were applied to the major questions of biology—the nature of cells, genetic mechanisms, and physiological and pathological processes at all levels of organization, from cells, to whole plants and animals, to man, to entire populations. A golden era of biological understanding was ushered in by the powerful insights thus gained. So armed, our national capability to utilize biology for societal purposes rapidly grew in sophistication and success.

This "system," by which the federal government became the principal patron of American science, grew rapidly but in a loose, uncoordinated manner. Each agency developed its own style, set its own priorities, established its own rules and regulations, and determined for itself what fraction of its resources should be devoted to fundamental and applied research and to development. Only the National Science Foundation was charged to assure the strength of the total national science endeavor, and its appropriations have never been adequate to this task. No federal agency has yet been fully charged with the welfare of the academic institutions in which so much of the research endeavor occurs, nor has any of them been given funds sufficient to ensure a continuing flow of scientifically trained manpower. Yet there have been important innovations in these regards.

Encouraged by the Congress, the National Science Foundation developed modest programs to permit individual colleges and universities to upgrade their capabilities for research and education in science, followed by an equivalent program for medical schools at the National Institutes of Health. The National Science Foundation and, later, the National Institutes of Health developed programs that conveyed block grants of modest size, calculated on a formula basis, to institutions or their medical schools to supplement research and research-training programs, while the Department of Agriculture continued to provide block-grant subvention to the state agricultural colleges.

Support of graduate students began with fellowship programs in several agencies. The major innovation in this regard was the institution of disciplinary training grants at the National Institutes of Health—grants that provide not only stipends for graduate students but also additional funds to strengthen the quality of their training—e.g., for equipment, consumable supplies, visiting lecturers, or, occasionally, additional faculty. The training grants of the National Science Foundation, however, have never been funded on a scale commensurate with such objectives and have been limited to what are, in effect, locally administered fellowship programs, while both agencies, and others, have managed predoctoral and postdoctoral fellowship programs. One additional innovation is noteworthy: the growth of large centers housing major facilities for the conduct of research, which are available to scientists based anywhere in the nation. While most such centers are concerned largely with aspects of the physical sciences, several phytotrons, biotrons, marine stations, research vessels, and controlled field areas may properly be regarded as national research facilities.

As we enter the fourth consecutive fiscal year in which the financial support of this patchwork of programs has either remained constant or declined in constant dollars, this system is experiencing a sense of crisis. Increasing numbers of universities, medical schools, departments, and scientists compete for these funds; the number of graduate students continues to increase; and scientific capabilities become ever more sophisticated, and hence more costly, while the purchasing power of the dollar has declined by about 30 percent during this four-year period. As signs of this retrenchment have become evident, morale has fallen and apprehension has risen. Neither institutions nor investigators can plan rationally for the future; graduate students and postdoctoral fellows become insecure and alarmed; undergraduates aspiring to graduate and professional schools can no longer sense the national purpose and, hence, falter in their decisions.

Unfortunately, the questionnaires providing the data base for this report were distributed and collected early in fiscal year 1968, and they reveal nothing of the impact of this alteration in the pattern of federal funding. Whereas we have some anecdotal evidence in this regard, there are no available statistics permitting us to assess the impact of this situation. It is evident, however, that most investigators carry on, although necessarily at a somewhat restricted pace; that large laboratories have been forced to dismiss some of their personnel; that young investigators find it increasingly difficult, if not impossible, to secure research support; and that, although the "system" is as yet intact, responsible investigators and academic administrators are increasingly apprehensive.

A recent survey by the National Academy of Sciences has established that only about 1 percent of all Ph.D. graduates of the classes of 1968 and 1969 are either unemployed or are employed in positions that do not permit them to utilize their graduate educations. Nothing is known of the fate of the postdoctoral fellows who should have emerged during this period, but mobility has been markedly restrained. And the consequences to the system of the depressed funding levels in the federal budgets for fiscal year 1971 will certainly be extensive and profound.

The studies reported here have exposed something of the nature and magnitude of the current endeavor in the life sciences. They provide guidelines for the societal support of this endeavor in the next decade; but they do not and cannot provide quantitative parameters adequate to establish the appropriate magnitude of that support. It is, rather, the opportunities for significant research, the national supply of talented life scientists, and the requirements for their support, combined with the national sense of urgency and purpose, that best define the appropriate magnitude of the research effort and determine its tempo. In turn, these must involve decisions concerning the support of educational institutions, the desired character of those institutions, and the provision of facilities and training of manpower, as well as decisions concerning a national system for delivery of health care—decisions that must necessarily be made in the public arena.

The acute financial plight of numerous medical schools has not gone unnoticed by the press, which expresses legitimate concern about the rising costs of health care and of hospitalization (largely due to wage increases for nonprofessional personnel); concern that care in university medical centers is superior to that in outlying hospitals; concern that the supply of physicians is inadequate to national needs and that, perhaps, medical schools train significantly fewer physicians than they might with their available resources; concern that disadvantaged citizens have little access to the present system for delivery of health care; and concern that many physicians and, indeed, many paramedical personnel are overtrained for their actual roles. The press has also evidenced concern about such allegations as, "research is performed at the expense of teaching," "research is misdirected to fundamental questions rather than the problems of disease," "research results, available even now, are not utilized in clinical practice," and "some significant fraction of research is mediocre or insignificant." In the milieu reflected in such criticisms the medical schools of the nation are confronted by a crisis of public confidence that is not abated by public concern about the parent institutions and their crises of student unrest. Until these concerns are resolved, recommendations regarding overall funding for research are futile. The American public, through its elected representatives, must soon decide what it wants and what it is willing to pay for.

Many of the recommendations presented here rest on the following premises, on which the Committee on Research in the Life Sciences is entirely agreed.

1. Although quite unplanned, the American biomedical enterprise is a unique and highly successful endeavor, the envy of the world, and, as such, should be a source of great national pride.

2. The rapidly unfolding understanding of life in molecular terms is not merely one of the gigantic intellectual accomplishments of man; it is the unique basis for hope that, in the future, we shall be able to cope with the major diseases to which man is subject, thereby lengthening the span of useful, enjoyable human life.

3. Again without a well-debated and accepted plan, the nation's universities, including their medical, agricultural, public health, veterinary, dental, and engineering schools, are both educational instruments and pri-

mary locales for the conduct of research. That research not only enriches the intellectual life of these institutions, it is also a prime national purpose in and of itself. In consequence, the quality of teaching and of education in these institutions is at an all-time high, incomparably superior to that which was available to the authors of this report when they were students. This is made evident by the competence of recent graduates of these institutions, whether undergraduate, graduate, or professional. And it is made no less true by student complaints about "relevance" or about insufficient contact with distinguished professors.

If support of research and research training is permitted to decline, even if simply eroded by inflation, responsible agencies will be confronted with a hard choice, and the public should participate in the decision process. One alternative is to maintain quality standards and to support only the best research, performed by the most competent investigators. Then, if current trends were to continue, within a few years, research funds would again be limited to a relatively small group of institutions, and great regions of the country would be without universities or medical schools with significant research endeavors. The general tone of life in these regions would suffer, as would the education of their students; this would not occur in areas still fortunate enough to be recipients of research funds. The second alternative is to cut the support of all projects in proportion to the reduction in funding. In time, research productivity in all institutions would be seriously injured, and all would sink into mediocrity.

We find neither alternative acceptable. Our students should be educated in the best settings we know how to provide. We are not content to train large numbers of "doctors" as slightly overeducated technicians unable to cope with any but the most routine clinical situations. We owe it to posterity to pursue research into the nature of life and into the nature of man and his disorders with all the imagination and tools at our command. This may appear expensive in total investment, but how expensive is it relatively to a nation that consumes tobacco, alcohol, cosmetics, automobiles, television sets, and outboard motors on an unprecedented scale and spends $60 billion, annually, on inadequate health care?

We need not speak here to the financial requirements for a vigorous research and development enterprise in those aspects of industry that depend upon progress in the life sciences. Indeed, because of the great difficulties in securing reliable information concerning the numbers and types of scientific manpower and their activities in industry, our studies shed little light on the nature of the life sciences endeavor in industry. The usual market forces should suffice, provided the educational system continues to yield an adequate flow of appropriately trained scientists and of new biological understanding.

Nor can we speak to the financial requirements of research in government, largely federal, laboratories. The scope and scale of such endeavors should always be appropriate to the current missions of those agencies that conduct in-house biomedical research, such as the National Institutes of Health, the Department of Agriculture, the Veterans' Administration, the Department of Defense, the Atomic Energy Commission, the National Aeronautics and Space Administration, and the Smithsonian Institution, each of which can boast of an illustrious history of distinguished research. *It is our deep conviction that each federal agency whose mission is science-based should continue to manage a vigorous intramural research program and support a commensurate extramural program of applied and fundamental research relevant to its mission.*

In light of our studies, we can, however, offer some prescriptions with respect to the major single component of the entire system—the educational and research activities of life scientists in the academic world.

We join many fellow citizens in urging that we more vigorously implement a variety of domestic programs; the federal government must undertake support of science on a scale commensurate with the magnitude and scope of those national aspirations that can be achieved only by further scientific understanding. *The present pause in the funding of science should be utilized as an opportunity for planning a complete system of support for the future, which should be not a haphazard patchwork but an orderly continuum.* Such planning should be undertaken with due regard for the financial stability and nurture of institutions of higher learning, which, collectively, have become a great national resource as well as for the specific requirements for research imperative to future national programs concerned with the magnitude of the population of both the United States and the world, the quality of the environment, the public health, and the world and American food supplies, while also assuring an adequate future flow of scientifically trained manpower. No other sector of the American economy can conceivably substitute for the federal government in these regards. Thus, it is essential that the federal government not only accept this responsibility but also plan accordingly in a thoroughly responsible manner. If this system is to survive and adequately serve our society, it cannot be repeatedly subject to sharp swings in funding levels. Funding must not be turned on or off on an annual basis. To the fullest extent possible, government programs should continue to develop in a carefully orchestrated manner, with the various major components of support for institutions, individuals in training, and the research endeavor itself appropriately synchronized.

We must re-emphasize the far-reaching benefits that will accrue to the citizens of the United States if the research endeavor in the life sciences is prosecuted vigorously, fully utilizing the talents of the existing pool of life scientists and of the vigorous young people who will respond to the excitement of this call. This enterprise, which cost about $2,473 million from all sources, public and private, in fiscal year 1968, should be recognized as a national purpose in and of itself, a component of the modern Enlightenment that gives luster and purpose to our civilization while providing the intellectual platform for the continuing improvement of the national enterprise for the delivery of health care and for our ever more productive agriculture. Moreover, only if this research endeavor is successful can we leave to future generations assurance of the quality of our national life, since only by utilizing the fruits of continuing research can this nation cope with the problems posed by a growing population, an advancing technology, a deteriorating environment, and dissipation of the bounty of great natural resources of our land.

We must guard against impatience; decades may elapse between appreciation of a new scientific observation and its intelligent application to some human or technical problem. The penalty for failure to prosecute a vigorous program of fundamental research will be paid by those yet unborn.

RECOMMENDATIONS

Population Problems

Growth of the world population and of the number of the American people, continued over historic time, will necessarily constitute a primary threat to the quality of human life. Containment of that problem will require vigorous educational, social, and political measures. But, to be effective, these must rest on enlarged understanding of human reproductive physiology. Only a broad program of research can provide the understanding necessary to make possible completely safe, reliable, effective, and reversible biological measures whereby appropriately educated couples may restrict the size of their families.

Achievement of these goals requires establishment of a responsible, vigorous, and highly visible mechanism, within the federal government, for the implementation of these diverse programs. We find no need for the constitution of a new agency but strongly urge the identification and appropriate funding of such a central mechanism within an existing federal agency. Such an agency should have statutory authority to engage in and financially support relevant educational and research programs at home and abroad. It should serve as a central repository of relevant scientific, social, educational, and political information and as a vehicle for dissemination thereof, but its identification need not exclude other agencies from contributing to the huge effort required.

The Environment

Evidence of the deterioration of the quality of the air, water, and fertile soil of the planet is reported daily by the press—deterioration that is proceeding most rapidly in the most technologically developed nations, but that worsens apace in virtually all nations. This growing threat to the quality of life and, indeed, to the habitability of the planet constitutes a profoundly human issue. Central to the effort to reverse this process is the capability of the life sciences. But neither the magnitude or criticality of that threat nor the standards to be sought can be estimated without a sufficient body of hard, reliable data. It is imperative that rational standards be established for the chemical and biological quality of the air and of both impounded and natural waters. It is imperative that new, degradable insecticides and pesticides with highly specific actions be devised and that their ecological consequences be understood, as it is imperative that the full ecological impact of the existing armamentarium of such agents be evaluated. Classical dose responses, evaluated only in terms of mortality or morbidity statistics, will not suffice; such data must also include assessment in terms of modern knowledge of cell physiology, metabolism, and cytogenetics. It is imperative that we acquire an inventory of the planet's useful renewable resources. And it is imperative that, as soon as possible, knowledgeable biologists cooperate with those who will plan the growth and redevelopment of urban and suburban communities in solving the complex problems generated by the perturbations of nature resulting from our diverse technologies. Cooperation among biological and social scientists as well as engineers in the design of new communities is required also in planning highways, in restricting natural areas for recreational use, and in evaluating proposals for changes in the national landscape.

Although the life sciences, even now, are capable of contributing significantly to this critical enterprise, the science of ecology, while crucial, is still developing; its capabilities are limited, as is the number of ecologists. It must be clear that ecological understanding rests upon the totality of all other biological understanding—upon our comprehension of physiological function, nutritional requirements, and reproductive mechanisms of plants, animals, and microbial forms as well as of the deleterious effects of chemical entities not normally present in the environment. Thus, continuing advancement of understanding along all biological fronts is essential to the development of ecological understanding.

As ecologists embark upon the essential exercise of extending their comprehension from relatively small managed plots and lakes to biomes, which are orders of magnitude greater, the data collection and analysis required are also manyfold greater in all aspects, including their cost. It is unfortunate that this youthful science has so rapidly become essential to maintaining the quality of our civilization, and it would be tragic indeed if its further development were limited by lack of the relatively modest funds required to assure its growth.

The multitudinous inputs required for management of the environment are such that no one federal agency will suffice to manage the various programs involved. In addition to the newly created agency, components of the Departments of Health, Education, and Welfare, Interior, and Agriculture, as well as the Department of Commerce, all have appropriate roles to play. Each should be encouraged to proceed with the utmost vigor and to seek support from the academic community wherever this is appropriate. Other agencies with no regulatory function but whose programs can affect the quality of the environment—e.g., the Atomic Energy Commission, the National Aeronautics and Space Administration, and the Department of Defense—should acquire the necessary biological capabilities to analyze the effects of their programs upon the environment and recommend accordingly. *In the face of a growing emergency, however, some one major agency should be designated as the planning focus for major action programs in this area and as manager of the relevant information system.* Presumably, this role will be assigned to the new agency.

Meanwhile, the National Science Foundation should be recognized as the principal federal agency for the support and strengthening of the national capability for research and education in ecology, and its support programs in this area should be augmented as rapidly as the capabilities of the academic ecological community will allow.

Finally, it is important to recognize that, with the inauguration of the research programs of the International Biological Program, investigations in ecology have begun to assume the proportions of "big science." *This survey committee recommends that there be brought into being a "National Institute of Environmental Studies," located at some central point in the United States.* The institute should have access to a large and powerful computer and to a variety of vehicles for use on land, on water, and in the air. The center should provide facilities for its central staff and for visiting ecologists and other life scientists, for physical and social scientists, for engineers, and for attorneys from universities across the nation; it should serve as a center for study of the sociotechnological aspects of the environment and as a source of advice to the many distinct entities in both the government and private sectors that can profitably utilize such assistance. In a sense, such an institute would be the converse of the National Institute for Environmental Health, which is concerned primarily with the effect on man of elements of his environment, although this term of reference should not be interpreted too narrowly. The proposed institute would be concerned, *inter alia,* with the effects of man on his environment.

Because the network of environmental responsibilities extends across so many agencies of government, *it is suggested that a working council composed of assistant secretaries of the appropriate departments of government, or their equivalents, specifically concerned with such problems, could serve as a vital coordination and communication mechanism,* working in concert with the White House Council of Environmental Advisers. Further, *we endorse the concept of an annual report on the state of the environment* to be transmitted to the Congress by the President.

Health

Advances achieved in the last decade in the understanding of man and the diseases to which he is subject have totally overhauled and remarkably enhanced the armamentarium available to the physician. For a variety of reasons, the quality of health care available at the great medical centers affiliated with our medical schools probably exceeds that available in most, but by no means all, other agencies for the provision of such care. We offer support and encouragement to all those who would utilize available medical technology to upgrade the quality of medical care wherever it is offered. Further, we recognize that the needs and enhanced expectations of disadvantaged citizens for medical care will necessarily make new demands on the federal treasury. At the same time, a substantial number of clinical scientists must direct their talents toward the development of more efficient and effective mechanisms for the delivery of optimal medical care to all citizens, regardless of income level.

Nevertheless, we deplore the fact that these new directions are already being sought within the static budget of the National Institutes of Health, and hence, necessarily at the expense of the biomedical research endeavor. It will be no service to our progeny if in this circumstance, as in so many others, the seemingly urgent drives out the important. Every physician knows how limited are the tools available for dealing with the endless variety of human illness, despite our recent progress. While learning and attempting to provide the benefits of that limited armamentarium to all citizens, it would be foolish in the extreme to do so at the expense of the very research enterprise that will fortify that armamentarium in the years to come. The total annual expenditure for health care is of the order of $60 billion, whereas about $2 billion is expended upon related research by all sectors of our society. Even if the entire $1 billion federal contribution to biomedical research were diverted to improvement of health care, it could have but trivial impact, while a vital national resource, our biomedical research capability, would be destroyed. *We warmly support an enlarged program of research directed at the improvement of our social instruments for the prevention of disease and the delivery of health care. But, equally strongly, we urge that such a program be funded in its own right, without injury to the national biomedical education and research capability.*

Moreover, the National Institutes of Health will be confronted with more and more fiscal and policy pressures that will tend to force the institutes into an ever narrower view of the "mission-relatedness" of the applications for research support that come before them. *We urge that these pressures be resisted.* The historic fact, illustrated in succeeding chapters, is that fundamental research has been a major contributor to the solution of eminently practical problems in the field of health. Additional research, designed to provide fundamental understanding, is imperative to the prevention and cure of the major killers of mankind. Until definitive preventive and therapeutic procedures for these disorders based on such understanding become available, we must continue to rely upon the costly—sometimes heroic but inadequate—measures that constitute the great bulk of current clinical practice. If historical precedent is any guide, when definitive procedures are provided by research, they are invariably not only more satisfactory but also decidedly simpler and cheaper than are the stopgap procedures that occupy most of the efforts of the health care system.

No less significantly, the majority of clinical investigators in the academic world, as well as those in both federal and industrial laboratories, characterize their own endeavors as "basic research." Yet these investigations, in their view, constitute the most rational, promising approaches to solution of the clinical problems that confront them daily. It would be tragic indeed if future medical research were confined to feckless attempts to apply the inapplicable, at the expense of fundamental research that may yet reveal the underlying causes of diseases and suggest effective approaches to their prevention and alleviation. For two decades *the National Institutes of Health has supported a balanced program of fundamental and applied biological and medical research. We urge that it continue to do so. We also recommend that, as its appropriations permit, the National Science Foundation expand its support of fundamental research in the life sciences,* watchful of restrictions in the support policies of the National Institutes of Health—e.g., in the diminished support of organic chemistry and studies of photosynthesis that occurred in fiscal year 1970—and of deficiencies in the programs of the Department of Agriculture. Already the major supporter of studies of systematic biology and fundamental ecology, the National Science Foundation should remain conscious of the need to pursue research along all the frontiers of the life sciences.

Agriculture

The American agricultural enterprise—the totality of "agribusiness"—is one of the most remarkable accomplishments of our civilization. Much of its success is owed to the network consisting of the United States Department of Agriculture, state departments of agriculture, and state university agricultural schools and agricultural extension services. For several decades, federal interest in the farmer was concerned in large part with the quality of rural life and the need to keep a reasonable number of farmers engaged in agricultural practice, while coping with the surpluses of agricultural productivity. But that era has passed. Our agricultural surpluses are well nigh gone, our own population increases, and we recognize a moral obligation to assist the poor and hungry in other lands. *Accordingly, we urge renewed and expanded attention to the classical problems of agricultural practice: (1) the development of new strains of a variety of crops resistant to disease and to adverse physical conditions; (2) the development of new strains of higher nutritional value and of greater productivity per acre; and (3) the discovery of new means for opening to agriculture of soils previously useless for this purpose.* The latter process should be avoided to the maximum extent possible in the United States, preserving such lands for wilderness and recreation, but it is very important in various developing nations, particularly those with laterite soils.

We take heart in the remarkable accomplishments in Mexico and in the Philippines, which have made available new strains of corn, wheat, and rice, which, for the nonce, have staved off the specter of world famine. But the need is still with us, and only a significant domestic research effort will permit both optimal nutrition of our own population and contribution to the world food supply. Entirely new plant species warrant investigation, and the time may well be at hand also for examining not only new disease-resistant animal strains but also new animal species not traditionally husbanded by man. Other experiments should be conducted with new technologies such as aquiculture in the nation's estuaries, ponds, and lakes, and even with the development of a technology for the eventual operation of inland "fish factories" analogous to the "chicken factories" of current practice. Indeed, catfish are already being bred successfully in this manner in several southern states, as are trout in Idaho. The totality of this effort can succeed only if the underlying biological principles are also given adequate attention.

For these reasons, it is our belief that the National Science Foundation, the National Institutes of Health, and the Department of Agriculture should constitute the major sources of support for the academic endeavor in the life sciences. It is regrettable that the Department of Agriculture has confined its support of academic research to the agricultural colleges. Such research necessarily rests on the much broader base of plant science and, indeed, on all of biology. Financial support of this broader effort by the Department of Agriculture would result in long-term strengthening of the scientific base of the department's programs and would engender useful liaisons between the department staff and some of the nation's leading scientists. This is not to deny significant roles to those other mission agencies with a stake in the progress of the life sciences, e.g., the Atomic Energy Commission, the National Aeronautics and Space Administration, the Department of Defense, the Department of the Interior, the Department of Commerce, and the Veterans' Administration. But if the first three agencies named provide the major sustaining support required, support from the second group could be limited to that research most germane to their mission requirements.

The Academic Endeavor in the Life Sciences

INSTITUTIONAL SUPPORT PROGRAMS

American universities and their graduate and professional schools have arisen over a period of more than two centuries, with diverse and often limited or local sponsorship. In the American tradition, each educational institution is both supported and controlled by the constituency it serves. This has long been evident in the hierarchy of institutions from grade school through the graduate universities and in the limited constituencies of certain private institutions. With the passage of time, the graduate and professional schools of both private and state universities have acquired national rather than local constituencies. Each receives young people from across the land, who later leave to seek careers wherever appropriate opportunities may beckon. Accordingly, *we consider that, of all elements of the educational system, the graduate and professional schools are most appropriately supported from the federal treasury.* Moreover, if that support were sufficient and delivered by appropriate mechanisms, it would provide urgently needed relief of the overall fiscal problems of these institutions of higher learning.

We must re-emphasize that, in the university setting, the research and educational endeavors are inextricably intertwined and cannot rationally be separated. Any effort to do so, for example, by time and effort reporting, is at best an artificial contrivance that will present a misleading and inappropriate picture of the nature of these endeavors. In designing instruments for financial support directed at either research or education in the universities, both aspects of the system must be strengthened, concomitantly, and such instruments should not be designed so as to distort the intrinsic nature of the relationship between education and research.

Our studies have revealed the heavy dependence of the universities on federal funding to meet their faculty payrolls, a dependence that is least apparent among the arts and sciences colleges of state universities, and much more apparent in private universities, most particularly in their medical schools. To charge federal contributions to these faculty salary payments against federal appropriations earmarked for research can be seriously misleading. Without those contributions, universities and their graduate and professional schools would find it impossible not only to pursue their research endeavors but also to meet their fundamental teaching obligations.

The increasing undergraduate, graduate, and professional enrollments expected in the decade ahead must necessarily aggravate these circumstances. As the traditional income sources of the universities become less adequate to offset mounting expenditures, in the absence of a research endeavor on the current scale, universities could undoubtedly meet their teaching obligations with smaller faculties. But then, not only would there be no research progress, but teaching at all levels would necessarily become dull, retrospective, and stultified as it becomes increasingly remote from the excitement of the present and the promise of the future. The scientific training of future citizens, practitioners, teachers, and investigators would be gravely compromised.

We believe support of the graduate and professional educational and research endeavor to be both a proper and a necessary function of the federal government. However, we also consider that payment of faculty salaries from grants and contracts in support of research constitutes a most undesirable distortion of the internal arrangements of academic institutions and of the relationship between institutions of higher learning and the federal government.

Hence, we strongly endorse recommendations, previously presented in a variety of other reports concerned with federal support of academic activities, that there soon be brought into being programs of block support to colleges and universities and their graduate and professional schools on a scale sufficient, when combined with their normal resources, to enable these institutions to meet all faculty salary payments from funds controlled by the central university administrators or those of the various colleges. Further, we believe that such support should be sufficient to provide opportunities for the development of new educational programs and services. These block grants distributed on the basis of appropriate formulae should be designed to provide for the specific needs of the medical, agricultural, engineering, and graduate schools, as well as the undergraduate components of both universities and liberal arts schools. Moreover, we hold that the variety of services essential to academic research activities, that, collectively, comprise those items allowable as "indirect costs" in the reckoning of government research grants and contracts, should be regarded as essential, intrinsic aspects of the operation of the university and should properly be defrayed from a distinct program of institutional grants dedicated to this purpose rather than related to individual grants and contracts. These various block grants to the universities and their component schools should alleviate current funding difficulties without altering the individual character of these institutions. Such a system would permit university administrators to plan rationally, based on the assurance of continuity of adequate funding for their continuing commitments, thus freeing them to concentrate on the magnitude and quality of their teaching and research endeavors.

GRADUATE EDUCATION IN THE LIFE SCIENCES

Graduate education in the life sciences is a large and expensive enterprise. The student proceeds through a series of didactic and laboratory courses, learns to communicate in smaller seminar groups, and, through a continuing dialogue with his peers, postdoctoral fellows, the faculty, and visiting lecturers, slowly acquires the values, the style, the history, and the experimental techniques of his discipline. These may be brought into sharper focus as he serves as a teaching assistant in an undergraduate course under the supervision of a more experienced member of the faculty and as he undertakes an original research project. Although such a project may be a totally independent endeavor, more frequently he conducts one complete phase of a larger research program in the laboratory of his faculty mentor where he works as a member of a coherent research group, utilizing all the laboratory's research facilities and conferring almost constantly with his mentor and his colleagues.

The process requires four to six years and, for the successful, culminates in receiving the degree of Doctor of Philosophy. The process is subject to a high degree of attrition, which may vary from 50 to 90 percent in various specific departments. These losses are somewhat inflated by the enrollment of students who intend only to expand their educations or acquire a master's degree and do not aspire to the Ph.D. Attrition among female students is significantly greater than that among male students. It will be evident that this highly individualized form of instruction is a great drain on the resources of the universities, demanding faculty time, space, research facilities and instrumentation, consumable supplies, and all the supporting services necessary to the conduct of research. In large measure, the costs of this enterprise are defrayed from research grants made to the faculty members, but these rarely suffice. The functional unit of graduate education is the disciplinary department or, occasionally, an organized multidisciplinary program. The quality and success of such a program is contingent upon the availability of funds to provide instrumentation used in common by the department staff and students, to support a vigorous seminar program, to operate communal facilities such as greenhouses, media preparation, animal care, and sterilization facilities, or instrument shops, as well as stipends for the graduate students. Without exception, these demands overtax the resources of the university.

Accordingly, *we strongly recommend the program of graduate-training grants instituted at the National Institutes of Health, particularly the disciplinary programs of the National Institute of General Medical Sciences.* These programs have clearly recognized the nature and requirements of modern graduate education in the life sciences as well as the supplemental training required if the graduate of a medical school is to become an accomplished investigator. The training-grant concept has been a remarkably successful educational innovation, upgrading the quality of education while contributing to an expansion commensurate with national needs. And we deplore the fact that, in recent times, these programs have failed to receive from either the executive or the legislative branch of the federal government the support that they merit. It is a matter of grave concern that appropriations for these programs are decreasing at the very time they should be increasing.

We are convinced that the budget for these programs for fiscal year 1971 seriously underestimates the magnitude of these needs, and it is our hope that appropriations for these programs will be doubled over the course of the next five years. Support of graduate education should not be turned on and off, nor should the effectiveness of graduate education be measured in enrollments. Only continuing support can assure the continued output of well-trained scientists. Reduction in training-grant support will necessarily result in a diminished output of the trained scientists required for a diversity of careers in research, teaching, administration, and public service, and, indeed, of those required to discharge national obligations to the developing countries overseas. Continuity of support to the graduate-educational endeavor is imperative to the national capability in the future. Indeed, we feel so strongly about this that, were it absolutely essential that some *small* decrease be made in appropriations for the research and graduate-research-training endeavors for the next year or two, we would consider it wise to delete these funds from research support rather than from training programs.

Patently, therefore, we greatly deplore the substantial reduction in the Administration's budget request for the training-grants program of the National Science Foundation for fiscal year 1971. Indeed, *we recommend a marked expansion of the program of training grants at the National Science Foundation, remodeled along the lines of the program of training grants conducted by the National Institute of General Medical Sciences.* Such a program should particularly support graduate education in those aspects of the life sciences seemingly more remote from the immediate problems of health, as well as in other scientific disciplines. The increment in lifetime earnings resulting from a Ph.D. in the life sciences is modest indeed, and it is unlikely that sufficient numbers of prospective life scientists from the lower income levels of society would undertake such study without personal stipend support. The nationally distributed traineeship program has demonstrated that it is better suited to the national purpose than are stipends defrayed from research grants or even direct fellowships.

Further, we recommend that the U.S. Department of Agriculture embark upon a program of analogous training grants for support of the graduate-educational endeavor, not only in the animal and plant sciences in the agricultural colleges but also in support of related education in graduate departments of the colleges of arts and sciences concerned with aspects of botany and zoology that necessarily underlie the future success of research in the somewhat more applied areas of animal and plant science.

STIPENDS

Until recently, relatively few qualified graduate students failed to find stipend support. Stipends are currently provided from training grants, national competitive fellowships, teaching assistantships, and research assistantships defrayed from research grants to members of faculties. We urge that it be made a matter of national policy, ultimately, to provide virtually all such stipends through the training-grant mechanism. Were such funds available on a sufficient scale, we would find little merit in a national competitive fellowship program wherein individual students bear a one-to-one relationship to an agency of the federal government, since these students must pass a screen of university and departmental admissions mechanisms in any case. We abhor the practice of supplementation of such stipends as a means of competitively attracting talented young people. That competition, which serves the national interest, should nevertheless rest on the research and educational opportunities of a given institution. *However, we do support local cost-of-living adjustments to a national basic stipend level, established by the sponsoring federal agencies. We urge also a substantial increment in the current basic stipend level, which has now remained unchanged during a period in which the purchasing power of the dollar has decreased by almost 30 percent.* Such stipends, intended to permit graduate students to pursue their educations relatively free of major personal financial problems, should certainly be set above the poverty level.

We regard teaching as an essential, normal component of the graduate student's personal career development, not only because it fits him potentially for a career as an educator, but also because it compels collecting, organizing, and presenting in orderly, rational fashion the facts and concepts of his discipline. Accordingly, we consider it inappropriate that graduate students should receive additional compensation for such services, regardless of the source of their basic stipends.

However, we are aware of the fact that increments above the basic stipend level can be utilized, with some success, to attract bright students into discipline areas that they might not have contemplated otherwise. *Accordingly, we do approve of limited supplementation from institutional funds of the stipends of students entering upon education in areas of specific national need in which manpower requirements are seriously limiting and an increase in manpower has been declared a matter of national policy by an appropriate sponsoring agency.*

We are aware that ultimate conversion to a system of stipend support almost exclusively from training-grant programs may generate a serious handicap for a student who, for personal reasons, must pursue his education in a science department that has not qualified for training-grant support. Although we cannot recommend to any student that he seek further education in a department deemed unqualified by a jury of peers, recognizing that compelling personal circumstances may occasionally prevail, we recommend that a modest fellowship program be inaugurated specifically to support such students.

CURRICULA

The pace of research has far outstripped the pace of educational reform in the life sciences. Scientists, who are always anxious to seize new experimental approaches to scientific problems, are as conservative as any other segment of the academic community with respect to curricular change. Yet changes are urgently needed; a few relevant suggestions follow.

1. Most university campuses offer a variety of courses in the life sciences for the undergraduate. Yet there is clearly a core of material central

to all biology, organized along the categorical lines by which biology is considered in this report. *Accordingly, a single core curriculum should be organized and presented to all undergraduate students who major in biology, enriched by elective opportunities in the senior year.*

2. *Every college and university should consider development of a one-year "humanistic" course in biology for students majoring in fields other than science.* In this vein, there is also need for humanistic reorientation of the one-year course in biology presented to high school students.

3. *A special one-year course in biology should be developed for under-graduate students majoring in mathematics, the physical sciences, or engineering.* Such a course could capitalize on the mathematical and scientific knowledge of this group of students, permitting an intensive, quantitative, and rigorous presentation of biology.

4. *The program leading to the Ph.D. should be more nearly standardized; the research goals of the dissertation program should be less ambitious* in view of the fact that so large a fraction of Ph.D. graduates go on to post-doctoral experience. As many students as possible should complete these requirements in four years.

5. Since this report is focused primarily on the national research effort in the life sciences, emphasis here has been placed on graduate education culminating in award of the Ph.D. degree. However, we fully recognize that a large fraction of those now educated in graduate schools do not go on to research careers, becoming, rather, teachers in undergraduate colleges or administrators, or entering one of a great variety of other occupations. We agree with those critics of graduate education who suggest that the formal requirements for the Ph.D. degree may not be appropriate for all such individuals. *Accordingly, we encourage experimentation with graduate curricula intended to lead to other advanced degrees, which would combine extensive training in biology with humanistic, social, engineering, or administrative studies, which would include some sophisticated laboratory experience but would not require an original research dissertation.* Success in such efforts would reduce the national bill for graduate education and would supply individuals better motivated for their actual careers than may be those Ph.D. graduates who do not find success as investigators. Also, the supply of Ph.D.-trained graduates would more closely approximate the foreseeable needs for such individuals.

6. The survey committee believes strongly that laboratory experience is an essential ingredient of an introductory biology course. But, across the country, the laboratories dedicated to this end are equipped only for a bygone era, lacking totally the instrumentation required for work in modern biology. *A national program to upgrade these teaching laboratories is highly*

desirable; the aggregate costs will be considerable, averaging perhaps $100,000 per institution, yet a start should be made on this task at the earliest possible date.

7. Little need be said concerning the curricula of medical schools. The dissatisfaction with the standardized curriculum adopted almost five decades ago is evident in every such school. Many are now in the early stages of major experiments with curricula, and these are summarized in a growing literature. It is evident that no one solution will suffice and that each school must establish for itself the means to cope with the fact that its graduates engage in a great diversity of medical specialties involving research, education, and administration, as well as clinical practice. Meanwhile, it is the fortunate student who is the subject of such an educational experiment, since the level of faculty interest in his personal progress is at a maximum during such an experience.

MEDICAL STUDENTS

The extended years of education and the high cost of tuition and of living presently combine to ensure that, overwhelmingly, physicians are drawn from the national upper-middle class. But the supply of physicians is insufficient for the nation's needs, and talent is to be found at all levels of society. If the future physician is to take his place in that society, unencumbered by the burden of severe indebtedness and great family sacrifice accumulated during his long years of training, some system of federal stipend support is essential. *We urge serious consideration of a significant program of personal stipend support for medical students, probably most suitably managed by the Bureau of Health Manpower at the National Institutes of Health.* A stipend level equivalent to that established for graduate students, combined with a cost-of-education allowance at least sufficient to meet tuition payments, would go far to broaden the pool of American families from which we may draw future physicians, and it would encourage the physician to view his career from the standpoint of public service rather than private gain.

We consider establishment of such a program much more important than the parallel decision concerning the mechanism by which it might be financed. *This Committee would look with favor upon inauguration of a program of National Medical Service wherein each medical graduate would serve in a civilian Medical Corps for two years, in exchange for his education.* Acceptable also would be a loan program whereby such loans could be repaid, at low interest, when the physician's income tax begins to exceed some appropriate minimum.

TECHNICAL ASSISTANTS

A major limitation of the national research capability is the supply of career technical assistants. *We encourage the National Science Foundation, through its educational and curriculum-development programs, to assist community colleges in the design of appropriate curricula and curriculum materials for training technical personnel.* The graduates of these programs should be assured of dignified, reasonably permanent positions in the laboratories of industry, government, and the academic world. This will require appropriate job titles and a salary scale commensurate with the contributions that are made to research by truly qualified technical assistants.

POSTDOCTORAL EDUCATION

Each year a larger fraction of those who receive the degree of Doctor of Philosophy in one of the life sciences have been going on to postdoctoral educational experience. A very large percentage of these seek such experience in a life science discipline other than that in which they received their graduate educations, and virtually all do so in laboratories other than those of their original research mentors, engaging in fields of research distinctly different from those in which they had originally been trained. *In view of the heterogeneity and scope of the life sciences and the great differences in style and experimental approach, we believe that postdoctoral education should constitute a normal experience for any life scientist who intends to pursue a career as an investigator; we recommend that all concerned recognize the propriety of postdoctoral education under these circumstances. We believe that postdoctoral stipends are appropriate items to be defrayed from both research and training grants. Further, we urge expansion by at least 50 percent within four years, assuming a supply of suitably qualified applicants, of the regular postdoctoral-fellowship programs managed by the federal agencies presently involved.*

Much attention has been given to the problem of the foreign postdoctoral in American laboratories. We find little reason to be concerned. Young foreign scientists bring to American laboratories new insights and skills and participate fully in the research programs of those laboratories, making effective contributions of ideas while performing laboratory procedures and assisting in the education of their graduate-student colleagues. We urge that our nation continue to welcome such individuals to our shores and continue to find it appropriate that they be supported from our funds. But we urge further that postdoctoral appointees from developing nations be encouraged by all means possible to return to their native lands in due course so that their own fellow citizens may reap the benefits of their educational experiences.

RESEARCH SUPPORT

We believe that the research-project-grant system is a remarkably successful social invention and that it should continue to serve as the backbone of the research-support system. While we have recommended above that other instruments be utilized to convey funds to be used, largely at the universities, for faculty salaries, student stipends, and provision of all those services included within "indirect costs," *research grants or contracts should convey those funds that, most appropriately, should be controlled by individual senior investigators—e.g., funds for equipment, research vehicles, consumable supplies, travel, computer costs, publication costs, and the salaries of individuals employed specifically for the purposes of the research project.*

We endorse the utilization of juries of qualified scientific peers for the evaluation of applications for such research support and of the qualifications of scientists making the applications. The scientific judgments made by such juries represent the best possible warranty to the tax-paying public that its funds are being utilized optimally and placed in the hands of individuals most likely to utilize such funds successfully in the prosecution of scientific research. *We urge that only the scientific credentials of accomplished scientists be considered in selecting personnel to serve on such juries.* Changes in this practice to give other qualifications for such service serious consideration can serve only to damage this system and destroy the confidence of the scientific community, the universities, the public, and their elected representatives in the operation of this system at a time when funds are already limited and constricting. It is all the more imperative that these limited funds be used to full advantage in the support of the most imaginative and important research of which the scientific community is capable, undistorted by political considerations.

From the best estimate we can make, in the current year (fiscal year 1970) appropriations for research, per se, are approximately 20 percent less than required to ensure that the nation's truly qualified academic life scientists are fully and usefully engaged. We urge that this deficit be eradicated as soon as possible and that, thereafter, the research-support system grow at a rate commensurate with the ability of the system to utilize such funds efficiently and wisely. The frequently proposed formula of an annual increment in research support of about 12–15 percent appears to us to be a rational approximation of desirable growth *as long as the system continues to expand to meet the perceived needs of society.* This would accommodate the increasing numbers of graduate and medical students, meet the increased costs of research due to increasing sophistication particularly noticeable in the life sciences as they lean ever more heavily on instrumentation, and compensate for the losses due to general inflation. Since graduate enrollments are expected almost to double in the next decade, since graduate study in the life sciences is proving attractive to an increasing number of all students interested in natural science, since a general course in the life sciences should be part of the education of all members of an enlightened citizenry, since there are 15 unstaffed medical schools now in advanced planning stages and at least as many others seriously contemplated, since even these proposed medical schools will not meet the national demand for trained physicians, and since a decreased flow of physicians from the developing nations could yet further aggravate the need for physicians trained in our own medical schools, generating yet further need for adequately trained faculty, it is abundantly evident that the academic endeavor in the life sciences must continue to expand by about 5 percent of the trained scientists per year and that employment opportunities for trained life scientists will exceed the supply for at least a decade. Since, further, we are confident that intriguing and important questions concerning the multitudinous aspects of life will continue to confront us and that the answers to these questions will be of great significance to human welfare, there will remain a broad scope for exercise of the research talents of the next generation of investigators. Expansion of the research-support system at a commensurate rate and by the mechanisms that have been suggested here seems entirely in accord with our national purpose.

If tax-derived research funds are to be efficiently utilized, we consider it imperative that senior investigators be assured of the continuity of their efforts. No useful end is served by arrangements that, by accident or design, compel senior investigators, at all times, to be concerned about whether supporting research funds will be available in the following year. *Research-grant awards should be of such character as to assure support for several years, conditioned only by reasonable progress and pursuit, in good faith, of the research that had been proposed, following where it leads.* Whether such assurance is provided by the "moral commitment" system of the National Institutes of Health, subject to annual budget negotiations, or by the "stepfunding" procedure of the National Aeronautics and Space Administration, is a matter for decision by agency administrators.

FEDERAL RESEARCH-SUPPORTING AGENCIES

As noted earlier, we believe that each agency with a science-based mission should both manage its own in-house research program and engage in support of research relevant to its mission that is conducted in the universities and in other nonprofit organizations. We regard it as unfortunate that, historically, the Department of Agriculture has not supported research in the animal and plant sciences outside of the agricultural schools and its own in-house laboratories. Collectively, the botany, zoology, and biology departments of the colleges of arts and sciences constitute a precious resource and include a large percentage of the talented individuals who contribute to the development of animal and plant science.

Accordingly, *we recommend that the Department of Agriculture inaugurate a research-grant program, analogous to that of the National Institutes of Health, with a sufficiently broad mandate to assure support of research in all germane aspects of animal and plant science.*

SPECIALIZED FACILITIES

Our studies revealed a heartening degree of use of existing specialized biological facilities. It is evident that these requirements will become even more extensive. Our questionnaire to department chairmen, which explored their understanding of current requirements for an extensive, but by no means totally comprehensive, group of specialized biological facilities and the priorities among them, revealed that there is a substantial and growing requirement for such facilities, each of which must be managed as an institutional, regional, or national resource. Moreover, this sampling of department chairmen does not necessarily adequately reflect genuine national needs. For example, a facility for experimentation in tropical agriculture or a relatively large conservancy within or in proximity to a city or a high-technology area may not appear to be the first-priority requirement of any single life science department chairman, and yet may be an important national need that should be recognized by an appropriate federal agency. Urban spread is rapidly consuming natural areas near cities and centers of learning that are needed for research and teaching. In many instances such an area could do double duty as a recreational park; such nature reserves are most suitably funded from local resources, although potential national values should be kept in view by federal agencies.

Appropriate programs that could fund the development and acquisition of such facilities already exist in the National Institutes of Health and the National Science Foundation, but they are inadequately financed to meet these requirements. *We recommend that both programs be substantially expanded* so that these needs may be satisfied in some measure as soon as possible. *We further recommend that the Department of Agriculture inaugurate such a program with specific relation to the need for such biological facilities as field areas and programmed climate-controlled rooms to facilitate the research of investigators whose studies are of high interest to the ultimate mission requirement of the department.*

INSTRUMENTS

Research in the life sciences is increasingly dependent upon complex and sensitive instrumentation, permitting a more highly specific and sophisticated approach to the primary problems confronting these disciplines. Yet more complex and costly instrumentation is, even now, under development. Our studies have documented the existence of a substantial backlog of requirements for the currently available major instrumentation needed for effective biological research.

We recommend the establishment at the National Institutes of Health and the National Science Foundation of programs specifically designed to meet the requirements for costly and occasionally unique instrumentation. An initial combined funding goal of approximately $25 million per year is suggested. We note further that the National Aeronautics and Space Administration has developed a realm of technology utilizing new types of sensors and telemetry, frequently but not necessarily in conjunction with satellites, which could be extremely useful to diverse areas of the life sciences, particularly behavioral biology and ecology. This agency is urged to exploit these possibilities by all necessary means, including collaboration with and support of academic life scientists in positions to capitalize on these opportunities.

Since these increasingly expensive instruments will generally be utilized by communities of scientists rather than by individuals, the administrators of these programs should negotiate with coherent disciplinary departments or multidisciplinary groups rather than with individual investigators. The latter should certainly still be free to justify acquisition of such instruments in their own research-grant applications where this is appropriate, but the suggested programs would permit acquisition of expensive instrumentation by groups of investigators who, collectively rather than individually, can justify expenditures on such a scale while ensuring full and efficient utilization of the instruments to be acquired. Moreover, in the long term, this approach would result in a much smaller expenditure than would the awarding of instruments to individual investigators.

COMPUTERS

Our study revealed a surprisingly large and diversified use of digital computers by the life sciences community, in which few scientists were encountering difficulties in securing funds for this purpose. The National Institutes of Health is converting from full to partial subsidy of computer centers for use in biomedical research. As the biological community becomes increasingly sophisticated in computer science and the uses of the computer, usage will grow rapidly. We approve the altered National Institutes of Health policy, believing that *the justification for computer time by investigators should be built into the process of research-grant evaluation.* Requests for funds to defray computer costs are certain to be a growing fraction of research budgets. Supporting agencies should recognize the propriety of such requests and be prepared to fund them. The annual bill for this activity will soon be of the order of $50 million.

LABORATORIES

Other than funds for the support of additional personnel, additional laboratory space is the primary requirement of the graduate, agricultural, and medical schools of the country and, perhaps to a lesser degree, of other professional schools such as dental, veterinary, and public-health schools. While the requirement for space continues to grow, support by federal agencies of construction of facilities has lagged seriously during the last four years. In fiscal years 1969 and 1970 no funds were appropriated under the authorization of $50 million per year for health-research facilities to the National Institutes of Health, and essentially no funds have been made available through the National Science Foundation program of graduate-research-facilities construction for the same period. Not only are new universities being brought into being and new medical schools and other health-professional schools planned and initiated, but the space requirements of existing institutions have grown with our society's expectations for the roles of these institutions.

Our studies reveal an acute backlog requirement for approximately $150 million for construction of research-laboratory space, already planned and partially funded, and required only for existing institutions to keep pace with their expanding graduate enrollments. Patently, that need would be considerably greater had we been able to estimate the real requirements of those institutions in which no plans exist for lack of matching funds—the requirements for newly planned institutions or for expansion of the student population of existing medical schools. It should be clear that these statements relate entirely to the requirements for laboratories for research and graduate education, quite apart from the equally serious failure to provide for the formal undergraduate and professional teaching activities of the same group of institutions.

We acknowledge the call on the public purse of new institutions, particularly the medical schools just coming into being, so that they may take their places in the educational world. But we also direct attention to the serious requirements of existing institutions, and *we recommend that both the National Institutes of Health and the National Science Foundation be en-*

abled to reactivate their dormant programs in support of facility construction at a rate that, between them, would provide no less than $50 million annually for the construction of laboratory facilities for research and teaching in the graduate departments of the life sciences and an equal sum for the research laboratories of medical schools. The planning and construction of a laboratory building consumes four to five years from inception to occupancy and utilization. Accordingly, if these programs are not inaugurated in the reasonably near future, they will not make buildings available in time to match the urgencies that will be generated by burgeoning graduate and professional enrollments.

Museums

The natural history museums of the United States constitute an invaluable and long-neglected resource for public education and research. Seemingly remote from some of the more attention-getting areas on the frontiers of science, they have been allowed to become dusty, lonely, and inadequately curated, with their collections lagging and their buildings sagging.

We recommend a vigorous program for upgrading the key museums of natural history across the country. Funds and management of such a program could be vested in either the Smithsonian Institution or the National Science Foundation. A specific program funded in the amount of about $10 million a year would be appropriate to this end. However, inauguration of such an effort should be preceded by appropriate national planning. It is unnecessary to upgrade, equivalently, all the nation's natural history museums. *A plan should be developed that identifies a limited number of general museums and a group of specialized repositories, the sum of which will satisfy requirements for research in modern systematic biology and for taxonomic identification services.* It is noteworthy that the latter function has been growing rapidly in volume and in importance. The National Academy of Sciences should take the initiative in developing the requisite plan.

Marine Biology Stations

Many of the most exciting chapters in the history of biology have been written at marine biology stations, largely because of the abundant, diverse, and remarkable organisms to be found in tidal waters, on the shore, and in the waters over the continental shelf. These still present great research opportunities, and we have yet to fashion a totally adequate scientific basis for large-scale marine agriculture. Again, however, it would be unwise to dissipate national resources by inadequate support of a substantial number of stations on all three coasts of the mainland, in Alaska, Hawaii, Puerto Rico, and various tropical locales. *A national plan identifying a network of principal laboratories with adequate representation of each major environment is essential so that each may, in time, be upgraded so as to be of optimal service.* As in the case of museums, this is not to deny the educational roles of many other stations, but it is necessary so that a reasonable number can be equipped to serve as major research centers.

Biological Information

The requirements for a national information system for the life sciences are distinct but not unique. *A national plan for such a system should be developed, utilizing the resources of the public and private sectors, of the three major operating units now functioning, and of the specialized information centers.*

For many years, scientific journals will continue to be the primary base of an information system. But most such journals are experiencing acute financial embarrassment. *A plan should be devised and a funding mechanism established to assist these journals in the near future, before some of them expire.*

Scientific meetings are an irreplaceable means of communication. Travel costs for delegates to such meetings should remain a legitimate item in research budgets.

The recommendations above, in our view, constitute a measured evaluation of the overall requirements to maintain the life sciences enterprise in the United States in the forefront of the worldwide scientific endeavor, to educate the next generation of citizens, scientists, and practitioners, and to construct the intellectual platform that will underlie future improvements in our public health, permit expansion of the economy, provide an adequate and wholesome food supply, and transmit to our progeny a bountiful land whose natural beauty and resources have been preserved and enhanced.

CHAPTER ONE

FRONTIERS
OF
BIOLOGY

Life, the most important and most fascinating phenomenon within the ken of man, is the subject of this chapter. It is our purpose to offer a brief overview of present understanding of life in its variegated manifestations. No attempt has been made to be comprehensive; examples and illustrations have been chosen because of the drama of certain findings or for the insight they provide. Necessarily, much has gone unsaid. The theme of this presentation is that life can be understood in terms of the laws that govern and the phenomena that characterize the inanimate, physical universe and, indeed, that living processes can be described only in the language of chemistry and physics.

Until the laws of physics and chemistry had been elucidated, it was not possible even to formulate most of the important, penetrating questions concerning the nature of life. For centuries, students of biology, in considering the diversity of life, its seeming distinction from inanimate phenomena, and its general inexplicability, found it necessary, in their imaginations, to invest all living objects with a mysterious life force, "vitalism." But in the late eighteenth century, Lavoisier and Laplace were able to show, within the considerable limits of error of the methods available to them, that the recently formulated laws of conservation of energy and mass were valid also in a living guinea pig. The endeavors of thousands of life scientists over the succeeding two centuries have gone far to document the thesis thus begun. Living phenomena are indeed intelligible in physical terms. And although much remains to be learned and understood and the details of many processes remain elusive, those engaged in such studies hold no doubt that answers will be forthcoming in the reasonably near future. Indeed, only two truly major questions remain enshrouded in not quite fathomable mystery: (1) the origin of life, the events that first gave rise to the remarkable cooperative functioning of nucleic acids and proteins that constitutes the genetic apparatus, and (2) the mind–body problem, the physical basis for self-awareness and personality. Great strides have been made in the approaches to both of these problems, but ultimate explanations are perceived very dimly indeed.

What follows, therefore, is a record of the present "state of the art." Treated are such questions as: Of what chemical compounds are living things composed? By what means are the materials of the environment converted into the compounds characteristic of life? What techniques have been employed to reveal the structures of the huge macromolecules of living cells? How are living cells organized to accomplish their diverse tasks? What is a gene and what does it do? What are the mechanisms that make possible cellular duplication? How does a single fertilized egg utilize its genetic information in the wondrous process by which it develops into a highly differentiated multicellular creature of many widely differing cell types? How do differentiated cell types, combined to form organs and tissues, cooperate to make their distinct contributions to the welfare of the organism? What is understood of the structure and function of the nervous system? What is a species? How does speciation occur? What factors give direction to evolution? Is evolution still occurring? Is man evolving still, and if so, can he control his own evolution? What relations obtain among the species in a given habitat? What governs the numbers of any one species in that habitat? Are there defined physiological bases for behavior? What is known of the bases for perception, emotions, cognition, learning, or memory, for hunger or satiety? For few of these questions, today, are there exact answers; yet the extent to which these are approximated, even now, constitutes a satisfying and exciting tale.

In the space available, there is little opportunity to describe *how* the facts and concepts considered have been garnered in the laboratory or field. The popular press occasionally presents descriptions of scientific "breakthroughs," while failing to indicate that no such event stands alone. Each research accomplishment is a bit of information in a large and growing multidimensional mosaic. Each investigator is aware of the history of the problem to which he has addressed himself and of the past and current contributions of others. Most successful demonstrations have occurred only when the time has been right and the stage set. Frequently, the idea had been discussed in one or another sense before a definitive demonstration was available. And each bit of information that illuminates a problem reveals a yet deeper layer of questioning to be explored.

Most particularly is it necessary to recognize the dependence of the investigator on his experimental tools. Indeed, the history of science, including the life sciences, is the history of the manner in which major problems have been attacked as more powerful and definitive tools have become available. Thus, living cells are invisible to the naked eye, appear as minute boxes or spheres with a denser nucleus by light microscopy, and exhibit an elaborate wealth of subcellular structural detail by electron microscopy. Techniques for isolation of pure proteins were developed in the 1930's and 1940's, but understanding of their structure seemed impossibly remote. Analytical tools such as electrophoresis (separation of molecules by virtue of differences in their electrical charge), ultracentrifugation (separation by virtue of differences in mass), chromatography (separation by virtue of their varying affinities for adsorption onto diverse solid surfaces combined with their varying solubility in diverse solvents), and appreciation of the specificity of action of certain hydrolytic enzymes permitted resolution of the linear sequences of amino acids along the strand that constitutes the protein chain. Without each of these tools, primary protein structure would remain a mystery. As they became available, the tools were rapidly applied to the problem by a waiting battalion of scientists, and a seemingly herculean task became almost routine. But there remained the problem of deciphering the three-dimensional structure of these large molecules. It was already known that x-ray crystallography could establish the structure of much smaller molecules; a series of refinements was required before this technique could be applied to proteins. But when these refinements were made, there remained a prodigious body of calculations required to convert the data into a model of a protein molecule. Fortunately, it was just at this time that the high-speed digital computer made its appearance. And the three-dimensional structures of proteins emerged as the triumphal accomplishment of this pyramid of scientific endeavor. Until all the bricks had been laid, the apex would have remained invisible and unattainable.

Biology has become a mature science as it has become precise and quantifiable. The biologist is no less dependent upon his apparatus than the physicist. Yet the biologist does not use distinctively biological tools; he is an opportunist who employs a nuclear magnetic resonance spectrometer, a telemetry assembly, or an airplane equipped for infrared photography, depending upon the biological problem he is attacking. In any case, he is always grateful to the physicists, chemists, and engineers who have provided the tools he has adapted to his trade.

Much as the biologist will employ whatever tool appears necessary to his inquiry, so too is he catholic in his use of biological forms for investigations. While mindful of the potential practical implications of his studies, the biologist does not restrict himself to the utilization of those species that appear closest to human concern. Rather does he select biological material particularly suitable for the problem at hand. Thus, the "alarm reaction" of the clam, among the least frantic of animals, reveals in slow motion the essentials of reactions that, in most other species, are too rapid to be readily analyzed. *Escherichia coli,* the innocuous colon bacillus, is utilized in a great variety of metabolic and genetic studies, largely because it can be safely handled in large quantities and is now a well-characterized organism. Other aspects of genetics are best observed in the huge chromosomes of fruit flies. The giant axon of the squid lends itself to studies of neurophysiology, particularly the mechanism of neuronal conduction, while the kidney tubules of the kangaroo rat have proved invaluable in elucidating the mechanism by which the kidney secretes an extraordinarily concentrated urine. Plant galls are simplified analogues of animal tumors that permit dissection of the elements of neoplasia. Domestic rodents—mice, rats, guinea pigs—have been employed on a large scale because they are cheap, breed relatively easily in captivity, are available in genetically homogeneous stocks, offer useful models of human disease states, and are now more thoroughly understood than are most other animal species. The cellular events of viral infection have been elucidated through detailed examination of infection of bacteria with bacteria-specific viruses, the bacteriophages. The simpler nervous systems of invertebrates such as insects and crabs offer models of neuronal integration, and their understanding paves the way for the immensely more complex nervous activity of the mammalian brain.

In each instance, the vast diversity of life permits selection of a particularly suitable study object. With the understanding so gained, subsequent examination of man or of the plants or animals important to his well-being is enormously simplified. In this sense, the approach to biological problems resembles the approach to many physical problems: the investigator seeks the most appropriate conditions for his experiments, conditions involving a minimum number of variables, all of which he may bring under control. It is regrettable that even well-intentioned critics have occasionally mistaken the biologists' search for the simplest possible model for dabbling with the inconsequential, failing to comprehend that, throughout the history of biology, far-reaching general principles have been deduced by comparative study of related structures or functions across a variety of species, genera, or phyla.

THE LANGUAGE OF LIFE

The last two decades have witnessed a prodigious gain in understanding of life at all levels. Undoubtedly, however, the crowning achievement of this era has been the spectacular growth of understanding of that process central to life itself—the chemical encoding of genetic information, the mechanism whereby it is read out to give direction to the life of the cell, and the mechanism whereby it is reproduced in the course of cell division. This area of understanding—variously termed molecular biology, biochemical genetics, the chemistry of reproduction, or the biochemistry of nucleic acids and proteins—flowered when the stage had been set. It could not have happened earlier and probably was inevitable when it did occur because of the centrality of the questions at issue.

Until 1940, genetics had been studied, in the main, with conventional higher species of plants and animals, the hereditary traits studied being those most readily observed—eye color, distribution of hair, flower pigments, etc. Through such studies, much of the language of formal genetics was generated, although the molecular mechanisms responsible were unknown. Biochemists had been concerned principally with identification and characterization of the chemical compounds characteristic of living forms and the pathways by which they are synthesized or degraded in cells. Microbiologists, long concerned with techniques for the identification and taxonomic classification of micro-organisms, then studied their susceptibility to sulfonamides and antibiotics as an adjunct to medical practice. Viruses had been a subject of study largely because of the diseases they engender in plant and animal species. By the mid-1940's it was possible to combine the understanding generated by these seemingly disparate disciplines into a concerted effort to understand the nature of the genetic apparatus.

It had long been apparent that the genetic complement of any individual must be encoded in some chemical form; and, although morphological traits had served the geneticist well, in fact, there can be no gene for height, or eye color, or number of teeth, or age of onset of baldness. Clearly, the genes that govern such parameters must actually govern specific chemical events, the consequences of which are evident in these more readily discerned characteristics. Slowly, the concept grew that each individual is a reflection of his complement of proteins; the latter serve both as structural materials and as the enzymes that synthesize all other types of biological chemicals. Each cell is whatever its proteins make possible, and the numbers of cells of each type and the manner in which they are distributed are, in some way, a consequence of genetic instructions with respect to which proteins to make and the relative amounts of each.

Belief in this concept began with the observations of Garrod in 1908, who assembled then-existing information concerning six hereditary diseases of man, indicating that each was the consequence of loss of some enzymic ability. This concept was solidified with studies of a bread mold, *Neurospora crassa,* which ordinarily can be grown on extremely simple nutritional media and synthesizes for itself all the usual amino acids, carbohydrates, purines, pyrimidines, vitamins, etc. Irradiated cultures of this organism were found to contain mutants that had lost the ability to make one or another of these vital components. By appropriate procedures it was ascertained which step in the sequence of chemical reactions by which such synthesis normally occurs (a "metabolic pathway" such as those shown in Figure 3) was actually blocked. In each instance, the mutant had lost the ability to catalyze one specific step in such a sequence. A large body of information accumulated from studies of a variety of bacterial forms confirmed this concept, encapsulated in the axiom, "A single gene determines the synthesis of a single enzyme."

The general relationship between the structure of genetic material and that of proteins, however, could not be established until the structural plan of proteins themselves had been revealed. The first protein to be studied appropriately was the pancreatic hormone, insulin, the structure of which is shown in Figure 4. Insulin proved to be constructed by the head-to-tail combination of 20 different kinds of amino acids; in all molecules of insulin, at each position along the chain, one and only one of the 20 possible amino acids does in fact exist. When the techniques developed for determination of this linear sequence became generally available, they were quickly applied to other proteins with similar results. Although a few differences were found among the insulins obtained from various species, in each species all the molecules of insulin are identical, as are all the molecules of cytochrome c, of myoglobin, and so on. These findings made explicit the nature of the information that must be encoded within the genetic material, i.e., instructions with respect to the linear sequence of amino acids in each of the proteins to be synthesized. If this concept is correct, the amino acid sequence of a given protein must, in some manner, be colinear with the instructions within the gene responsible for its synthesis.

The special advantage of utilizing micro-organisms is that it is possible to screen for mutants in populations of billions of individuals at one time and, by applying to them appropriate modifications of the classical tech-

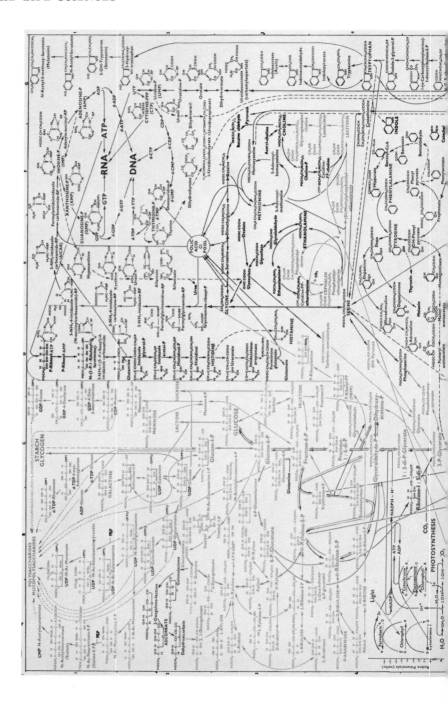

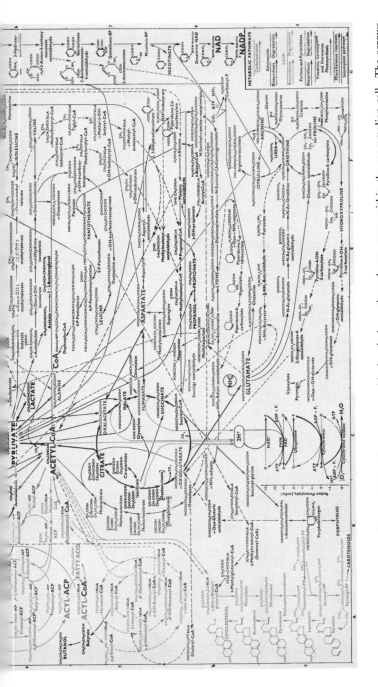

FIGURE 3 Metabolic Pathways 1969. The numerous biochemical reactions shown occur within the mammalian cell. The arrows indicate their diversity and interrelatedness. Details of the individual reactions are beyond the scope of this report. (Reprinted with permission of D. E. Nicholson, Department of Bacteriology, The University, Leeds, England, and Koch-Light Laboratories Ltd., Colnbrook, Bucks, England, with the knowledge of the U.S. distributor, General Biochemicals, Laboratory Park, Chagrin Falls, Ohio. Copyright © 1969 D. E. Nicholson.)

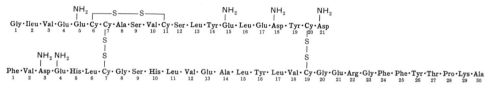

FIGURE 4   The amino acid sequence of bovine insulin. (From *Principles of Bio-chemistry*, 4th ed., A. White, P. Handler, and E. L. Smith. Copyright © 1968 McGraw-Hill, Inc. Used with permission of McGraw-Hill Book Company.)

niques worked out by geneticists for larger organisms, construct "maps" of the genes. The earliest such maps, of the genes of fruit flies, simply related the position of each gene to that of other genes along a chromosome. The refinements possible with bacterial genetics enabled construction of detailed maps indicating the relative positions of individual mutations along the length of a single gene. Such techniques have been applied to a considerable variety of bacterial and viral genomes (the totality of genetic material in a cell). A rather thoroughly studied gene in this regard is that which directs the synthesis of an enzyme in *E. coli* called "tryptophan synthetase." The positions along the map of hundreds of mutants of this gene have been scrutinized and related to the accompanying change in the amino acid sequence of the protein itself. A partial summary of these studies is shown in Figure 5, which illustrates the colinearity of the gene that provides the instructions for making tryptophan synthetase and the amino acid sequence of that protein itself.

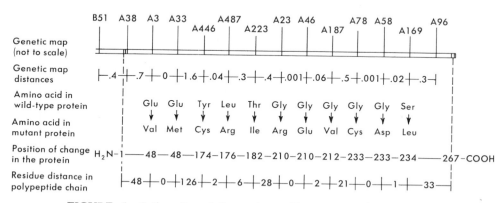

FIGURE 5   Colinearity of the amino acid sequence of the enzyme tryptophan synthetase and the substructure of the gene that governs its synthesis. (From C. Yanofsky, G. R. Drapeau, J. R. Guest, and B. C. Carlton, "The Complete Amino Acid Sequence of the Tryptophan Synthetase A Protein (α subunit) and Its Colinear Relationship with the Genetic Map of the A Gene," *Proc. Natl. Acad. Sci. U.S.* 57:296, 1967.)

Figure 5 also reveals a concept first made clear by an understanding of the defect in sickle cell anemia. This disorder is the consequence of an alteration in the structure of hemoglobin, in which, at only one specific position ($\beta6$) along a chain of 146 amino acid residues, there occurs a substitution of the amino acid valine for glutamic acid. This observation, with its profound implications for the understanding of genetic disease, reveals the nature of the simplest possible kind of mutation: A change in the structure of the genetic material at one point along the strand of genetic instruction results in substitution of one amino acid for another in the strand of amino acids.

The Genetic Material

Meanwhile, an ever-growing body of evidence indicated that the genetic material itself must be the polymer called deoxyribonucleic acid (DNA), which is peculiar to the cell nucleus and is the stuff of which chromosomes are made. The ultraviolet-absorption maximum of this material occurs at the wavelength most effective in creating mutants. As techniques for the purification of viruses accumulated, each in turn was found to contain a nucleic acid as a major component. The capstone in this argument, which was not truly recognized as such at the time it occurred, was a study undertaken for rather practical clinical purposes. Two strains of *Pneumococci* were known, one of which was characterized by an outer coat of a carbohydrate polymer that the other strain lacked. When, however, cell-free preparations of cultures of the former were added to cultures of the latter, they were "transformed," acquiring the ability to make the carbohydrate polymer and retaining that ability through an indefinite number of subsequent cell divisions. In effect, these cells had acquired a gene they formerly lacked.

The material in the cell-free culture that made this possible was found to be DNA—in retrospect, categorical evidence that DNA is indeed the material of which genes are made. The lack of immediate appreciation of the profound implications of that finding was a consequence of earlier studies of the structure of DNA, which were misleading in that they suggested it was a dull repetition of a fundamental repeating polymer unit without variation. In light of such studies it had seemed unlikely that the structure of DNA could be the basis for genetic instruction, which was known to require immense variation. Only as the structure of DNA was re-examined was this incorrect impression rectified (Figure 6).

More careful analyses, using newer techniques, demonstrated great variability even in the gross structure, the relative composition varying from species to species, yet with one pair of cardinal rules. All DNA's

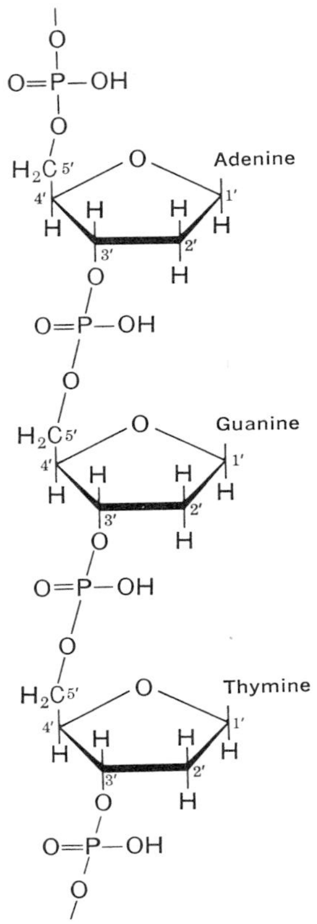

FIGURE 6 A segment of the backbone structure of a single strand of DNA. Representation of a portion of a DNA chain, showing the position of the inter-nucleotide linkage between C-3' and C-5'. (From *Principles of Biochemistry,* 4th ed., A. White, P. Handler, and E. L. Smith. Copyright © 1968 Mc-Graw-Hill, Inc. Used with permission of McGraw-Hill Book Company.)

are constructed of only four fundamental units called nucleotides: adenylic acid (A), guanylic acid (G), thymidylic acid (T), and cytidylic acid (C). This small number of units, nevertheless, permits encoding of a vast amount of information; indeed, in Morse code, with only two symbols, it is possible to transmit all the works of Shakespeare. In all specimens of DNA, A = T and G = C, whereas there is no consistent relationship between A and G. The meaning of this constancy was not apparent until combination of this information with studies of the x-ray-diffraction pattern of nucleic acids led to the now well-known depiction of DNA as two very long strands wrapped about each other in the familiar double helix, and so aligned that, on the strands, every A is opposed by a T, and each G is opposed by a C (Figure 7). In each case the pair of bases is linked by the relatively weak forces of hydrogen bonds, as illustrated in Figure 8. The great stability of the double helix is the consequence of the sum of thousands of such unions. Importantly, there is no rule with respect to the actual consecutive sequence along one strand. If one knows the sequence along one strand, one automatically knows the complementary sequence along the opposing strand, as shown in Figure 7.

This structure immediately solved the two basic questions concerning the chemical structure of genetic material. This material must, as an intrinsic property, both achieve its own self-duplication as cells divide and provide for the great variability that would permit the encoding of instructions to make the enormous diversity of proteins found in nature. Self-duplication is achieved in a most ingenious way. Decades of bafflement concerning how any chemical could achieve its own self-duplication were resolved by the recognition that as cells divide, the double strand is, in some manner, disengaged and each strand then governs the synthesis not of itself but of its complement. Where one double strand existed, two double strands are brought into being, each of which contains one of the original strands and a complementary new partner (Figure 9). A great body of evidence now supports that concept, although some of the details remain obscure. The base pairing described is the feature of the structure that determines how the new strands shall be formed. Occasionally mistakes are made, and when this occurs there is poor fitting of the strands. Under those circumstances a set of additional enzymes comes into play. One snips out the ill-fitting sections, thereby permitting another to reinsert

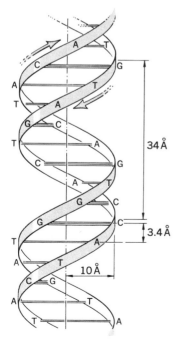

34 Å

3.4 Å

10 Å

FIGURE 7 Schematic representation of the double helix of DNA. The ribbons represent the deoxyribosephosphate backbone chains. The opposing arrows indicate that one strand is running from the 5′ position of one sugar to the 3′ position of the next, while the other strand is running in the opposite sense. The horizontal lines represent hydrogen bonds between opposing pairs, two for each AT couple, three for each GC couple. (From I. Herskowitz, *Genetics*, 2nd ed., 1962. Copyright © 1962 Little, Brown and Company.)

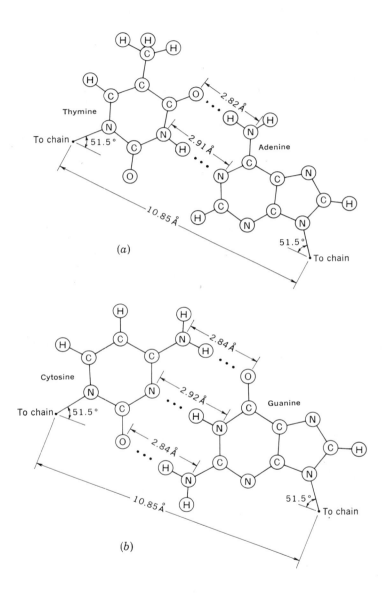

FIGURE 8   Molecular dimensions and hydrogen bonding of base pairs of DNA. (Adapted from S. Arnott, M. H. F. Wilkins, L. D. Hamilton, and R. Langridge, "Fourier Synthesis Studies of Lithium DNA, Part III, Hoogsteen Models," *J. Mol. Biol.*, 11:391–402, 1965, p. 392.)

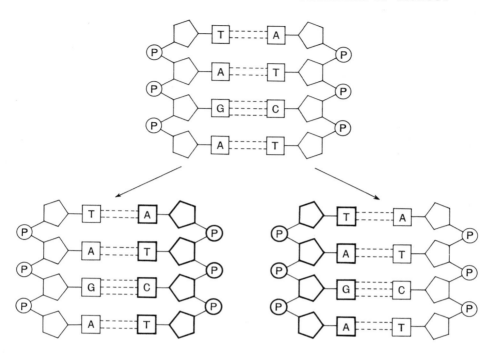

FIGURE 9   Scheme of replication of a DNA model. Boldface chains represent the newly synthesized strands of the two daughter molecules. (From J. Josse, A. D. Kaiser, and A. Kornberg, "Enzymatic Synthesis of Deoxyribonucleic Acid," *J. Biol. Chem.,* *236*:864, 1961. Copyright © 1961 The American Society of Biological Chemists, Inc.)

the proper bases, which then make the normal tight fit of the double helix. Interestingly, the double helix of a bacterial chromosome, like that of most viruses, is a circular molecule, the head, as it were, being joined to the tail. This was originally recognized by the mapping procedures noted above, and then it was visualized by electron microscopy. Duplication of the DNA, therefore, must commence by opening this circle.

In this concept, instructions for protein synthesis must be provided by the linear sequence of bases along the DNA strand; given the fact that the chromosome of *E. coli* consists of a single helix of about 10 million consecutive base pairs, and that any one of the four bases may lie to left or to right of any other base, there is essentially an unlimited number of statistical possibilities, only one of which is the actual structure of a specific DNA chromosome. And the possibilities become astronomical in a human cell, the DNA complement of which is 1,000 times as great as that of *E. coli.* Subsequent research efforts concentrated on the mechanisms by which the instructions encoded in the DNA are utilized to give direction to protein synthesis. This is now understood in remarkable detail, although there are a few serious gaps in this knowledge. The summary that follows encompasses the work of hundreds of investigators over a period of two decades.

PROTEIN SYNTHESIS

At an early stage it was recognized that, although the DNA holds the primary instructions for protein synthesis, it does not itself directly participate in that process. In the cells of higher organisms, DNA is locked in the nucleus, whereas protein synthesis occurs in small bodies called "ribosomes" stippled throughout the cytoplasm. It followed, therefore, that the instructions for protein synthesis in the DNA must be dispatched from the cell nucleus to the ribosomes. Ribosomes were found to be composed of a mixture of about 20 different proteins plus several forms of ribonucleic acid (rRNA); RNA differs from DNA in that the sugar component is ribose rather than deoxyribose, and most RNA is single-stranded rather than double-stranded. Ribosomes are leaflets constructed of a small component and a larger component, disposed much as a partially open clam. The messages from nucleus to ribosome were shown to consist of yet another form of single-stranded ribonucleic acid, messenger RNA (mRNA), which is fabricated in the nucleus by a specific enzyme called the DNA-dependent RNA synthetase. This form of nucleic acid is transcribed from one strand of the DNA helix by the same base-pairing rules, except that the pyrimidine nucleotide uridylic acid (U) is utilized, rather than the thymidylic acid (T) of DNA, so the four letters of the RNA alphabet are A, G, U, and C. Just as in DNA itself, the growing mRNA is formed on one of the DNA strands in an antiparallel manner (see Figure 9). In a living cell the long fibers of mRNA can be seen threaded through as many as 10 ribosomes at once (a polysome), so different areas of the message are being "read" by each of the ribosomes consecutively.

Assuming some kind of colinearity of the mRNA and the protein to be synthesized, it must be the base sequence of the mRNA that specifies the amino acid sequence of the protein. Since there are 20 different amino acids in proteins and only four letters in the RNA alphabet (A, G, U, C), obviously these cannot bear a one-to-one correspondence; moreover, one can form only 12 two-letter words with four letters. Hence, the minimal number of letters that would suffice is three per "word," i.e., the "codon" that specifies the exact amino acid next to be incorporated in a growing protein chain. Indeed, one can form 64 three-letter words with a four-letter alphabet. The problem then, was to establish the relationship between the four-letter alphabet of the RNA and the 20 words in the amino acid dictionary.

If three-letter words are utilized—and a large body of evidence now indicates that the code words for amino acids are built of three letters each—there seemed no way physically to relate the structure of the amino acids to a sequence of three nucleotides in RNA. Accordingly, it was postulated that some form of "adapter" would be required. That adapter proved to be yet a third general type of RNA, termed "transfer RNA" (tRNA), the smallest kind of RNA known. Several pure tRNA's have been isolated, each of which serves as the adapter for one specific amino acid; Figure 10 shows a complete structure of one tRNA. All tRNA's appear to be built along the same general plan.

Within the cell there is a family of "amino acid-activating enzymes," and the fact that the entire apparatus actually works successfully rests on the remarkable properties of these enzymes. In absolutely specific fashion, each such enzyme esterifies one and only one of the 20 amino acids to the hydroxyl group at the 2-position of the ribose at one end of one specific form of tRNA; it is imperative that the enzyme make no error since any such error would necessarily become an error in protein synthesis. What is shown in Figure 10 as the projecting round knob of the tRNA molecule is the "anticodon," a sequence of three bases which fit, by the usual base-pairing rules, against three consecutive bases in the messenger RNA, the codon for an amino acid. The amino acid is attached to the tRNA at a position quite remote from the anticodon. In a cell engaging in protein synthesis, there is a pool of all 20 amino acids, each affixed to its specific tRNA by virtue of the activity of the appropriate activating enzymes. As the long mRNA (500 to 10,000 nucleotide units) threads through the ribosome leaflet, it is attached to the smaller ribosomal component, while an amino acylated tRNA, which can achieve the necessary complementary base pairing to the message, is fixed in position on the larger member.

The first amino acid to be laid down is that at the amino terminus of the chain. The protein chain grows by the reaction

$$
\begin{array}{c}
\overset{O}{\underset{\parallel}{}} \quad R_1 \quad \overset{O}{\underset{\parallel}{}} \qquad\qquad R_2 \quad \overset{O}{\underset{\parallel}{}}\\
\text{------C—N—CH—C—tRNA}_1 + \text{H}_2\text{N—CH—C—tRNA}_2\\
\overset{}{\underset{}{\text{H}}}
\end{array}
$$

$$
\begin{array}{c}
\overset{O}{\underset{\parallel}{}} \quad R_1 \quad \overset{O}{\underset{\parallel}{}} \quad R_2\\
\text{------C—N—CH—C—N—CH—tRNA}_2 + \text{tRNA}_1.\\
\overset{}{\underset{}{\text{H}}}\overset{}{\underset{}{\text{H}}}
\end{array}
$$

As each such reaction is completed, the freed tRNA departs, and the mRNA must move through the ribosome so that the next three "letters"

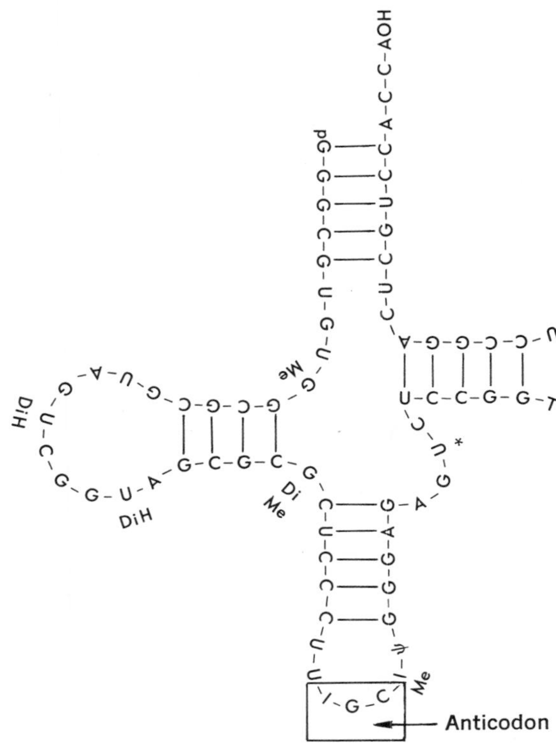

FIGURE 10 Base sequence and general structure of a tRNA for alanine. The anticodon, the three bases that pair with the three-base codon of mRNA, are shown at the bottom. As in DNA, and DNA–RNA hybrids the strand of tRNA is running in the opposite sense to that of the mRNA, to which it must attach on the ribosome surface. (From J. T. Madison and H. K. Kung, "Large Oligonucleotides Isolated from Yeast Tyrosine Transfer Ribonucleic Acid after Partial Digestion with Ribonuclease T1," *J. Biol. Chem.*, *242*:1324–1330, March, 1967, p. 1329. Copyright © 1967 by The American Society of Biological Chemists, Inc.)

are aligned at the working site. The mechanism by which this ratchet-like process is accomplished is totally unknown. These events are schematically shown in Figure 11.

There remained the task of establishing the dictionary; as the decade of the 1960's began this appeared to be a herculean task. A fortunate combination of accident and experimental virtuosity rapidly broke through this problem, the solution to which is shown in Figure 12. All 64 possible three-letter words are utilized and, hence, show considerable redundancy. Where more than one code word is utilized to signify the insertion of a given amino acid, an equal number of tRNA's must also be available to the cell; in several instances this has been shown to be the case. As will be seen, three of the three-letter words do not relate to any amino acid; where these occur in the mRNA sequence, no amino acid can be inserted, and synthesis of the protein chain terminates and the tRNA at the end of the chain is removed by hydrolysis. This, therefore, is the "punctuation" in the message, the "period that ends the sentence."

Understanding of exactly how chain initiation is accomplished is less satisfactory. It is all-important that reading of the message begin at a precise point; if the reading frame were to shift by one letter, the entire message would be garbled and a completely different set of amino acids would be assembled. In bacterial forms, it would appear that the first amino acid in the chain is always the same—methionine, the amino group of which bears a formyl group. Message reading, therefore, begins by utilizing formyl methionine tRNA as the first "word" and is terminated when any one of the three nonsense words shows in the message. Acetyl serine tRNA may serve the same role in animal systems.

One of the most powerful tools in the multitude of studies that gave rise to the remarkable picture just described was the experimental demonstration of the effectiveness of base pairing. The force that holds the two strands of DNA together is the sum of a multitude of tiny forces—hydrogen bonds between nitrogen and oxygen atoms. These are the forces that also hold

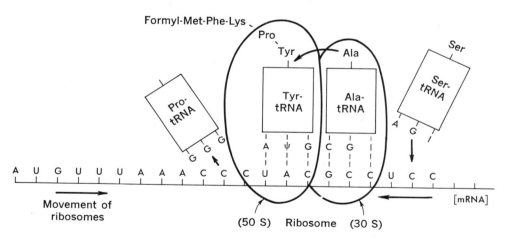

FIGURE 11   Schematic representation of protein synthesis on a ribosome. The protein tRNA is departing, having been used in the previous step; the bond between tyrosine and its tRNA is being displaced by the amino group of alanine; and a seryl-tRNA is just coming into position. (From *Principles in Biochemistry*, 4th ed., A. White, P. Handler, and E. L. Smith. Copyright © 1968 McGraw-Hill, Inc. Used with permission of McGraw-Hill Book Company.)

2nd POSITION

1st POSITION		U	C	A	G	3rd POSITION
	U	PHE PHE LEU LEU	SER SER SER SER	TYR TYR CT-1 CT-2	CYS CYS CT-3 TRY	U C A G
	C	LEU LEU LEU LEU	PRO PRO PRO PRO	HIS HIS GLN GLN	ARG ARG ARG ARG	U C A G
	A	ILU ILU ILU MET	THR THR THR THR	ASN ASN LYS LYS	SER SER ARG ARG	U C A G
	G	VAL VAL VAL VAL	ALA ALA ALA ALA	ASP ASP GLU GLU	GLY GLY GLY GLY	U C A G

FIGURE 12 The genetic code. Each amino acid in a protein is specified by a nucleotide triplet in RNA, e.g., aspartic acid (asp) is specified by the triplets GAU and GAC. UAA, UAG, and UGA are utilized for punctuation, viz., to indicate when to terminate the chain.

crystals of water, i.e., ice, together and, like ice crystals, they can be melted by warming. If double-stranded DNA is brought to an elevated temperature, the double helix comes apart and the DNA becomes individual random flopping coils. If the temperature is then lowered very slowly, the coils find each other and the double helix is restored. By the same technique one can prepare in the laboratory hybrid DNA–RNA complexes, but only when the latter can be aligned and can join by multitudinous base pairing. In this way it was shown that a rather substantial fraction of DNA of E. coli and a much larger fraction of mammalian DNA are used to code for the preparation of rRNA, but only a tiny fraction, less than 0.1 percent, specifies the formation of all the tRNA's. And by the same procedure one can artificially achieve attachment of tRNA onto RNA. Indeed, this technique was the basis for the most successful procedure for determination of the genetic code. Thus, a trinucleotide of RNA of known base sequence can be added to a ribosome suspension. To a sample thereof is added a tRNA charged with its amino acid, the latter labeled with ^{14}C. If the radioactivity adheres to the ribosome–RNA complex, the tRNA has paired its anticodon with the trinucleotide and the code word is established.

Much of these concepts had been elaborated by deduction from a great variety of experimental observation. A capstone on this intellectual structure was provided by a series of maneuvers, conducted with great technical skill, which achieved the isolation, in pure form, of a single gene from among the many of the genome of *E. coli,* that which directs the synthesis of any enzyme called β-galactosidase. The final proof of this overall picture has been provided by a remarkable *tour de force.* From knowledge of the structure of the tRNA for alanine, an antiparallel, complementary length of DNA was synthesized chemically. The usual DNA-synthesizing enzyme was utilized to form its complementary DNA strand. This relatively short double-stranded DNA was then used with the DNA-dependent RNA-synthesizing enzyme to form the tRNA for the amino acid alanine and yielded the predicted structure. This constitutes true chemical synthesis of a gene! Similarly, synthetic lengths of RNA have been used with ribosomes as mRNA and have yielded small polypeptides of the structure predicted by the code.

Self-Assembly Such studies demonstrated that base pairing is the primary mechanism involved in DNA duplication, in the synthesis of RNA on DNA, and in message reading in the ribosome. But they also demonstrated a cardinal principle of the biological world, *the principle of self-assembly.* Organisms are assemblages of cells, and within the cells there are myriad organized sub-cellular bodies, within which, in turn, are macromolecules that are aggregates of smaller molecules. Yet, as we have seen, all that genetic instructions can provide is information descriptive of the synthesis of protein chains. Are other types of instructions required, or does all else follow from the fact of protein-strand synthesis? The answer, unquestionably, is that all else is derivative, that all other structures combine, because they do indeed fit together and are held together by a collection of small forces, much as are the two strands of DNA that make such a remarkably tight fit.

For example, the hemoglobin molecule consists of four subunits, two α-chains, and two β-chains. The two chain types can be separated by appropriate means but, when remixed, the normal tetrameric units reconstitute themselves without assistance. The enzyme ribonuclease is a single protein chain within which are three internal disulfide bridges (—S—S—). These can be opened by appropriate chemical means (reduction). In this form the enzyme is a random flopping coil and lacks catalytic activity. When allowed to reoxidize slowly, virtually every molecule regains its enzymic activity, despite the fact that there are eight different ways in which the disulfides might have recombined, only one of which would have assured catalytic activity. Accordingly, the manner in which the sulfhydryl groups "find each other" and are reoxidized is an inherent property of the primary amino acid sequence.

Yet another illustration of this principle is the manner in which whole viruses are constituted. For example, a bacteriophage virus consists of a core of nucleic acid surrounded by a coat protein, a base-plate protein, at least one enzyme, and tail fibers. Assembly of the virus normally occurs by an orderly, stepwise process. But mutants are known that are unable to conduct one or another of the syntheses required and, as shown in Figure 13, accumulate incomplete parts of the virus. When such pieces are simply mixed, the complete virus forms without any guiding agent, enzyme, or energy source, much as simpler molecules assemble to form orderly crystals. The final structure of the virus is the obligatory consequence of the three-dimensional structures of its various parts. By extension of this concept, the variety of structures common to a cell—membranes, ribosomes, chloroplasts, mitochondria, etc.—are all considered to come into being by spontaneous self-assembly in consequence of the existence in the cell of the requisite subunits, the system attaining in every instance the state of lowest free energy. This concept can be extended to the manner in which cells "recognize each other" and form the cell aggregates that are tissues and organs. It is an intriguing stretch of the imagination to project this still further to the fact that biological individuals recognize each other and dwell communally as colonies of bacteria, schools of fish, herds of mice, or clans of human beings.

THE LIFE AND TIMES OF A CELL

A living cell is a remarkable and rather unlikely object. Within itself it fabricates, from a few low-molecular-weight materials received from the environment, a great variety of additional molecular species, some of which are then utilized for the further synthesis of various high-molecular-weight polymeric materials, including nucleic acids, proteins, and polysaccharides. And these continue to exist within the cell despite the presence therein of enzymes capable of degrading them back to their original monomeric forms. Numerous low-molecular-weight compounds and mineral ions are maintained in concentrations decidedly greater than those that prevail in the cell's environment. In general, the medium surrounding the cell is rich in sodium ions and low in potassium ions, whereas the interior of the cell is rich in potassium and low in sodium, despite the fact that the outer mem-

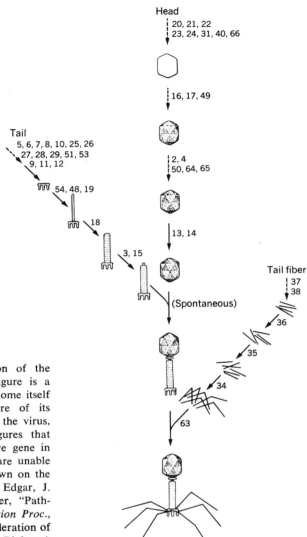

FIGURE 13 Sequential construction of the
T-4 bacteriophage. At bottom of figure is a
complete virus particle. The viral genome itself
conveys instructions for manufacture of its
parts. Specific individual mutants of the virus,
denoted by the numbers in the figures that
represent the position of the defective gene in
the chromosomal map of the virus, are unable
to perform the specific operation shown on the
arrow. (From W. B. Wood, R. S. Edgar, J.
King, I. Lielausis, and M. Henninger, "Pathway of T-4 Morphogenesis," *Federation Proc.*,
27:1163, 1968. Copyright © 1968 Federation of
American Societies for Experimental Biology.)

brane of the cell permits passage of both substances. This remarkable
dynamic steady state, so remote from equilibrium with the environment,
is maintained by the cell through the continual expenditure of energy.

Still more remarkably, such a structure is capable of accepting additional
materials from the environment and proceeding through a complex set of
operations that results in the formation of two daughter cells identical in

structure and potential to the original parent cell. Conceivably, the earliest "cell" was rather like some currently existing bacteria, although it may be assumed that it did not possess an outer coating or cell wall. Presumably, no such coating was required because the osmotic pressure of the environment was much the same as that of the cell interior. With the passage of eons, such single-celled objects found themselves in environments inimical to their existence, where extracellular concentrations differed from those of the original primordial soup. Under these circumstances, mutations that resulted in greater capability for meeting the environmental challenge imparted survival value. Among these was acquisition of the ability to surround the cell with a tough outer wall, which is evident in all current plant cells and in bacteria. The latter show a great diversity of such structures, all of which share one aspect: The entire wall surrounding a single cell is a single "bag-shaped" molecule made of several different types of repeating units. In higher plants, the wall is fashioned of cellulose that imparts not only stability against changes in the salt concentration of the environment, but also vertical rigidity to the growing plant.

The Energy for Cell Work

From all available evidence, at the time these events occurred the atmosphere of the planet was rather different from what it is today and contained such gases as methane, hydrogen, and carbon monoxide, as well as some carbon dioxide, but little if any oxygen. One can only assume that the earliest cells "learned" to utilize as their energy source the energy available in anhydrides of phosphoric acid, particularly that of adenosine triphosphate (ATP) (Figure 14), since it is the energy available from hydrolysis of the bonds between the phosphoric acid components of this compound that, in all currently living cells, is utilized to drive all synthetic chemical processes and all other events requiring a source of energy, e.g., secretion, contraction, conduction of electrical impulses, or emission of light (Figure 14). As the supply of this compound, available from the primordial soup, disappeared, an advantage accrued to those cells that had "learned" to synthesize ATP for themselves by utilizing the energy potentially available in the chemical structure of glucose. Such processes have persisted in all living cells to the present time. When performed by bacteria, they are called fermentations; in animal cells the related process is called glycolysis. In human muscle, the process can be summarized by the following equation, which requires the operation of 12 distinct enzymes:

$$1 \text{ glucose} + 2\text{ADP} + 2\text{P}_i \longrightarrow 2 \text{ lactic acid} + 2\text{ATP}.$$

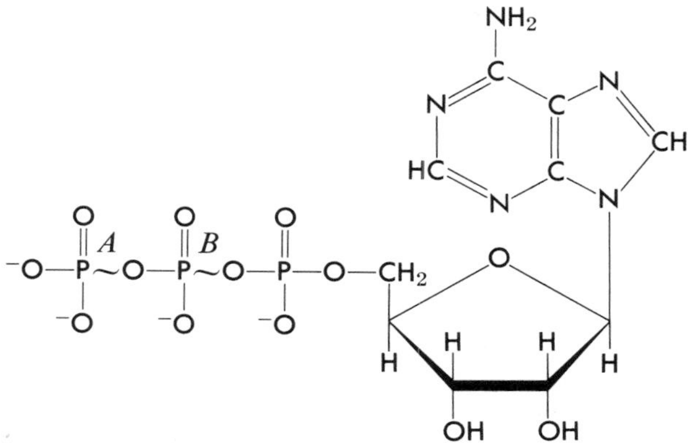

FIGURE 14 Structure of adenosine triphosphate (ATP). The wavy lines represent bonds whose hydrolytic rupture is accompanied by a relatively large change in free energy. Hydrolysis of bond A yields adenosine diphosphate (ADP); hydrolysis of bond B yields adenosine monophosphate (AMP).

P_i in these equations denotes inorganic phosphate. The process may be more familiar in the manufacture of beer, in which yeast conducts an essentially similar process:

$$1 \text{ glucose} + 2ADP + 2P_i \longrightarrow 2 \text{ ethanol} + 2CO_2 + 2ATP.$$

PHOTOSYNTHESIS

As long as the supply of glucose in the medium sufficed, this would have permitted generation of ATP to meet the requirement of the cell for an energy source. But the time came when the sugar in the cellular environment was consumed; at that time an advantage accrued to any cell that "learned" to utilize some other means of trapping energy—e.g., solar energy—in such a way as to generate a supply of ATP. This may be presumed to have happened in the early progenitors of current forms of algae or purple bacteria. The particular novelty required was a pigment that could absorb the energy of light and in some manner take advantage of that circumstance. That pigment was, and still is, chlorophyll. In somewhat loosely organized minute membranous bodies, called "chromatophores," the chlorophyll can accept the energy of sunlight. (It looks green because it is absorbing the red light of sunlight and reflecting green light back to the observer.) From the light-activated chlorophyll molecule is ejected an electron that is transferred to some acceptor and then to an iron-protein called ferredoxin, the chlorophyll being left as a free radical. In the most primitive instance, this electron is transferred in turn from one carrier to another and ultimately returns to the original chlorophyll; in the course of this electron passage along the consecutive carriers, one or two molecules of ATP can be synthesized.

In a cell to which there was still available a reasonable abundance of diverse organic chemicals, this process would suffice as a source of energy because the cell still could utilize substances taken from its environment for building materials. But, in time, the organic chemical supply of that environment must have dwindled; then those cells gained an advantage that could not only make ATP by the photosynthetic process but also manage to utilize photosynthetic energy for the reductive formation of carbohydrate from CO_2 in the environment. Even now, such processes are evident in photosynthetic bacteria and algae.

The actual reactions by which CO_2 is fixed into carbohydrate are not, strictly speaking, *photo*synthetic; they readily occur in the dark in a sequence of reactions made possible by an assortment of 12 different enzymes. The overall process may be summarized as:

$$6CO_2 + 18ATP + 12TPNH + 12H^+ \longrightarrow$$
$$\text{glucose} + 18ADP + 18P_i + 12TPN^+. \qquad (1)$$

TPN is a small molecule (triphosphopyridine nucleotide) that can accept a pair of electrons and is thereby reduced to TPNH. But Equation 1 thus defines the role of the true photosynthetic process, which is to provide the requisite ATP and TPNH. The overall photosynthetic process may be summarized as:

$$MH_2 + TPN^+ + xADP + xP_i \xrightarrow{h\nu} M + TPNH + H^+ + xATP. \qquad (2)$$

Some details of this mechanism are shown in Figure 15. The reaction expressed in Equation 2 would be thermodynamically impossible were it not for the energy of sunlight coupled into the system through the action of chlorophyll.

As shown in Figure 15, the ultimate source of reducing power, the source of the electrons to reduce the TPN to TPNH, is an organic compound, like succinic acid. Electrons flow over a series of carriers to the chlorophyll free radical that is formed when photoactivated chlorophyll releases its electrons to the iron-protein, ferredoxin, and thence to TPN. But again we may imagine that with the passage of time the supply of compounds like succinic acid was seriously depleted. Ultimate advantage then accrued

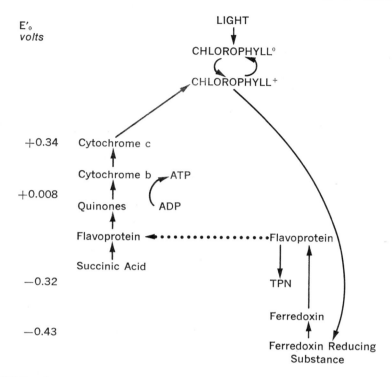

FIGURE 15   Path of electron flow in bacterial photosynthesis.

to those plant cells that acquired the capability to replace the succinic acid of Equation 2 (MH$_2$) with water (H$_2$O). This remarkable capability appeared in the higher plants.

$$2H_2O + 2TPN^+ + xADP + xP_i \xrightarrow{h\nu} O_2 + 2TPNH + 2H^+ + xATP. \tag{3}$$

This process is even less likely, thermodynamically, than that summarized in Equation 2. The overall reaction, glucose + oxygen to yield water and CO$_2$, is by far the more likely event, familiar as the combustion of wood. The reverse process, (3 + 1), is made possible by the extraordinary organization of molecules in the chloroplast, summarized in Figure 16. Two distinct chlorophyll centers collaborate in the process. The electrons ejected from the first chlorophyll (on the left side of the figure, as drawn) are replaced from water molecules with the evolution of oxygen gas. Electrons ejected from that activated chlorophyll traverse a set of carriers until they reach a second chlorophyll center (P$_{700}$). This must be irradiated in

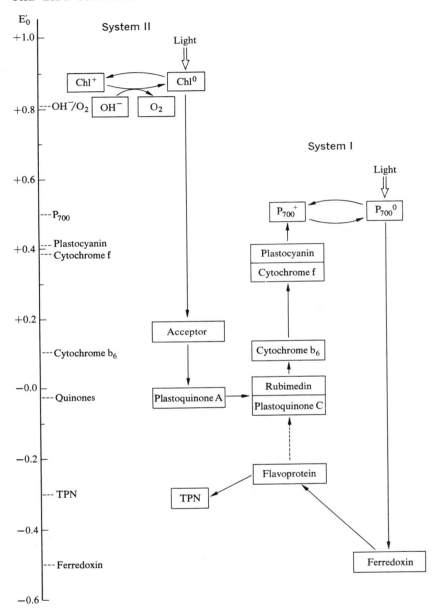

FIGURE 16   Path of electron flow in photosynthesis by higher plants. (From *Principles of Biochemistry*, 4th ed., A. White, P. Handler, and E. L. Smith. Copyright ©
1968 McGraw-Hill, Inc. Used with permission of McGraw-Hill Book Company.)

turn; at the second center an electron is ejected and accepted by ferredoxin as in simpler systems, and the electron ejected from this second chlorophyll is replaced by the one that was ejected from the first chlorophyll. About 15 different kinds of molecules are required in this overall process, all of which must be located quite specifically with respect to each other in the working machine. Given a supply of TPNH and ATP made by this marvelous machinery, these can be utilized elsewhere in the cell to "fix" carbon dioxide into glucose. This is the process upon which all life on this planet ultimately depends. This is the primary source of organic materials generated by the chemistry of higher plants and made available to animals by ingestion, as well as the source of the oxygen that animals require in order to metabolize the organic compounds they have eaten. It was this process in times past that generated the organic compounds left to us as the carbon and hydrocarbons of our fossil fuels. These concepts of the evolution of the photosynthetic machinery, based entirely upon living forms now available, were entirely unknown a decade ago. Not only the overall outlines, but all the intimate details of these processes, have been revealed in the years since World War II, utilizing a great variety of techniques and instruments developed during this period.

The appearance of animal cells necessarily had to wait until plant-photosynthetic activities were well advanced and had enriched the atmosphere with oxygen while generating a supply of organic compounds. Animal cells developed their own miniature power plants, termed "mitochondria," in which electrons taken from ingested foodstuffs similarly pass, hand to hand, as it were, over a series of intermediary carriers and are delivered to oxygen with the formation of water. As in chloroplasts, this passage of electrons is coupled to the operation of a molecular machinery that can generate ATP, so that, in animals, the overall process is described by the following reaction:

$$\text{glucose} + 6O_2 + 38ADP + 38P_i \longrightarrow 6CO_2 + 38ATP + 6H_2O.$$

Metabolism

METABOLIC PATHWAYS

In animal cells, as in plant cells, it is the energy intrinsically available in the structure of ATP that is utilized to drive all other processes associated with the life of the cell. In both series of cells, function is intimately associated with the architecture of the minute bodies—chloroplasts or mitochondria—in which these events occur. The "life" of the cell then consists of the utilization of ATP and other raw materials to synthesize the variety of compounds required for cell function and for cell growth and division, as well as to maintain the constancy of the interior of the cell in the face of the challenges of its environment. Each cell, therefore, is a miniature chemical factory. Plant, algal, and bacterial cells have wider capacities for synthetic activity than do animal cells, which rely on the previous chemical activity of the plant material that they ingest as the source of many of their raw materials. The incredible complexity and diversity of this chemical activity is shown in Figure 3, pages 38 and 39, in outline form only. Starting with a trivial number of precursors in the environment—inorganic elements and carbon dioxide—cells synthesize the huge number of compounds shown in this "metabolic map." As the 1940's opened, only a handful of those compounds had been identified. By the mid-1960's, the major features of the metabolic map were well-nigh complete and, where gaps existed, intelligent guesses could be made with respect to the missing compounds. The sequential events from a common starting material such as glucose to an end product—e.g., an amino acid required for protein synthesis—is termed a metabolic pathway. A few generalities in this regard are noteworthy.

METABOLIC CONTROLS

A pathway commences with a reasonably readily available material, e.g., glucose, made by the photosynthetic apparatus

$$
\begin{array}{c}
\text{m} \\
\uparrow \\
\text{glucose}\rightarrow a\rightarrow b\rightarrow c\rightarrow d\rightarrow e\rightarrow f\rightarrow g\rightarrow \text{end product} \\
\downarrow \\
\text{n}
\end{array}
$$

Each step thereafter is made possible by the catalytic activity of a specific protein called an enzyme, which serves no other function in the life of the cell. Early in the pathway, there may be branches in which intermediates can be utilized for entry into other pathways, but one step, in our example the reaction b→c, is called the "committed step" in the sense that, thereafter, the intermediates serve no purpose in the life of the cell but as stages in the ultimate formation of the desired end product.

Most importantly, the rate at which the overall process proceeds, in most cells, is dictated in turn by the requirements of the cell for the end product. This is achieved in a variety of ways. In mitochondrial oxidation, the formation of ATP is tightly coupled, as if by a clutch, to the passage of electrons. If no ADP is available for synthesis into ATP, electron transfer comes to a halt, *viz.,* the cell oxidizes glucose only when the energy so released can be utilized for the formation of ATP, a process strikingly evident in muscle cells, in which the requirement for ATP can rise spectacularly when the muscle begins to contract.

In a synthetic pathway, the end product itself, when present in sufficient amount, can "turn off" the pathway, so that no more is manufactured until it is needed. This is brought about in two quite different ways, both of which are examples of "negative feedback." The most thoroughly studied example of negative feedback is the formation of cytidylic acid (CTP) (related to the "C" of the genetic code) in a reaction sequence that begins with aspartic acid, an amino acid. This is schematically represented in Figure 17. The committed step consists of the carbamylation of aspartic acid to form N-carbamyl aspartic acid, catalyzed by aspartyl transcarbamylase. The rate of this enzymic reaction is markedly reduced by the presence of CTP. This enzyme can be dissected into two types of subunits. The larger variety, even in the absence of the smaller, can catalyze the reaction in question but is not affected by CTP. The smaller subunits tightly bind CTP but have no catalytic properties. When both types of units are recombined, catalytic property is retained, but CTP binding to the smaller units so alters the three-dimensional structure of the enzyme complex that its

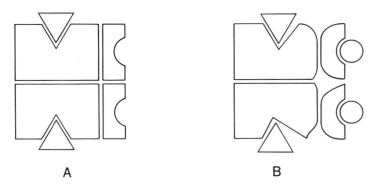

A   B

FIGURE 17   Schematic representation of allosteric inhibition. *A* shows the enzyme, constructed of two enzymic subunits with active sites at the angular clefts, which exactly fit the substrate (triangles). The smaller units have no substrate binding sites but can bind the final product (circles) of the reaction sequence in which the enzyme participates. When the latter binding occurs (*B*), the structure of the reacting subunit is deformed, in turn deforming the enzymic units so that they no longer snugly fit the substrate. Hence, enzymic function is inhibited by the end product of its catalysis. (From *Biology and the Future of Man,* P. Handler, ed. Copyright © 1970 by Oxford University Press, Inc.)

catalytic properties are markedly inhibited, a process termed "allosteric inhibition." Numerous examples of such allosteric inhibition have been observed in the last decade; all conform to this general pattern.

An additional form of feedback control is evident in bacteria. It has long been known that bacteria such as *Escherichia coli* are capable of synthesizing for themselves all 20 of the amino acids that they require. However, when placed in a medium rich in those amino acids, they rapidly cease to manufacture any and grow exclusively by use of the amino acids from the medium. In part, this is accounted for by the kind of feedback control described above, *viz.,* the amino acids inhibit the enzymes responsible for the committed steps in their own syntheses. In addition, however, with time, as the cells multiply, the actual amounts of those enzymes dwindle. Such cells actually cease to make the enzymes responsible for the synthesis of the amino acids already present in the medium, a phenomenon called "repression." Understanding of this process came from studies of the converse of this process, which has also been observed. When *E. coli* are placed in a medium that contains milk sugar (lactose) instead of glucose, they very rapidly begin to produce enzymes that permit them to metabolize the milk sugar; this process is called "induction." In both repression and induction, control is exercised by regulation of the expression of the genetic apparatus. When *E. coli* is placed in a medium containing lactose, rapid accumulation of three enzymes is observed. One, β-galactosidase, makes possible the hydrolysis of lactose, a double sugar of glucose linked to galactose, into its two components; a second is a protein, galactose permease, which serves in the membrane of the cell to facilitate entry of galactose into the cell interior; and the third catalyzes the acetylation of galactose, a process whose metabolic meaning is obscure.

Detailed analysis of these events has led to the following formulation as summarized in Figure 18. Within the circular chromosome of *E. coli,* there is a region that can be mapped as five consecutive genes. The terminal three are the genes that when expressed, give rise to the synthesis of the three proteins cited above. The first gene, however, directs synthesis of a protein, the repressor, which, when formed, attaches itself to the second gene. When it is so attached (the manner of attachment is unknown at this writing), the subsequent three genes cannot be transcribed; no mRNA is made on their surfaces, and accordingly the cell is unable to metabolize lactose. If, however, a small amount of lactose enters the cell, it preferentially binds to the repressor protein made by the activity of the regulatory gene, removing it from its attachment to the operator gene, thus freeing the rest of this locus for its expression, and the cell rapidly gains ability to metabolize milk sugar. That this is the case has been demonstrated by the actual isolation of the repressor protein, by the finding that it does indeed bind to some portion of the chromosome, but not as tightly as it does to lactose, thereby giving direct experimental confirmation to postulates deduced from the general behavior of the system. Thus, "induction" is more properly regarded as "derepression."

The process of repression typical of the cell growing in a medium rich in amino acids is visualized as a modification of the process described above.

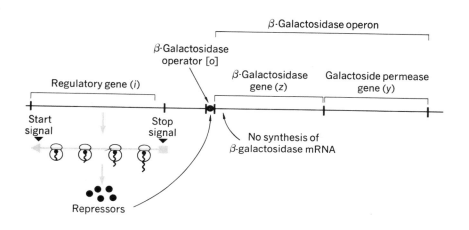

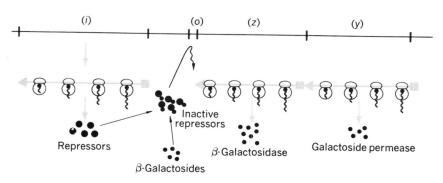

FIGURE 18 Structural relationships and function of the lactose operon. In the top figure, repressor protein, made by action of regulatory gene, binds to operator gene, and the other genes of the operon are not functional. Newly arrived substrate, below, binds to repressor protein, which departs from regulator gene, and the structural genes of the operon go to work, as indicated by the growing protein chains on the ribosomes. (From J. D. Watson, *Molecular Biology of the Gene*, W. A. Benjamin, Inc., New York, 1965. Copyright © 1965 W. A. Benjamin, Inc.)

The repressor gene is thought to make a protein that can lock onto a regulator gene, but the repressor protein is successful in repressing only when it is in the proper three-dimensional conformation that results from binding to it of the amino acid—purine, pyrimidine, etc.—which is the repressing agent. By a combination of such modifications of the genetic apparatus and allosteric inhibition of enzymes, the cell's chemical activities are brought into a harmonious whole, all activities proceeding at rates commensurate with the cell's requirements at any time.

ACTIVE TRANSPORT

Every cell is in contact with its environment at its interface, the plasma membrane. But movement across this membrane is not the mere consequence of passage through little holes. Although a few components—e.g., water and lipid-soluble gases—can move freely, most materials—*viz.*, amino acids, sugars, and charged ions—cross the membrane only in consequence of a process termed "active transport." As noted earlier, galactose permease takes its place in the bacterial cell membrane when the cell has been derepressed by the presence of milk sugar. Available information indicates that on the surface of this protein there is a site that tightly binds the sugar galactose. Thereafter, the protein rotates within the membrane so that the galactose so bound now faces inward. A second site on the protein then binds a molecule of ATP. When the latter attaches, the three-dimensional structure of the protein is so altered that the galactose falls off and becomes available for the metabolic activities of the cell interior. Hydrolysis (rupture of a covalent bond by addition of the elements of water), by that same protein, of the ATP into ADP and P_i discharges these from its surface, and the lactose permease molecule becomes free to rotate in the membrane to repeat the cycle. Only presumptive evidence suggests that similar events account for the active transport of the potassium ions into cells and sodium ions of cells, of amino acids, of other sugars, and so on. But for the moment, this appears to be an acceptable working model.

ENZYMES

The working machinery of the cell is its complement of enzymes. Enzymes have been studied for more than 50 years, but the pace of this endeavor has accelerated enormously in recent times. The primary task has always been to relate the structure of an enzyme to its catalytic function. Many types of enzymes have been under continuous study; the best understood of all enzymes is ribonuclease, made in the pancreas and secreted into the intestine, where it serves for digestive purposes by hydrolyzing long-chain ribonucleic acids into the smaller nucleotide units of which they are composed.

The initial approach to this problem was to establish the linear sequence of amino acids along the chain, an accomplishment that was completed by the early 1960's (Figure 19). A great body of evidence was then assembled that indicated that the catalytic properties of this enzyme are dependent upon the coordinated, concerted activity of the histidine residues in positions 119 and 12 of the chain. It was then assumed that the polypeptide chain must be so coiled as to offer a crevice into which the substrate, ribonucleic acid, would fit, such that these two histidine residues would then sandwich the bond to be subjected to hydrolysis. Convincing proof that this was so awaited the advent of x-ray crystallography, which provided a total three-dimensional representation of the ribonuclease model (Figure 20). In complete accord with expectations, in this as in every other hydrolytic enzyme examined to date, there is indeed a crevice into which the substrate molecule can fit, and, as predicted, the two working histidine residues are so located as to sandwich the portion of the substrate molecule to be attached.

This description also accounts for the catalytic activities of enzymes generally. Each must be so arranged that the substrate is attracted to its surface and tied down by appropriate groups on that surface so as to permit

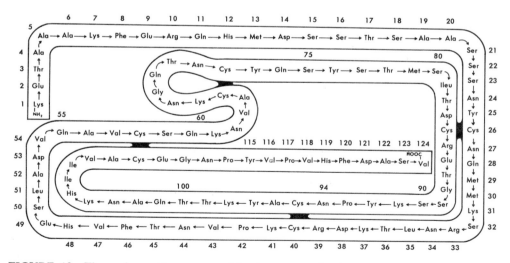

FIGURE 19   The amino acid sequence of bovine pancreatic ribonuclease. (From D. G. Smyth, W. H. Stein, and S. Moore, "Sequence of Amino Acid Residues in Bovine Pancreatic Ribonuclease: Revisions and Confirmations," *J. Biol. Chem.*, *238*:227, 1963. Copyright © 1963 The American Society of Biological Chemists, Inc.)

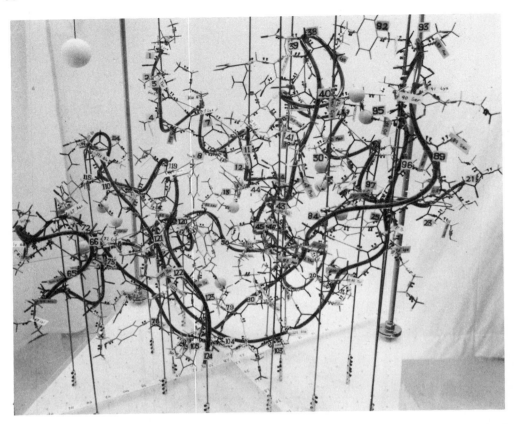

FIGURE 20   Three-dimensional model of ribonuclease. The continuous strand of tubing follows the peptide chain. α-Helix constitutes about 15 percent of the molecule: residues 2 to 13 (partly visible), residues 26 to 33, and residues 50 to 58 (in the back of the molecule). There is a section of antiparallel β-structures comprising residues 71 to 92 and 94 to 110. The pairs of balls represent sulfur atoms in disulfide bridges, e.g., between 40 and 95 at the upper right. The single balls represent sulfur atoms of methionine residues. The imidazole rings of histidines 12 and 119 are, as predicted from the chemical data, near one another in a groove. The competitive inhibitor, 5-iodouridine-2'(3')-phosphate, is bound in this groove. (From H. W. Wyckoff, K. D. Hardman, N. M. Allewell, T. Inagami, L. N. Johnson, and F. M. Richards, "The Structure of Ribonuclease-S at 3.5 A Resolution," *J. Biol. Chem.*, 242:3984, 1967. Copyright © 1967 The American Society of Biological Chemists, Inc.)

attack by specific "working" residues on the enzyme surface. Such residues include phenolic groups from tyrosine, carboxyl groups from glutamic acid, and sulfhydryl groups from cysteine, as well as the imidazole group of histidines. It is the combination of specific binding and "concerted attack" that gives enzymes both their high degree of specificity for substrates and their remarkable catalytic rates.

The final proof in the present instance was the total synthesis of ribonuclease from component amino acids by completely chemical procedures. When a chain of amino acids was synthesized with the sequence that had been ascertained by analytical procedures, it spontaneously assumed the three-dimensional conformation that results in normal catalytic activity of this enzyme, the ultimate triumph capping five decades of research. At the same time, this new capability offers a bold vista for the future, the prospect of synthetic proteins that will serve a variety of purposes. Perhaps the most important immediate prospect is the availability of the polypeptide hormones of the anterior and posterior pituitary glands, lack of which results in a variety of human disorders. In several cases—e.g., growth hormone—specificity for man is absolute in that material obtained from bovine or porcine sources is ineffective; patently, the supply of human hormones is limited.

SUBCELLULAR ORGANELLES

The advent of electron microscopy ushered in a new era of understanding of the structure and functioning of living cells; the seemingly simple little box with an enclosed nucleus is, in fact, a highly detailed, organized structure (Figure 21). The nucleus itself was found to be enclosed in a double membrane. Occasionally, that membrane is interrupted and becomes continuous with membranous sheets that continue through the cell, some of which find their way to the plasma membrane at the cell surface and thence to the cell exterior. These sheets of membranes through the cell are termed the "endoplasmic reticulum." In cells engaged in protein synthesis, these membranes are stippled with ribosomes, the complex little machines that achieve protein synthesis. It is noteworthy that ribosomes also come into being by self-assembly. By appropriate means, separated ribosomes can be made to fall apart into a collection of several different kinds of nucleic acids and approximately 18 different proteins, the roles of which are not yet evident. When the medium is changed, all these components reaggregate and assemble spontaneously into working ribosomes. Thus the genetic machinery does not have to give additional instructions to assure construction of these little chemical factories. Nor is there reason to think that any additional instructions are needed for the synthesis of the membranes themselves. These are constructed of lipids (fatty materials) and proteins; some are enzymes or participate in the active transport process and some may be purely structural. As the cell grows, these membranes continue to increase by the incorporation of additional proteins and lipids, and no

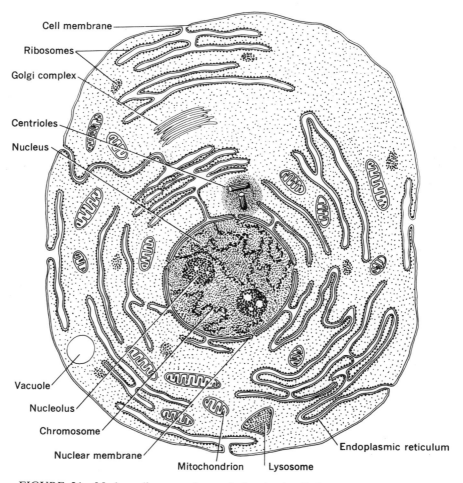

Cell membrane
Ribosomes
Golgi complex
Centrioles
Nucleus
Vacuole
Nucleolus
Chromosome
Nuclear membrane
Mitochondrion   Lysosome
Endoplasmic reticulum

FIGURE 21   Modern diagram of a typical animal cell, based on what is seen in electron micrographs. The mitochondria are the sites of the oxidative reactions that provide the cell with energy. The dots that line the endoplasmic reticulum are ribosomes: the sites of protein synthesis. In cell division the pair of centrosomes, shown in longitudinal section as rods, part to form poles of the apparatus that separate two duplicate sets of chromosomes. (Reproduced from *Biological Science* by William T. Keeton, with the permission of the publisher, W. W. Norton & Company, Inc. Copyright © 1967 by W. W. Norton & Company, Inc.)

instructions need be given because this process is a consequence of the structure of the lipids and proteins themselves. Like the membranous sheets through the cell and the one surrounding the nucleus, the outer membrane of the cell consists of two layers, one facing the environment and the other facing inward toward the cell.

A similar double-membrane structure is evident in mitochondria, the small powerhouses to which reference was made earlier. Their outer membranes are exposed to the cell cytoplasm; the inner membrane consists of the working parts of the electron-transfer and ATP-generating machinery. Chloroplasts in plant cells are rather similarly constructed. Withal, the great advantage to the cell of this double-membranous structure remains to be understood, although it is a primary characteristic of life itself. Some portions of the endoplasmic reticulum are particularly concentrated with respect to enzymes that engage in the metabolism of lipid-soluble materials—e.g., the synthesis of cholesterol and steroid hormones—presumably because these water-insoluble materials can be managed only when dissolved in an appropriate medium, the lipid of the membranous structure.

Unexpectedly, both mitochondria and chloroplasts have been found to contain DNA. The latter is insufficient to specify all the proteins of these structures, so they are not fully autonomous bodies. But these DNA's do find expression; in the living cell, mRNA is made on their surfaces and is translated into proteins on ribosomes in the usual manner. But the special virtues of this process are not apparent.

Cell Division

The great marvel of the living cell is the cycle of events whereby one parent cell gives rise to two identical daughter offspring. Indeed, for unicellular organisms and for the early stages of the embryonic growth of more complex organisms, cell division is itself the very essence of life. The life cycle of most cells can be viewed as a sequence of events shown as follows:

Stage S is the period of DNA synthesis, during which the total DNA complement of the cells is doubled. Stage D is the process of mitosis or gene segregation, in which the total DNA is separated into two identical sets of genetic material, after which cell division occurs. Stage G_1, the gap in time between cell division and the next onset of DNA synthesis, varies from only a few seconds in some cells to many hours in others. During this period the cell engages in all the synthetic activities required for its own growth and metabolism, reaching its mature size, form, and function. If the cell is not to divide again, it remains in this stage.

Nothing is known of the trigger that initiates the next burst of DNA synthesis, nor is it clear how this actually occurs in animal cells with complex chromosomes. The DNA content of a human cell is thousands of times greater than that of a simple bacterium. It is organized in chromosomes, structures which, at greatest magnifications in some cells, somewhat resemble a bottle brush. Whereas a bacterial chromosome consists of a single circular DNA molecule, it is unknown whether mammalian chromosomal DNA is circular, whether each chromosome is a single molecule, or whether each of the protruding filaments from the main axis is a separate discrete molecule of DNA. During the period of DNA synthesis, the chromosomes are not recognizable morphologically, losing their characteristic structure.

When the total DNA content has doubled, a second resting stage, G_2, is apparent. During this period the marvelous mitotic apparatus is generated. Centrioles appear at either side of the cell, and spindle fibers, made of an unknown protein, are organized. Phase D, mitosis itself, then commences, and the spindle fibers attach to the paired chromosomes that are aligned in parallel at mid-cell. The fibers somehow shorten and contract, drawing one of each chromosome pair to the opposite sides of the cell. A membrane begins to grow at the mid-line; when this is completed, there are two cells. The chromosomes in each daughter cell compact themselves into a dense nucleus within which it is almost impossible to recognize specific chromosomes. Each daughter cell then enters Stage G_1.

It will be evident that some chemical event must initiate each of the four phases of cell life and serve as a signal inaugurating the processes characteristic of that phase. But in no case is the nature of the signal known. The mitotic apparatus requires a signal for fiber production, some mode for fiber attachment to the chromosomes, a mechanism for spindle shortening, a signal that occasions formation of new membrane, and the disappearance of the spindle fibers after two cells are fully formed, as well as the subsequent compacting of chromosomes into nuclei. These triggers and many other details governing the process are unknown. A decade of careful electron microscopy and study of cell division has served only somewhat to fill in details of this remarkable machinery. But the cardinal aspects remain elusive and warrant intensive investigation because the potential of control of these events by external manipulation would provide a means for an approach both to cancer therapy and to the alleviation of a multitude of other disorders.

A powerful tool for the study of these events was made available by the discovery of techniques for the cultivation of mammalian cells in disassociated suspended culture, much like the growth of bacteria or algae. By such techniques it has been possible to learn the nutritional requirements of mammalian cells of various cell types and to observe some of the chemical events associated with each of the four major phases of cell life.

Discovery of the consequences of cell–cell interactions has been particularly significant. The free-living mammalian cell is motile and proceeds through round after round of the cell cycle, but when cells in culture find each other and establish physical contact, motility abruptly ceases and the cells are likely to remain in phase G_1. Whatever may initiate the cancerous transformation, the consequences are exactly the reverse, regaining of motility and re-entry into Stage S. Cell–cell interactions are of immense significance in cell life, but the molecular bases for such events remain totally obscure.

Development of an Organism

Sexual reproduction, the addition of two sets of genes, makes possible the great variety of individuals that constitute a species, and also underlies the processes that have made possible the origin of new species. No phenomenon within the experience of man is more wondrous than the events following the union of sperm and egg to form a complete zygote. The fertilized egg divides again and again. The embryo grows and proceeds through an invariant series of shapes. Early in the process, all cells appear outwardly to be virtually identical, although in fact this is never true.

With the passage of time, however, some cells begin to take on the characteristics of the specialized cells of adult forms. Primitive organs develop; here and there cells die while other cells migrate from relatively distant portions of the embryo to take new places. In time, there is a fully formed organism, with each cell seemingly in a foreordained place and exhibiting the special properties of the cells of muscle, nerve, liver, kidney, bone marrow, connective tissue, etc. The heart of the mystery lies in the fact that all the information directing this process must, necessarily, have been encoded in the nucleus of the zygote. Moreover, all the genetic information of the zygote is later to be found in the nuclei of all the cells of the fully formed organism. How then does the process of differentiation occur?

In the main, understanding of the basis for differentiation remains woefully meager. Two models are already before us. The first comes from the understanding, originally gained from studies of bacteria, that, at any time in the life of a cell, only a portion of its genetic information is being expressed; much of the genome remains repressed until some specific event can achieve derepression. Similarly, in each of the four major phases of the life of any dividing cell in the normal cell cycle, different aspects of the genome find expression. We can only assume that essentially equivalent events are operative in the process of differentiation. In the adult form, only cells that become erythrocytes engage in the synthesis of hemoglobin, but the genes for hemoglobin synthesis are to be found in all cells. Only cells destined to be muscle synthesize the contractile protein myosin; only fibroblasts make collagen; only cells in the retina elaborate and organize the visual pigments; and so on. This is a thoroughly satisfying concept, but it provides no insight into the actual nature of the repressors and derepressors; it fails to account for the events that give direction to the future of any specific cell. An intense effort is currently under way to ascertain the intrinsic nature of these events now that a suitable set of models permits rational questions.

Patently, each cell in the developing embryo is sensitive to its environment. Long suspected, the definitive evidence came from studies of bits of embryo in tissue culture. Tiny embryo sections taken from an area that, from past experience, should go on to become working muscle have been grown in tissue culture; they go through repeated cell divisions, and, at a characteristic time, the cells gather together to form a primitive "myotube." Cell membranes become obscure as the cells blend into a single syncytium, which then begins to contract spontaneously. But the initial cell samples were found to have contained two types of cells: One type normally goes on to become muscle, and one becomes the fibroblasts responsible for the synthesis of the connective tissue protein, collagen, as well as the variety of carbohydrate polymers that are secreted into the medium. If the same experiment is repeated, starting with only a single cell capable of becoming muscle, the cells do indeed proliferate but the myotube does not form and contractions are not evident. If, however, the flask in which the experiment is conducted is coated with some pure collagen, then the ensuing events are much as might have occurred were fibroblasts present in the culture. Thus, only when the potential muscle cells "sense" collagen in their environment can they complete their normal development.

Moreover, such experiments need not be conducted by starting with relatively primitive cells. Fully differentiated cells—lens or retina cells of the eye, primitive pancreas cells, or those of the kidney—can be employed in such experiments and will divide as many as 50 consecutive times, each division yielding differentiated cells resembling the one cell with which the experiment began. But again, such experiments are feasible only when conducted in a medium that provides fluid from the area originally surrounding the cell with which the experiment began. Indeed, that cell can be separated from the fluid by membranes of varying porosities, and the experiment can be successfully repeated. It is clear that the dividing cells are receiving "information" from their environment, but in no instance, except in the requirement of dividing muscle cells for collagen, is there understanding of the nature of that chemical message, much less how that message affects the genetic apparatus.

Cell–cell interaction apparently lies at the heart of the differentiation process, but again its nature is obscure. The chemical aspects of the cell surface remain to be established. Examination by the techniques of immunochemistry shows that each differentiated cell type has a defined chemical surface specific to that type. More than that, within a given type, there must be subtle distinctions that account for the mechanism by which a given nerve cell takes up residence in such a position as always to form a synapse with a specific other nerve cell or to innervate a specific structure. That the latter does indeed occur is shown by a few relatively simple experiments. For example, if the nerve to an area of skin is sectioned and the patch of skin is rotated by 180 degrees—e.g., the skin that used to be on the abdomen now lies on the back—the nerve grows out new endings to the skin. But if now one lightly touches the animal's back, he scratches his abdomen. As the nerve grew back, it found the original types of cells with which it was supposed to make contact, but the brain records those as having been in their original position. In certain lower forms like the frog, one can successfully section the optic nerve and then rotate the globe of the eye by 180 degrees. The optic nerve regenerates, and each nerve sends out endings until it finds the same portion of the eye to which it had originally been attached. In consequence, when a fly appears overhead, the frog's tongue darts down instead of up. The nerves regenerated and found the original portions of the eye to which they had always been attached, and hence in the frog's forebrain his entire optical field has been inverted with respect to reality. What chemical attraction leads the growing nerve cell to the portion of the retina to which it was "foreordained"?

If man is to understand himself, to find new bases for contraceptive techniques, to find diagnostic procedures that can detect improperly fashioned fetuses early in pregnancy, perhaps one day to undertake surgical repair of such fetuses, it is imperative that an intensive effort be made to understand the fundamentals of the process of differentiation. Only a beginning has been made in this direction. Perhaps the experimental limitation has been failure to identify the most suitable biological study object. Much of earlier embryology rested on studies of the development of sea urchins, frogs, salamanders, and chicks. None has proved quite as rewarding as the use of *E. coli* for genetic studies or use of the squid axon in neurophysiology studies. Many investigators have been seeking simple models of the differentiation process—for example, spore formation in bacteria and fungi or flagellum formation in motile bacteria—systems that show great promise but that have not yet revealed new generalities. Rather, have they simply afforded additional examples of the validity of the concept of genetic repression and derepression as the underlying mechanism of differentiation.

Consider, for example, the initial event of fusion of sperm and egg. This event has been studied for generations; electron microscopy has now revealed the process in great detail, but, withal, the process remains mysterious. The developed egg, before fertilization, is unique among cells. It is extremely large, as cells go, and is provided not only with storage carbohydrate and fat, but also with an abundance of nucleoli, the small bodies within the nucleus within which are made the specific forms of RNA that serve as the structural, and perhaps functional, material of ribosomes. The mature egg already has a very considerable complement of ribosomes on which RNA messages are lodged, but these are not finding expression—protein synthesis is not occurring. Upon the approach of the sperm, its "acrosome," the bulge at the leading edge, burrows through the egg cell membrane, the sperm nucleus seeks and finds the egg cell nucleus, and these fuse. Shortly thereafter, the ribosomes become "activated," the nucleoli begin to disappear, more ribosomes appear in their stead, and the preformed messages begin to be read. Thus, the initial events in embryo formation are turned on by the introduction of the sperm nucleus, but the messages which are then read were prepared in the developing egg before ever the sperm entered. The chemical nature of this activation process is totally unknown. In mammals, the egg then travels down the horn of the uterus and almost invariably lodges in the same spot on the thin wall of the uterus. Neither the mechanism nor the virtue of this "choice" of location is known. But it is there that placenta formation is to occur. Cell division occurs all the while the fertilized egg travels down the uterus to its location and may exist in the 100-cell stage of the blastocyst at the time of nidation attachment to the uterine wall. In species that prepare more than one egg at a time, the eggs then space themselves equidistantly in the uterus. The basis for this spacing is unclear. Thus description of the early stages of mammalian life has become much more detailed, but understanding of the process is surprisingly meager.

Development of the Nervous System

Through interactions of embryonic cells and tissues, some cells on the dorsal surface of the early embryo are induced to organize themselves into a thickened plate, which rolls into a tube and sinks beneath the surface. From this tube and from associated cellular masses, the entire nervous system arises. The essential events during this key phase of development involve individual cells that, typically, form elongated cytoplasmic strands that become the adult nerve fibers. Before the embryonic nerve cells first send out their processes toward the periphery or to some station within the nerve centers, four interrelated events may be discerned: cell divisions; migrations of cells; changes in the shape, size, and content of individual cells; and the death and disappearance of partly differentiated cells. Closely timed and precisely interlocking, these events produce all the major features that finally characterize the adult nervous system—e.g., a recognizable neural axis with appropriately segregated parts, rudiments of cranial and spinal nerves, and enlargements, foldings, and outpocketings that foretell both the final gross form and the inner detail of localized cell groupings. The key point is that the final unity emerges from the coordinated activity of thousands of individual cells, each engaging in one or more of the four activities mentioned, each behaving as if it knew its place in the final structure, yet each subordinating itself to that structure.

Just as these developmental events are to be understood best in cellular terms, so the function of the completed nervous system demands expression in the same terms. The human brain and spinal cord consist of billions of cells, all arranged in an orderly manner with precise interconnections, structural and functional. Communication among nerve cells depends on the passage of impulses. These arise in individual cells and are effectively transferred from cell to cell either by one cell directly exciting another or by an intermediate step involving a chemical transmitter. Small groups or very large numbers of nerve cells are thus brought into activity in highly selective patterns. Out of this complex of neural events emerges the coordinated product we call behavior.

In the early embryo the young neurons develop sharp affinities and disaffinities for one another and for peripheral tissues. They reveal these properties both by entering selectively into specific cellular associations that become the interknit pathways and centers of the adult nervous system and by selectively re-establishing some of these associations during nerve regeneration. The capacity of developing cells generally to form intercomplementary groupings lies at the basis of essentially all embryonic events at cellular and higher levels, but the origin of these abilities and their nature are largely unknown.

Plant Embryogenesis

The developmental process in plants presents a unique set of puzzles. One attribute of this process—the photoperiodic control of flowering—is particularly fascinating. In many plants the length of the daily exposure to

light determines whether flowers and their associated structures can develop and thus, in effect, controls sexual reproduction. This process is
often quite striking.

For example, some varieties of soybeans are "short-day" plants, which
flower only when the daily period of illumination is shorter than some
critical day length. In consequence, although plant sowings may have been
made over the spring and summer and the plants are thus of varying size,
when the days begin to grow short, all flower at once. If kept under artificial
illumination, they may never flower at all. Mysteriously, in many such
short-day plants the effect of a few days' exposure to an appropriate day
length may persist long afterward. For example, in the cocklebur, a common weed that will never flower if kept under continuous light, flowering
will be initiated if the plant is exposed only once to a dark period more
than nine hours long, even if it is immediately replaced in continuous
light. No anatomical or biochemical test has yet distinguished between
induced and noninduced plants immediately after the inducing dark period.
Yet clearly, some basic change has been brought about, some self-main-
taining or steady-state condition that governs the entire subsequent development without additional external stimulus. Obviously, insight into the
nature of this induced state would be valuable to commercial plant growers.
Even more mysterious is the fact that the effects of the continuous dark
period can be utterly abolished if it is interrupted by only one minute of
exposure to light! Red light is the most effective region of the spectrum,
and even this can be annulled if the brief exposure to red light is followed
immediately by exposure to light of a wavelength in the "far red." Discovery
of these properties led to a search for a pigment with an appropriate
absorption spectrum that might account for those effects. Such a material,
a protein called "phytochrome," has indeed been isolated. But the function
of this material in plant physiology continues to remain obscure.

One other facet of photoperiodism—the fact that it can be translocated—
is of interest. For short-day plants, flowering can be induced if only a
portion, indeed if only one leaf, of the plant is maintained in the dark for
a sufficient period. Presumably, this implies the formation of some compound that can serve as a plant hormone and, leaving the site of formation,
will trigger the events that lead to the initiation of flowering. All attempts
to find such a hormone have thus far failed, yet such a compound must
exist. For example, if a branch of a plant maintained in continuous light
is grafted onto a plant that has been maintained for a sufficient period in
the dark, to induce its own flowering, flowering is also initiated in the
grafted branch, which has never been out of the daylight. These events,
of obvious commercial interest to plant growers, are also of profound
significance to the whole field of developmental biology because they may

well serve as clues to the nature of the chemical messages that induce differentiation in all developing embryonic forms.

Animal Viruses

An understanding of the cellular processes induced by infection with animal viruses is important for progress in several areas of biological and medical sciences. About 60 percent of all illnesses are estimated to be caused by viral infection. The meager results of over 20 years of empirical searching for clinically useful antiviral agents indicate that further knowledge of the fine structure of viral components and the molecular events involved in viral replication and virus-induced cell damage is required in order to design effective agents for the treatment of viral diseases.

Understanding of the cancer cell and the neoplastic process in molecular terms, a hopeless pursuit several years ago, is now a realistic research goal for those studying tumorigenic viruses and the mechanism of virus-induced cell transformation. Because tumor-producing viruses can now be obtained in highly purified form and suitable "normal" cells in culture can be transformed into "neoplastic" cells, experimental systems are becoming available for a rational analysis of the molecular basis of neoplasia.

Animal viruses provide unique experimental footholds for attacking these complex problems of mammalian cell function for which other approaches may be virtually nonexistent. Virus infection is the only experimental procedure available by which a defined segment of genetic material can be introduced into a mammalian cell. Since viruses contain only a limited number of genes, from five to several hundred, it is technically feasible to analyze in detail the factors governing the synthesis of specific viral macromolecules employing the virus-infected cell for experimental analysis.

More than 500 animal viruses of various sizes and degrees of chemical complexity have been discovered, each containing either DNA or RNA and multiplying or maturing in different parts of the host cells; over 300 of these can infect man. Animal viruses have been tentatively classified into eight groups on the basis of biochemical and biophysical properties of the virion (the extracellular mature virus particle): four DNA-containing groups, the papovaviruses, adenoviruses, herpesviruses, and poxviruses; and four RNA-containing groups, the picornaviruses, reoviruses, arboviruses, and myxoviruses. Viral DNA's range from 3 million to 160 million in molecular weight, are double-stranded, and either circular (papovaviruses) or linear. Viral RNA's have molecular weights ranging from 2 million for the picornaviruses to over 10 million for the reoviruses, the latter uniquely being double-stranded. Their great variety of chemical compositions, structures, and sites of replication, presumably also reflecting differences in replicative patterns and induced cellular modifications, makes animal viruses unique tools in experimentally dissecting cellular function in molecular terms. By infecting homogeneous cell cultures, one can insert viral genetic material of different types and sizes into defined intracellular regions and study the ensuing biosynthetic events.

Studies with representative members of the four DNA virus groups, during the past five years, have revealed the following series of events during an infectious cycle: (1) attachment of virus to specific receptor sites on the host cell; (2) uptake of intact virus into phagocytic vesicles and transport to cytoplasmic or, possibly, nuclear sites; (3) intracellular un-coating of viral DNA; (4) transcription of specific regions of viral DNA; (5) attachment of the transcribed product, viral mRNA, to cytoplasmic or, possibly, nuclear ribosomes; (6) synthesis of viral-specific enzymes and other "early" proteins, utilizing the viral message on cellular ribosomes; (7) replication of viral DNA by enzymes of uncertain origin; (8) a second wave of transcription involving parental or progeny viral DNA or both; (9) translation of these viral mRNA's to viral-structural proteins and other viral-specific proteins, some of which engage in regulatory functions such as "switching off" the synthesis of virus-induced "early" enzymes, presumably as repressors; and (10) the final construction of the virion, presumably by self-assembly.

While the overall replication patterns of the various DNA viruses are similar, the individual biosynthetic steps can differ considerably. Three interesting general patterns involved in the replication of most DNA viruses are the early inhibition imposed upon host-cell macromolecular synthesis, the mechanism of which is unknown; stimulation of the activity of DNA-synthesizing enzymes; and the regulatory role of "late" viral gene products on "early" viral gene functions. Further analysis of these phenomena should provide insight into the mechanisms that regulate cellular and viral macromolecule synthesis. Hopefully, information concerning regulatory mechanisms that operate specifically in viral infections may provide clues to the control of virus disease.

The biosynthesis of RNA animal viruses, especially poliovirus, has been studied intensively during the past few years and, although similar to that of DNA viruses in many respects, it differs in several significant ways. RNA viruses attach to cells, penetrate, and are then uncoated. But the DNA–RNA transcription step, characteristic of DNA viruses, is bypassed. The parental viral RNA strand itself serves as a messenger RNA. It forms viral polyribosomes and directs the synthesis of viral-specific proteins, among which are: (1) regulatory proteins that, in some manner, inhibit normal host-cell synthesis of RNA and protein, (2) a unique RNA-dependent RNA polymerase that catalyzes synthesis of new viral RNA on the surface of the invading RNA, and (3) viral-structural proteins. RNAase-resistant RNA structures, thought to be double-stranded intermediates in viral RNA replication, have been demonstrated with many RNA viruses.

Interferon is the name given to a cell-coded protein formed in response to infection with most DNA and RNA animal viruses. In sufficient quantity, interferon inhibits virus multiplication and, thus, may play a role in recovery from viral disease. Recent studies indicate that interferon acts by inducing the synthesis of a second cell protein that, in turn, somehow prevents the association of viral mRNA with ribosomes to form viral polyribosomes. Clearly, if a sufficient supply were available, interferon would have great therapeutic potential. Current research is directed toward means of eliciting maximal interferon synthesis in cell cultures. A double-stranded, synthetic RNA has proved quite effective in this regard and has afforded protection against inoculation of mice with large doses of hoof-and-mouth disease virus.

Conversion of a normal cell to a cancer cell is thought to involve a permanent genetic alteration in the affected somatic cells. But the mammalian cell contains at least a million genes, and identification of the specific genes that are altered, in terms of their products (i.e., proteins and enzymes), without some suggestive clues has posed insurmountable technical difficulties. However, the analysis of viral carcinogenesis greatly simplifies this problem. Cancer-producing viruses contain only between 7 and 50 genes, yet one or several of these genes can induce cancer in animals and transform normal cells to malignant cells in culture. Understanding of the functions of these relatively few viral genes should greatly facilitate elucidation of the carcinogenic process.

Indeed, significant progress has been made in understanding the mechanism of tumor induction and cell transformation by DNA and RNA oncogenic viruses. Shortly after they invade their host cells, the DNA tumor viruses, including eight human adenoviruses (31 human adenoviruses are known), virus SV40, and polyoma virus, are no longer demonstrable as infectious in virus-induced tumor cells. However, the presence in and on these cells of viral-specific proteins ("tumor antigens") distinct from virion structural proteins argues for the persistence and functioning of at least part of the viral genome in the tumor cell. The direct demonstration of viral DNA in virus-induced tumor cells is a technically formidable problem since a single viral gene would represent a very small portion of total cellular DNA, about one part in a million. However, it has recently been shown that tumor cells, induced in animals by polyoma virus, SV40, and seven different human adenoviruses, continue to synthesize viral-specific mRNA, as demonstrated by the formation of DNA–RNA hybrids when their pooled mRNA is mixed with pure viral DNA. A surprisingly large amount of the total mRNA in the polyribosomes of adenovirus-transformed cells is viral-specific, 2 to 5 percent, suggesting that a small amount of viral DNA present in the tumor cell is preferentially transcribed while most of the host RNA goes unexpressed. This selective transcription during viral carcinogenesis may represent only a specialized case of the more general phenomenon occurring during cell differentiation, or it may occur by an entirely different mechanism. It seems likely that these viral mRNA molecules are translated into proteins, some of which account for the altered growth and antigenic properties of the tumor cell.

Host-cell DNA synthesis is inhibited during the cycle of infection with most DNA viruses. However, infection of nondividing cells with polyoma or SV40 virus induces the synthesis of host-cell DNA, a phenomenon thought to be of importance in viral transformation.

The oncogenic RNA viruses, including the avian and murine leukemia and sarcoma viruses, are capable both of transforming cells and of replicating within the same cell. Particularly intriguing are those viruses that are defective in the genes for synthesis of viral coat protein; they transform cells without the production of infectious virus. Others are defective only within certain host cells. Co-infection of tumor cells induced by such a defective virus in conjunction with a second virus ("helper virus") is required for the synthesis of infectious virus; newly synthesized virus then contains the genome of the transforming virus and the coat protein of the helper virus.

FORM AND FUNCTION

For the isolated cell, structural form is correlated with its simplest needs—to remain alive in the face of adversity, to grow, and to reproduce by fission. To accomplish these simple ends, cells possess a variety of substructures, each specialized for the performance of a specific chemical task. Multicellular organisms—man himself—are made possible by the collective structures, and the functions they permit, of organized groups of cells, organs that serve the entire organism much as cellular organelles serve the cell. One need consider only the brain, the gastrointestinal tract, the cardiovascular system, the kidney, the musculoskeletal system, and the genitalia to recognize the extent of this subdivision of labor. The goal of practitioners of the anatomical sciences and of physiologists has been to obtain detailed understanding of these correlations of form and function—to understand how nerves conduct, muscles contract, kidneys regulate the composition of the extracellular fluid, the gastrointestinal tract degrades foodstuffs into metabolizable nutrients; how joints permit motion; how gases are taken into the body or removed; and how the organism is integrated into a harmonious whole through the operation of the nervous and endocrine systems. The last two decades have witnessed an immense expansion of research in these areas, which has been made possible by new awareness of the nature of the problems, by new experimental tools such as the electron and phase-contrast microscopes, fast multichannel recorders, microspectrophotometers, the flying-spot microscope, x-ray analysis, and radioisotopes and techniques for their detection and measurement, as well as by the entire armamentarium of the biochemist. In no area of the life sciences is the melding of such classical disciplines as anatomy, physiology, pharmacology, anthropology, and biochemistry so clearly evident. Review here of the multitude of accomplishments at this level of biological consideration is impossible; accordingly, a few examples will be cited only as illustrations of current approaches to a few classical problems.

Muscular Contraction

For decades, muscular contraction has been an attractive object of study. Although it is easily amenable to experimental approach, no useful working concept of the fundamental mechanism was developed until two quite independent approaches revealed complementary information that, combined, led to a highly satisfying model of the nature of this process.

As shown in Figure 22, phase-contrast and electron microscopy revealed the presence in muscle of two quite distinct types of fibers: In skeletal muscle, filaments of about 200 Å are each surrounded by six filaments about 100 Å in diameter. When muscle is stretched, the two sets of filaments pull away from each other, and when muscle is contracted, they telescope into each other. The individual filaments do not themselves shorten but appear to interact with each other by a "clawing" action that pulls the filaments past each other. The thick filaments are constructed of the protein, myosin; each molecule of myosin (mol. wt. 480,000) is an elongated multiple-stranded helix, about 1,800 Å by about 20 Å, with a "knob," oriented at right angles to the fiber axis at one end. The knob portion has the properties of an enzyme capable of catalyzing hydrolysis of ATP. The thin filaments are formed of a second protein, actin, the fundamental unit of which is a simple globular molecule (mol. wt. 60,000), which polymerizes into a filament analogous to a string of beads

Skeletal Muscle

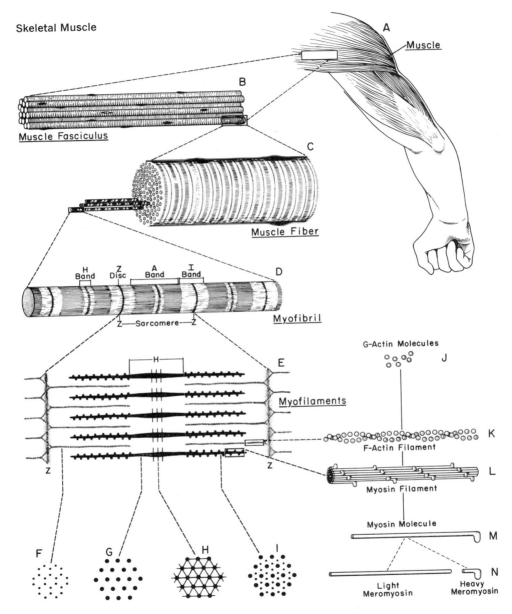

FIGURE 22   Structure and function of skeletal muscle. F, G, H, and I are cross sections at the levels indicated. (From W. Bloom and D. W. Fawcett, *A Textbook of Histology*, 9th ed., 1968. Copyright © 1968 W. B. Saunders Company. Drawing by Sylvia Collard Keene.)

with two such strands wrapped around each other in a double helix. In solution, purified actin and myosin react rapidly with each other to form a tightly bonded complex, bond formation occurring between the knobs of the myosin and some aspect of the actin molecule. Addition of ATP permits transient separation of the actin–myosin complex, which is re-formed as the myosin hydrolyzes the ATP. Presumably, some equivalent interaction is responsible for the ratchet-like action of actin and myosin in contracting or relaxing muscle, although the details of this process remain obscure.

How then does an intact muscle contract? The process begins with the arrival of a nerve impulse, itself the consequence of a rapid change in the surface properties of the nerve such that sodium ions leave and potassium ions enter. At the muscle–nerve junction a similar wave of excitation commences and rapidly sweeps over the muscle surface. This wave is essentially of similar character—a change in the distribution of sodium and potassium ions near the muscle cell surface. A few milliseconds later the entire cell begins to contract.

In a brilliant series of experiments, microelectrodes with tips less than a micron in diameter were touched to the muscle surface, applying shocks too small to cause a general excitation. Local contraction responses were obtained, but only at particular spots on the cell surface. Electron micros-copy revealed these activating spots to be narrow indentations of the surface membrane, which penetrate deep into the cell. These structures, "transverse tubules," carry the change in surface charge into the cell in proximity with the contractile elements—the actin and myosin filaments. Contraction itself therefore occurs in consequence of a change in the local ionic environment of these proteins. To some degree, it is the change in sodium and potassium concentration that is meaningful, but more im-portantly, these changes act to release calcium ions from some bound form, and it is the increase in calcium ions, specifically, that stimulates the ATP-hydrolyzing activity of the myosin and makes contraction possible. Unless the ionic changes were reversed, the cell would continue to remain con-tracted until all available ATP had been utilized. However, in close rela-tion to the transverse tubules is a network of extremely tiny tubes, the "sarcoplasmic reticulum," a system that, utilizing the energy of ATP, sequesters the calcium inside the tubules, thereby bringing contraction to a halt and permitting the muscle to relax until the next wave of excitation arrives. Numerous details of this fundamental life process remain to be unraveled. But, in the main, the totality is a satisfying concept, consistent with all known evidence, and represents the ultimate convergence of physiological, biochemical, and anatomical studies.

THE CONSTANCY OF THE *Milieu Interieur*

The free and independent life of vertebrate animals became possible when their ancestral forms first developed closed circulations, which assured that all cells of the body, no matter how remote from the external environment, had a nutrient supply, removal of waste products, and an environmental composition compatible with life. In turn, such an arrangement renders imperative some device for assuring the constancy of that internal environment, despite the vicissitudes of life. On some days, one may drink little or nothing; on others, rather prodigious quantities. Salt intake may vary equally widely. Some diets are effectively alkaline; some are acid. Changes in the rate of respiration may affect not only the oxygen supply but also the rate of removal of carbon dioxide and hence the acidity of the blood. That we survive and scarcely notice such variation is a tribute to a remarkable organ, the kidney.

Until relatively recently, thoughts concerning renal function rested on speculations based largely on the appearance of the anatomical structure of the kidney. More than a century ago it was recognized that the fundamental operating unit of the kidney is the "nephron," of which millions are arrayed, in parallel, in the kidney cortex (Figure 23). Each commences with a small arteriole that is branched rapidly from the main renal artery and becomes a tuft of capillaries (the glomerulus) that coalesce in a venule. Surrounding the tuft of capillaries is a structure made of connective tissue that leads into a miniature tubule. The latter goes straight down toward the renal medulla, makes a hairpin turn, returns toward the outer surface, and then descends again into a thicker channel, the collecting duct, which drains into the hilum of the kidney. Particularly noteworthy is the fact that the blood in the venule formed by coalescence of the glomerular capillaries also surrounds the ascending tubule and the collecting duct before entering into the larger veins for exit from the kidney. From the appearance of this structure it was deduced that the glomerulus must be a filter through which passes an "ultrafiltrate" containing all the plasma constituents except the relatively large plasma proteins. It was further assumed that, as the glomerular filtrate traverses the tubules, materials are removed from it, ultimately leaving the presumptive urine.

Slowly, over the first two thirds of this century, evidence accumulated suggesting the essential validity of this concept. Indirect techniques that permitted measurement of the magnitude of these operations were devised. In a 160-lb man, the overall rate of glomerular filtration is about 125 milliliters per minute, or about 180 liters per day, a volume 65 times that of the entire volume of plasma. Almost all (99.5 percent) the water and

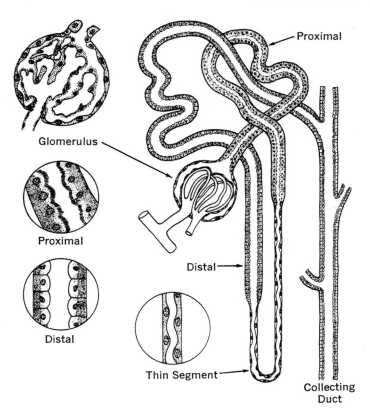

FIGURE 23   The nephron—fundamental unit of the kidney. (Adapted from H. W. Smith, *The Kidney*, Oxford University Press, New York, 1935.)

its solutes are reabsorbed as this filtrate passes along the tubules, for urine output is only about one liter per day.

The experimental challenge has been to ascertain the mechanisms by which this versatile organ so alters its behavior as to assure excretion of very dilute urine when water intake has been copious or extremely concentrated urine when water intake has been meager or salt intake excessive, to excrete alkaline or acid urine as may be appropriate to physiological circumstance, and to assure that none of the material in the glomerular filtrate that is valuable to the body, e.g., glucose, is lost. Although indirect evidence in these regards was accumulated over a long period, the detailed picture now available has been the consequence of recent development of techniques for micropuncture, originally used in frogs and mud puppies but since extended to mammals, permitting direct sampling from precise locales within the minute renal tubules under a variety of conditions. Ten years were then expended in perfecting adequate microanalytical techniques for the examination of these tiny specimens (0.01–0.1 μl). Once the techniques were developed, thousands of such determinations were undertaken, and from them has emerged a detailed blueprint of the *modus operandi* of this organ. A few aspects of this operation warrant summary.

First it was established that the pressure in the glomerular capillaries is unusually high (about 60 mm Hg) and thus sufficient to ensure filtration through the capillaries. As the fluid passes through the descending proximal tubule, a large fraction of the total is removed; with it most of the desirable organic compounds, e.g., glucose, are removed by processes of active transport similar to those operating in most living cells. Only about 10 percent of fluid remains at the level of the hairpin turn. The fluid in this region is decidedly more concentrated than earlier in the proximal tubule, largely due to removal of water by simple osmotic forces because of the high concentration of salt in the surrounding region. As the fluid rises in the ascending limb, specific facultative adjustments are made in the sodium, potassium, and hydrogen content. In large measure, sodium ions removed in this region are exchanged either for hydrogen or potassium ions. It is in this region also that the tubular cells manufacture ammonia (NH_3), which is secreted into the duct fluid, to combine with hydrogen ions that the same cells have secreted into that fluid. By this process a considerable amount of acid can be secreted without unduly acidifying the presumptive urine. Final adjustments are made in the early portion of the collecting duct, where the final salt concentration is achieved.

A series of controls assures that the composition of the final urine is commensurate with physiological requirements of the moment. Foremost among these is the ingenious mechanism that makes salt concentration possible. It was long known that man can excrete urine about four times as concentrated in salt as his own blood plasma. Other animals, particularly desert-dwellers, are considerably more adept at this task than are we; for example, the kangaroo rat need never drink water and survives by excreting urine 14 times as concentrated as his own plasma! The principal feature that distinguishes the kidney of the kangaroo rat from that of man is that in the kangaroo rat the descending tubule dips much farther down into the cortical and medullary tissue of the kidney; i.e., the tubule is decidedly longer relative to kidney size than is that of man. Inspection of the kidneys of a variety of species indicated that, in general, concentrating ability is a function of tubular length. It was this observation that suggested that the entire apparatus is patterned after the principle used by engineers in the design of heat-exchanging apparatus—i.e., as a countercurrent multiplier.

This notion has been subject to a variety of tests and has indeed proved to be valid, operating as shown in Figure 24. Since then, other examples of the same process have been sought, and it was learned that the circulation of large pelagic fish utilizes the same principle to conserve heat, a feature particularly evident in the tuna.

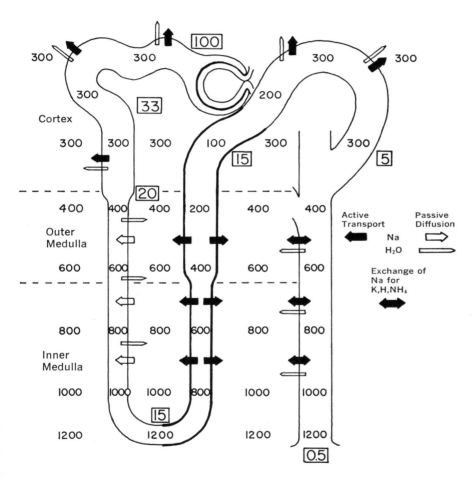

FIGURE 24   Salt concentrations in various regions of the nephron, basis for the countercurrent concentration device. Summary of passive and active exchanges of water and ions in the nephron in the course of elaboration of hypertonic urine. Concentrations of tubular urine and peritubular fluid in mOsm/L; large boxed numerals show estimated percent of glomerular filtrate remaining within the tubule at each level. (From *Physiology of the Kidney and Body Fluids,* Second Edition, by Robert F. Pitts. Copyright © 1968 Year Book Medical Publishers, Inc. Used by permission.)

Several hormones also control renal function. Individuals with Addison's disease were long known to become dehydrated due to excessive urinary excretion of sodium and chloride, with retention of potassium and acid. Such individuals fail to secrete the normal hormones of the adrenal cortical gland. The specific hormone in question has been identified as aldosterone, a steroid that is necessary to the process whereby sodium is reabsorbed in the distal tubule. Aldosterone secretion, in turn, is sensitive to the sodium concentration of blood passing through the adrenal cortex and through the pituitary gland.

A second hormone was also recognized by virtue of a disorder of man—diabetes insipidus—a syndrome characterized by excretion of as much as six gallons of urine per day. Individuals suffering from this disorder fail to elaborate in their posterior pituitary glands antidiuretic hormone, a relatively small polypeptide or protein, which determines the ease with which water may penetrate the walls of the collecting duct. The amount of hormone in contact with that duct determines the volume and final salt concentration of urine.

A third hormone, that of the parathyroid gland, governs the fate of phosphate as it traverses the tubule. This ion is removed in the descending tubule by active transport; the hormone blocks this process and phosphate continues into the urine. Secretion of this hormone, however, is sensitive not to the phosphate concentration but to the calcium concentration of blood plasma. When the latter is below the physiologically desirable level, hormone is secreted. It stimulates resorption of bone mineral; both calcium and phosphate concentrations rise in blood plasma, and the phosphate is then discarded in the urine under the action of the same hormone.

Thus the intrinsic architecture of the kidney, the special adaptation of ordinary mechanisms for cellular active transport, the countercurrent apparatus, and the superimposed hormonal control all combine to permit survival under a wide variety of environmental circumstances.

For many years, hypertension has been associated with renal disease. Forty years of research addressed to the nature of this relationship culminated in the demonstration of the following set of events. When circulation to the kidney is impaired, for whatever reason, e.g., a failing heart or an occluded renal artery, kidney cells release into the circulation a protein, renin, which has proved to be a proteolytic enzyme. The latter, in turn, digests from a normal plasma protein a medium-sized polypeptide that is somewhat further digested by a second enzyme, already present in plasma, to form angiotensin, the most powerful of known vasoconstrictive agents. The presence of this material in blood accounts for the disease of a large group of patients with hypertension. Moreover, angiotensin has been found not only to raise arterial blood pressure, but also directly to stimulate the cortex of the adrenal gland to secrete aldosterone, which, as we have seen, stimulates the mechanism for tubular reabsorption of sodium. In consequence, the kidney reabsorbs an inappropriately large amount of sodium, sodium excretion falls behind daily intake, and the body content of sodium, and with it water, increases. It is for this reason that the patient with congestive heart failure becomes waterlogged with edema fluid, usually evident as swelling of the extremities. Elucidation of this sequence of events was possible only after sophisticated techniques had been developed for the purification of proteins and of polypeptides.

The actual molecular basis for the influence of aldosterone, antidiuretic hormone, parathormone, and angiotensin on the structures of the kidney and of the adrenal gland remains totally unknown. Each is secreted in response to changes in the internal ionic environment. The kidney is a versatile machine whose activities, therefore, are programmed by these hormones. The result is the remarkable constancy of the *milieu interieur*.

Endocrines

By 1945, students of endocrinology thought that the science of endocrinology was well developed. The list of endocrine glands was thought to be essentially complete, and the gross consequences of under- or over-secretion of each were thought to be known, particularly as manifest in diseases of human beings. Yet the list of endocrines continues to proliferate, understanding of the mode of action of these regulatory secretions becomes increasingly detailed, and the subtle manner in which they integrate the life of the organism becomes increasingly evident.

The most surprising discovery of recent years was that certain nerve cells respond to neural signals by secreting hormones into the bloodstream, and in this way the two major integrative systems of the body are themselves functionally interlocked. The best demonstration of this is the picture that has emerged of the regulation of the female reproductive system. Suggestions of such interlocking came from old observations of the influence of the environment on the success of mating and pregnancy in domestic animals. Exciting stimuli related to photoperiodicity, ambient temperature, visual perception, odors, and even sounds were recognized. Among wild birds, the female generally requires the attention of the male before she will engage in nest building and egg laying. The female pigeon must perceive the visual image of her mate in order to produce crop milk for the young. Crowding can completely disrupt the reproductive processes of rats and mice. A female mouse who has mated with a male of her own strain does not become pregnant if, within 24 hours, she merely senses the odor of a male mouse of a different strain; her pregnancy is blocked by suppression of pituitary support of the *corpora lutea*. In all these situations the environmental stimuli perceived by sensory receptors in the brain are translated into chemical signals secreted by nerve cells in a small region at the base of the brain—the hypothalamus. From this small group of cells, neuroendocrine secretions, simply termed "releasing factors," are transmitted to the nearby pituitary gland via the very short hypophyseal portal vein.

At least two specific releasing factors are involved, causing the pituitary to release, respectively, its follicle-stimulating hormone and its luteinizing hormone. The former stimulates the development of ovarian follicles, while the latter effects the rupture of mature follicles and the release of eggs and causes the ruptured follicle to develop into the *corpus luteum*. Earlier, it had been supposed that the pituitary gland and the ovary were linked in a closed feedback regulatory mechanism. It was presumed that ovarian hormones act on the pituitary to suppress the secretion of its gonadotropins. But it has recently been demonstrated that the ovarian hormones act not on the pituitary but on the hypothalamus, suppressing the production of releasing factors. The latter, of unknown structure, are of enormous potency and are made and secreted only in the most minute quantities.

Other releasing factors that serve to stimulate the pituitary to release its thyrotrophic, growth, and adrenocorticotrophic hormones are thought to be elaborated in this same small region of the brain. Thus, the notion is implied that the cells of this area are equipped with a set of chemical sensors exquisitely sensitive to many aspects of their environment, because it is here, also, that one experiences thirst, satiety, hunger, and external temperature changes, and responds accordingly.

Another new addition to the family of hormones is thyrocalcitonin, a hormone secreted by the parafollicular cells of the thyroid gland and which, together with the long-known hormone of the parathyroid gland, serves to regulate calcium metabolism. Preliminary observations suggest that this new hormone may be precisely what is required to prevent the postmenopausal osteoporosis responsible for bone fractures in a large number of women of advanced years.

Surprisingly late in the history of endocrinology was the discovery of a compound that appears to deserve the appellation "intracellular hormone" and may be, in the evolutionary sense, the first hormone. This compound, 3',5'-cyclic adenosine monophosphate (cyclic adenylate), is formed from ATP under the influence of the enzyme phosphoadenosine cyclase. Remarkably, this substance has been independently discovered in several connections. It was first noted as the material formed when the hormone epinephrine, from the adrenal gland, stimulates the release of glucose from storage as liver glycogen. Epinephrine operates by triggering the action of the liver cyclase, and the resultant cyclic adenylate then trips off a series of events that lead to the activation of the enzyme responsible for the degradation of glycogen. But cyclic adenylate formation was also discovered as the primary event in the activation of the cells of the adrenal cortex by the trophic hormone of the pituitary; after arrival of the latter hormone, cyclic adenylate is formed, which somehow makes possible the formation and secretion of adrenocortical hormones. The same compound is made when pituitary antidiuretic hormone acts upon the collecting duct of the renal tubule; somehow its presence permits movement of water across that structure.

The effects of cyclic adenylate are not confined to animal cells. It has been found to stimulate synthesis of enzymes in *E. coli,* normally controlled by catabolite repression, to induce aggregation in starved slime molds and to derepress the synthesis of several enzymes of the mold *Neurospora*— enzymes that usually appear in the course of spore formation. It seems remarkable that the same compound can serve so many functions. The manner in which it does so is obscure in each instance, but from this brief account it is clear why this newly discovered material has been dubbed an "intracellular hormone."

PLANT AND INSECT HORMONES

At least five major hormones or groups of hormones have now been discerned in plants, including the classical auxins, gibberellins, cytokinins, abscisin, and even the simple gas ethylene. Unlike the highly specific effects of animal hormones, plant hormones appear to lack single clearly defined functions. Each is involved in a variety of different processes of growth or development. The gibberellins stimulate stem and leaf growth, seed determination, growth of some fruits, and in some cases, the formation of flowers and even the sex of flowers. Cytokinins affect root and bud formation; like the gibberellins, they are capable of overcoming dormancy in some plants, and they have the interesting property of rendering plants less sensitive to all kinds of unfavorable conditions, such as shortage of water, extreme temperatures, and even the effects of weed killers. Abscisin generally seems to reduce the growth activities of plants and to induce dormancy, as observed in a variety of plant parts. Accordingly, it would appear that, at any time, the behavior of a given portion of a plant is determined not by the absolute amount of any one of these hormones, but by the ratio of one to another.

Insects elaborate at least three and probably several more hormones. The two that have commanded attention are large steroid-like molecules, ecdysone and the "juvenile hormone." The former stimulates the processes of metamorphosis in insects. The chromosomes of insects treated with ecdysone have been observed to exhibit puffed, swollen areas where the protective coating of protein around DNA has been removed and transcription of RNA is taking place. In contrast, juvenile hormone causes insect larva to remain in the larval stage, preventing them from progressing to the pupate stage and becoming adults. The mechanism of this effect is without explanation. These relations are summarized in Figure 25. Although insect endocrinology may seem a specialty far removed from the cares of man, studies in this area have led to important advances in understanding· of the relation between nerve cells and hormones, of the control of cellular growth and death, and of the mechanism of hormone action, while at the same time suggesting, as recounted in Chapter 2, how, one day, we can better control insect pests.

Enough has been said above to indicate the main thrust of future research in the field of endocrines, *viz.,* intensive scrutiny of the mechanism of action of these chemical messages. Clues are available in some instances, but in most cases only morphological response or relatively gross chemical changes have been observed to result from endocrine activity. Indeed, it is not clear that any hormone actually enters the cells that it affects. Conceivably, all effects result from binding to cell membranes. Release of cyclic adenylate in some instances, or activation of portions of the genome in others, may be direct or indirect effects of the hormone. Clearly, the effect of insulin in facilitating penetration of glucose into muscle and kidney cells is the consequence of binding of insulin to cell membranes, but other effects of insulin, for example those on amino acid metabolism, must find an alternative explanation. Increasing evidence indicates that many hormones effect a change in the function of some portions of the genome, as we have already noted for cyclic adenylate. Insulin, thyroxin, sex steroids, and the plant kinins have all been shown to elicit a burst of RNA synthesis, presumably of messages previously repressed. Whether they directly affect the genes or operate through altered cellular metabolism remains to be established. When such molecular understanding is at hand, a new era of pharmacology may dawn.

THE NERVOUS SYSTEM

What strikes one most forcibly in thinking about the human brain is that it is made up of more than a trillion cells and that the interconnections between those cells are many times more numerous still. The nervous

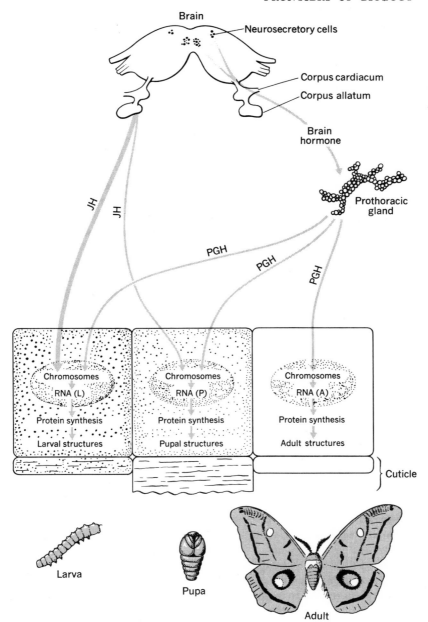

FIGURE 25 Endocrine relations in insects. JH, juvenile hormone; PGH, ecdysone. (From H. A. Schneiderman and L. I. Gilbert, "Control of Growth and Development in Insects," *Science, 143*:325–333, January 1964. Copyright © 1964 by the American Association for the Advancement of Science.)

system continually receives, transforms, stores, and updates information concerning the world about us. It is through the interactions of nerve cells that we are aware of the world around us and are capable of learning and remembering, of feeling and acting. The task of studying brain function is, at first thought, staggering. Fortunately, it has been experimentally simplified in a number of ways, and these simplifications have permitted increasing insight into brain function.

The most profound simplification has resulted from development of techniques for studying the activity of individual nerve cells, a simplification that is successful because the building blocks of the different regions of the vertebrate nervous system, and indeed of all nervous systems, are everywhere about the same. What distinguishes one region from another or one brain from the next is the way the building blocks are put together. By this cellular approach some insight has been gained into the complex problems of how sensory stimuli are sorted out at various levels in the brain. It is now apparent that cells in the cerebral cortex are organized into elementary groups arranged in narrow vertical columns, whose function is to perform a particular "transformation" of the neural message. But this explains why the cerebral cortex requires so many cells. It is not that the actions mediated by individual cells are trivial; rather, it is because the cortex receives such an enormous amount and variety of information from the sensory apparatus at the body surface (eyes, ears, nose, skin) that huge numbers of cells are required to handle and to "transform" this information so it can be used for perception and for action.

A second simplification has emerged from the study of the brains of relatively simple animals, or even of ganglia, organized bundles of a few thousand nerve cells, since these manifest a variety of interesting characteristics that makes them useful as models for study of certain elementary features of behavior in higher animals.

Study of neural function has engaged man for many decades. The rewards of the last 20 years have been rich indeed, but it is apparent that we have merely crossed the threshold of significant understanding. In the years ahead, neural science will be wed to molecular biology, biochemistry, and genetics, on the one hand, and to the concepts and techniques of experimental psychology on the other. Only bare beginnings have been made in these directions, but it is already clear that interdisciplinary effort will be essential to bridge the gap between the function of individual cells or groups of cells and psychologically meaningful behavior. The task is vast indeed, and dramatic rapid progress is not likely. New tools must be fashioned, and new approaches must be developed. Nevertheless, those engaged in the study of nervous systems are aware of an unusual excitement and great promise.

The Neuron

Neurons have irregular shapes; each consists of a cell body with numerous processes, axons, and dendrites, with varying prominence from one group of neurons to another. An example of a motor neuron is shown in Figure 26. The dendrite-cell body region most commonly receives the contacts, or synapses, projected to it by other neurons. These contacts vary from relatively few to several thousand. Each contact can influence the excitability of the recipient neuron. The essence of this process derives from the fact that the dendrite-cell body region can sum or integrate all the influences that converge upon it via these contacts.

Compared with most other cells, the neuronal nucleus is extraordinarily large; although it has the normal species component of DNA, it has an unusually large nucleolus engaged in making ribosomal RNA and possesses a variety of enzymes that in other cell types are restricted to the cytoplasm. Within the cytoplasm, masses of ribosomes (the Nissl bodies) are actively engaged in protein synthesis, the purpose of which is not apparent. Each day a nerve cell manufactures an amount of protein equal to approximately one third of the total protein of the cell. This protein is prepared for export, since it travels down the cell and along the axon and some of it must leave the cell. The function served by this impressive axoplasmic flow remains unknown. The narrow cylindrical axon varies from a fraction of a millimeter to more than a meter long and conducts the action potential generated in the cell body to the terminal contacts in a repetitive code of all-or-none impulses.

SIGNALING IN NEURONS—THE TRANSFER OF INFORMATION

The complex organization of the nervous system is based upon a precise, selective interconnection of neurons, many of which are present and operative at birth. Functional connections may not be permanent, and conceivably they can be unmade and others formed, but the mechanism and circumstances of this change, if it occurs, are unknown. Each nerve cell possesses, in miniature, the integrative capacity of an entire nervous system, but the intricacies of the latter should not be understood to be simple addition of large numbers of the former. Each nerve cell evaluates the totality of its inhibitory and excitatory input, and then either fires or does not fire, the only alternatives available to it. Figure 27 summarizes the electrical changes that occur when a nerve fires.

As measured with intracellular microelectrodes, at rest, a neuron maintains across its surface a voltage difference of 60–70 millivolts by maintaining an unequal distribution of sodium and potassium ions. This distribution

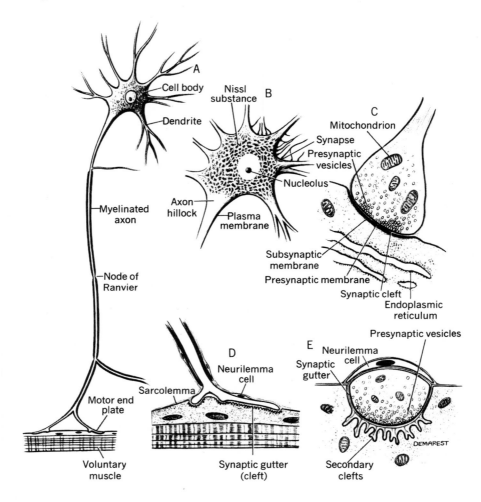

FIGURE 26   The neuron. Diagrams of a motor neuron of the ventral horn of the spinal cord. (A) The neuron includes a cell body and its processes (dendrites and axons). Within the cell body, there is Nissl substance, nucleus, nucleolus, and axon hillock. The axon collateral process branches at a node of Ranvier. (B) Several axons form synapses on the cell body and base of dendrites; motor neurons are densely covered with such synapses. Some have an excitative, others an inhibitory action. (C) A chemical synapse as reconstructed from electron micrographs, including synaptic vesicles in bulbous ending, mitochondria, presynaptic membrane, thick subsynaptic membrane on the postsynaptic side, and the synaptic cleft, which is 200–300 angstrom units wide. (D) Enlargement of a portion of the motor end plate region of (A). (E) Cross section through (D). (From *The Human Nervous System* by C. R. Noback. Copyright © 1967 McGraw-Hill, Inc. Used with permission of McGraw-Hill Book Company. Drawn by R. J. Demarest.)

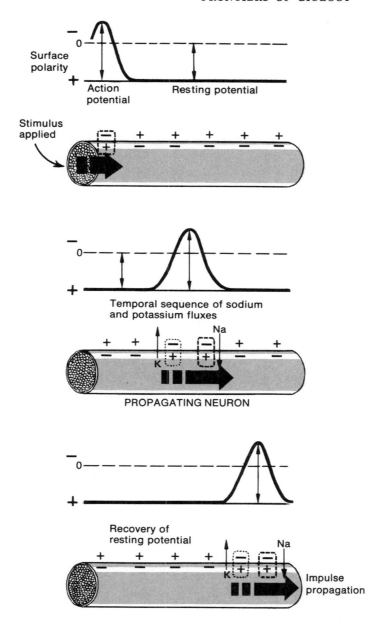

FIGURE 27   Monophasic nerve action potential. Diagram shows overshoot above zero and temporal sequence of sodium ion influx and potassium ion efflux during propagation of an impulse. (From Tuttle, W. W., and Schottelius, Byron A.: *Textbook of physiology,* ed. 16, St. Louis, 1969, The C. V. Mosby Co.)

is achieved by the sodium-transporting pump characteristic of most cells. If the resting potential is reduced by 10 or 15 millivolts, the nerve-action potential is initiated because of a sudden reversal of the characteristic sodium–potassium selectivity of the membrane. Sodium ions enter rapidly, and potassium ions leave the cell somewhat more slowly. As this happens in a given patch of membrane, it results in lowering of the potential in the adjoining patch, where the same events then ensue and the wave is propagated in this manner along the axon and dendrites. During the subsequent period, while the original sodium and potassium concentrations are being restored, the nerve cannot fire.

INITIATION OF IMPULSE ACTIVITY AT SENSE ORGANS

Entering the central nervous system is a vast network of nervous fibers whose peripheral endings constitute the sensory apparatus, distributed in sheets such as those of the retina of the eye, the cochlea of the ear, or the skin. At the periphery, some transduction process is required to convert pressure or temperature on the skin surface, absorption of light by the visual purple of the eye, or vibration of the minute hairs of the ear, into depolarization (reduction of the potential) at the nerve terminal, resulting in initiation of an impulse that then travels away from the initiating event. Details of these transducer mechanisms are scanty, but it is evident that the spatial and temporal patterns of such stimuli are transformed into impulses of varying patterns and frequency. The transduction process and the tropic effect of the nerve itself upon the tissue surrounding it constitute a subject of intensive interest.

In no case does the signal go directly from the initiating site to the central nervous system as a single impulse. Interposed is a set of relay stations, or interneurons. The junction between two neurons occurs at the synapse, where a dendrite from one neuron is brought into close apposition to the cell body, the axon hillock, or even a dendrite of a second neuron. In most instances, they do not make actual contact. Between them is a cleft, perhaps 250 Å wide. In the peripheral nervous system, arrival at the dendrite of an impulse from the afferent neuron results in the discharge into the cleft of a "chemical transmitter," either acetylcholine or norepinephrine, found in minute vesicles just at the tips of the dendrites. Diffusing across this slender space, they result in depolarization of the apposing neuron and in initiation of a new impulse to travel along its axon. The mechanism of such cell–cell interaction in the central nervous system has not been established with certainty. But since the same chemical species are to be found in the brain proper, it seems likely that the peripheral system may offer a fair model of transmission in the central nervous system. For sensory signals, successive excitation occurs in the direction of the central nervous system; successive motor signals move in the opposite direction. In rare instances, the cleft is missing, and the two dendrites are fused so that the electrical impulse simply continues from neuron to neuron.

In the neurons of the brain, each presynaptic cell either excites, i.e., lowers the voltage difference between inside and outside, and leads to initiation of an impulse, or inhibits, i.e., increases the resting potential, thereby preventing or depressing the ability of the receiving cell to discharge. On most central interneurons, as is the case for other central neurons, there are both inhibitory and excitatory synapses; the balance of their influences determines whether a neuron will or will not generate an action potential and whether its frequency of discharge will increase or decrease. In both types of chemical synapses, the transmitter substance increases the permeability of the postsynaptic membrane for selected ions only. But at excitatory synapses the transmitter increases the permeability to sodium ions, which reduces the membrane potential and causes current flow, which would initiate an impulse, while at inhibitory synapses permeability is increased for potassium ions and chloride ions, which prevent impulse propagation. The basis for this distinction is not understood, but it resides in the structure of the synapse, not in the intrinsic chemistry of the transmitter substance. Whether an interneuron does or does not fire, and how frequently it does discharge, depends entirely upon the balance of inhibition and excitation that it receives at any instant. There is no reason to believe that all transmitter substances are known. In invertebrates, γ-amino butyric acid serves as an inhibitory transmitter, and serotonin has been found to act as a chemical transmitter. Since both are present in high concentration in the human brain, as are acetylcholine and norepinephrine, in all likelihood these chemicals serve there as transmitter substances also. Termination of the activity of the transmitter is accomplished by its removal from the field of action; acetylcholine is hydrolyzed, and norepinephrine can be actively transported back into the neuron from which it was released, setting the stage for the next such event. Since norepinephrine also serves as a more conventional hormone when released by the adrenal medulla and may participate in the interaction between the hypothalamus and the pituitary, it will be interesting to learn whether the releasing factors of the hypothalamus, which governs pituitary secretory activity, also are, in effect, neurotransmitter substances.

The Central Nervous System

While built of essentially identical subunits, different brains have greatly different capacities for generating behavior. They differ in the number of neurons that compose them and in their interconnections—i.e., not in the nature of their components but in the manner in which these components are organized. As a result of their many types of interconnections, even simple neural systems manifest emergent properties that cannot be intuitively predicted. To achieve greater understanding, two stratagems have been adopted: (1) the study of small brains and (2) the study of subsystems of known interconnections in large brains. In each approach, the exercise consists of sequentially examining the responses of a large number of the elements at each level to a given stimulus pattern and making inferences concerning transformations that have occurred from one level to another.

SMALL BRAINS

The central nervous systems of most invertebrates contain only 10^4 or 10^5 cells, and the more primitive contain no more than a few hundred. Even more propitious is the fact that some invertebrate nervous systems contain small organized groups of cells capable of generating rather specific behavior. For example, a ganglion of only nine interconnected cells is responsible for the rhythmic beat of the heart of crustaceans. More complex motor reflexes are controlled by segmental ganglia that may contain but 500 cells; an excellent beginning has been made in tracing the connection of the individual nerve cells in such an apparatus. In insects, rather simple types of avoidance learning can be accomplished by a segmented ganglion containing only 3,000 cells. Among these, neurons vary in their size and in the efficacy—the influence on the receptor neuron—of individual connections. One of the major tasks of the next few years is achievement of further understanding of some of these simpler systems by combined anatomical, biochemical, physiological, and psychological approaches. But a beginning has been made. The primary tool in all such studies is the electrical detection of passage of a nervous impulse. To illustrate the nature of such progress, we will consider some findings with respect to the visual systems of several types of brains.

Visual phenomena commence with the absorption of light by specially adapted nerve cells in the retina at the back of the eye. Rod cells "see" only in black and white and are effective in dim light; cones "see" in colors but only at relatively high light intensity. The functional material in rods is visual purple or rhodopsin, a dye that is bleached when light of the proper wavelength is absorbed. This material consists of a protein (an opsin) to which is bound the aldehyde form of vitamin A (retinal). In the latter, all bonds are in the *trans*-configuration. (See diagram at top of page 101.) When a quantum of light is absorbed, the bond at position 11–12 is isomerized to the *cis*-configuration, swinging the bulk of the molecule away from

All-*trans*-Retinal

Δ^{11}-*cis*-Retinal

the protein surface, resulting secondarily in a subtle change in the three-dimensional conformation of the opsin. It is this event that somehow initiates the nervous impulse, which then passes over the surface of the rod cell. Although each rod contains about 10 million molecules of rhodopsin, absorption of only one light quantum suffices to cause the nerve to fire, and if five to seven rods fire, this is perceived by the dark-adapted individual as a faint light flash!

Much the same arrangement serves in the cones, except that there appear to be three categories of cone cells. In each, there is a visual pigment utilizing retinal, as in the rods. But each category of cone has a different opsin protein, so, when the retinal is bound, one appears to be red, one blue, and one yellow, thus permitting color discrimination. Color blindness consists of the genetic inability to make one of these three opsins.

At all levels of evolutionary development a few general principles appear to be applicable. The afferent input is distributed widely among various regions of the central nervous system, and a great amount of sensory transformation occurs at the earliest points in the sensory pathways as a result of lateral interaction between adjacent elements. When excitatory, lateral interactions lead to facilitation and sychronization of the activity in adjacent members of a neural population. When inhibitory, lateral interactions lead to spatial contrast. Lateral interaction has been extensively studied in the visual system of the horseshoe crab, *Limulus*. The eye of *Limulus* is constructed of a multitude of individual receptor units (ommatidia), which are not independent in their action. Although the activity of a given receptor unit is principally determined by the light shining on its facet, this activity is significantly modified when light is shone upon neighboring receptor units, causing them to become less active. The receptor units are mutually inhibitory; excitation in one unit produces inhibition in all the surrounding units. The spatial spread of inhibition is such that it is most effective for the nearest units and falls off sharply with distance. The strength of the inhibitory influence exerted by a particular receptor channel on neighboring receptor channels depends both on the effects of the stimulus on the reference channel and on the inhibitory influences from its neighbors. The strength of this influence depends in turn on the neighbors' level of activity, which is partially determined by the inhibition that the reference receptor elements exert on them.

This anatomical arrangement provides an example of the major principle in the dynamic organization of neural populations: balanced opposition of excitatory and inhibitory tendencies in molding patterns of neural activity. In the *Limulus* eye, these inhibitory influences, exerted quite indiscriminately by receptor channels on their neighbors, result in enhancing contrast in the visual image. If each unit in the mosaic of receptor channels inhibits the activity of its neighbors to a greater degree as it is more strongly excited, then brightly lighted elements will exert a stronger suppressing action on dimly lighted neighbors than the dimly lighted neighbors can exert on the sharply lighted elements. Consequently, the disparity in the activities of the two channels will be exaggerated and brightness contrast enhanced. If the inhibitory interaction is stronger for near neighbors in the retinal mosaic than for more distant ones, such contrast effects will be greatest in the vicinity of sharp light discontinuities in the retinal image, and the outline of objects imaged on the retina, their resolution, will be sharpened. Thus the pattern of neuronal activity generated by the action of light and by boundaries between light and dark is transformed in a physiologically important way, by the mosaic of receptor channels, at a very early point in the analysis of sensory input.

By using small brains, it has been possible to describe in similar fashion the interrelationships and properties of interneuronal populations, of "command interneurons" that operate at a stage beyond initial interneuronal sensory processing, and to sum up all the influences that lead them to make the equivalent of a decision. By either firing or not firing they then determine whether efferent impulses will then flow along axons that lead to the musculature.

LARGER BRAINS

Early studies of the visual systems of larger brains made it evident that the retina projects in an orderly fashion upon the visual cortex region of the central nervous system. But changes in the configuration of the receptive field occur at various levels between the retina and its final representation, as can be observed with microelectrodes implanted at various levels along the optic tracts. The first neural elements in the mammalian visual system whose receptive fields have been successfully studied are the ganglion cells of the retina, each of which is receiving information from a considerable number of rod or cone cells in the retina proper, as shown in Figure 28. The fields of these cells—the collection of rods or cones to which they are connected—are circular, with an "on" center and an annular antagonistic surround region, or the converse. The most effective excitatory stimulus for cells with an "on" center receptive field is a circular light spot covering the entire central "on" region of the field. If the stimulus is enlarged to include any of the annular surround region, the effectiveness of the stimulus is reduced because of the mutual antagonism between the center and surround regions. Accordingly, a retinal ganglion cell does not primarily signal the intensity of light impinging on a given part of the retina, but signals the contrast between the intensity of illumination in the center of its receptive field and that of its surround region. At the next synapse (in the lateral geniculate region of the brain, Figure 29), the effective excitatory receptive field resembles that of the retinal ganglion cells.

However, at the first cortical synapse beyond, the receptive field changes dramatically. Small spots of light, which are effective in stimulating retinal ganglion and geniculate cells, are practically ineffective. To drive cortical cells, the stimulus impinging on the retina must have linear properties (bars, lines, rectangles). Within this cortical area there are two general classes of receptive fields. The simpler receptive fields resemble geniculate cells in that they can be described in terms of discrete excitatory and inhibitory zones, although the receptive field must be rectangular and not circular, and the effective stimulus is not a spot of light but a bar with a specific inclination, e.g., a vertical, horizontal, or oblique axis of orientation. For example, the most effective excitatory stimulus for cortical cells with a simple receptive field may be a bar with a receptive-field orientation from 12 to 6 o'clock projected upon some retinal position. This rectangular excitatory zone is framed by a rectangular inhibitory zone, and it must be built up by receiving input from a set of geniculate cells having appropriate properties ("on" centers) and similar retinal positions. Thus, for a stimulus to be effective on a cortical cell, it must have the proper axis or orientation. Since all areas of the retina must be presented in all orientations and for several stimulus types, one can understand why the cortex needs so many cells for the normal functioning of the visual apparatus. Importantly, although geniculate cells respond to stimulation of only one eye, cortical cells tend to respond to both eyes; so it is at the level of the first cortical synapse that one finds the first evidence for the binocular-fusion character-

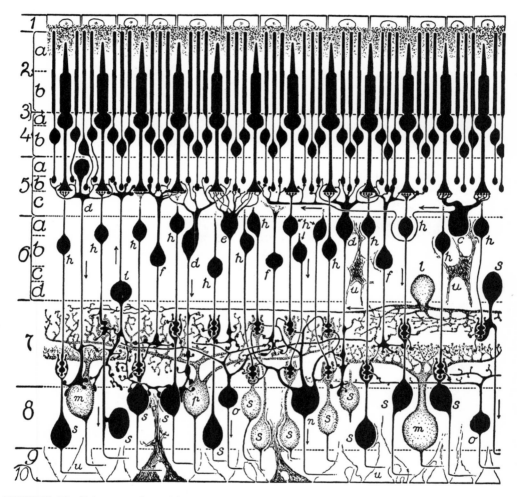

FIGURE 28 Primate retina. Diagram representing the structures of the primate retina, based on numerous Golgi-stained preparations of man, chimpanzee, and macaque. In the upper part, the slender structures are the rod cells (*a*), the thicker ones, the cone cells (*b*); *c*, horizontal cell; *d, e, f, h,* and *i,* centripetal bipolar cells; *l,* inner horizontal or association cell; *m, n, o, p,* and *s,* ganglion cells; *u,* parts of the radial fibers of Müller, with their nuclei in 6 and their lower or inner ends forming the inner limiting membrane (10). Note the various synaptic relations between different neurons, reciprocal overlapping of expansion or its absence, the probable direction of the nervous impulses indicated by arrows, and other details. (After Polyak. From W. Bloom and D. W. Fawcett, *A Textbook of Histology,* 9th ed., 1968. Copyright © 1968 W. B. Saunders Company.)

istic of the vision of higher animals. These simpler cortical cells are found in vertical columns within the visual cortex, running from the surface of the brain to the white matter (Figure 29B). Within these columns are also found complex cells for which the effective stimulus is again linear and must have the correct orientation, but its exact position in the receptive field is unimportant. Clearly, these complex cells are receiving their input from a variety of the simpler cells, and the column becomes the elementary unit of neural organization, bringing together cells that are appropriately interconnected to generate a cell with a higher order of receptive field.

In the adjoining area of the brain the visual message undergoes still further processing. Here some cells have "hypercomplex receptive-field properties" and respond only to a highly specific stimulus such as an angle or corner. In this region, complex cells feed into hypercomplex cells, which serve to detect curvature or changes in the direction of a line. The totality of these levels of activity permits estimation of such a change in light intensity as contrast between light and dark, as well as changes in contour. Elsewhere in the system are cells that recognize only change or movement.

From this analysis it is not yet possible to grasp how a total visual image—a room and its contents—is built and perceived almost instantaneously, but the elements are now available. One can now say that the task of understanding the neural basis of perception is no longer impossible or incomprehensible. It is only immense. Indeed, what has been learned about neural mechanisms of perception indicates that the details are not only elegant, they are beautifully simple. Equivalent analyses are available for somaesthetic perception and for the control of motor activity. These differ significantly in detail, but the principle of the experimental approach is equivalent and has yielded, to date, about the same degree of understanding.

INTERCALATED SYSTEMS: HOMEOSTATIC REGULATION

Between the major sensory and motor systems there exists the great mass of the central nervous system, composed of subsystems organized for the regulation of intrinsic brain functions and for the control of the function of other organs. These are particularly voluminous and diverse in the forebrain, where they are represented by the large association areas of the cerebral cortex, the limbic system, a heterogeneous array of large neural structures in the medial and basal walls of the cerebral hemispheres and the corpus striatum. Although this portion of the brain has been undergoing analysis for many years, understanding is not comparable to that provided by studies of sensory perception or the governance of motor function,

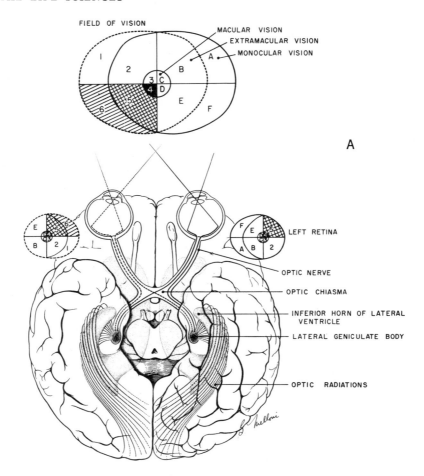

FIELD OF VISION

MACULAR VISION
EXTRAMACULAR VISION
MONOCULAR VISION

LEFT RETINA

OPTIC NERVE

OPTIC CHIASMA

INFERIOR HORN OF LATERAL
VENTRICLE

LATERAL GENICULATE BODY

OPTIC RADIATIONS

A

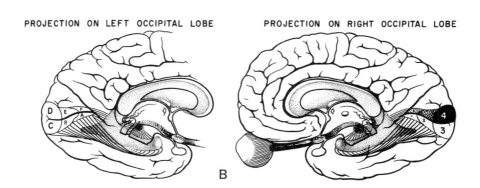

PROJECTION ON LEFT OCCIPITAL LOBE

PROJECTION ON RIGHT OCCIPITAL LOBE

B

largely because of the experimental difficulties in establishing the nature of effective input stimuli to the cells of these structures. Most current understanding has come from clinical and behavioral studies of the deficits that result from pathological destruction or surgical ablation of specific regions within this vast part of the brain. These have permitted somewhat detailed mapping of the responsibilities of specific regions of this portion of the brain, and it is clear that the association areas are involved in the elaboration of specific sensory information into percepts of varying complexity. Further, reference has been made earlier to the role of the hypothalamus (part of the limbic system) in the regulation of the pituitary gland. This small region is sensitive to diverse other stimuli and responds by driving appropriate motor neurons controlling autonomic functions such as respiration, heart rate, and blood pressure.

One more aspect of the study of the central nervous system may be cited merely to indicate the patterns of current research. No problem is older than man's concern with himself—with the relation between the physical entity of the brain and "mind" or "self-awareness." Little progress can be recorded, but one aspect of self-awareness, the phenomenon of sleep, is subject to experimental approach. Even now it is unclear what the underlying physiological function of sleep may be. One can only ask how it occurs and in what phenomena arousal consists. The extensive changes in cerebral action associated with sleep–wakefulness transitions are not limited to the level of consciousness. There are also changes in muscle tone and reflex thresholds, for example, which are signs of central nervous system excitability. Yet there is no general diminution in the overall activity of the sleeping brain—no change in cerebral blood flow or oxygen consumption, no apparent overall decrease in the activity of cortical neurons.

Two important discoveries underlie modern concepts of the mechanisms involved. The first is that slow (eight per second) rhythmic electrical

---

JURE 29   Primate vision. (A) Diagram of the central visual pathways showing the rse of impulses from the retinal quadrants to the visual cortex. Small inserts beside ball show projection of image within the visual field on the retina. Images on the right of the subject are projected on the left side of the retina; those above, on the lower half he retina, and so on; i.e., the visual field is inverted on the retina. There is normally a on of maximum acuity in the central field, with a decline in the resolving power periphly. The visual field of one eye overlaps that of the other, providing the basis for eoscopic sight (notion of depth). The notion of depth can be given in monocular vision is far more perfect in binocular vision. (B) Projection of optic radiations to the occipital al cortex. Corresponding visual quadrants, 1, 2, 3, 4, 5, 6, and A, B, C, D, E, F of the al field. (From S. Grollman, *The Human Body, Its Structure and Physiology,* 2nd ed., Macmillan Co., New York, 1969, © copyright Sigmund Grollman 1969.)

stimulation of the general thalamocortical portion of the brain, entrains the electroencephalogram (EEG) in rhythmic oscillation and provokes all the behavioral signs of sleep, whereas more rapid electrical stimulation of the same region, or of the ascending reticular system of the brain that impinges upon it, wakes a sleeping animal. The second observation is that deafferentation of the forebrain by a midbrain transection leaves it with electrical and behavioral signs of continued sleep, while transection of the brain below the medulla leaves the sleep–wakefulness cycle intact. These observations and the studies that have followed from them have led to the proposition that sleep is due both to a withdrawal of driving electrical input and to an active process, originating in the brain stem, which tends to drive the forebrain electrical activity in slow rhythmic oscillation that produces sleep. According to this concept, sleep is neither exclusively active nor a passive process, but a combination of both. Much current research is aimed at clarifying the nature and mode of operation of these reciprocal mechanisms.

An alternative approach, which appears to warrant aggressive continued study, rests on the observation that spinal fluid taken from a goat that has been kept awake for several days contains a material of low molecular weight that induces deep natural sleep in other animals. If confirmed, this observation offers enormous potential, not only for understanding of the mechanisms involved in sleep, but also for the development of the ideal sedative or anesthetic.

A quite independent set of problems originates in the fact of the existence of the two major cerebral hemispheres. These are connected through the corpus callosum, a bundle of perhaps 100 million nerve fibers. If this structure is severed, the functioning of each hemisphere remains intact, but quite independent. Learning in each hemisphere becomes dependent upon its own sensory input, with no crossover unless the hemispheres receive identical input. For example, an animal with a severed corpus callosum can be taught, with one eye, that a given signal represents the availability of food, and subsequently, with the other eye, taught that quite a different signal indicates such availability. Thereafter, it will continue to respond to these signals entirely in accord with whichever eye is left uncovered. In a dramatic instance, prefrontal lobotomy was performed on only the right hemisphere of a "split-brain" monkey. When introduced into a cage containing a snake, this subject cowered with fear when one eye was covered and the snake was perceived with the left hemisphere, but completely ignored the snake when it was perceived only with the right hemisphere. These studies have had practical utility. Surgical separation of the hemispheres has been accomplished in individuals with certain forms of epilepsy who were thereby returned to a semblance of normal life. These patients also are clearly under the control of two independent half-brains, without fusion of individual inputs. Accordingly, in man, each hemisphere is capable of independent conscious awareness. Surgery leaves "split-brain" people with two separate minds, indicating that the higher functions are not the exclusive domain of a major hemisphere. Such studies offer great promise of revealing the highest orders of organization of the brain.

BEHAVIOR

Scientists of diverse backgrounds investigate the ways animals behave toward each other and toward their environment. These "behavioral biologists" include zoologists, physiologists, psychologists, psychiatrists, and anthropologists, among others. Some observe whole animals; others, small parts of their nervous systems. Some study the physiological mechanisms of behavior in adults; others, the development of behavior in embryos. Some, concerned with specific problems, use any particularly appropriate animal as a tool; others are fascinated by particular kinds of animals. Some approach animal investigation primarily as a means of understanding human behavior; others are not primarily motivated by the potential social utility of their work. In sum, however, their interrelated work illuminates in diverse ways the behavior of organisms.

This field of endeavor is in its infancy. Nevertheless, starting from different points of view, studying quite different animals, formulating different problems, a coherent picture is emerging of the way in which behavior has historically evolved and the way it develops in individuals, and of the physiological mechanisms that make behavior possible. Clearly, this endeavor shades into the totality of experience in observational and experimental psychology. So large is the field and so disparate the parts that no adequate summary is possible here. We can, however, offer a small sampling of the kinds of studies that, in recent times, have contributed to growing understanding of the physiological bases for behavior.

Evolution, Inheritance, and the Development of Behavior

The biologist assumes that behavior is not an accidental feature of the existence of a species—that it has, itself, been subject to selection in the evolutionary process. In some instances, unique behavioral patterns clearly relate to the ecological niche in which a species has lodged, as, for example, various aspects of the behavior of the kittiwake, a cliff-nesting gull, which are shared with the swallow-tailed gull of the Galápagos, thousands of miles distant. Both of these birds, however, engage specifically in cliff-nesting. Not all aspects of behavior can be so identified, and indeed some may be irrelevant, but surely the remarkable formalized courtship and mating patterns of many species have served to preserve their genetic distinctiveness and to assure their continuity. There can be little doubt that behavioral qualities can be treated as genetic traits and followed in longitudinal breeding experiments. Few are simple Mendelian traits governed by single genes. Nevertheless, maze performance in rats, aggressiveness in many species, geotropic responses in flies, can all be manipulated by breeding, although the chemical functions of the genes involved have not been identified. At the same time, the genetic contribution to behavior can be modified by individual experience; e.g., producing mild behavior in genetically aggressive animals by frequent fondling during their early lives.

However, assorted allegations of causality with respect to behavior associated with genes known to affect specific biochemical functions have, in the main, proved instead to be associated only by chromosomal linkage with genes of unknown biochemical function. A clear correlate of general behavior with genetic endowment is illustrated by individuals afflicted with Down's syndrome, the result of trisomy (the relatively rare occurrence of three rather than two chromosomes) and by recent suggestions of excessive violence in some proportion of human males whose sex chromosomes are XYY.

In the main, the nervous system is "prewired" in the sense that most of its connections are the consequence of genetic endowment. Few connections, if any, are formed simply on the basis of reaction to the environment. In this sense the newborn child is not simply a "clean slate," but has built-in attributes that are subject to modification with experience. Examples of such environmental influences include the modified behavior of the genetically aggressive rats cited above; the impaired visual–motor coordination of an adult monkey if, during the early months of life, it is prevented from visually perceiving the consequences of voluntary movements; the strikingly abnormal social and sexual behavior of monkeys that are reared without normal social association with other monkeys of their own age; and the well-known "imprinting," as the maternal object, of the first moving object to come within the ken of a newborn duckling.

Physiological Analysis of Behavior

A concerted and many-sided attack on the ways in which the nervous and endocrine systems influence behavior is now under way. One challenging problem is the "clockwork" responsible for a large group of biological rhythms with a range of frequencies from a few seconds to several years, occurring at all levels of organization. Some occur in single cells or small cell aggregations, such as the pacemaker of the heart or rhythmic impulses from the brain stem; some involve almost the totality of the endocrine system, as in the ovarian cycle; some are the basis of diurnal or annual rhythmic behavior, as in the breeding and migration patterns of many species; some are responsible for cycles in the size of entire populations. Clearly, some of these have been imposed by the environment—by the length of daylight or seasons of the year. But most circadian rhythms, such as sleeping patterns, are basically of internal origin in the individual and become entrained in some aspect of the environment. In no instance to date has the intrinsic clockwork actually been established.

Simple motivational behavior lends itself readily to experimental approach, permitting analysis of questions concerning such behavioral attributes as drive and satiation, reward and punishment, pleasure and pain, and, particularly, physiological regulations of hunger, thirst, sexual and maternal behavior, and temperature regulation. In each of these the activity of the hypothalamus at the base of the brain has proved to be central, exhibiting excitatory and inhibitory mechanisms that accelerate or brake behavior. For example, experimental destruction of the excitatory mechanism can result in fatal refusal to eat or drink in the presence of an abundance of food and water. Conversely, destruction of the inhibitory mechanism elsewhere in the hypothalamus leads to gross hyperphagia and inordinately obese animals. These regions of the brain can be manipulated by electrical or chemical stimuli, causing animals to eat or to starve themselves on signal. Evidently, the normal hypothalamus is responsive to internal signals such as the blood concentrations of sugar, sodium ions, estrogens, and androgens. Normal man regulates these behaviors with remarkable precision; thus, for example, one can maintain relatively constant weight for 30 or 40 years of adult life. At the next level of understanding, it should be possible to obtain sufficient information to permit clinical management of the obese, the anorexic, the depressed, and persons with damaged brains in whom these normal controls have failed.

ORIENTATION AND HOMING

Spatial orientation to the surroundings is one of the most characteristic behavioral attainments of animals, including man. Many animals, however, orient themselves by sensing mechanisms or behavioral discriminations that pose perplexing mysteries because they differ considerably from those that affect human behavior. Echolocation enables a bat with a brain weighing less than a gram to locate and intercept minute flying insects within a fraction of a second, but only very limited echolocation has been achieved by blind humans. Similar abilities are obviously exercised by marine mammals in murky waters.

Migratory animals pose elementary questions for which, despite great effort, there are as yet no adequate answers. The aspects of their environment that provide directional information and the sensory channels that convey that information to the central nervous system remain obscure.

A catalog of the achievements of migratory animals never fails to astonish. Birds such as Arctic terns and some of the shearwaters migrate thousands of miles. On land, some mammals such as caribou migrate great distances under conditions so difficult as to pose extremely complex questions concerning how they manage their orientation and sustain their motivation. Sea turtles swim across hundreds of miles of ocean to locate relatively small islands where they lay their eggs on a few beaches. Many fish and eels are known to migrate similarly over thousands of miles of ocean. Even some marine invertebrates go through life cycles that include extensive migratory behavior. The migrations of whole populations of insects such as locusts have been notorious since biblical times because of their destruction of human crops. Tagging programs have revealed that monarch butterflies migrate for more than a thousand miles and in directions that cannot be accounted for by passive transport by the wind. Not only do the sensory mechanisms involved in their orientation require analysis, but also attention must be given to the equally significant questions of motivation, genetic programming in relation to individual learning, and the ecological advantages of migration, which suffice to offset the exertion and hazards of such extensive travel.

The homing behavior evident in numerous species is a related phenomenon. It is usually studied by artificial displacement of animals by considerable distances to unfamiliar surroundings. Success in their performance is even more remarkable than that of long-distance migrations. Conceivably, the latter are guided by sun or star compass orientation, which also requires compensation for the apparent motion of celestial bodies due to the earth's rotation. But homing after arbitrary displacement requires that the animal select the homeward direction as well as keep moving in that direction. We know how some animals obtain the equivalent of a compass, but we know almost nothing about how, in homing, they obtain the equivalent of a map. The heart of this problem is the search for sensory cues not apparent to man. Bees orient themselves by the polarization of light, and some fish detect distortion of the electric fields generated by their own electrical organs. Conceivably, other attributes of the environment are as important as the star map appears to be to migrating birds.

The effect of the endocrine system on behavior has been extensively explored for some time, largely because it lends itself readily to such experimentation. Well known are the profound consequences on male and female sexual behavior that result from castration and restoration of normal behavior by administration of appropriate sex hormones. More recently, it has been shown that if radioactive hormones are administered, they bind to a relatively small group of cells in the brain stem, and it appears that subsequent behavior is the consequence of such binding. As we have already seen, a converse set of relationships is mediated by the hypothalamus and pituitary glands. Some such relations require more complex interactions. For example, stimulation of the mother's nipples by the suckling young is recognized in the brain and, in turn, through the hypothalamus, initiates the secretion of pituitary lactogenic hormone, which, in its turn, stimulates the mammary glands to manufacture milk. In a mother rat even the sound or smell of hungry young sets off this train of events.

The nervous system of a newborn infant is already so constructed as to exhibit either male or female character. Injection of the opposite sex hormone into adolescent or adult animals does not elicit a striking change in sexual behavior. However, if the hormone is administered during embryonic life, the nervous system takes on the character associated with the sex of the administered hormone, at least as measured by the ability of the adult animal to respond to the sex hormone when it is administered later. As yet, there is no clue as to the nature of the "wiring diagram" or any other attribute of the nervous system that gives it its male or female character. Such studies have deep relevance for that unfortunate group of individuals whose genetically intrinsic sex is mistaken during early life.

Learning and Memory

Learning is the process whereby behavior is modified by experience; it is a measurable and lasting change in behavior produced by a specifiable set of environmental circumstances. All learning can be reduced to specific parameters operating under the following principles: (1) There must be continuity in time and space of the items learned (stimuli to be associated with other stimuli or response). (2) There must be repetition of the associated items, although occasionally there is successful one-try learning. (3) There must be reinforcement, whether reward or punishment, for with reinforcement the learned response grows in force and, in its absence, decays (forgetting). (4) An interference or intermission process is required to produce forgetting, presumably by covering over memory rather than destroying it, because there is evidence that the basic process of learning is essentially permanent. But having made these general statements, one must admit that the nature of the learning process, in its essence, remains obscure. Primitive organisms can engage in the simplest learning, if only as avoidance responses. Ascending the phylogenetic ladder, there is a continual increase in the complexity that the organism can handle in learning. A wide variety of experimental approaches have been brought to bear on this process, but without genuine success.

It is apparent that no one area of the brain constitutes its memory. Memory is diffused over much of the total cortex. No one lesion in the brain, even a sizable one, produces permanent amnesia or inability to learn, although such lesions can produce amnesia for recently learned material. A variety of investigative techniques have yielded similar findings. Lesions placed in various areas of the brain, violent electric shocks, epileptic seizures, and drugs that inhibit protein synthesis, specifically puromycin, can all abolish a very recent learning experience, but none removes well-established memory. Investigators engaged in such studies have contemplated that memory may take one of several forms. Among these are electrical circuits in the brain that are self-sustaining for long periods; connections between nerve cells, *viz.,* synapses, which may either be brought into being or broken; and new outgrowths or sproutings of such connecting terminals, which might be generated. Finally, there may be specific chemical changes, conceivably in the transmitter molecules, but, more likely, experience could be encoded within macromolecules of the nerve cells.

The favorite candidates for the latter, if this possibility is valid, have been RNA, proteins, or even the complex polysaccharides. Each affords sufficient variability to give rise to speculation that learning is an experience that can be coded within its structure. The antibiotic puromycin, which inhibits protein synthesis, can totally eliminate memory of a recent learning experience in mice or fish. In view of the wholesale protein-synthesis characteristic of cells of the nervous system, some relationship between protein synthesis and learning may not be entirely unexpected. But the matter is not simple, because cycloheximide, which also inhibits protein synthesis, has no effect on memory. An important clue may reside in the fact that puromycin blocks protein synthesis by interrupting growth of polypeptide chains, whereas cycloheximide somehow prevents the normal breakdown of polysomes so that the mRNA remains in the cell, inaccessible to the enzymes that normally degrade it after about 30 minutes. These are the most recent observations, and the fact that the experiments are at all productive gives hope of the possibility that the intrinsic nature of the learning process may yet come to light.

A vast literature deals with sensation and perception of the environment and the behavioral consequences of these processes. Since these are thoroughly summarized elsewhere, no attempt will be made to do so here. But we would direct attention to a growing body of information that stems from analysis of relatively primitive nervous systems. For example, caterpillars are relatively highly specific in their choice of plant food, yet there are only four taste-receptor cells on each of their maxillary palps. An acceptance response that initiates feeding, based on the chemical structure of the available food, is determined by one unique pattern of impulse discharge coming from the four axons of these cells; all other discharge patterns lead to rejection of the "tasted" material. Not only is this of great theoretical significance for behavior, but it can also be a practical lead to the control of plant pests.

An elegant example is found in the adults of certain moths, which are normally preyed upon by insectivorous bats. The ears of these moths can detect the ultrasonic cries made by their predators when they are echolocating. Each ear of the moth is supplied with only two acoustic sensory cells, but arrival of sound of the wavelength used by the bats automatically triggers evasive behavior on the part of the moth. Through these and a host of similar examples, the elements of behavior may be observed in primitive animals. From such understanding, ultimately, we may be able to build a platform from which to view the behavior of that most complex of all animals, man.

Ecology

The objective of man's struggle with the environment is not to win but to keep on playing. The biosphere as a whole is certainly not simple, but it has been remarkably reliable up to now. The objective of applied ecology is to keep it so.

Ecologists are generally concerned with the interactions among living forms and between them and their environments. Our society has placed upon the ecologists the fearful responsibility of safeguarding the planet for human habitation. Patently, the working ecologist, on a day-to-day basis, does not live with that concern. On any day he may be on a lake or the ocean, examining bottom mud with the use of carbon-dating procedures; following migratory patterns of birds, mammals, or fish by telemetry; computing the energy balance of a lake, of a terrarium in the laboratory, or of the southwestern desert; examining the history of a peat bog as recorded in its fossils; or engaging in a computer simulation of the total ecology of a rain forest. His tools are not uniquely his; they are whatever applicable tools science has made available. Clearly, so all-encompassing a set of concerns cannot be adequately summarized here, and it must suffice to provide only some indication of the problems that give ecologists concern and the manner in which these problems are approached.

Some Areas of Ecological Research

ENVIRONMENTAL CHALLENGES TO INDIVIDUAL ORGANISMS

Although the concern of the ecologist is with populations or species, understanding must rest on the behavior of individual organisms as well. No one who has seen a 40-pound salmon fling itself into the air again and again in a vain effort to surmount a waterfall can fail to marvel at the strength of the "instinct" that draws the salmon upriver to the stream in which it was born. How do salmon "remember" their birthplace and find their way back, sometimes from thousands of miles? The answer has economic and political interest because dams athwart the routes of the salmon have cut heavily into their reproduction, and the diminished numbers of salmon have affected the fisheries of many nations. Salmon literally smell their way home. Each home stream, presumably from the plant and mineral oils of its drainage basin, acquires a unique organic fragrance. Young salmon "learn" this fragrance in the early weeks of life and retain recognition of it thereafter. The chemical nature of these odors has not been identified. How the salmon navigates long ocean distances is unclear, but, as it finds the main river, it selects that tributary that carries the home odor and continues upstream.

The energy requirements for this extraordinary journey are prodigious. A group of salmon, observed while migrating 600 miles up the Fraser River in British Columbia, took almost three weeks to make the traverse. By the time the female salmon, which do not feed during migration, had reached the spawning grounds, they had expended 69 percent of their body fat and over half of their protein. Energy had been expended throughout this period at 80 percent of the maximum rate possible for these animals! With only a small reserve remaining for emergencies, one can readily understand why only one additional small dam on a river may cause a run of salmon to collapse.

The continual pressure for survival led those organisms that had adequately adapted to inhabit a remarkable variety of ecological niches. Of these, none is more remarkable than the animals that inhabit the desert, where water is scarce. How they survive under those circumstances is a matter of considerable import. As we have already seen, desert rodents evolved kidneys capable of secretion of extraordinarily concentrated urine.

At the same time, they "learned" to avoid the harshest aspects of life in the desert by remaining in their thermostatic, humid burrows during the day, emerging to feed and explore only at night. The camel and the ostrich each solved this problem, in part, by acquisition of a heavy coat that serves as an extremely effective insulator. Others learned to permit their body temperatures simply to approach that of the environment, rising and falling with the days and nights as with the seasons.

Because of its novelty, one should take note of the mechanism that permits marine birds to spend virtually the entirety of their lives at sea. The least saline water available to them is that in the flesh of the fish that constitute their diet. But each such encounter also entails ingestion of some seawater as well. Yet the salt concentration of the blood of these birds is essentially identical with that of terrestrial forms. This is accomplished by a unique adaptation, the presence in their nasal passages of a "salt gland," an organ that, on sensing an increase in the sodium concentration of the plasma, secretes a highly concentrated salt solution of fixed concentration and continues to do so until the plasma sodium level has returned to normal. The mechanism responsible for this function is not yet known.

Plant species are equally sensitive to the water supply, and the record of annual rainfall is to be found in the width and density of the annual deposition of wood. When midday temperatures and radiation exceed a critical level, which varies with the species, the plant's conducting system can no longer transfer water from the soil; evaporation from the leaves continues and water tensions are established within the plants. A suction force equal to 16 atmospheres has been measured in the tops of redwood trees under such circumstances. When internal water stress becomes excessive, the stomata of the leaves close, minimizing water loss but also shutting off uptake of carbon dioxide and release of oxygen, and hence halting photosynthesis. Records of the past found in the rings of trees require careful study, combined with associated fossil analyses and examination of current climatic conditions, to permit predicting the more important biological effects of future weather modifications when this becomes feasible.

THE ABUNDANCE OF LIVING THINGS

The numbers of animals and plants in a given area are primary concerns of the ecologist. Both absolute numbers, or density, and the ratios among species are primary variables. Both are subject to external and internal controls. External controls are exemplified by the weather, application of pesticides, natural catastrophes, and variation in food supply or in numbers of predators. Internal controls are exemplified by the changes in breeding patterns by which the subject populations respond to all aspects of their environment. Both have been studied in detail, in the wild and under laboratory conditions. In natural habitat, both absolute densities and interspecies ratios are maintained over remarkably long periods; the objective of population ecology is to understand the play of forces that achieves this. These forces vary from species to species, but for all species the product of the probability of death and the probability of birth, summed for all ages over a generation, must come to an average value of 1.0 if constancy is to be maintained. A few examples will suffice. *Daphnia,* a small freshwater crustacean, can maintain remarkably dense populations on an adequate food supply. If a modest grade of removal (fishing) is applied, the density falls sharply but not linearly. Even with removal rates as high as 30 percent per day, a small but vigorous population persists. Patently, the effect of fishing mainly for the young is different from that of fishing for adults only. This displays the resilience that man exploits; if we somewhat rarefy a population, we increase its rate of growth, so a certain degree of rarefaction actually increases the steady yield of animals harvested. This is a benefit that works both ways; pest controllers are less than enchanted to find that the more rats they kill the more there are.

Some forms of predation can completely exclude an organism from an otherwise suitable environment. Native silkworms can be successfully raised on wild cherry trees as long as each is enclosed in a protective net, with yields as high as 80 percent from egg to adult. But if the trees are left unprotected, not one silkworm survives through the larval stage. The breeding response to adversity is illustrated by the size of egg clutches in bird species. For example, the habitat of the European robin extends over a great area, yet these birds maintain a fairly constant density over 35 degrees of latitude. Since the winter death rate of adults is higher in the colder northern latitudes, the population can remain constant only if this is offset by an increase in the birth rate. And, indeed, per degree of latitude, this species adds about 0.1 egg to the annual setting.

Some forms of external control are not always obvious. Red tides, the sporadic blooms of a microscopic dinoflagellate that is highly toxic to fish, occur occasionally on our eastern coast. The circumstance that leads to such blooms is not predictable or constant. It usually begins with the formation of a pool of nutrient-rich brackish water in an estuary that has been temporarily prevented from tidal flushing and becomes enriched with stream-borne nutrients. When flushed to sea, the toxic water mass maintains its integrity for days before dissipation. The nutrients that touch off such an event are small quantities of primary elements and of a few organic compounds. A few micrograms of one of these per cubic meter of water appears to make the difference between bloom and no bloom, between environmental health and disaster.

SPECIES INTERACTIONS

Much of the biology of each species is devoted to accommodation to other species, including such phenomena as defense against disease and predators, patterns of courtship and mating, body size, life-span, and reproductive potential. The number of interspecies interactions among the several million species is vast; systematic studies of these are certain to reveal profitable ways of utilizing more of the earth's biota for man's benefit. Biological control tests, new drugs, and repellents are obvious potential applications. It is this largely untapped potential and not sentimentalism that makes ecologists protectors of threatened species and of dwindling habitats that harbor unique combinations of species. It would be tragic if potentially easy solutions to future major problems were lost through ignorance or indifference.

It is repeatedly impressive that seemingly competitive species manage to accommodate in given ecosystems. For example, study of a spruce forest containing five species of wild warblers, at first glance much alike in their habits, revealed their actual ecologies to be surprisingly diverse. Although they all fed on the insects in the spruce trees, each species had a unique combination of behavior patterns based on the proportion of the time it spent hovering, whether it tended to feed on peripheral or central parts of the trees, how frequently it flew from one place to another, and so on. Thus, in effect, by exposure to different items of food, they share the resources of the forest.

Such understanding can be utilized in practice. For example, the Klamath weed was accidentally introduced into the livestock ranges of northern California about 1900. This plant not only replaces valuable foliage, but it is also highly toxic. But in 1947 a beetle was found in Europe that fed upon this plant, and it was introduced into the affected areas. Within a few years, this beetle achieved mass destruction of the Klamath weed population. However, along highways and under the shade of trees, the beetle population is ineffective in controlling the plant. As a result, beetle and weed populations have achieved an equilibrium with rather low weed infestation. Meanwhile, the range has been repopulated with useful grasses.

No episode in the deliberate manipulation of ecological systems is more dramatic than that which began with the introduction of the European wild rabbit into Britain by the Normans at the beginning of the twelfth century. With it came the flea, which is the only known host for the myxoma virus. The virus itself had long been established in the native rabbits of South America, but it is not pathogenic for that species; yet it induces a rapidly debilitating disease in European rabbits. This virus was deliberately introduced into France in 1952 and into England in 1953. By 1955 rabbits were practically extinct over most of Britain. As the rabbits disappeared, plant species that had previously been highly restricted in distribution by the grazing of rabbits now flourished and spread. Areas that had previously supported a covering of low mosses and turf became covered with deep mats of grass and strands of heather. In turn, the entire insect population changed with it. Thus, the total ecosystem was profoundly affected by removal of only one link in the food chain.

ENERGY FLOW IN ECOSYSTEMS

Only recently has ecology become sufficiently sophisticated to be concerned with the flow of matter and energy in ecosystems. A growing body of information reveals the efficiency of the conversion of solar energy into organic matter in ponds, ocean areas, open fields, and cultivated farms. In turn, these provide lessons for man in his future management of the earth. For example, the average efficiency of an Iowa cornfield is only about 2 percent, whereas the conversion of solar energy into organic matter by algae in a fertilized pond may be as high as 20 percent. The use of radioactive carbon dioxide as a tracer has permitted the construction of balance sheets of carbon flow, particularly in small lakes; chemical analysis reveals the total flow of mineral elements as well. Similar techniques have been applied to small patches of forests.

The consequences of removal of tree cover are extremely dramatic. For a long period thereafter, the impinging energy is wasted, and the bulk of the available mineral matter is removed by the leaching action of rainfall. Studies of these effects must be continued both on a small scale and on the larger scale of total drainage basins, forestlands, and deserts. What has already become apparent is that, although 70 percent of the surface of the globe is covered by ocean, the total biological yield of the oceans is approximately of the same order as that of the land area.

STABILITY AND DIVERSITY

Much ecological concern has been addressed to those factors that make for a stable ecosystem—a system whose numbers and balance of species remain relatively constant over prolonged periods. Examinations of such systems have involved description of the patterns of the food web, the distribution and arrangement of different species in space and time, and the grouping of individuals of several species into higher taxonomic units with varying functional roles. Such studies have led to a few generalities.

In general, the greater the number of species at any level in the food chain, the greater the community stability at that level and at lower and higher levels as well. Amplitudes of fluctuation of herbivore populations are determined by the number of species of plants they eat and by the number of predators and parasites that attack them.

The density of food plants determines the extent to which herbivore populations fluctuate, and the size of the herbivore population, in turn, determines population fluctuation among predators.

The more diverse the species at any taxonomic level, the more stable the system. A system of ten equally abundant species is more diverse and successful than one with a single very common species and nine relatively rare species.

Although stability may be hard to measure satisfactorily under any circumstances, it is easier to measure after the fact, when disturbance, the inverse of stability, has occurred. Such events usually occur spontaneously or are produced accidentally by advancing civilization. Relevant studies have recently been performed deliberately. For example, an hourglass-shaped bog lake was deliberately separated into two lakes by an earthen barrier across the constriction. Lime was added to one lake, raising the pH from 5.9 to 7.3. Within a year the transparency of the latter increased remarkably and, after two years, the well-lighted zone had increased from 2.7 to 7 meters in depth. The *Daphnia* population was found to replace itself in one third the time in the lime-treated lake, and new species of phytoplankton and of rooted aquatic plants began to thrive. Thus the initial disturbance that had led to acidification of the lake could, in fact, be successfully reversed, and a more stable, more diverse, and, for man, more attractive ecosystem was restored.

Many illustrations of the practical applications of ecological understanding are to be found in Chapter 2. The foregoing discussion has served only to indicate some of the kinds of problems addressed by ecologists. Only now has the science advanced to a point at which ecologists consider that they can usefully and successfully construct mathematical models of large ecosystems—e.g., the coniferous forest, the western grasslands, the southwestern desert. The success of these efforts remains to be ascertained. Meanwhile, applied ecology is man's greatest hope both for protection of the natural environment and for human survival. Ecological understanding is required to guide intelligent use of pesticides and of biological mechanisms for pest control; to indicate when clear-cutting or slash-and-burn approaches should be used as forests are put to new use; to give guidance to the appropriate scheduling for plowing, planting, burning, and normal agricultural practice; to predict the consequences of the introduction of new strains of crop plants (already it is clear that, on a single farm, genetic diversity of a single crop is preferable to reliance on a single genetic strain); to indicate the possibilities of utilization of new species of plants and animals as major foodstuffs for man (e.g., the many ungulates of Africa, the Saiga antelope, originally from Alaska, which is much more efficient than sheep and goats at cropping the tundra, or red deer, which are more successful on rocky islands than are sheep); to maximize the food harvest from the oceans and larger bodies of fresh water (the manatee, consumer of water hyacinth, one of the most productive of crops, looks particularly promising); to enable us to predict adequately the consequences of increase in the consumption of fossil fuels as opposed to increase in the utilization of nuclear power; and to enable us to protect naturally or deliberately impounded bodies of fresh water—examples without end.

Man has claimed this planet as his own. In so doing, he must accept responsibility for the multitudes of species that he has displaced and that he husbands. The planet can never again return to the circumstances that obtained when *Homo sapiens* was a small wandering clan of hunters. Nor is there any reason to think that desirable. But it can be preserved in beauty with an immense and diverse flora and fauna, while supporting its human population, provided sufficient ecological sophistication is brought to bear. It is regrettable that the need for such understanding has become imperative so early in the life of this young science, which warrants all the support our society can provide.

THE ORIGIN OF LIFE

The origin of life is the least understood aspect of biological evolution. Significant progress in untangling this puzzle has been made in recent years. This progress, stemming from advances in such diverse fields as cosmology, geochemistry, and molecular genetics, together with the search for extraterrestrial life, which is one of the prime objectives of the national space program, has now heightened interest in this central problem.

In the past, discussions of the origin of life tended to be entirely speculative exercises, often tinged with superstition. But this topic has now become a problem for legitimate inquiry, subject to the same intellectual discipline as other attempts to understand evolutionary processes, including the requirement for logical elaboration of hypotheses, avoidance of arbitrary assumptions, and recourse to observation and experiment. Unfortunately, knowledge of the terrestrial environment in the remote past is uncertain and, as the history of this question shows, is liable to drastic revision from time to time as new evidence accumulates. In any case, it is impossible to duplicate or approximate the geological time scale, as well as the variety of conditions and the secular changes in these conditions that have occurred during the Earth's history. Because of such constraints, the most one can hope to claim for conclusions on this subject is a high degree of plausibility. In this respect, however, studies of the origin of life differ in degree but not in kind from other scientific investigations.

Life is not one of the fundamental attributes of the universe, like matter, energy, or time, but is a manifestation of certain molecular combinations. These combinations cannot have existed forever, since even the elements of which they are composed have not always existed. Therefore, life must have had a beginning. Current views of the origin of life differ fundamentally from those of preceding centuries in that they are concerned with the origin of these molecular combinations rather than of organisms endowed with mysterious properties. From this standpoint, the origin of life must be viewed as a historical incident in the evolution of our planet, i.e., as an event limited in place and time by prevailing physical and chemical conditions.

The unique attribute of living matter, from which all its other remarkable features derive, is its capacity for self-duplication and mutation. Living systems reproduce, mutate, and reproduce their mutations. The endless variety and complexity of living organisms and the seeming purposefulness of their structure and behavior are consequences of their mutability. Any system that has the capacity to mutate randomly in many directions and to reproduce those mutations must evolve.

On the basis of various geological dating methods, it is estimated that the earth was formed about 4.5 billion years ago. The first hard-shelled animals in the fossil record appear at the beginning of the Cambrian, about 0.7 billion years ago. It is clear that life was present well into the Pre-Cambrian, a period lasting 3.8 billion years, but one cannot yet say how far back. The time when life started is an important parameter because, by difference, it provides the time scale for the organic synthetic reactions leading up to the origin of life. Paleontological evidence comes from examination of various Pre-Cambrian rocks for their fossil remains and organic content. These include the Nonesuch shale of northern Michigan, the Gunflint chert of Ontario, the Soudan shale of northern Minnesota, Bulawayan limestone of Southern Rhodesia, and the Fig-Tree chert of the Transvaal. All contain isoprenoid hydrocarbons, particularly pristane and phytane, both of which are breakdown products of chlorophyll, and perhaps of other biological molecules. The oldest, the Fig-Tree chert, is three billion years old and contains small amounts of these hydrocarbons as well as microfossils of bacterial and algal size and form. If these are genuine fossils and residues of biological activity and are not the consequence of subsequent contamination, then life started between 3 and 4 billion years ago.

The conditions on the earth's surface at that time are not entirely certain. But all the indications are that the atmosphere was unlike that of the moment, that in place of oxygen, nitrogen, water, and carbon dioxide, there were methane, ammonia, carbon monoxide, hydrogen, and lesser quantities of hydrogen cyanide and formaldehyde. This is a reducing atmosphere, in contrast to the oxidizing atmosphere of the moment, and was probably generated largely by outgassing of the initial solid matter of the earth. On this assumption, a variety of experiments have been conducted to ascertain what circumstances might have led to the kinds of organic compounds characteristic of living material. In the first such successful experiment, an electric discharge was passed through a mixture of ammonia (or nitrogen), methane, and water vapor above boiling water. Aldehydes and hydrogen cyanide were formed under these conditions, and these in turn reacted to form detectable amounts of the amino acids, glycine, alanine, serine, and aspartic and glutamic acids. Remarkably, these are the very amino acids that, to the present day, are the most abundant amino acids of proteins. By increasing the amount of hydrogen cyanide in such preparations, and by varying the energy source to simple application of heat or ultraviolet radiation, the number of products formed has been extended significantly. Among these products are hydrogen cyanide and its polymers such as dicyandiamide, which in turn have led to the formation of the major purines, guanine and adenine, and, under the correct conditions, to guanylic and adenylic acids. It was more difficult to find conditions that would lead to the formation of pyrimidines, but both cytosine and uracil have been obtained in impressive yields. In yet other experiments, at $100°$ C clays such as kaolin catalyze formation of such sugars as glucose and ribose in good yields from dilute solutions of formaldehyde. Moreover, cyanide and its polymers have been found to catalyze synthesis of random polypeptides from relatively concentrated solutions of amino acids, and of random polynucleotides from mononucleotides. In sum, therefore, conditions that may or may not mimic those of the prebiotic era on this earth, but that are as close as current theory can suggest, result in the formation of a large number of the primary building blocks of biological macromolecules as well as primitive macromolecules, thereby setting the stage for the origin of life.

Beyond those facts, all is speculation. The problem is to ascertain how these noninformational proteins and polynucleotides combined to form a self-duplicating system. No point is served by repetition here of current highly speculative hypotheses. It must suffice to indicate that the necessary raw material would have accumulated somewhere on earth as a consequence of the physical and chemical conditions on the earth's surface.

If, over the passage of 1 or 2 billion years, molecules with a very low order of accidental catalytic activity served to catalyze further syntheses either on their own surfaces or on the surfaces of other molecules, the earliest beginnings would have been made. According to this concept, life is not necessarily a highly unlikely event, but rather the almost obligatory consequence of the zero-time conditions on the earth's surface.

One further set of concepts warrants recital. From the notion described above, once primitive life began, and the crudest membrane surrounded such macromolecular packages, the materials available for further trans-formation, like those that contributed to the primitive genetic apparatus, must have been those that were present in the original "primordial soup." It is in that context that one finds explanation of the fact that the nucleo-tides not only serve as the building blocks of all forms of nucleic acids, but also participate in intermediary metabolism as the coenzymes for hundreds of metabolic processes—presumably because "they were there." ATP be-came the energy source for cellular reactions because it was there. In time, as the original supply of organic materials began to dwindle, selective ad-vantage would come to those primitive cells that "learned" to synthesize what they required from other organic materials still present in the medium. In a metabolic pathway such as those we have considered, there is invari-ably a set of intermediates that serve no purpose but as substrate for the transformations that lead to the desired end product. Patently, a cell that had "learned" to convert what we currently recognize as a starting com-pound—e.g., glucose—to one of the first stages in the current biosynthetic pathway would not have benefited at all by such an event. Synthetic path-ways as we now know them must have evolved backward in the sense that the next capability when guanine and adenine began to disappear should have been an enzyme that could make use of the hypoxanthine also formed in the primordial reactions. When that was gone, utilization of some other substrate to make hypoxanthine would again confer high sur-vival value. It is of interest, therefore, that aminoimidazole carboxamide, even today an intermediate in purine biosynthesis, was also formed under the same rudimentary circumstances that have been shown to lead to the formation of adenine, guanine, and hypoxanthine.

Patently, the information at hand is a far cry from genuine understanding of the origin of life, but it may well be that an important beginning has been made.

HEREDITY AND EVOLUTION

When the *Origin of Species* was published in 1859, Darwin convinced biologists that all living forms are related to life in the past, from which they evolved by small changes preserved by natural selection. But the fundamental mechanism of this process eluded him. This was provided in the formulation of the rules of inheritance by Gregor Mendel, based on his observations with garden peas, which led to understanding that heredity is particulate in nature. Shortly thereafter, observations of the behavior of chromosomes, particularly the manner in which the chromosome number is halved during the formation of sperm and egg, made it apparent that the chromosomes were the carriers of these particulate genetic factors, later called "genes." In the 1920's, as a result of a large body of ingenious, elegant experimentation with the fruit fly *Drosophila melanogaster,* genes were shown to be arranged in single file on the chromosomes. By 1940, the chromosomal mechanisms of genetic transmission were thoroughly understood, and the concept of mutation—a heritable change in a gene or chromosome so that it produces a new effect—was well established. The rules of mutation were well understood by 1950—that any individual occurrence is essentially unpredictable although the average can be measured, that most mutations are harmful, and that the rate of mutation can be increased by raising the temperature, by radiation, and by a variety of chemicals. But the chemical nature of the gene itself and of the mutation process was not understood. The chemical structure of DNA, and the concept of mutation as a process involving the substitution of one of the four bases of DNA for one of the others, or of deletion of some portion of the DNA, or inversion of some portion of the molecule, or crossing-over by exchange between two genes or chromosomes, was established in the last two decades. Fundamental theory and the detailed mechanisms of genetics now rest on rather secure grounds.

Population Genetics—An Extension of Mendelism

A new field, population genetics, is addressed to more detailed understanding of how mutation, selection, migration, population size, environmental conditions, and other factors influence the distribution of individuals, the character of populations, and the evolutionary process. To date, however, available methods permit the study of only microevolutionary change— changes within species—rather than the study of the origin of higher categories such as genera or families. The first step was to determine the relationship between the proportions of genes in the population and the proportions of the various genetic types of individuals (genotypes) that arise by random combinations of these genes. Thus, if P_1, P_2, P_3 . . . represent the proportions of alternative genes at a single chromosomal locus, the genotype frequencies are given by appropriate terms in the quadratic expression $(P_1 + P_2 + P_3 \ldots)^2$. Many natural populations are sufficiently large, and mating is sufficiently random for a number of traits, to demonstrate that this principle applies remarkably well. It has been a powerful tool for analyzing natural populations and determining the mode of inheritance of genetic traits in populations in which experimental matings are not possible, *viz.*, man. One of the earliest demonstrations of the validity of the principle was the finding that the distribution of the major blood groups of man, A–B–O, conforms to the above rule. Because the potential number of genotypes in the population is vastly larger than the number of genes, a considerable simplification is achieved by considering changes in gene frequencies rather than genotype frequencies. In any case, in a sexually reproducing population, the genes are reassorted by the Mendelian shuffle that occurs in each generation; the effects of such reassortment are transitory because gene combinations are put together and taken apart with each generation. Thus, the value of considering gene frequencies rather than genotype frequencies is that the genes perpetuate themselves as intact units, whereas the genotypes are constituted anew each generation. The weakness of this simplification relates to the fact that individual genes are not completely randomized in each generation. Genes close to one another on the same chromosome often tend to be linked together in inheritance. Accordingly, a truly sufficient mathematical theory requires extremely complex models.

One of the earliest contributions of population genetics was to provide understanding of the disadvantage of inbreeding. Sibling mating had long been discouraged in human tradition, and many biologists had noted the weakening effect of inbreeding. It was recognition of the fact that harmful recessive genes usually produce their effects only when present in duplicate and that the occurrence of such duplication is enormously increased by inbreeding, as evident in the occurrence of human hereditary disorders, that led to objections to inbreeding on scientific rather than social grounds.

Observational and experimental population genetics started with field studies of natural populations. Although most evolutionary change is too slow to be witnessed in a human lifetime, some examples of unusually rapid evolution have been described. Among these is the evolution of "industrial melanism." As smoke from factories gradually darkened tree trunks, the light-colored forms of moths that rest on these trunks were replaced by darker forms, a phenomenon possible only under rather intensive selection pressure. This pressure was demonstrated experimentally by the observation that birds selectively caught light moths from the darker tree trunks found in industrial areas and dark moths from the lighter tree trunks found in rural areas. A more recent and more unfortunate demonstration of the operation of such processes is the development of insects resistant to common insecticides and of micro-organisms resistant to antibiotics. The large numbers of individual organisms of varied genotypes and the rapid rates of reproduction make possible rapid selection of such resistant mutants, which then grow out and replace the original susceptible strains.

One of the greatest contributions of experimental population genetics has been the demonstration of the large amount of hidden genetic variability, which is concealed in relatively constant natural populations.

Cytogenetics

Not long after discovery of the chromosomal basis of heredity it became apparent that, for normal development, there must be exactly the right number of chromosomes in the fertilized egg. Most species are diploid, having two representatives of each chromosome in the fertilized egg and in all the body cells derived from it. Some are polyploid having four, six, or eight chromosomes of each type. For the diploid species in particular, when there are too many or too few chromosomes, or even if there is an excess or deficiency of only a part of a chromosome, the resulting animal or plant will exhibit abnormalities and frequently cannot survive. A classic demonstration of this phenomenon was obtained by breeding studies with the common jimsonweed, *Datura stramonium,* which normally has 12 pairs of chromosomes in each cell. If an organism has three representatives of a particular chromosome instead of two, it is said to be *trisomic.* Each of the 12 possible trisomic types of jimsonweed was found to have a characteristic abnormal appearance. *Monosomic* plants with only a single representative of a given chromosome were also discovered, and these, too, had characteristic abnormalities, usually considerably more severe than the corresponding trisomics. Such data were not expected. From them came understanding of the great importance of the correct balance among genes on each chromosome. If an organism is to develop normally, not only must the genes be normal, but there must be the correct numerical balance among them. Unbalanced chromosome combinations arise because of an accident in the process of meiosis, the reduction division that yields egg or sperm cells containing half the normal complement of chromosomes of somatic cells. Such accidents are not common but neither are they exceedingly rare, and whenever they have been systematically sought they have been found.

In agricultural practice trisomic and monosomic types have been of great practical utility. They permit the determination of both the consequences of a specific gene and its chromosomal location. Application of this knowledge facilitates the transfer of useful genes—e.g., those for resistance to diseases, insects, or poor soil—from one variety to another. It was with such techniques that the gene for rust resistance was transferred from the common weed, goat grass, into commercial wheat and, in a recent *tour de force,* that a useful gene was transferred from rye to wheat. Similarly, polyploidy can be exploited as in the creation of a new cereal crop, *Triticale,* an octoploid with six sets of chromosomes from wheat and two from rye.

HUMAN CYTOGENETICS

Adequate techniques for the study of human chromosomes were invented only a dozen years ago. In fact, they are quite unsophisticated, and it is difficult to understand the long delay in their development; even the correct number of human chromosomes was unknown until a little more than a decade ago. But because of the large body of material and great interest, human cytogenetics has advanced extremely rapidly. With the new techniques, a few drops of blood suffice, and knowledge of human cytology is now comparable to that of the best-studied plants. The overriding question was to what extent chromosome anomalies are responsible for human disease, malformation, or early death.

The first trisomic disease, discovered in 1959, turned out to be the already familiar "mongolism" or Down's syndrome, characterized by severe mental retardation, a number of characteristic physical abnormalities, and a face that appeared to Down to resemble the mongoloid peoples, a fact that resulted in his unfortunate choice of a name for this disease. About one infant in 700 is born with this disorder, and such patients constitute as much as one third of the population in institutions for the most severely retarded. A long-standing puzzle was created by the fact that mongolism is associated with a remarkable variety of seemingly unrelated abnormalities. Now that the cause is known, this is not surprising, for there must be many genes on the responsible chromosome and any gene that produces an abnormal result if present three times in the genome rather than the normal two would produce this effect.

This suggested an approach to finding other chromosomal diseases, and the study of patients with mental retardation and other superficially unrelated abnormalities led almost immediately to the discovery of two other trisomic types, both new conditions previously unknown to clinical medicine. But no other examples of trisomic types have been found in several years of subsequent study. Since the three known types all involve quite small chromosomes, it appears likely that trisomy for the other chromosomes of man is simply incompatible with the development of the embryo and fetus. Experience with experimental plants and animals indicated that monosomy is more harmful than trisomy for the same chromosomes. In confirmation, a study of miscarried human embryos indicated that more than a third of spontaneous abortions are caused by trisomy, monosomy, and more complex chromosome anomalies.

An exception to the rule that chromosomes occur in identical pairs is found in the sex chromosomes X and Y (Figure 30). Moreover, these are also the basis for the exception to the rule that monosomy is lethal since monosomic individuals with a single X chromosome and no Y chromosome turned out to have the well-known but previously not understood

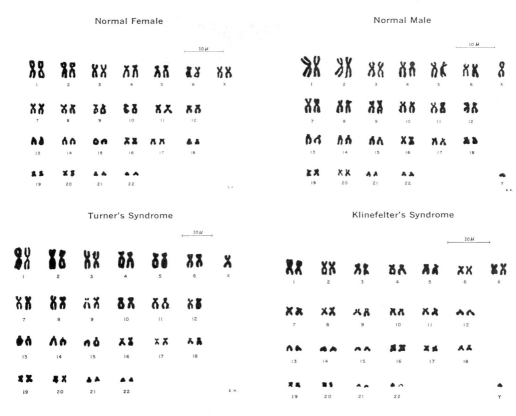

FIGURE 30   Human chromosomes of the normal female and the normal male, and in Turner's syndrome and Klinefelter's syndrome. (From *Metabolic and Endocrine Physiology* by Jay Tepperman. Copyright © 1968, Year Book Medical Publishers, Inc. Used by permission. Photograph courtesy of Mary Voorhess and Lytt Gardner.)

Turner's syndrome (small stature, amenorrhea, absence of secondary sex characteristics), while the trisomic XXY type has turned out to be the known condition, Klinefelter's syndrome (apparent male with feminine stigmata and very small gonads). In individuals with chromosomes that have broken and been reattached, there may be no immediate abnormality, but such abnormal chromosomes complicate the normal process of sperm and egg formation, markedly increasing the likelihood of unbalanced chromosome combinations in the next generation. This imposes a serious burden on the prospective parents, who should be made clearly aware of the potential consequences.

Polymorphism

Naturalists have long noted that some species of animals are polymorphic, i.e., they exist in two or more quite distinct forms such as black and cinnamon bears or red and silver foxes in the same litters. Population geneticists have sought to understand this process in a general way for any gene locus. If the heterozygous individual (A_1A_2) is somehow selectively advantageous in comparison with either form of homozygote (A_1A_1 or A_2A_2), then both types of genes persist in the population in a stable equilibrium determined by the relative fitnesses of the two homozygotes. This has been demonstrated amply with many forms; fruit flies have been most thoroughly studied, but polymorphisms are found throughout the animal and plant kingdoms. Sickle cell anemia is perhaps the classical human example. Individuals with "normal" hemoglobin are susceptible to malaria, those with two genes for the sickle cell trait are subject to a profoundly debilitating anemia, while the heterozygote is resistant to malaria and only slightly incapacitated by having half his hemoglobin abnormal; hence, the heterozygote has a selective advantage. Accordingly, both genes flourish in African populations. As man is investigated more intensively, an increasing number of polymorphisms, usually detectable only at the molecular level, have become apparent; presumably all are maintained by the same principle.

Some Recent Accomplishments

A primary accomplishment of population genetics is the understanding it has provided of the nature of the evolutionary process, of the value of the sexual reproduction mechanism, and of the relation of man to all other living organisms.

No chapter in modern biochemistry is of greater interest than the accumulated demonstration of the kinship of man to other living forms. The primary datum is the universality of the genetic code and of the operation of the genetic apparatus. The fact that the genetic code is a constant from bacteria to man to higher plants is the most cogent available argument supporting the concept that all living forms derive from a single common ancestor. At the same time, by tracing the amino acid sequences of specific proteins with essentially identical functions, it has been possible to illustrate how the process of mutation causes divergence from an original form. The most thoroughly studied such protein is the electron carrier cytochrome c. Amino acid sequences of this protein from more than 30 species are now available, and of the 105 residues in this protein, only a quarter of the positions remained filled by the same amino acid in all species. Strikingly, at what is considered to be the binding site of cytochrome c to the enzyme responsible for its oxidation, there is a run of 12 amino acids where no substitutions have occurred in recorded history. Moreover, in a general way, the number of differences in amino acid sequence between the cytochromes c of any two species is a function of the time since the two species separated from a common ancestral form, as indicated by the paleontological record. Accordingly, the cytochromes c of the primates, like the hemoglobins of the primates, are virtually identical, which is convincing proof on the molecular level of our common origin. In contrast to the great variability of most of the amino acids of the proteins studied to date is the remarkable constancy of structure of a histone (a basic protein bound to DNA in cell nuclei). Of the 115 amino acids in this molecule, only at one position does the histone of the garden pea differ from that of the cow, testifying to their common ancestry.

One other aspect of such studies is particularly noteworthy. As one ascends the phylogenetic ladder, the DNA content of cells increases, as does the number of proteins that such organisms synthesize. How did this increase in DNA come about? The answer derives from comparative studies of amino acid sequences of diverse proteins within a given species. For example, it is apparent that much of the structure of the digestive enzymes trypsin, chymotrypsin, and elastase, as well as one enzyme that operates in the blood-clotting process, is identical. Similarly, the polypeptide hormones of the pituitary are very much alike, and a protein hormone produced in the gastrointestinal tract closely resembles one produced in the pancreas. Yet all of these are quite distinct proteins at the present time. The first observation in this area was the fact that although four different types of protein chains are involved in the structure of normal human hemoglobins, these are essentially of equal length and show very large areas of identity. For each such related group, one can only conclude that the genes now responsible for their synthesis derive from a common ancestral gene that, by a process of duplication (presumably independent of normal mitotic division), doubled and redoubled at some time in the past, with the progeny genes free to mutate in different directions. Thus, the answer to our question is that the increased DNA found in cells of higher organisms resulted from duplication of the old DNA, followed by an independent evolutionary pattern occurring as a result of mutation and selection.

Population genetics has also demonstrated that evolution need not be observed only over geologic time. It can happen, and perhaps it is happening, in current human populations. The rates of change that could arise from the crossing of human races, from increasingly high correlation in intelligence of husband and wife, from changes in the mutation rate as a result of radioactivity in the atmosphere, from a possible law requiring sterilization of persons with mental retardation, can all be predicted with reasonable accuracy on the basis of current understanding of the way in which natural selection acts on genetic variation. It also enables prediction of some of the ecological consequences of the introduction of insecticides, pesticides, or other new agricultural practices.

Clearly the most striking practical achievements based on genetic understanding have been in agriculture, particularly the creation of fertile hybrids between divergent plants, e.g., the radish–cabbage hybrid or the wheat–rye hybrid mentioned above, resistant strains of almost all primary food crops and, of course, the development of hybrid corn. The well-known success of that endeavor has led to intensive efforts to generate equally successful hybrid wheat and rice; these studies are not yet complete. In medicine, the most important practical result has been the deeper insight into disease-producing mechanisms related to genetic variability. These will be discussed in greater detail in Chapter 2.

THE DIVERSITY OF LIFE

Systematic biology was begun by amateur students of natural history who catalogued the animal and plant life in the world about them. During the eighteenth century, Linnaeus was able to treat both animal and plant kingdoms in their entirety, but he was the last person to do so. In 1818, Lamarck published an account of all invertebrate animals, and again he was the last person able to do so. By now, two centuries after Linnaeus, systematists have discovered, described, and taxonomically classified about 1.5 million different species of organisms; yet each year more than 10,000 previously unknown species are discovered, described, and catalogued. Withal, this may represent no more than one third of actually existing species, and it is clear that at least 99 percent of all species that have appeared on the face of the Earth are now totally extinct.

One of the tasks of the systematic biologist, in addition to the labor of love entailed in finding, observing, describing, and cataloguing new species, is to determine how speciation actually comes about, what the essential attribute of a species really is, how current species relate to those of the past, what mechanisms account for extinction, and how this multitude of species cohabits our planet. This is by no means a purely academic task. It has been just such knowledge that has made possible many of the breeding programs of modern agriculture, that underlies all attempts at biological control of pest species, and that must suggest future attempts at large-scale cultivation of species not currently utilized as foodstuffs for man or for the production of new drugs, fibers, fermentations, etc. Previous successes in such activities are treated in greater detail in Chapter 2.

What Is a Species?

With increasingly sophisticated biological understanding, with introduction of chemical techniques for examination of proteins, polysaccharides, and lipids, the description of a given set of organisms tentatively considered a "species" has become increasingly detailed and has permitted discriminations impossible in an earlier era. Currently, a species of animal or plant is characterized by two major properties: a gene pool adapted to occupy a particular niche in nature and protection mechanisms that prevent mixing with other gene pools.

The gene pool cannot yet be described by direct examination of the genes themselves; it is described rather by phenotypic expression in the structure, physiological function, biochemical composition, breeding and mating patterns, choice of food, annual life cycle, etc. This mode of analysis is a sophisticated extension of the endeavor, once totally summarized by Linnaeus. The processes by which cross-breeding with other species is prevented are called "isolating mechanisms." In a general way the protection mechanisms operate to prevent the mixing of two incompatible genetic systems that could lead to the production of disharmonious and selectively inferior hybrids. Classically, this was exemplified by failure of the sperm of one species to fertilize the egg of another. And for distantly related organisms this is indeed the case. But distantly related organisms have little tendency to attempt such breeding in the first instance. The more important mechanisms are those that prevent cross-breeding between gene pools where, indeed, the biological event might be possible. Among animals, behavioral barriers that prevent mating are far more important than genetic incompatibility. Two main classes of premating isolating mechanisms are commonly employed. Signals by which individuals attract and identify other individuals as being reproductively ready members of the opposite sex are most effective because they are misunderstood only rarely by members of the same species and are likely to be ignored by members of other species. In birds, frogs, and insects, vocalizations are important signals of this type. In other kinds of animals such as mammals, salamanders, many insects, and invertebrate species, specific secretions (scents, sex-attractant chemicals, etc.) serve as primary means of communication among potential mates. In still other animals, such as insects, fishes, reptiles, and birds, visual signals involving color patterns and ritualized displays and postures serve as premating isolating mechanisms. Another category of such mechanisms is separation of the reproductive activities of sympatric (cohabiting) species by space or time. They may reproduce at different times—e.g., one species of elm that flowers in the spring and a different species in the same area that flowers in the late summer—or they may select slightly different breeding sites, as do related species of toads, one of which breeds in running water and the other in rainpools of the same small valley. Isolating mechanisms of this category are widespread in plants as well as animals.

Origins of Species

From examination of both living and fossil forms, it is clear that geographical speciation has been the principal mechanism by which new gene pools have originated protective isolating mechanisms. This involves the spatial separation of a subpopulation from the gene pool of a parental species and the gradual building up of isolating mechanisms in the isolated population. When these have been reasonably perfected through repeated mutational changes, the external barrier can break down and the daughter species can now coexist protected by the isolating mechanisms. In a general way, small founder populations have been found to speciate far more rapidly than have large populations, particularly when the small group has either invaded a new territory or been subject to imposed climatic change. The great representatives of these generalizations are to be found in the manner in which marsupials in Australia have so adapted as to fill most of the ecological niches that, on other continents, are claimed by placental mammalia. Darwin's original example, the finches of the Galápagos, were a prime example because the descendants of the original finches now show beak forms that vary from one like that of the grosbeak to long, slender needles for feeding on nectar.

In general, much remains to be learned about geographical isolation mechanisms. It is unclear how these have operated in the sea, in large freshwater bodies, or in the intertidal zones of shorelines, which are remarkably rich in fauna and flora. Nor is it clear what factors determine the actual number of species found in a given locality. Why is a tropical reef ten times as rich in species as a rocky reef in cool waters? Why is a tropical forest so very rich in the number of species of trees, insects, and birds, while each species is represented by relatively few individuals? Documentation of geographical speciation is overwhelming; witness changes in the protectively colored coat of mice with changes in soil color, or seemingly identical grasses that may flower in March along the Gulf Coast but in midsummer in Kansas. No matter how well populations respond to the demands of local environments, there is a line—the geographical species border—which is the limit of tolerance of the genetic system of that species.

A tantalizing question remains as to why the frontier populations of a species do not respond by the selection of new genotypes that would permit expanding the species border over time. Some aspects of the genetic system of the species render it geographically cohesive and place limits on expansion. Nor is the situation entirely clear in any instance. Patently, the entering of a new niche has resulted with great frequency in a budding off of the phyletic line by the formation of new gene pools (speciating). Yet, with at least equal frequency, entering of a new niche can be effected by a minor reconstruction of the genotype; witness the diversity of niches occupied by the woodpeckers, hummingbirds, albatrosses, ducks, and swifts, despite the fact that these birds are anatomically, physiologically, and in most aspects of their life histories remarkably similar wherever they are found. In some degree, the answer is to be found in the fact that the more closely adapted a given species is to life in a restricted environmental circumstance, the less well it can radiate into other circumstances.

The influence of the environment is equally apparent in the phenomenon of evolutionary convergence in which similar selective pressures drive unrelated genotypes into similar forms. Thus, the cactus growth form has appeared in several distinct families of plants quite distantly remote from the true cacti. In Australia, where the ordinary frog is absent, a tree frog has evolved with habits, body size, shape, and appearance of the common leopard frog of North America, even though the two species are only very remotely related genetically.

Origin of Higher Groups

Macroevolution, the development of new broad assemblages of species such as birds, beetles, and ferns, is less understood than is speciation itself. But the first step must always be the entrance of a population of some species into a new adaptive zone. Once this step is taken, a new selection pressure arises favoring all individuals that by their genetic constitution are superior in adapting to the new niche. Presumably this is how the earliest ancestors of all terrestrial animals became amphibious and how the earliest ancestors of all flying animals began to glide. It probably is extremely rare that any particular genotype has the potential to initiate such a new major group. But, with millions of species in existence simultaneously, this must have happened many times in the course of the 3 billion years of life on earth. Even when the first adaptation was behavioral, like air-swallowing in some ancestor of the present lung fish, it set the stage for additional structural and functional adaptations that made the occupation of the new adaptive zone more effective. New organs or structures have come into existence by means of two mechanisms. Sometimes it has been merely the intensification of the function of the previously existing organ, as when lungs developed as sinuses of the esophagus in air-gulping fishes living in stagnating swamps. In other instances, a pre-existing structure has taken on a second function without interference with the original function until the new function became the primary one. Thus, the initial acquisition of bird feathers almost certainly facilitated maintenance of constant body temperature, and only secondarily acquired their function as enlargements of the gliding surface of the wing. What is not clear is why certain groups have remained virtually unchanged for hundreds of millions of years—indeed appear to be immortal—while others simultaneously have undergone radical changes in what would appear to have been the same environment. Indeed, the entire fossil record is punctuated by instances in which one or another group entered upon almost explosive diversification leading to a simultaneous invasion of all sorts of new adaptive zones, referred to as adaptive radiation. But the circumstances that prompted such events are unclear.

Understanding of such events requires reconstruction of the intermediary stages that led from one kind of organism to another. Such reconstruction is immensely aided when there is available a key fossil that actually appears to be a linking organism—e.g., *Archaeopteryx* between reptiles and birds, the mammal-like reptiles (therapsids) between reptiles and mammals, and *Ichthyostega* between certain fishes and amphibians. But most such transitions are still shrouded in complete mystery.

Transitions among the plants are even less well understood. The higher plants (angiosperms), now the dominant group of plants in most of the world, have not been clearly traced to any ancestral group among the lower plants. Some 25 major phyla are recognized for all the animals, and in virtually no case is there fossil evidence to demonstrate what the common ancestor of any two phyla looked like. Nevertheless, there is a possibility of finding these ancestors because only a tiny fraction of the fossil record of the Cambrian and Pre-Cambrian eras, when the first animals turned up in the fossil record, has been subjected to careful examination.

Extinction

Extinction, the termination of an evolutionary line without descendants, is even less well understood than the origins of new species. Although there has been progress in explaining the extinction of certain species owing to changes in the physical or biotic environment, explanations offered for the decline and final extinction of entire major groups, as has occurred many times in the geological past, are totally unsatisfactory. Of 60 trilobite families, the dominant group of animals at the close of the Cambrian period, 40 disappeared from the subsequent fossil record and, after flowering in the Ordovician, the entire phylum disappeared before the end of the Paleozoic. Near the end of the Permian, about 24 orders, half of the then-current phyla, became extinct; again, during the last part of the Cretaceous, one quarter of all animal families were eliminated. (See Table 1.)

TABLE 1   Geologic Time Scale

ERA	PERIOD	EPOCH	(BEGAN) YEARS AGO
Cenozoic	Quaternary	Holocene	11,000
		Pleistocene	2,000,000
	Tertiary	Pliocene	13,000,000
		Miocene	25,000,000
		Oligocene	36,000,000
		Eocene	58,000,000
		Paleocene	63,000,000
Mesozoic	Cretaceous		135,000,000
	Jurassic		180,000,000
	Triassic		230,000,000
Paleozoic	Permian		280,000,000
	Carboniferous		345,000,000
	Devonian		400,000,000
	Silurian		425,000,000
	Ordovician		500,000,000
	Cambrian		600,000,000
Pre-Cambrian			over 600,000,000

Such periods of extinction have been equally devastating for dominant groups in the oceans, e.g., the ammonites, and on land, e.g., the dinosaurs. None of the theories advanced to explain these catastrophes is convincing. By contrast, plant extinctions have been far more gradual, and the important floristic changes have not coincided with the major extinction of animal groups. Neither changes in climate nor competition from evolutionary newcomers can explain the disappearance of these large, dominant, widespread groups. With so many species in existence, one would have expected at least some of them to shift into new adaptive zones and thus escape the fate of their relatives.

It may well be that the reasons were subtle, as in the rise of the angiosperm plants in the Cretaceous, which undoubtedly had an adverse effect on the ruling herbivorous reptiles, this, in turn, bringing about the decline of the carnivorous types. The important point is that the entire biota at any one time has been interrelated and interdependent in an extremely complicated way. Whatever factor affects the primary producers in such a system will have a profound and selective effect on the primary and secondary consumers, which may lead to partial or complete extinction and to the formation of new ecosystems. No species has so profoundly affected the distribution of others as man. His appearance in the Pleistocene coincides with the disappearance of a great variety of large mammals; some were hunted, and others lost in the competition for common food sources. And this competition between man and the other species continues even in the modern era.

Diversity and the Conceptual Framework of Biology

The study of diversity, with its emphasis on evolution, the creativeness of the selective process, the history of genetic programs, the uniqueness of individuals in species, and the statistical properties of populations, places emphasis on the most strictly "biological" of all phenomena. The evolutionist is constantly aware that organisms are the products of individual genetic programs carefully adjusted by hundreds of millions of years of natural selection. The remarkable successes of the reductionist approach of the biochemist have permitted understanding of life at its essence. But understanding of DNA, of enzyme mechanisms, of membranes, and of nervous conduction could not have permitted prediction that there would be butterflies, orchids, or porpoises—much less man.

Evolutionary biology has produced fundamental generalizations of concern to every thinking human being. Perhaps the most important contribution after the general theory of organic evolution itself is the development of "population thinking." This view stresses that all classes of objects and phenomena in the living world are composed of uniquely different individuals and that all statements concerning such populations of individuals must be taken in a statistical sense. One cannot understand the working of natural selection or the phenomena of race or fitness unless one appreciates the populational nature of biological phenomena. The gradual replacement during the past century of the ideology of essentialist philosophy, dominant from Plato and Aristotle to Kant, by population thinking has been one of the most important, although scarcely noticed, conceptual revolutions.

Although man is unique, he is part of the evolutionary stream and cannot be understood as an isolated phenomenon. He must be viewed within the context of the remainder of the organic world, a comparison that reveals respects in which man resembles other organisms as a consequence of his evolutionary heritage and in which he is indeed unique. *Evolutionary biology has succeeded in dealing with man objectively and scientifically, rather than as an object of ideology or dogma.* The evolutionary process applies to man as to all other organisms. Every problem faced by man, whether it be disease prevention, the life-span, population control, ethical decisions, or other, will be better understood as our comprehension of the evolutionary process by which man evolved and acquired his present characteristics increases.

Man's knowledge of his long history gives him a different perspective about his future. He has changed greatly in the past, and it is in his nature to continue to change. He understands his history and the origin of human diversities and similarities. These diversities are not unnatural and thus must be seen as part of a continuing process. It is perhaps humbling to realize that man is in this sense only a part of nature. But evolutionary biology has also shown us the central role that man is destined to play in evolution from now on, unless, of course, he engineers his own extinction. Although man arose through an evolutionary process that he did not understand and over which he had no control, he must now realize that he is unique in the living world and that the responsibility for continuance of the evolutionary process is his. The future evolution of the orangutan and the whooping crane and of most other species will be determined by human decisions and hardly at all by anything done by those species themselves. Thus the evolutionary view gives man not only a sense of humility but also a sense of responsibility. The question is not whether man is to influence evolution; he is already doing so, and indeed is changing things so that evolution is taking place more rapidly than at any time in recent history. He now has not only the opportunity to influence the other species as he has done in the past with domestic plants and animals, but also the opportunity—perhaps the obligation—to influence his own future evolution. That he will continue to evolve is the inevitable consequence of genetic variability and differential reproduction. The capacity of biologists to develop ways by which man can determine his future evolution is undoubted. The more difficult questions are whether man will choose to take full advantage of that capacity and with what wisdom.

CHAPTER TWO

BIOLOGY IN
THE SERVICE
OF MAN

Progress in biological understanding has proceeded at a spectacular rate for two decades. The deepening insights into the nature of man and his diverse living kin could well be reward enough for the large investment of effort and funds. Such understanding is more than a highlight of our culture; it is a primary tool of our working civilization. In the pages that follow we shall seek to illustrate and document that statement. Only a small sampling can be offered here, but it should become evident that the life sciences have dramatically altered our life style, contributing to our security, our health, our comfort, and our enjoyment.

BIOLOGICAL RESEARCH AND MEDICAL PRACTICE

The impressive and rapidly growing, though fragmentary, conceptual structure of biology has greatly increased understanding of disease mechanisms; presumably, as the conceptual framework becomes more general and more coherent, comprehension of disease will grow correspondingly, thereby enlarging opportunity for the alleviation and prevention of many disorders.

In considerable measure, the history of biology is the history of attempts to cope with disease. Many disorders have fruitfully been viewed as "nature's experiments" and, as such, have proved to be cardinal clues in elucidation of major fundamental phenomena. Thus, vitamin-deficiency diseases—e.g., pellagra, beriberi, sprue, and scurvy—were the clues to the very existence of vitamins and, hence, to the coenzymes of metabolism; investigations of diabetes and glycogen-storage diseases revealed the hormonal control of carbohydrate metabolism and, indeed, the pathways of that metabolism; the prevalence of pernicious anemia revealed the existence of vitamin B_{12} and of the unique biochemical reactions it makes possible; the requirement for agents to manage infectious diseases stimulated the discovery of antibiotics, and these, in turn, proved to be powerful tools in the elucidation of the mechanism of operation of the genetic apparatus and the synthesis of bacterial cell walls; the dramatic changes in the volume, pH, and salt concentrations of blood plasma in such disorders as infantile diarrhea, pernicious vomiting, diabetic coma, and Addison's disease have been both the primary stimuli and the major "experiments" in revealing the complex homeostatic mechanisms that control the volume, acidity, and electrolyte composition of the body fluids of both the intracellular and extracellular compartments; the variety of cardiac disorders has revealed the fine mechanisms and neural control of the cardiovascular system; and the existence of sickle cell anemia and other instances of altered hemoglobin structure were the first demonstration that a "point" mutation results in a specific amino acid replacement in a protein, as well as the demonstration that the genetic code in man must be identical with that in the bacterial species in which it was first determined. In each instance, the knowledge so gained, abetted by insights from other areas of biology, has resulted in expansion and improvement of the therapeutic armamentarium to the great benefit of those afflicted with the very disorders that served as clues.

This mutual feedback has characterized much biomedical practice. Advances in practice have come only when the intellectual stage was set and suitable methods were in hand. Painstaking analyses of the electrolyte composition of the blood in health and disease, over a period of 40 years, contributed much to current understanding. But the analytical methods required were tedious, slow, and unreliable in the hands of any but highly qualified experts. In the last decade, these were replaced by a variety of thoroughly reliable, semiautomated procedures, allowing the benefits of this understanding to be brought to virtually all those requiring it. The precise control thus afforded, symbolized in the bottles of intravenous infusions so common in modern hospital practice, has dramatically reduced mortality in a variety of illnesses and has been a major contribution to the success of current heroic surgical procedures. Hence, one no longer encounters the once painfully exact irony, "The operation was successful but the patient died." The new analytical methods and their use in guiding parenteral fluid therapy are the fruit of thousands of painstaking investigations. This chapter cannot hope to provide a comprehensive summary of such contributions but will, rather, describe a few recent noteworthy illustrations.

The National Health

The dramatically altered national health picture since the turn of the century broadly illustrates the changes man has wrought through his science and suggests those yet to be accomplished. Fifty years ago the major medical problems afflicting individuals in the United States were similar to those now facing developing nations. In 1900, both influenza and pneumonia killed more persons than any other disease. Tuberculosis came next. The combined death rate (deaths per 100,000 of population) from these diseases was greater than that from heart disease today, a malady that killed more than 712,000 persons in 1965, when cancer took the lives of an additional 300,000 individuals. In the early 1900's, the death rate from tuberculosis exceeded that from either of these causes, while diphtheria, now almost unknown, was the tenth leading cause of death. For three decades, pellagra—deficiency of the vitamin nicotinic acid—was the leading cause of death in eight southeastern states, whereas cases of this disease have rarely been reported since 1945, and mortality is zero.

Diagnosis, Disease, and Drugs

SULFONAMIDES AND ANTIMETABOLITES

Quite evidently, many people who would have succumbed to infectious disease in an earlier day now survive to die, at a later age, of degenerative disease or cancer. The advent of antibiotics deserves major credit for current ability to cope with infections. Moreover, antibiotics have played a major role in the development of drugs and approaches to the treatment of other diseases, including cancer, by illuminating the broad principle of drug design, which is fundamental to much current research.

From understanding of the mechanism by which sulfonamides inhibit the growth of bacteria came the concept of antimetabolites and new insights into the essential relationship between molecular form and physiological function. Simply stated, an antimetabolite inhibits the activity of an enzyme

that cells need for growth or other normal activity because it closely resembles the natural substrate of that enzyme. However, the antimetabolite
cannot be affected by the enzyme and remains attached to its surface; in
consequence, the enzyme cannot perform its normal function. The discovery
of specific bacterial inhibition has a long history. In 1904, Paul Ehrlich,
a German scientist, postulated that infectious diseases could be treated if
chemicals could be found with a greater affinity and toxicity for parasite
organisms than for host cells. Using dyes against trypanosomes and arseni-
cals against spirochetes, he demonstrated the validity of his hypothesis and
provided the earliest useful treatment for syphilis. In 1935, a dye called
Prontosil was shown to be effective in treating streptococcal infections in
patients, though it had no effect on bacteria in a test tube. The demonstration that individuals treated with Prontosil excrete sulfanilamide, a degradation product of the dye in the body, was soon followed by observation that
this compound inhibits both infection in patients and the growth of organisms in laboratory test media. A vigorous program of chemical modification
of the basic structure led to a new class of drugs, the sulfonamides. Even
now, these are the drugs of choice in the treatment of gastrointestinal and
urinary-tract infections.

Early empirical success with sulfanilamide rendered it imperative that
the mechanism of its effect be understood, so as to permit design of even
more effective congeners. A lengthy series of observations, conducted in
a multitude of laboratories at home and abroad, yielded the following conclusions:

Sulfonamides inhibit bacterial growth by preventing the organisms from
synthesizing folic acid, a vitamin for man, lack of which results in sprue.
Normal synthesis of folic acid by bacteria and plants commences with the
incorporation of p-aminobenzoic acid. In molecular structure, p-aminobenzoic acid and the sulfonamides are distinctly similar.

p-aminobenzoic acid   sulfanilamide

When a sulfonamide attaches itself to the enzyme responsible for the
normal reaction with p-aminobenzoic acid, synthesis is blocked, and, for
lack of folic acid, the bacterium cannot survive. Because man is unable to

synthesize his own folic acid, the sulfonamides do his metabolism no harm, selectively attacking bacteria while leaving human cells undamaged. It was these observations that gave rise to the concept of antimetabolite drugs. Many have since been usefully synthesized, but no better example of the concept is yet available.

ANTIBIOTICS

Penicillin was discovered in 1929 when a British bacteriologist observed the inhibitory properties of the fungus *Penicillium notatum,* which secretes penicillin into surrounding media. This substance, destined to become the most widely used antibiotic, was, however, originally discounted as impractical because of its seeming chemical instability. But by 1940 other British scientists showed that it was reasonably stable when partially purified and dried. Their material, only 50 percent pure, proved to be nontoxic to man and very active against susceptible micro-organisms, including staphylococci. Although effective, penicillin was tedious to purify, and problems of mass production seemed insurmountable when the calamity of war prompted members of the British group to look across the Atlantic for help.

The mass outbreak of typhus during World War I and the loss of countless wounded to secondary bacterial infection, followed in quick succession by the influenza pandemic of 1917–1918, gave urgency to the search for an effective antibacterial agent as we entered World War II. It took the crisis of the Second World War, which harnessed the potential of the American drug industry, until then running a distant second to Europe as a source of new drugs, plus the resources of the Department of Agriculture, to create the antibiotic age. The results were nothing less than spectacular. Success was based upon already developed techniques for large-scale cultivation of micro-organisms, the isolation of *Penicillium* strains that secreted large quantities of penicillin, and the development of suitable growth media. By September 1943, there was enough of this drug to supply all the Allied forces. This phenomenal accomplishment not only markedly reduced mortality among the wounded but also launched a new and fruitful search for other antibiotics.

After elucidation of the chemical structure of penicillin, in due course natural penicillin was replaced by semisynthetic penicillins, which are comparatively simple to manufacture and which retain the essential molecular configuration of the parent molecule, which is so effective against Gram-positive organisms.

The attempts to prepare semisynthetic penicillins bore an additional fruit. The earliest such attempts, which seemed entirely rational, failed. When the explanation was found, it proved to be an important extension of the antimetabolite principle. Sulfanilamide and p-aminobenzoic acid are essentially planar molecules; thus the analogy suggested by the two-dimensional formulae above is indeed valid. But the unsuccessful semi-synthetic penicillins, which appeared to be reasonable analogs of natural penicillin—as these structures are conventionally represented on paper—differed significantly when three-dimensional models, based on x-ray evidence, were constructed. Since then, chemists engaged in the synthesis of new drugs have been acutely aware of the fact that, to be effective, the drug must attach properly to the surface of the enzyme or membrane to be affected, and this must be a property of its three-dimensional conformation.

Extensive screening of soil samples, largely by drug manufacturers, then led to the discovery of an ever-increasing family of antibiotic agents, among them streptomycin, chloromycetin, aureomycin, and terramycin. Although there is as yet no universally effective agent, one or another of these drugs can mitigate virtually all known infections.

Antibiotics have drastically altered the patterns of medical practice. Prior to 1940, thousands of hospital beds were occupied by patients with infectious diseases. Today, in the main, these patients receive a prescription for antibiotics and return home. The morbidity associated with postoperative infections has dropped sharply. And the damaging, once frequent, chain of events that began with a "strep throat" and went on to scarlet fever, rheumatic fever, and serious heart disease has been broken. The search for new and better antibiotics continues in an effort that counts on both rationally exploited chance and accumulated skills and understanding. New antibiotics are still discovered by screening methods in which activity is sought in extracts of thousands of yeasts and fungi and soil samples of unknown microflora from around the globe. Modified, improved semisynthetic compounds then follow as drug designers attempt to deal with the two most critical problems posed by these drugs.

As predicted by scientists familiar with the physiology and genetics of bacteria, as use of antibiotics spread throughout the population, so, unfortunately, did bacteria that are antibiotic-resistant. The antibiotic boom fostered selective processes that bred resistant organisms. Among a normal population of bacteria there are, almost invariably, a few organisms that have spontaneously mutated, the mutation rendering them immune to the bactericidal action of a given antibiotic. As the drug suppresses the growth of sensitive members of the colony, resistant mutants flourish. In some cases, simultaneous use of two antibiotics with differing *modi operandi* is effective to a limited degree. But the problem is compounded by the fact that resistance, like an infectious disease, is catching. Both by sexual mating and by transduction, a process in which a virus carries a bacterial gene from one cell to another, bacteria can spread their resistance among related strains, and some organisms have been isolated that are resistant to several antibiotics at once. Recent work describing transduction may open the way to "outwitting" this threatening phenomenon, as should continuing improvement of semisynthetic antibiotics that are of greater potency and specificity than natural antibiotics but that are insensitive to the enzymes that destroy the latter.

The spectacular success of these antibiotics gave sharp stimulus to inquiry into their mode of action, an inquiry that continues with increasing intensity. In a few instances, partial answers are already available. Thus, penicillin selectively inhibits one specific enzymatic step in the complex process whereby the cell walls of Gram-positive bacteria are fabricated. Each such wall is a single "bag-shaped" macromolecule built of 10 different kinds of subunits. As the cell grows, or divides, linkages must be broken and additional subunits inserted. Interruption of this process leaves the cell without a casing and, hence, renders it susceptible to damage by diverse physical or chemical changes in its environment. Since mammalian cells employ no such casing, they are unaffected by penicillin. Actinomycin D, which has found only limited use as an antibiotic, operates by interference with the mechanism by which RNA is made on the surface of DNA. Because it affects mammalian cells in the same way, it has found little clinical use as an antibiotic. Streptomycin in some manner so affects the ribosomes of Gram-negative bacteria that they make mistakes in translating RNA into protein, and hence make useless, nonfunctional proteins. As this field progresses—as the secrets of naturally occurring antibiotics are revealed— it should be possible to improve on antibiotics, permitting synthesis of chemical entities that are lethal for invading organisms yet relatively innocuous for man. In each case, the new drug must be so constructed as to fit, sterically, onto an enzyme or a membrane surface in such fashion that it will seriously limit normal function, presumably by extension of the antimetabolite principle.

A more sophisticated understanding of the operation of the pathways by which products are synthesized in the body has offered a new approach to drug design. Early attempts to block the synthesis of a given product, e.g., cholesterol, sought to inhibit an enzyme known to be vital to its biosynthesis. Research generally was directed at finding a drug that mimicked the substrate with which a specific enzyme reacted, as noted earlier. However, a new avenue of pursuit was opened by the understanding that the "committed step" in most synthetic metabolic pathways (pathways that involve a series of consecutive reactions) is subject to allosteric feedback inhibition by the final product, which bears little resemblance to the substrate of the enzyme responsible for the committed step. It is clear that ingestion of cholesterol drastically inhibits its own biosynthesis. Patently, a foreign molecule that, in low concentration, could accomplish the same event might serve as a potent drug for prevention of atherosclerosis, and a series of other such possibilities is also under active investigation. But until the principle of allosteric feedback inhibition had been revealed in studies of bacterial metabolism, this approach could not have been conceived.

There is good reason to expect a considerable increase in the sophistication of drug synthesis in the near future. In addition to the factors considered above, it is evident that many drugs—e.g., morphine and digitalis—work by attachment to specific loci on cell membranes or intracellular membranous structures. Partial understanding of how a drug interacts with a cell membrane at the molecular level has only evolved in recent years. As this field matures—as the structures of membranes are revealed—it may well become possible to alter them usefully in specific states. Quantitative information about the biochemical events in metabolic disease is badly needed, permitting construction of mathematical models of metabolic events in a form manageable in a computer. Such information can be applied in testing new drugs for a given disorder and in determining suitable dosage regimens. For years, the interrelationships between levels of blood glucose and secretion of insulin after the administration of sugar to normal volunteers and to diabetics have been crudely understood. More recently, a carefully constructed mathematical model describes the effect of administered insulin on the uptake of glucose by the tissues, with resultant changes in blood-glucose levels and in the release of insulin from the pancreas. Use of this model permits more nearly normal regulation of the blood-sugar levels of diabetics. The benefits to man to be derived from this advance are not yet certain, but the potential is huge. The insulin regimens available since 1920 have sufficed to maintain the lives of hundreds of thousands of diabetics. In time, however, they progress to a series of highly undesirable sequelae—cataract, peripheral vascular disease, hypertension, atherosclerosis, and a disease of the lining of the minute filters of the kidneys. A generation will be required to establish whether the dosage schedules suggested by the new mathematical model, which, far more than in the past, mimics the release of pancreatic insulin by normal individuals, will also prevent the physical deterioration that is characteristic of diabetics treated with insulin for the last half century.

As understanding of disease has dramatically increased, so have demands for better comprehension of what disease is on the molecular level. Simultaneously, the development of a new drug has become a considerably more complex operation due to the effort to meet increased requirements for specific details about mode of action, specificity of action, safety, and effectiveness in man. From the time a scientist arrives at an idea of the pharmacological potential of a new compound to the time that compound actually reaches the market—a period of five to ten years—a pharmaceutical house must invest between $5 million and $10 million. Yet, this is our ultimate hope for useful new drugs, and increasingly such developments must rest on sound fundamental studies.

VIRAL DISEASES

In contrast to the great success of antibiotic therapy for bacterial infections, only trivial progress has been achieved in coping with viral diseases. A virus consists of a relatively small amount of genetic information, as either DNA or RNA, with a protein coat. This coat is shed as the nucleic acid enters the cell, where it usurps the normal genetic apparatus, shutting off normal production of cellular RNA and proteins so as to turn out many copies of the virus itself. Patently, any drug or procedure calculated to interfere with this process must also similarly interfere with normal operation of the genetic apparatus. Although this is probably tolerable for brief periods in a tissue such as muscle, it could be highly injurious to such rapidly dividing tissue as that of the bone marrow or the intestinal tract. Clearly, drugs intended to serve these ends must possess a very high degree of specificity and, despite much work, only a few useful leads are available. One noteworthy example is the treatment of viral eye infections, e.g., the herpesvirus, with a halogenated pyrimidine compound, 5-iododeoxyuridine. Although quite toxic systemically, it can be safely applied as eye drops. In the eye this compound is incorporated into the new viral nucleic acid, which then, as if mutated, directs the synthesis of inappropriate proteins, and the infection cannot sustain itself.

A recent finding of considerable promise is that an antibacterial antibiotic, rifampicin (rifantin), also has very significant antiviral potency. Its mechanism of action is highly interesting; for unknown reasons, in the presence of this compound, the coat proteins of several viruses cannot assemble on the viral nucleic acid surface. Hence, although both nucleic acid and coat proteins are made in the infected cell, the full virus cannot be assembled and fails to leave the cell in which its components were synthesized, and thus the infection is terminated. Since there is no analogous assemblage in the metabolism of mammalian cells, the antibiotic can be used in animals, in adequate dosage, as an antiviral agent without concern that it will interfere with any vital process in the host animal cells.

For the present, the major defense against virus infection must remain man's own principal defense mechanism, the immune system. The efficacy of this system was long evident in the list of diseases that strike but once in a lifetime, e.g., smallpox, measles, mumps. In each case, the "antigen," the foreign material that is "recognized" as foreign and that both elicits formation of antibodies and combines with them, is the viral protein coat.

Effective defense is possible either by deliberate immunization in advance, or by enhancing the immune response early in natural infection. Deliberate immunization has long been practiced, as in smallpox vaccination, while enhancement of the immune response has, until recently, consisted largely of administration of antibodies from someone who has already had the disease, as in administration of pooled γ-globulin to prevent a suspected case of measles.

Understanding of the nature and behavior of viruses, coupled with methods for culturing them, lay behind the development of the polio vaccine. As recently as 1954, this crippling disease struck 20,000 Americans annually. Eleven years later, only 61 cases were reported in the United States, the dramatic achievement of a mass-immunization campaign. Although the general principles of immunization had been known since Jenner introduced smallpox vaccine, much fundamental knowledge had to be acquired before it could be applied with impunity to the polio virus. It was first necessary to develop a cell system—monkey tissues grown in culture—in which polio viruses could be grown. Initially, viruses grown in this way were chemically inactivated and then administered. The subsequent perfection of live polio vaccines depended upon an independent line of research and the discovery of three mutant forms of the virus that could no longer cause disease but retained their immunizing effect.

A development of molecular biology that may yet offer large dividends is the recently acquired knowledge of a material called "interferon." This is a protein, perhaps an enzyme, that is produced in small amount by animal cells infected with a virus. In sufficient quantity it increases remarkably the efficacy of the immune response. Until techniques become available for its large-scale production, the best hope has appeared to be stimulation of the mechanism by which one's own cells engage in interferon synthesis. The primary trigger seemed to be the viral nucleic acid. Following this clue, it was found that synthetic double-stranded RNA (a simple polymer devoid of meaningful genetic information) is at least as efficient a stimulus as viral nucleic acid. When given early in an infection—e.g., mice given sufficient virus of hoof-and-mouth disease to assure 100 percent lethality—such material has offered complete protection, not only sparing lives but preventing the disease. This may yet prove to be the basis of a truly useful clinical approach to viral infection with the happy property of being generally useful without regard to the specific virus in question in any given patient.

CANCER THERAPY

Insights into the nature of DNA, its biosynthetic processes, and its role in cell growth and development have had wide application in recent cancer research. Coupled with recognition of the antimetabolite principle, these insights stand behind the development of a series of anticancer drugs that are able to check the growth of tumor cells, prolonging the lives of some patients by several years.

Cancer, second only to heart disease in the mortality tables, is not one but many diseases. Slow-growing solid tumors, such as lung cancer, are extremely difficult to treat unless the tumor is localized so that it can be removed by surgery or destroyed by irradiation. Significant progress has been made in treating by chemotherapy fast-growing tumors such as leukemia or other blood or lymph cancers. Cancer of both types is characterized by abnormal, uncontrolled growth of cells. Successful therapy depends upon an understanding of the metabolism and synthetic activities of those cells and rests on the principle of attacking them when they are in a vulnerable state.

The process of cell replication occurs in four stages: two pauses or resting states, a period of DNA synthesis, and one of mitosis and cell division. Different chemical agents selectively inhibit cell metabolism at different stages in this cycle and, because the cancer cells in an individual are not all synchronized—that is, they are not all in the same phase at the same time—judicious use of a combination of antimetabolites is necessary to destroy the maximum number of tumor cells.

Folic acid is used by man as a coenzyme in the process of synthesis of DNA precursors. Therefore, it was reasoned, an antimetabolite that could disrupt this sequence would inhibit the growth of tumor cells in the DNA-synthetic phase. Of a series of structural analogs that were tested, one called methotrexate (amethopterin) is clinically useful. Its drawback is its lack of specificity; it acts against all cells in the DNA-synthetic phase, whether they are cancerous or not. The turnover of normal cells, however, is distinctly less than that of rapidly dividing cancer cells; in weighing risk versus benefit, it was concluded that the toxic effects of methotrexate are less than its benefits, particularly in the treatment of leukemia.

Another agent in the arsenal of anticancer agents is actinomycin, originally found as an antibiotic, which checks cell growth by limiting RNA synthesis on DNA. When used against choriocarcinoma, an all too frequently fatal cancer of young women of childbearing age, it effects a 50 percent cure rate (remission of symptoms for five years). In combination with methotrexate, cures are achieved in close to 80 percent of cases. The same combination effects a 70 percent cure rate in Wilm's tumor, a kidney cancer, and a 25 percent cure rate in cases of Burkitt's lymphoma, a malignancy of the lymph glands first identified in children in Central Africa but now known to be widespread. Patently, without knowledge of the mode of action of actinomycin as an antibiotic, there could have been no reason to consider it as a potential anticancer agent.

Progressively more complete knowledge of DNA and RNA metabolism has opened other encouraging new avenues of cancer therapy. One clue came from the observation that rat tumors metabolize excess quantities of the pyrimidine uracil in the synthesis of RNA. Laboratory production of a series of uracil analogs resulted in two compounds that have proved valuable against leukemia cells: 5-fluorouracil and 5-fluorouridine. In addition to their considerable antileukemia action, these agents are effective in about 20 percent of cases of cancer of the colon.

Another drug, cytosine arabinoside, induces remissions in almost 40 percent of leukemia cases and hence is the most successful antileukemic agent yet assayed. Recently licensed by the Food and Drug Administration, this drug was developed after zoologists discovered that a species of sponge contains a class of nucleosides (subunits of RNA) that contain the 5-carbon sugar, arabinose, instead of the normal sugar, ribose. This structural analogy suggested the desirability of testing cytosine arabinoside in an anticancer screen. Until recently, the supply was limited to the very small amounts available from sponges. For several years, no enzymatic or chemical synthesis proved useful. This bottleneck was broken by a group of organic chemists studying chemical events under what are presumed to be the circumstances prevalent on earth when biological macromolecules first appeared, who found that under such circumstances cytosine arabinoside is readily formed. Thus, thanks to a zoologist interested in the biochemistry of sponges, an organic chemist interested in the origin of life, and biochemists curious about the metabolism of this unusual compound, the most useful of all antileukemic drugs tested to date was made available. How long would one have had to await this achievement if only "targeted" or "directed" research were supported?

One more drug, of limited utility in the treatment of leukemia, affords an excellent illustration of the manner in which fundamental understanding leads to practical application. A continuing theme in cancer research has been the thought that a somatic mutation underlies the malignant transformation. If such a mutation led to a metabolic or nutritional difference between normal and neoplastic cells, one might be enabled to utilize that difference in the design of a therapeutic approach. This has been realized in one instance. Of the 20 amino acids found in proteins, 10 cannot be synthesized by man, and hence are nutritionally essential. The other 10 can be synthesized by most human cells, but some cells rely on receipt of such amino acids in the blood after they have been synthesized by the liver. Asparagine is one such amino acid. All cells made in bone marrow are normally capable of synthesis of this amino acid, but screening of leukemic cells revealed that the cells of some patients are incapable of such synthesis and are dependent upon blood plasma for a supply of asparagine. Several micro-organisms make an enzyme, asparaginase, that hydrolyzes asparagine. This enzyme has been highly purified and injected into such patients; their leukemic cells, starved for asparagine, then succumb. The consequence has been a gratifying, sustained remission of symptoms in this small population of patients.

Today, carefully controlled combination drug therapy offers to leukemia victims a survival time of two to five years, whereas only a few years ago, they would have died in a few months. True, permanent cures must await a form of therapy addressed to the not-yet-comprehended underlying cause of these disorders.

Comprehension of the intricacies of nucleic metabolism and of specific metabolic inhibitors of various kinds have applications in other areas of medicine as well. Seldom is an advance confined to a single field. In psoriasis, a disease characterized by excessive growth of portions of the skin, methotrexate, particularly in combination with 6-azauridine, an inhibitor of purine metabolism, brings the disfiguring disease under control, though it is not a lasting cure. In organ transplantation and immunological research, as we shall see, drugs originally developed in anticancer programs have played a primary role in scientific progress by serving to suppress the immune system, the essential step in preventing the rejection of a foreign organ. Yet another drug, 5-iododeoxyuridine, originally synthesized, tested, and discarded as an anticancer agent, as we have seen, turned out to be highly effective in curing herpes keratitis, a virus infection of the eye, thereby preventing what was previously a major cause of blindness in the United States. This advantageous action of the drug has encouraged trials of its effectiveness in obliterating other viral infections, including meningitis and smallpox.

The major question before all those concerned with cancer therapy is the underlying nature of the neoplastic transformation of previously normal cells. It has long been known that many physical agents—e.g., chronic mechanical irritation, carcinogenic hydrocarbons, various dyes—predispose to such transformation. But their role remained unclear. Half a century ago it was shown that papillomas (warts) on rabbit skin contained a virus that, when administered to another rabbit, resulted in formation of more virus-containing papillomas. Other examples followed, perhaps most notably avian leukosis, a disease analogous to but not identical with human leukemias, in which neoplasia followed viral infection. The most dramatic stimulus to this general theorem was given by the demonstration that a remarkable variety of mouse tumors are all consequences of infection by one agent, the polyoma virus. In consequence, an intensive search is in progress for analogous carcinogenic viruses in man. The clearest success to date is the positive identification of a virus in the etiology of Burkitt's lymphoma. But virus-like particles have also been found in the cells of a wide variety of other malignant and nonmalignant tumors; it remains to show their causality.

The viral theory of cancer would seem less plausible were it not for a readily available model in bacterial life. Bacteria are subject to infection by their own specific viruses, bacteriophages. Some of these, the "temperate" phages, enter a cell and disappear, their nucleic acid seemingly becoming an integral portion of the bacterial genome, reproduced only when the entire genome is doubled prior to cell division. However, a sudden change in the environment can result in rapid multiplication of only the viral nucleic acid in the cell genome, with formation of a multitude of virus particles and rupture of the host cell. By analogy, then, carcinogenic viruses could be carried in the genomes of mammalian cells and yet be invisible and of no consequence until some change—e.g., cigarette smoking—accumulated a sufficient challenge to produce specific virus duplication and carcinogenic transformation. The nature of this process is discussed in Chapter 1.

Finally, it is apparent that this generalization, if valid, has only slight impact on the strategy of anticancer programs. Whether it be the nucleic acid of the host cell or of the virus, all available chemotherapeutic approaches, like x irradiation, must affect the biosynthesis of this component of the system. If the generalization proves valid, the design of anticancer drugs will be more clearly delineated in the future and will become decidedly less empirical.

GOUT

A significant bonus from cancer research has been the development of a drug for the treatment of gout. Originally synthesized as a potential anticancer agent, in the last few years allopurinol has become the treatment of choice for gout, a disease marked by unusually severe, acute arthritis and, in many patients, by the presence of deposits of chalk-like material that lead to grotesque deformities and serious crippling. Familial in distribution, gout afflicts about 275 of every 100,000 persons in the United States. Allopurinol is effective in a majority of gouty patients, preventing crippling and alleviating pain. It has also become standard therapy in the treatment of patients who form uric acid stones, whether or not they have gout.

In the course of essentially negative experimental trials with allopurinol in cancer victims, it was observed that, during treatment with this drug, patients excreted unusually small amounts of uric acid. That the measurement was made at all derived from the fact that allopurinol was synthesized as an antimetabolite of the purines required for synthesis of RNA and DNA;

in normal and gouty individuals, uric acid is the ultimate end product of purine metabolism.

This observation prompted what then proved to be highly successful trials of the agent in patients with gout, a disease in which uric acid deposits accumulate in the joints. Indeed, as long ago as 1850, gouty patients were discovered to have elevated levels of uric acid in the blood.

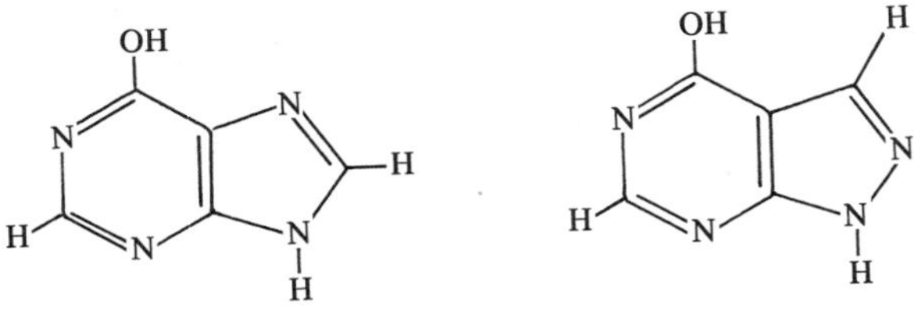

Hypoxanthine
(6-oxypurine)

Allopurinol
[4-hydroxypyrazole (3,4-d)
pyrimidine]

Approximately two thirds of the uric acid formed each day is excreted through the kidney, and one third by way of the gastrointestinal tract. Because it is so sparingly soluble, it tends to form crystals when its level is elevated in the blood or urine. Allopurinol is structurally similar to hypoxanthine, one of the purines formed by degradation of the nucleic acids and the precursor of uric acid. Because of this structural similarity to hypoxanthine, allopurinol can attach to xanthine oxidase, the enzyme that catalyzes formation of uric acid and can inhibit its normal activity. In consequence, daily formation of uric acid is much reduced, while the hypoxanthine is disposed of by an alternative process, viz., reutilization for nucleic acid synthesis.

A by-product of this work with allopurinol has been elucidation of the genetic basis of a rare form of gout observed in children who also exhibit cerebral palsy, mental deficiency, and self-destructive biting. In these children, while allopurinol effectively inhibits xanthine oxidase, total purine excretion is not reduced as in other gouty individuals. This suggested that hypoxanthine could not be reutilized for nucleic acid synthesis in these children—a postulate that proved to be correct. Thus, the precise metabolic defect in this form of gout, transmitted as a genetic recessive trait by a gene located on the X chromosome, was identified. It may be hoped that such understanding may one day lead to rational therapy. For now, it must suffice to incorporate knowledge of this dreadful disease into sophisticated genetic counseling.

We cannot refrain from an additional note. In all mammals but the primates, uric acid is subject to a further metabolic degradation that leads to highly soluble products. Hence, gout is a disease that can occur only in man and the other primates. But loss of uricase (the uric acid-destroying enzyme) is the only specific biological alteration one can temporally associate with the evolutionary appearance of the primates, and one cannot help but wonder whether the presence of uric acid in the blood is in some manner related to the subsequent rapid evolutionary development of the brain.

GENETIC DISEASES

The first scientist to document the fact that some diseases tend to run in families was A. E. Garrod, physician to the British royal family. In 1908, not long after the resurrection of Mendel's work, Garrod published a remarkable treatise entitled "Inborn Errors of Metabolism," at a time when almost nothing was known about metabolism. He listed six inborn errors that are transmitted as recessive Mendelian traits. Now, several hundred such genetically transmitted errors have been identified; in many cases (Table 2) the specific missing enzyme or other protein is known; in other cases (Table 3) it has not yet been identified. It is to be emphasized that, with few exceptions, these genetically transmitted abnormalities are detected because they do, in fact, occasion disease. This comes about, generally, in one of three ways:

1. Because of the blocked pathway, a desirable end product is lacking, e.g., in albinism, lack of one of the enzymes responsible for metabolism of the amino acid, tyrosine, renders synthesis of the pigment melanin impossible.

2. Accumulation of an intermediate that normally is further metabolized may precipitate the difficulty, e.g., the arthritis of alkaptonuric individuals in whom a block in tyrosine metabolism is responsible for accumulation of homogentisic acid.

3. The blocked pathway may result in diversion of a normal intermediate into an alternative but normally little-used metabolic channel, forming intermediates that themselves cause difficulty, e.g., mental deficiency caused by accumulation of phenylpyruvic acid in phenylketonuria.

In the main, type (1) is predominant, since it also includes the host of situations in which the genetic defect results in a great variety of structural and functional failures. It will be clear that such diseases are the consequence of genetic alteration of important but not vital processes. Un-

TABLE 2   Some Hereditary Disorders in Man in Which the Specific Lacking or Modified Enzyme or Protein Has Been Identified

DISORDER	AFFECTED ENZYME OR PROTEIN
Acanthocytosis	β-Lipoproteins (Low Density)
Acatalasia	Catalase
Afibrinogenemia	Fibrinogen
Agammaglobulinemia	γ-Globulin
Albinism	Tyrosinase
Alkaptonuria	Homogentisic Acid Oxidase
Analbuminemia	Serum Albumin
Argininosuccinic Acidemia	Argininosuccinase
Crigler-Najjar Syndrome	Uridine Diphosphate Glucuronate Transferase
Favism	Glucose-6-Phosphate Dehydrogenase
Fructose Intolerance	Fructose-1-Phosphate Aldolase
Fructosuria	Fructokinase
Galactosemia	Galactose-1-Phosphate Uridyl Transferase
Goiter (Familial)	Iodotyrosine Dehalogenase
Gout	Hypoxanthine Guanine Phosphoribosyl Transferase
Hartnup's Disease	Tryptophan Pyrrolase
Hemoglobinopathies	Hemoglobins
Hemolytic Anemia	Pyruvate Kinase
Hemophilia A	Antihemophilic Factor A
Hemophilia B	Antihemophilic Factor B
Histidinemia	Histidase
Homocystinuria	Cystathionine Synthetase
Hypophosphatasia	Alkaline Phosphatase
Isovaleric Acidemia	Isovaleryl CoA Dehydrogenase
Maple Syrup Urine Disease	Amino Acid Decarboxylase
Methemoglobinemia	Methemoglobin Reductase
Orotic Aciduria	Orotidine 5'-Phosphate Pyrophosphorylase
Parahemophilia	Accelerator Globulin
Pentosuria	L-Xylulose Dehydrogenase
Phenylketonuria	Phenylalanine Hydroxylase
Sulfite Oxidase Deficiency	Sulfite Oxidase
Wilson's Disease	Ceruloplasmin
Xanthinuria	Xanthine Oxidase

doubtedly genetic alteration of vital processes exists, but is expressed not as defined disease but as complete lethality with failure of the fertilized egg to develop. With few exceptions, all these diseases reflect the presence of the mutant gene in the chromosomes of both parents; hence there is the great potential of genetic counseling in the future to limit such diseases. Finally, we must note again that, without the huge advances in fundamental

TABLE 3   Some Hereditary Disorders in Which the Affected Protein Has Not Been Identified

DISORDER	BIOCHEMICAL MANIFESTATION
Congenital Steatorrhea	Failure to digest and/or absorb lipid
Cystic Fibrosis	Thick viscous mucous secretion, high sodium content of all secretions
Cystinuria	Excretion of cystine, lysine, arginine, and ornithine
Cystinosis	Inability to utilize amino acids, notably cystine; aberration of amino acid transport into cells
Fanconi's Syndrome	Increased excretion of amino acids
Gargoylism (Hurler's Syndrome)	Excessive excretion of chondroitin sulfate B
Gaucher's Disease	Accumulation of cerebrosides in tissues
Niemann-Pick Disease	Accumulation of sphingomyelin in tissues
Porphyria	Increased excretion of uroporphyrins
Tangier Disease	Lack of plasma high-density lipoproteins
Tay-Sachs Disease	Accumulation of gangliosides in tissues

understanding in the last few decades, it would have been impossible to detect and define this list of defects in man's own essential biology or to arm genetic counselors in the future.

Examination of individuals with inborn errors has contributed significantly to understanding of normal metabolism. The accumulation of a compound not normally observed indicates that it must be an intermediate in some metabolic pathway. Such information has contributed significantly to construction of "metabolic maps" of the multitude of synthetic and degradative chemical reactions that constitute the activities of normal cells. (See Figure 3, page 38.)

Moreover, in a few instances therapeutic regimens have been developed that markedly mitigate the genetic disorder. Phenylketonuric infants may be detected by a simple test applied to soiled diapers. When placed on diets very low in the essential amino acid phenylalanine, they grow to maturity with essentially normal mental capacity—at least sufficient to obviate the necessity for institutionalization. Galactosemic infants also are now readily detected. A feeding formula containing cane sugar rather than milk sugar prevents the cataracts and bone malformation caused by this disease. Both these procedures get around the genetic disorder by avoiding the problem—eliminating from the diet the material the child is genetically unable to metabolize. But this approach is open in only a few such diseases.

An alternative plan is to replace some of the defective cells of the affected individual with normal cells from some donor. For more than a year and a half, a boy with agammaglobulinemia—a total lack of immune competence—and one with the Wiskott-Aldrich syndrome—a partial immune deficiency—have survived with immunologically competent marrow cells grafted from genetically related siblings. In the United States and abroad, a long-standing embargo on marrow grafting has been lifted because new understanding of the immune mechanism, advances in tissue typing, and methods of controlling immune responses through immunosuppressive drugs have markedly raised the level of knowledge.

In theory, a similar approach might be useful for sickle cell anemia, for example, or for the nongenetic aplastic anemia. However, the massive scale of marrow transplantation then required renders such plans rather unlikely.

The final alternative is repair of the genetic defect itself, *viz.,* introduction of the normal gene into the body cells of affected individuals, an approach that has been called "genetic engineering." To date, this has not been tried. The proposal is to learn how to transduce the appropriate gene with an innocuous virus—analogous to transduction of antibiotic resistance in bacteria. It is impossible to assay the chances of success in this effort, which will be long and difficult. But there are, at least, no patent theoretical obstacles to success, and mankind would be immensely rewarded.

In this regard, it may be noted that sophisticated knowledge of enzyme systems and microtechniques for identifying them while the fetus is *in utero* are enabling geneticists to identify an increasing number of metabolic disorders in infants during the first trimester of pregnancy. Most recently, it has become possible to detect Tay-Sachs disease, a fatal neurological disorder resulting from impaired lipid metabolism, *in utero.* Such techniques offer the parents the option of terminating pregnancy in such cases but raise ethical questions outside the scope of this technical summary.

THE IMMUNE SYSTEM

Mechanisms for resisting infection began evolving in early biological time. Immunity is a highly specific condition in which, having produced antibodies to the toxins produced by bacteria or to the exterior of bacteria and viruses, animals are not adversely affected by their invasion.

The system responsible for antibody formation occupies the current scientific limelight because, if it could be understood and controlled, the possibility of victory over cancer and infectious diseases would be markedly enhanced while one of the major stumbling blocks to organ transplantation would be overcome, and because, intrinsically, it is a fascinating process. An antigen must simply be a foreign molecule of sufficient size to trigger the formation of antibodies. To this day, it remains unclear how a given antigen initiates the process of formation of antibodies that specifically combine only with molecules identical with the initiating antigen.

The first time the body confronts an antigen, it produces a small amount of antibody that slowly disappears from the blood. The response may involve only a few antibody-making cells, but it paves the way for a significantly more powerful response if the same antigen is introduced a month or more later. The immune system, in short, "remembers" the foreign molecules it has encountered before and is prepared to defend itself against them, responding with 50 to 100 times the vigor exhibited during the first encounter. This biological memory is specific, greatly fortifying resistance to a second attack of disease. Infections that come and go suddenly are those in which the attack is successfully thwarted by antibodies. Chronic infections are those in which specific resistance fails.

One of the principal questions before immunologists is the nature of this sophisticated memory. Biochemically, in what kind of compound or mechanism is it stored? Antibodies and antigens fit together like a hand and a tightly fitting glove. Is there, perhaps, a biochemical library in which the shapes of previously encountered antigens are shelved? If so, where? Current hypotheses suggest that storage must be in a protein, perhaps in a variety of antibody protein that may be attached to the surface of a cell, such as a lymphocyte coursing through the blood or a macrophage or a plasma cell precursor in the bone marrow, spleen, or thymus gland. A recent discovery of great importance indicates that antibody formation begins by cooperation between two families of cells—scavenger cells called macrophages, and antibody-synthesizing cells. The macrophage engulfs the antigen, perhaps changing it into a different form, which is a highly active stimulus to antibody formation by plasma cells in lymph nodes and bone marrow. The extent to which this process occurs is dictated by hormones elaborated by the thymus gland, but the exact role of this gland remains obscure.

Particularly intriguing is the mechanism whereby the antigen, or a derivative generated in a macrophage, causes the formation of the highly specific antibody. Antibodies are large protein molecules constructed, as usual, of amino acids, according to the plan shown in Figure 31. Each such molecule has two combining sites for antigen. Much of the molecule is invariable among all antibodies, while a significant fraction, in two different sections of the molecule, shows variation in amino acid sequence, depending upon the initiating antigen. The problem, then, is whether there are already genes for all possible antigens, with a given antigen serving somehow as a derepressor for that gene which corresponds to the antibody, which combines with itself, or whether, in some manner, there is a set of plastic genes whose expression is altered by presence of the antigen. These two hypoth-

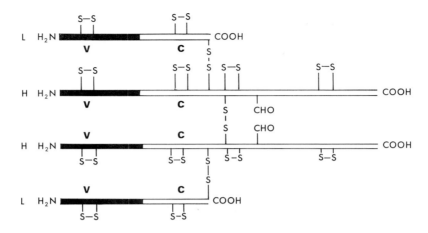

FIGURE 31   Schematic structure of plasma immunoglobin. The structure given is
that of the major antibody protein of the class IgG.   L = light chains; H = heavy
chains; CHO = carbohydrate unit. The variable and constant portions of the chains,
with respect to amino acid sequences, are indicated by the labels V and C, respectively. It is not known whether V is the same length in H and L chains. (From
Principles of Biochemistry, 4th ed., by A. White, P. Handler, and E. L. Smith. Copyright © 1968 McGraw-Hill, Inc. Used with permission of McGraw-Hill Book Company.)

eses seem almost equally unreasonable, and the true mechanism may well
differ from either. Until such understanding is gained, knowledge, and hence
control, of the immune system must remain essentially empirical.

Antibody function, which may take several forms, is only partially understood. The simplest of its activities is to neutralize, by binding at its active
site, a toxic molecule, thus preventing its toxicity until it can be destroyed
by scavenger cells. With viruses, interaction with the protein coat of the
virus may suffice to prevent its penetration into a susceptible cell, and hence
account for the usual immunity produced by viral infections and by vaccines.
In a more complicated maneuver, antibody may react with the surface
molecules of invading cells—e.g., bacteria. Cellular destruction then requires the operation of "complement." The latter is a group of substances
in normal serum that becomes functional only after antibody–antigen interactions take place. Apparently, in a group of hydrolytic enzymes, each
activates the next like a row of falling dominoes, the final activated member
being the enzyme that attacks and destroys the cell. It is possible that
complement, or a deficiency of its naturally occurring inhibitors, plays a
role in the development of some autoimmune diseases—instances in which
a person makes antibodies to his own tissue, with very serious consequences.

Aberrant functioning of immune mechanisms is associated with a number
of important chronic disorders. Common allergies, including hay fever and

hypersensitivity to some protein in shellfish, are among these. So are more serious maladies, including an inability to make antibody or the destructive process of making antibodies against oneself. Studies of children with the former disorder show it to be inherited as a recessive trait; the consequence is a continuing series of infections, inevitably ending in death. In contrast are the autoimmune diseases. Acute and chronic kidney disease develops in persons who make antibodies against their own kidneys. Rheumatic fever and damaged heart valves appear in patients whose antibodies fight their own myocardial muscle fibers. Hemolytic anemia can result from antibodies against red blood cells. Multiple sclerosis may well have a related etiology. To date, this understanding has done little to assist in the management of these disorders, but it is the essential first step.

TISSUE TRANSPLANTATION

As was brought to public attention so dramatically in recent attempts at cardiac transplantation, management of the immune system is a key factor in the success or failure of homografts. The problem is still but partially solved, while investigators search for ideal immunosuppressive drugs and regimens for their use. (Too little does not work; too much so paralyzes the immune system that the patient is exposed to repeated infections.) Currently, the search is for means to induce specific tolerance, that is, to induce the immune system specifically to accept a transplanted organ from a given donor while retaining its normal vigor against infection.

When an organ, e.g., a kidney, from a random donor is transplanted, it serves as a source of antigens. The antibodies produced in response can attach to a variety of sites in the transplanted organ and effectively destroy it. Moreover, the transplant may bring with it immunologically competent cells that make antibodies against the host, damaging diverse normal tissues. In the earliest such studies it was shown that transplants involving identical twins posed no such problems. The question, then, is whether a vast number of operative antigens is involved or a lesser, perhaps manageable, number. Although the chemical nature of the antigens remains largely unknown, "typing" procedures have been developed that indicate the presence of perhaps 30 tissue antigens—procedures analogous to typing procedures for classical blood transfusion. Accordingly, it has now become possible to type prospective donors and recipients, thereby permitting identification of suitable donors who are not necessarily identical twins. This is a momentous achievement that, already in use, should go far toward reducing the serious-ness of management of the immune system after such procedures, but it has by no means yet obviated this problem.

At the same time, one may well ask, "What purpose, if any, is served by the mechanisms involved in these rejection reactions? Why should cells from one human being be rejected by another?" There is no obvious evolutionary explanation. Since transplantation is entirely man-made, there is no known selective mechanism that would account for one man's biochemical refusal to tolerate the tissues of another. One hypothesis under consideration is that the mechanism evolved and has been maintained to eliminate mutant cells not under normal growth controls, *viz.,* cancer. Indeed, there is evidence to support this view; if true, it suggests that the neoplastic transformation of normal cells may be a frequent event, but that the normal immune system quickly destroys them. Established cancer, then, would be the consequence of failure of the rejection reaction—a distinct possibility.

CARDIAC DISORDERS

Heart diseases take the lives of more than 700,000 Americans annually and disable millions more. Acquired and congenital anomalies need surgical repair. Malfunctioning valves need replacement. Hearts too weak or diseased to beat regularly require regulation or stimulation. And some hearts are beyond repair. These needs have motivated cardiac research for decades. Some have been met; others may be met before long.

Although heart transplantation is an experimental and controversial measure, heart surgery includes a variety of well-established procedures that have saved thousands of lives. Pacemakers save many more, as do drugs that control irregular heartbeats. For these achievements we are indebted to the physiologists who have explored the workings of the heart, to the engineers who designed the heart–lung machine and the miniaturized transistors that power pacemakers, and to pharmacologists for their fundamental research into the chemistry of a heartbeat.

The primary problem of cardiac physiology is accurate estimation of blood flow through the heart. Studies to overcome this problem, begun in the 1920's, and continued ever since, have been richly rewarded. First came procedures whereby, from the behavior of an injected pulse of an indicator dye, one could calculate blood flow. Twenty years later, it was shown that thin, radiopaque catheters could be introduced into the heart chambers via peripheral blood vessels and carefully positioned visually by fluoroscopy, permitting direct sampling of blood in the cardiac chambers. Before complex heart surgery could be attempted, it was imperative that the surgeon have, in advance, precise information about specific functional and anatomical abnormalities. Catheterization, radiopaque dyes, and angiocardiography (x-ray visualization of the heart and associated vessels as the dye passed through them) provided some of this information. Recently, valuable knowledge has been obtained describing the response of the heart to various types of abnormal mechanical overloads from obstruction and insufficiency of the several heart valves; this is especially useful in selecting candidates for surgery.

In short, a variety of increasingly sophisticated and reliable techniques have been developed over the course of half a century. With them the physician can make a highly precise diagnosis, establish the quantitative as well as the qualitative nature of the problem, and rationally decide upon a therapeutic or surgical course. One of the more remarkable aspects of the surgical technique is the recent capability to literally patch major blood vessels and cardiac valves, thanks to the availability of suitable, nonreactive plastics from the chemical industry.

Pacemakers, first used to control cardiac rates in physiological studies on animals, have been employed in animal and human studies ever since the finding that electrical stimulation by high-voltage direct-current shocks, delivered directly to the heart during episodes of fibrillation, induce the return of properly coordinated rhythms with return of normal cardiac function. In fibrillation, the muscular fibrils of the heart twitch rapidly, independently, and irregularly so that coordinated contraction and pumping cannot occur. During the 1950's cardiologists studying patients suffering Stokes-Adams attacks (syncope—bouts of loss of consciousness due to insufficient blood flow to the brain) developed pacemakers that trigger heartbeats by means of small electrical shocks. The electrodes of these little generators are directly implanted in the patient's heart. This procedure has evolved as the optimal therapy for patients with heart block, prevents Stokes-Adams attacks, and also serves persons in cardiac failure characterized by extremely slow ventricular rates.

Antiarrhythmic drugs have also found a valuable place in the cardiologist's arsenal of weapons against irregular heartbeats. The first clinical trial of such drugs took place in 1912, when the effect of quinine alkaloids was observed. Since then, extensive pharmacological studies have attempted to explain the mechanism of the quinine alkaloids in this regard. (Quinidine from the cinchona plant is the most effective.) Not until 1951 did another antiarrhythmic agent, procaine amide, become available. Like quinidine, it depresses contractility of the heart and similarly affects its electrical activity, decreasing the formation of impulses, slowing conductivity and excitability, and prolonging the lag time between beats. These drugs have a like range of clinical use; both have similar toxic effects; both remain the most frequently used agents to control a wildly beating heart.

For most of this period, the hunt for such drugs was entirely empirical. The failure to produce more effective, more specific, and less toxic drugs to control cardiac rhythm stems from the facts that the underlying mechanisms responsible for many arrythmias were, and are, unknown, and that the pharmacology of the existing drugs is imperfectly understood. Knowledge of the electrical basis for the formation and conduction of impulses within the heart gained impetus when it became possible to record the transmembrane potentials of single cardiac fibers by implanting microelectrodes within cells. By means of this technique and associated studies, it became possible to characterize the ionic basis of cardiac electrical activity, to identify the unique properties of certain specialized cells, and to observe the influence of, for example, quinidine and digitalis on these parameters. Now, from studies of the electricity of the heart and of the ionic processes associated with it, highly detailed, though not yet complete, pictures have been drawn of each of the major clinical types of arrhythmia, a new beginning is under way, and suitable test systems are available for the search for specific antiarrhythmic drugs.

Critical to ultimate management of these disturbances is improved understanding of the underlying electrical activity, which is the result of the operation of the cellular "electrolyte pump." It must be more than fortuitous that the "cardiac glycosides," particularly ouabain, which can assist a failing heart are the most effective known inhibitors of the cellular transport system, which, in cardiac muscle as in all other cells, achieves the outward movement of sodium ions and the inward movement of potassium ions using the energy of ATP. The responsible protein is associated with cell membranes; the model proposed in Chapter 1 for transport processes seems an adequate description of its function, $viz.$, binding of $3Na^+$ and $1ATP$, a conformational change that permits rotation in the membrane, hydrolysis of the ATP, discharge of the Na^+, binding of K^+, and rotation to the original position. Control of this basic life function, which is adapted to the special purpose of "electrical" conduction in nerve and muscle fibers, appears central to progress in a variety of cardiac disorders. Parenthetically, one may note that hyperactivity of this system in the various secretory glands is one of the manifestations, now used as a definitive diagnostic sign, of cystic fibrosis. In yet another context, it is the genetically controlled synthesis of this same transport protein in kidney tubules that is regulated by aldosterone, the adrenal hormone that, in excess, causes Cushing's disease and lack of which occasions Addison's disease.

The advent of cardiac surgery (and indeed, of cardiac transplantation) is one of the most dramatic episodes in the history of medicine. Clearly, these heroic procedures could not have been attempted until all the necessary knowledge, skills, materials, and tools were at hand. Illustrative is the instrument that has become the $sine\ qua\ non$ of modern heart surgery—the heart–lung machine.

When the heart is opened for repair of a valve or closure of a hole between the two ventricles (pumping chambers), the heart–lung machine is temporarily employed to assume the function of the heart and lungs, to pump blood, supply oxygen, and eliminate carbon dioxide. First used successfully in man in 1953, its origins can be traced through preceding centuries to the 1500's and 1600's, when double-valve, one-way pumps were designed to draw water from deep mines. These, in turn, apparently inspired William Harvey to recognize the true nature of the heart, which he likened to a water pump. In the 1800's, physiologists attempted to duplicate the work of the heart by perfusion of various animal organs such as the liver and kidney, using a pump to better understand the functions of these organs. In time, physiologists tried to add oxygen and remove carbon dioxide from the blood used in perfusions, thus experimenting with crude forerunners of the heart–lung machine. Their glass, rubber, or metal parts, however, severely damaged the delicate red blood cells, a fact of little consequence in short-term experiments on isolated organs but of obvious import for human application. The plastics industry solved this problem by offering virtually inert, smooth plastics with nonwetting surfaces, which minimize damage to the blood cells as they pass through the pump.

Once developed, successful application of the heart–lung machine awaited solution of one other problem. When blood comes into contact with surfaces other than normal blood vessels, it clots. Indeed, this would happen in the blood vessels themselves were they not coated with natural anti-clotting compounds. Among these is heparin, which can be obtained in quantity from beef lungs. Heparin inhibits the clotting mechanisms, permitting blood to course through the tubes and chambers of the machine for hours.

Many refinements in recent years have made the heart–lung machine safer and more readily available. Artificial heart valves and plastic blood vessels, developed in collaboration with engineers, are available to surgeons. Even totally implantable artificial hearts have been tried in man. And long years of animal experimentation, coupled with the availability of immuno-suppressive drugs and techniques for determining tissue matching, make human heart transplantation a feasible, though still highly experimental, procedure. Yet the road ahead is long. Oxygenators causing less damage to blood than those currently available are needed if heart–lung bypass is to be applied for long periods of time. Such an instrument is essential to save patients with serious but reversible lung diseases, such as hyaline membrane disease in newborns. The use of artificial pumps either partially or completely to support the circulation of patients during a heart attack is a logical move that has already been attempted, but it is far from routine and demands considerable refinement. Significant progress toward production of totally artificial hearts is thwarted by our inability to produce compact, long-lasting power sources capable of responding to the biochemical signals that control muscle blood flow. But there is reason to hope.

Remarkable as all these accomplishments are, it must not be forgotten that a large fraction of the conditions that impose these requirements for drastic surgery are the consequence of one process, atherosclerosis, the deposition of mushy lipids on the surface and within the walls of the arteries, which then calcify, become brittle, and serve as foci for clot formation and infection. In the long run, it is to be hoped that understanding of this process will permit its prevention, thereby obviating the need for many current surgical and therapeutic procedures. The alternative, more than a thousand cardiac transplants per day in the United States alone, is scarcely an appealing prospect. Meanwhile, the efforts of thousands of scientists have brought surgery to this remarkable peak.

If the physiologist originated the idea of a heart–lung machine, he also has greatly benefited by its sophisticated use in the hands of surgeons and engineers, for today he uses the same instrument as a tool for learning still more about the intricate mechanisms of the heart and lungs. And, eventually, the information he gathers will further enlighten the physicians and surgeons in their battle against disease.

Diuretics   A serious, occasionally life-threatening, complication of heart failure, liver and kidney diseases, and hypertension is edema, the excessive accumulation of salt and water in body tissues at large. Today, diuretics, drugs that interfere with the mechanisms by which kidneys retain sodium, and hence chloride and water, control edema rather successfully in most patients. A major class of modern diuretics was made possible by observations during the early history of sulfanilamide. Ironically, no one would have predicted that the background essential to the rational development of diuretics would be supplied by research quite unrelated to the function of the kidney or to the need for such agents.

Early in the clinical use of sulfonamides, it was noted that such patients excreted an alkaline urine and developed a mild acidosis (acidification of blood plasma). Then, biochemists observed that sulfanilamide inhibits the enzyme carbonic anhydrase that catalyzes the simple hydration and dehydration of carbon dioxide, a process necessary to the escape of carbon dioxide from the blood as it travels through the lungs. When carbonic anhydrase was then found to be present in quantity in the kidney, it became apparent that this enzyme plays a role in kidney mechanisms for excretion of acid and that sulfanilamide's inhibitory effect on the kidney enzyme accounted for its effect on urinary secretion, with consequent acidosis. Accumulation of acid is the consequence of excretion of sodium ions. In otherwise normal individuals, e.g., the sulfanilamide-treated patients, this effect is undesirable. But in patients whose kidneys are failing to excrete salt (sodium ions) normally, the same process could be decidedly beneficial. At that point, chemists had a rational test system for fashioning a drug that would be a more effective inhibitor of carbonic anhydrase than sulfanilamide by modifying the structure of sulfanilamide, and designed acetazolamide (Diamox), which, in 1950, became the first useful oral diuretic. Incidentally, it also became a remarkably successful agent for treatment of glaucoma, excessive secretion of fluid into the anterior chamber of the eye, by interfering with the carbonic anhydrase of the overactive secretory cells.

While Diamox was safe, it was not an ideal agent because it could not be used continuously. Five years later, by continuing modification of the basic structure, another diuretic, chlorothiazide (Diuril), was constructed and proved useful not only for treatment of water and salt retention in ambulatory, nonhospitalized patients but also in lowering their blood pressures. In the five-year period following its introduction, prescriptions for diuretics in the United States increased sixfold. But Diuril, too, had limitations, particularly limited ability to cope with massive edema and excessive stimulation of urinary excretion of potassium and sodium. Pharmaceutical chemists then looked to aldosterone, the adrenal hormone that normally occasions retention of sodium and loss of potassium from the body. Several compounds that structurally resemble aldosterone, yet are sufficiently different that they do not possess its pharmacological actions, were synthesized to displace the hormone from sites where it is normally bound in the kidney. By thus occupying the effector sites of the natural hormone, they function as antimetabolites and prevent excessive secretion of potassium.

Meanwhile, another approach resulted in a diuretic of an entirely different class. It had long been known that mercurial compounds are diuretic, but their toxicity precludes their use. These agents were known to work by reacting with sulfhydryl groups of proteins in the lining of the kidney tubules. Accordingly, a compound was sought that also reacts with such groups but lacks the toxicity of mercurials. The result, ethacrynic acid, is so effective that it must be used with great caution. Happily, like Diuril it can be taken orally. With this armamentarium it is now possible to treat successfully virtually all forms of salt retention except for those that reflect primary disease of the kidney itself. Such cases can be managed only by dialysis with an "artificial kidney," itself the product of two decades of research, entirely dependent upon growing understanding of the role of a normal kidney.

As is so often true in science, new discoveries seldom have only a single application. Investigations of sulfanilamide culminated in establishment of the antimetabolite principle and development of Diamox, and Diamox, in turn, became an important tool for fundamental research. First, it represented the beginning of a rational scientific approach to seeking diuretics by relating chemical structures to kidney mechanisms. Second, it became extremely useful as a device enabling renal physiologists to evaluate the role of carbonic anhydrase in kidney-transport processes. It was of prime importance in elucidating the renal mechanisms of bicarbonate reabsorption and hydrogen secretion. The concepts thus developed were then amply supported by direct renal-micropuncture experiments. This development had an important influence on clinical care of patients because the new understanding of physiology enabled scientists to predict the specific electrolyte losses in the urine produced by various drugs.

Much public concern and attention is directed to the problem of providing the best of medical care to all Americans, a concern we fully share. But the nature of medical care and its relation to research should be clearly understood. The component of medical practice that makes the greatest demands on our resources—measured in the time of physicians, nurses, paramedical personnel, hospital beds, and the ever more complex technology of intensive medical care—is the management of those disorders for which research has, to date, made possible only palliative or physiologically corrective measures, termed by some "half-way medical technologies." When research has provided a definitive therapeutic or preventive regimen, it is invariably cheaper and simpler than the palliative treatment previously available for the same disease. This is surely true for a wide range of infectious diseases such as lobar pneumonia, poliomyelitis, tuberculosis, bacterial endocarditis, typhus, typhoid fever, and diphtheria, to name but a few. Almost all nutritional diseases—e.g., pellagra, beriberi, rickets, and scurvy—and a variety of other ailments such as pernicious anemia, Addison's disease, goiter, juvenile diabetes, Parkinsonism, and glaucoma fall within this category. Only a few years ago, it was these disorders that dominated the efforts of the health care system. Most remain serious, but they are but a minor aspect of medical practice. The diseases that now overwhelm the health care system are those for which research has not yet provided the understanding required to design truly definitive procedures. It is not lack of physicians, nurses, technicians, or hospitals that limits our capability to manage such problems as most forms of cancer, coronary occlusion, myocardial infarction, stroke, acute rheumatic fever, osteoarthritis, pyelonephritis, bronchial asthma, schizophrenia, muscular dystrophy, cystic fibrosis, and multiple sclerosis; it is lack of understanding sufficient to permit development of a really therapeutic procedure. Biomedical research, which represents only 1.5 percent of total expenditures for health, is, therefore, both the biggest health bargain one can purchase and the only hope for future progress. If this opportunity is neglected or minimized for shortsighted fiscal reasons, then, by the turn of this century, our nation must double the number of physicians, nurses, technicians, hospital beds, and sanitaria and learn to live with the equivalent increment in human suffering. Grim prospect indeed!

Population Control

While biomedical scientists pursue greater sophistication in the understanding and treatment of disease, this attempt must be matched by a concerted effort to solve the crisis being brought on by the continuing increase in human population. No matter what contributions scientific investigation and new technologies make in the coming decades, it is hard to imagine that they will come quickly enough or be sufficient to meet man's needs if his sheer numbers continue to mount unchecked. The problems of population control are both biological and sociological. From studies in reproductive biology must come new and better contraceptive procedures, which must then be put into general use.

The oral contraceptives that became widely available in 1961 are consumed by millions of women the world over; they symbolize society's recognition of the need for birth control. They also illustrate the beneficial results of concentrated, deliberate research. Birth-control pills in current use are usually a combination of the two hormones that regulate the reproductive cycle—a synthetic estrogen and a progestin, a synthetic version of natural progesterone. If taken as prescribed, they appear to be almost invariably effective, although reproductive biologists are not entirely certain why. That these agents strikingly alter the output of the related regulatory pituitary hormones is certain. Beyond that, explanations of their mechanism of action are tentative. They may not actually prevent ovulation each month, yet exert their contraceptive effect nonetheless. The appearance of the endometrium that lines the uterus is somewhat altered in women taking these drugs; perhaps this relates to the failure to conceive. Another possibility is that the progestin in the combination products stimulates the release of cervical secretions so viscous that they effectively entrap spermatozoa. Indeed, there is some evidence that progestin alone is an effective contraceptive, and various experiments with low-dose progestational compounds are under way. Unfortunately, it was one of this class of compounds that was recently shown to induce tumor formation in dogs; hence, the future of this program is uncertain.

The availability of the current pill is the culmination of 70 years of study of the operation of the mammalian reproductive apparatus. Step by tedious step, understanding of the nature and function of the two pituitary hormones—the estrogen of the ovary and the hormone of its corpus luteum —as well as the progesterone of the uterus—was achieved. The accumulated information found its way into pregnancy tests, diagnoses of abnormal pregnancies, and correction of faulty development of secondary sex characteristics. Natural sources of estrogens and progesterones were inadequate; substitutes were synthesized that were more effective than the natural forms and that could be taken by mouth. Detailed studies revealed the precise cellular changes occasioned by each natural and synthetic hormone; slowly, the precise clockwork that governs the menstrual cycle and the stabilization and climax of pregnancy was elucidated.

With such knowledge came successful diagnosis of the cause of a large fraction of all instances of sterility—imbalance of the two pituitary hormones. Therapeutic trials failed until it was realized that only human hormone is effective. This is available in urine, and a modest supply now permits pregnancy for many childless wives. But the supply is limited and one must await precise establishment of the amino acid sequence of this hormone, followed by synthesis using the recently developed methods for polypeptide synthesis, to overcome this shortage.

It was with this slow and difficult accumulation of understanding that the search for a contraceptive pill began, both estrogens and progestins being tested separately before it became clear that a combination might be required. Most important is the realization that, until the whole stage had been set, the final undertaking could not have been possible. There has been no better illustration of the culmination of many years of interaction between clinical observation and clinical and basic research. As use of the pill increased, reports of clotting disorders and breast tumors became more frequent. Even at this writing, the validity of such claims is somewhat uncertain, and the adverse effects of the pill remain to be established with certainty. Assuming the reality of such effects, there remains the societal decision of weighing hazard against benefit—the death rate due to pregnancy versus that associated with the pill and the risk entailed versus the societal imperative that population growth be brought under control. Meanwhile, the search for other, less hazardous but still effective, measures must be prosecuted vigorously.

The search for contraceptive drugs began with animal studies and now returns to the laboratory to create the next generation of pills. In addition to seeking an explanation of the mechanisms of current agents, investigators must explore the phenomenon of conception itself even further. A quite subtle interruption in this delicately balanced sequence of biological events may well prevent conception just as surely as the grosser effects of present agents.

The newly established Center for Population Research at the National Institutes of Health has initiated a program focusing on four targets:

1. The reproductive physiology of the male, particularly the processes that permit the maturation of sperm cells. Only a mature sperm can penetrate and fertilize an egg. If the biochemical events surrounding this process could be controlled, a new approach to contraception would be available.

2. The structure and function of the oviduct through which an egg travels from the ovary to the uterus.

3. The function of the corpus luteum, the yellow body, formed after ovulation, that produces progesterone for the maintenance of pregnancy.

4. The biology of the fertilized egg cell before and during implantation in the uterine wall.

Prior to the introduction of oral contraceptives, reasonably satisfactory methods of mechanical or physiochemical contraception existed. Human nature limits the success of these methods; all too often they are used improperly or not at all. They are not, however, to be discarded, nor are the increasingly satisfactory intrauterine devices. If population control is to be achieved on an acceptable scale, a variety of contraceptive methods will be required. This will be possible only in the light of additional knowledge.

If indeed a promising lead for the development of a new contraceptive drug does emerge from research, there will remain an extremely lengthy process, prescribed by the Food and Drug Administration (FDA), before it can be brought to market. Such research and development is performed in the laboratories and under the auspices of pharmaceutical companies, which spend collectively, even now, more than half of all funds devoted to research on reproductive physiology, quite apart from the high costs of prolonged toxicity testing and development. Precisely because such a drug would be taken by "normal" women over many years, the FDA procedures are conservative, demanding prolonged test trials to establish safety, side reactions, and so on. At best, there can be no way to shorten the testing trials in women—and the world's population will have increased by at least one billion before widespread, unrestricted use of such a new drug could be considered, even if the structure of the compound were known, its synthesis worked out, and its general biological properties known at this writing. The great expense of the necessary prolonged procedures is a serious deterrent to the undertaking of such activity by the drug manufacturers, who must somehow be assured that they will at least recover their investment. *Meanwhile, the needs of humanity are so great that we suggest that the Secretary of Health, Education, and Welfare develop some new set of relationships wherein the government joins with the drug manufacturers in funding such research activities, utilizing to the full the organized multidisciplinary capabilities of these organizations, underwriting their costs, and pooling their competence.* Confronted by the crisis of population growth, the government is justified in taking emergency measures.

The Early and Latter Years of Life

Half the individuals born today will die before their seventieth birthdays; yet, for all the hazards that beset man during his middle years, the gravest threat remains with the first year of life. Infant mortality (deaths in the first year of life) has been declining steadily in the last half century as a result of significant advances in infant care, but it is still higher in the United States than in several other countries—22.1 deaths per 1,000 live births in 1967.

Human biological potential is conditioned, in large measure, by the events of prenatal and early postnatal life. The quality of adult life is predetermined by such phenomena as inherited defects, environmental influences, including disease, exposure to radiation or drugs, and the quality of nutrition. The first stages of man's life are the object of growing scientific attention, yet there are few areas in which clinical applications are as severely handicapped by lack of fundamental understanding.

There is, as yet, no precise description in biochemical terms of the mating of sperm and egg. The fetus, in the protective environment of its mother's womb, nourished through the placenta, is particularly susceptible to environmental influences as its cells differentiate and become specialized tissues, and it is subject thereafter to the health of its mother. Diabetes, toxemia of pregnancy, and blood-group incompatibility can threaten its health, and even its survival. Parturition must come neither too early nor too late, and the newborn must then adjust to his world. Whereas a significant fraction of infant mortality may be eliminated by applying available understanding, further progress will be entirely dependent on improved knowledge of the entire process from conception to the early years of life.

No problem appears more urgent than definitive establishment of the consequences in later life of early nutrition. This problem first came to attention with respect to peoples of developing nations as it became evident that the apathy, stunting, susceptibility to infection, lack of energy, and, perhaps, limited intelligence of certain tropical populations were related to their nutritional status, since this characteristic is particularly obvious among those groups in which kwashiorkor (generalized protein deficiency) is rife. Significantly, the data also indicate that there is no genetic basis for this problem. Accordingly, there is urgent need to learn how protein deficiency results in these sequelae, whether there are key amino acids, what level of nutrition is required to prevent the process, etc. Early evidence strongly indicates that the brain of the protein-deficient individual may contain as many as 30 percent fewer than the normal number of neurons (nerve cells). Since the process of neuron generation is completed within the first two years of life, this deficit can never be overcome. Solution of these problems could go a long way toward helping protein-deficient people to help themselves. Equally important is the need to establish the extent to which similar nutritional influences are at work in the United States. Animal studies will continue to be revealing, but safe and sensitive techniques for monitoring the physiological state of the fetus as it develops are sorely required.

To know the mechanisms of genetic and environmental effects and to comprehend the role of nutrition, the influence of hormones in fetal life, and the interactions of tissues, these factors must be measured and charted. Efforts to accomplish this are under way. New methods are now being applied to measurement of maternal excretion of hormones, particularly estriol (an estrogen), and relating it to fetal development, to monitoring of fetal heart rates and correlating these with the fetal condition, and to analysis of fetal blood, even during labor itself, by obtaining microsamples that are examined by new microchemical procedures.

At the opposite end of life's scale, the process of aging is even less well understood. Indeed, it has yet to be described adequately. What processes are responsible for the progressive decline in the structure and function of an adult organism? What aspects of the process are intrinsic to the organism, i.e., the consequence of its initial genetic complement, and what aspects result from environmental assaults? How would we age in the absence of intercurrent trauma or infectious disease? In both young and old organisms, muscles contract, nerves conduct, glands secrete, and so on. The changes occurring in tissues that distinguish youth from age are too subtle to be detected by currently available techniques. How does deterioration in structure and function become incompatible with life? Has anyone ever died of "natural causes"?

One aspect of aging seems incontrovertible. With the passage of time, cells die in certain organs—the brain, the muscles, the lymphatic system— and are not replaced. Is aging merely the consequence of this one-way process? If so, what clockwork fixes the norm for a mouse at one year, for man at three score and ten, for the giant sea tortoise at 500 years, and for the sequoia at several millennia? Can this clockwork be reset?

One prominent theory of aging holds that it reflects a developed instability of the genetic apparatus of individual cells, i.e., that aging occurs because of highly specific deviations within single cells rather than among whole cell populations. Perhaps, for example, in the course of time, subtle errors in the self-duplicating process of DNA accumulate. Perhaps the accuracy of transcription fades, though it does not fail completely. To date it has been possible only to refine these questions, not to subject them to rigorous test, for lack of a reasonably short-lived but acceptable model. Current efforts utilize mammalian cells in tissue culture and such organisms as the thousand-celled rotifer. A suitable test model should have a short life-span and well-established standardized nutritional requirements, should be maintained in freedom from infections and other external insults, and must possess genetic uniformity.

An alternative hypothesis suggests that, whereas cell death and failure of replacement do indeed lie at the heart of the aging process, the reason may not be intrinsic in the cells themselves but may be secondary to changes in their environment. Certainly, with the passage of time, connective tissue becomes tougher, thicker, and less elastic. If such changes also occur on a minute scale at the level of capillaries, this could result in local nutritional failure or intoxication by the products of the cells' own metabolism.

Regrettably, all such studies are in their infancy. Only when they have produced sufficient understanding will it be clear whether man may aspire to a prolonged span of enjoyable, fruitful years.

This brief summary has only touched upon the approaches to biomedical research that may be anticipated in the next decade. Predictions of the future direction of clinical investigation, like those of other human affairs, are hazardous, but the record suggests that the greatest benefit will accrue from the slow accumulation of basic knowledge concerning the nature of normal and pathological physiological and chemical processes. Obviously, one cannot apply knowledge to the prevention and treatment of disease until that knowledge exists.

Biomedical research has come of age. In the intensively managed, highly instrumented clinical research units of our great hospitals, clinical investigation has become a legitimate science. Human biology is being explored with unprecedented vigor and sophistication, and the information net of the biomedical community assures that scientific discovery in all disciplines is readily applied to human disease.

This endeavor, the focal activity of university medical centers, is less than two decades old. How, then, shall one measure its success? Not alone by the large and small insights into the nature of life or the pathogenesis of disease, nor by the pain alleviated or the lives saved. We are all too aware of the woeful limitations of medicine, of the anguish of suffering, tortured humanity, including those who are left behind after death. The true measure of this research enterprise is to be found in the hopeful spirit of the biomedical research community as it faces the future. The increasing wealth of information and insight provided by molecular understanding of normal structure and function and their pathological aberrations render this community confident that it will be armed with ever more powerful tools with which to undertake its noble task.

On Feeding Man

In ten years' time, human beings will eat human beings in Pakistan.

— President Mohammed Ayub Khan, 1964.

If man is to control his own destiny, he must understand his world as profoundly as he must understand himself. Only when there is a balance between the human population and its food supply will the threat of mass starvation be lifted. But calories alone will not suffice; the protein and vitamin content of the food supply must also be adequate to human need. Moreover, a balance in planetary terms could well be misleading. In the long term, each major population group must feed itself.

Accomplishment of these goals is a major challenge to the human race, but it is feasible. Indeed it is in prospect, although that seemed unlikely only a few years ago. Mild optimism in this regard rests on the facts that:

1. On a worldwide basis, food supply has been increasing faster than has population for several years.

2. Recently introduced strains of wheat, rice, maize, sorghum, and millet have dramatically increased food production in areas in which food shortage has been traditional.

3. Population control is gaining worldwide acceptance and its practice is increasing, albeit less rapidly than might be hoped.

4. A sound scientific basis has been constructed for agricultural practice; its extension, worldwide, coupled with provision of the necessary capital, could undoubtedly assure an adequate food supply for a world population that can limit its numbers to only moderate growth in the future.

Modern agricultural practice is one of the greatest of scientific triumphs. Since the turn of the century, agriculturists have been quick to utilize the most recent applicable understanding of genetics, plant physiology, soil chemistry, and physics. The result, combined with generous use of fertilizer in the developed nations, is that an ever-diminishing fraction of the working population is required to feed the remainder, who enjoy the most diverse and nutritious food supply in history.

Crop Yields

GENETICS AND AGRICULTURAL PRACTICE

The primary challenge to the farmer is to achieve the greatest possible yield of his crops. His actual choice of crop rests on market forces—price and regional eating habits—coupled with the suitability of his farm for specific forms of tillage. Thereafter, the result depends upon the genetic strain employed, application of fertilizer, soil and cultivation management, control of pests and weeds, water supply, and harvest. The first, choice of genetic strain, is undoubtedly the most successful of all applications of genetic understanding, and in the United States is the basis of a significant industry. For example, dozens of strains of wheat, tomatoes, and hybrid corn are under cultivation in this country. They have been developed to maximize crop return under diverse local circumstances—mean temperature, temperature maxima and minima, amount of rainfall, soil structure, and resistance to infection by specific viruses and fungi—and also to take advantage of heavy applications of fertilizer. Mutants tested in research stations are selected and improved by commercial breeders, who make stocks available to seedsmen, who in turn make them commercially available. A few examples will suffice.

Photosynthesis is the function of the leafy structure of the plant, and crop yields can be significantly increased by maximizing that fraction of solar energy per acre that is absorbed by the crop leaves. This is achieved in part by the spacing of rows, but a major limitation is imposed by the shading of lower by upper leaves. This problem can be minimized by increasing the verticality of the upper leaves. Strains of all major grains are now being bred to achieve this; already it is clear that substantial gains will thus be realized, particularly in semitropical regions where intense sunlight exceeds the light-absorptive capacity of the upper-leaf canopy.

Selective breeding has also improved the desirable intrinsic properties of many major crops. Tomatoes now under commercial cultivation contain several times the vitamin C concentration of older strains. Sugar beets were made competitive with cane by raising their sugar content from 6 to 18 percent, while, at the same time, new strains of cane were developed that do not flower for several months, thereby doubling their sugar content. Success in the breeding of maize affords examples of the use of genetics and breeding methods in agriculture. Maize, or Indian corn, originated in prehistoric times in the highlands of southern Mexico. It has remained to become the most productive of grains, sometimes referred to as the backbone of American agriculture. At first, it was improved, even by prehistoric man, by field selection of outstanding plants for seed. But early in this century, the possibility of greatly increased productivity by hybridization of inbred lines was realized. Most corn now produced in the United States is grown from hybrids that best fit the many demands of local conditions. When the opportunity is taken to aid corn production in developing countries, field selection of varieties is first resorted to for expediency before undertaking the slower development of hybrids.

Corn has the drawback of being deficient in the content of the nutritionally required amino acids lysine and tryptophan. Fortunately, the corn plant was chosen early for detailed genetic mapping and study of the functioning of inherited characteristics. Among the genes studied were two designated as opaque-2 and floury-2. The action of these genes determines in part the degree to which synthesis of the protein zein in the grain is replaced by glutelin of higher lysine content. Incorporation by breeding methods of one or both of these genes in the chromosomes of desirable varieties and hybrids is now in progress. This promises to alleviate, in part, the protein deficiency of world diet, particularly in some Latin American and African nations.

An improvement in protein quality similar to that realized in corn is now in progress for rice. Rice protein is nutritionally excellent, qualitatively, but the amount of protein per serving is rather limited. The results to date indicate that a gain of more than 25 percent in protein content of the rice grain over that of varieties now in use can be attained. The situation for wheat and sorghum is also promising.

Wheat and rice, like sorghum, corn, and other grains, can be bred as dwarf varieties. The wheat and rice dwarf varieties have short, stout stems, allowing greater numbers of plants per acre and use of fertilizers at the higher rates necessary for attaining high yields. Resistance to some of the prevalent diseases can be incorporated into the dwarf varieties, and the protein content of the grain can be changed. All these endeavors are in progress in the strikingly successful programs of the Rockefeller and Ford Foundations for improvement of agriculture in developing countries.

The soybean, which is valuable as a crop because of its high protein and oil content, presents considerations somewhat different from those pertaining to the grains. Although its use was recorded in Chinese *materia medica* as early as 2838 BC, it was not used as a crop in the United States until about 1900. The United States is now the leading producer, with an annual crop of about one billion bushels.

The soybean plant was at first poorly adapted to growth in many areas. It was discovered that its flowering and yield depended on the length of the day and consequently varied strongly with latitude. Varieties were therefore bred for restricted latitude regions. A variety suited for culture in Arkansas would be killed in bloom in Iowa, where the season is too advanced for maturing when days become short enough for blooming. This property of the plant is a display of the endogenous biological rhythm that is also important for reproduction of many animal species and is present in man.

The quality of the soybean can be varied by breeding to vary the oil or protein content. The slowly changing economic need for the one or the other allows adequate time for development of appropriate strains. Yields of fields, however, even now are relatively low in terms of maximum known yields; the reasons for this are now being sought, with interest centering on factors controlling the extent of flowering and retention of fruit.

Like animals, higher plants are subject to infection by many micro-organisms—by bacteria, fungi, and viruses. In a field unmanaged by man, an equilibrium is achieved among all these, as well as insects and predators. But man-managed monoculture, with great acreages planted with a single strain of one crop, are far more susceptible to such infections, which are not tolerable in agricultural practice. Moreover, few useful therapeutic measures are either available or desirable in view of the low value of in-dividual plants. Accordingly, the success of monoculture rests on the breed-ing of resistant strains.

Although, even now, 10 to 15 percent of each major crop is lost to infectious disease, virtually the whole of American agriculture consists of plantings of strains especially bred for resistance to specific pathogens. In this way, the wheat crop was saved from attack by bunt (stinking smut) and rusts. Genetically based resistance to both of these diseases exists and is constantly being exploited. Varieties are bred that have resistance to the dominant strains of rust in a particular region. After a few years, however, mutant strains of the rust fungus develop that are capable of invading the wheat varieties in use. Meanwhile, other wheat varieties selected for re-sistance to the mutant rust are developed and introduced. There is yet little hope of breaking this cycle of resistant plant, mutant fungus, and back again.

Beans and cotton have been protected, albeit only in part, from fungi and root rot. The sugar-beet industry was almost abandoned because of the huge losses to "curly top" virus until resistant strains were developed. And the oranges of Southern California were almost lost to the virus causing "quick decline," which was transferred by aphids through the sour-orange rootstocks in common use. Only the last-minute discovery of resistant rootstocks saved this industry. Numerous other instances of the application of genetics to practical agriculture could be described, but these should suffice to indicate the scientific sophistication of these endeavors.

Agricultural Practice

Once a suitable strain of crop plant is available, adapted to local climate, and as resistant as possible to serious disease, successful agriculture then requires an adequate water supply, intelligent management of the soil, and minimization of ravages by pests. The soil is the farmer's principal resource, and its conservation is imperative. Optimal tilth depends upon a suitable combination of sand, silt, and clays maintained in miniature aggregate by the degradation products of plants formed many years before that lead to the formation of humus. The combination should prevent puddling or compacting, permit easy penetration by root hairs, water, and air. Salinity and lack of drainage must be avoided. Only recently has soil received the close attention it warrants so that crop production can be maximized by application of rather precisely formulated fertilizers, addition of nitrogen-fixing bacteria, pesticides, and herbicides. In combination, these have been responsible for the increments in crop yields of recent decades. Moreover, attention must be paid not only to the major minerals—potassium, nitrogen, phosphorus, calcium—but to trace elements as well. The nutrition of citrus trees growing on sandy soils in particular must be watched carefully. In many of the western states particular difficulties are met by assuring adequate iron supply for many crops and taking care with interaction of iron, copper, zinc, and phosphate nutrition. If legumes such as soybeans, alfalfa, or clover are raised, the presence of suitable strains of nitrogen-fixing bacteria—the *Rhizobia*—is necessary. Few agricultural triumphs exceed the rich reward gained by application of traces of cobalt to Florida grazing ranges. This metal, then found in the grasses, is utilized by rumen bacteria for synthesis of vitamin B_{12}. For grazing cattle, rumen bacteria are the only source of this vitamin, which is entirely essential to them as to all animals. The entire Florida cattle industry rests on the scientific detective work that elucidated the basis for the emaciation of Florida cattle in early years and provided this almost absurdly inexpensive solution.

Soils are living microcosms; if permitted to die, their useful rejuvenation is extremely difficult. Below the surface are bacteria, actinomycetes, fungi, and algae in prodigious numbers. An equal living mass of animals—nematodes, mites, springtails, earthworms, potworms, ants, insect larvae, and even the larger burrowing animals—cohabit this domain. Successful agriculture requires continuation of this equilibrium. But it can be seriously altered by monoculture; crop rotation was long ago recognized as a partial answer to this problem, although devised entirely empirically. It is an expensive way to use valuable land and today need be done only for specific reasons. For example, rotation with barley can be used to reduce the population of saprophytic fungi in the soil, which otherwise cause "take-all" disease of wheat, or rotation of corn with beans in areas of Michigan can be used to reduce the fungal population responsible for bean-root rot. But it is no longer necessary to rotate other crops with legumes in order to enrich the soil with nitrogen. When these considerations are added to well-understood aspects of plant physiology, which dictate the manner of seedbed preparation, use of irrigation, density of planting, and soil salinity and acidity, soil management becomes an increasingly scientific enterprise, which feeds us today and assures that we will transmit this paramount heritage, the soil on which life depends, to future generations.

Meanwhile, the standing crop must compete with weed plants and survive its predators long enough to come to harvest. For centuries, manual and then mechanical hoeing required much of the farmer's labor. The discovery of a cheap chemical analog of the natural plant hormone auxin, 2,4-dichlorophenoxy acetic acid (2,4-D), in 1941 ushered in a new era. In low doses, this compound is highly toxic to some plants and innocuous for others. Ideally, of course, the former is the weed and the latter the crop. A great diversity of effective compounds allows the ideal to be approached, particularly where a broadleaf weed infects a gramineous crop or a grass infests a broadleaf crop, as is common. A series of congeners have since become available to spare the farmer in this classical task, because spraying with appropriate herbicides, tailored both to the major weeds and the crop to be spared, is even more successful than mechanical procedures.

Similarly, recognition of the insecticidal properties of DDT in 1939, initially used against insects directly injurious to man, indicated that intelligent application of understanding of insect physiology, entomology, pharmacology, and the arts of the organic chemist could prevent crop destruction by insects. To date, the use of 2,4-D has increased yearly even though it has been replaced in part, and DDT is being withdrawn because of concern for its potentially adverse effects on man, transfer to the general environment, prolonged persistence, destruction of beneficial insects and possibly other wildlife, and stimulation of resistance in the target insects. These are now matters of broad general concern, and it is regrettable that public decisions must be made on the basis of our limited knowledge. But these compounds paved the way for modern agriculture. Without their equivalent, modern intensive agriculture is not possible, and, just as the continual breeding of new crop strains is imperative, so too is a continuing search for effective herbicides and pesticides, optimally with specific effects on offending organisms, degradable in the soil and nontoxic to man and animals. Attainment of these goals will require continuously increasing understanding of plant and insect physiology and life cycles.

Control of undesirable species by biological means is, in many ways, the most attractive possibility for future exploration. The notion is by no means new; attempts at such control began late in the nineteenth century. Indeed, some 650 species of beneficial insects have been deliberately introduced into the United States from overseas, of which perhaps 100 are established. These are now major factors in the control of aphids and a variety of scale insects and mealybugs. More recently, microbes and viruses have been considered for these purposes, a few of which are being used; for example, spores of the bacterium *B. thuringiensis* are used to control the cabbage looper and the alfalfa caterpillar. Some insects have been utilized for control of weeds—e.g., prickly pear in Australia and the Klamath weed in the western United States—while a combination of the cinnabar moth and the ragwort seed fly is required to keep down the population of the toxic range weed, the tansy ragwort.

Still more imaginative and dramatic are such special procedures as the elimination of insect species, e.g., the screwworm, by introduction of sterile males (sterilized chemically or by gamma irradiation), thereby eliminating this longtime scourge of southern cattle; the use of minute quantities of synthetic or natural attractants, combined with an insecticide, have been used to eliminate the oriental fruit fly; and the setting of traps with flashing ultraviolet lights, which reduced the population of tobacco hornworm in some southern regions of the United States. The sterile-male approach is now being attempted for control of other insects, including some fruit flies and such devastating species to man and stock as the tsetse fly. Despite the successes in control of crop diseases and pests, losses are still serious, as can be seen from Table 4. In the United States, much labor could be saved and product quality enhanced through better controls. In underdeveloped countries, the margin of safety between an adequate diet and malnutrition is so narrow that an unusual loss, as in a locust plague, is a disaster. At this time, it seems likely that current biological research will have some success even against the locust.

Each of these biological and chemical procedures has necessarily been the result of many years of investigation. Each is put into practice only when its consequences appear to be adequately understood. Yet each must necessarily alter the ecology of the affected region. When the pest species is successfully eliminated, some other species will probably take its place and may also require control. For the foreseeable future, the prospects are bright if the requisite research effort is maintained.

TABLE 4   Crop Loss Due to Pests and Disease

CROP	PERCENTAGE OF TOTAL CROP LOST TO			
	Diseases	Nematodes	Insects	Weeds
Corn	12	3	12	10
Rice	7	—	4	17
Wheat	14	—	6	12
Potatoes	19	4	14	3
Cotton	12	2	19	8

Source: *Scientific Aspects of Pest Control,* NAS Publ. 1402, National Academy of Sciences, Washington, D.C., 1966, p. 27. (Data from USDA Agricultural Handbook Number 291, *Losses in Agriculture.*)

Animal Science

Meat and milk are important components of the total food supply. They provide protein of high quality and make otherwise uninteresting diets attractive. The herbivores, moreover, harvest the cellulose of plants, which man cannot digest. They can graze sparsely vegetated rangeland and can be brought to very high efficiency through management and use of biology. In the United States, more than two thirds of the total crop production is fed to animals.

Cattle, hogs, and sheep in the United States, taken together, total about 180 million, and there are twice that many chickens. The domestic animals of the Western World are the highly specialized results of careful breeding and selection. Biological understanding has been crucial to achievement of the desired goals of these breeding programs and will continue to be so.

No enterprise applies more science to the problems of breeding, nutrition, disease, and economics than does the poultry industry. Genetic improvement of poultry has yielded superlative results and holds even greater promise for the future. There is no reason to believe that the growth rate, efficiency of feed utilization, meat quality, and egg-laying capacity have reached the highest possible levels. In spite of much progress in disease control and nutrition, the mortality of older fowl is often as high as 50 percent during the first laying year; since these animals are genetically capable of high egg productivity for two years, reduction of this mortality rate would yield great economic benefit. While chickens have been bred for high laying capacity and meat production, deliberate adaptation to diverse environments is only beginning. On a worldwide scale, geographical conditions of day length, temperature, humidity, and altitude warrant consideration in breeding programs if other nations are to share the boon of abundant inexpensive eggs and chicken meat.

In light of detailed understanding of chick nutrition and of the environment conducive to maximal growth and to egg laying, chicken growing has passed from an aspect of farming to a large industry, in which individual "chicken factories" grow tens of thousands of chickens simultaneously, each in its own enclosure, automatically fed, watered, and cleaned. In the most advanced practice, a computer program, containing the nutritional requirements of chickens and the composition of food grains, recomputes the most economic satisfactory mixture of cereal grains, based on daily or even hourly changes in grain prices and directs the mixing machinery accordingly. To be sure, this conversion to a chicken industry is not without cost. Such factories are deliberately located close to their urban markets. Grains are transported from their sources, but whereas the chicken farmer once returned the manure to the soil, it is uneconomic to transport manure back to Midwestern grain fields. The latter use chemical fertilizer while the manure accumulates outside the chicken factories, a problem not yet adequately managed.

When dependence on milk fat in the American diet was greater than at present, dairy-cattle breeders selected simultaneously for milk volume and high butterfat content. With increasing "calorie watching," avoidance of saturated animal fats and acceptance of products based on plant oils, breeding attention has turned to emphasis on protein quality and content. Many genes control the various characteristics of cattle and, unless their heritability is reasonably high, selective breeding presents difficulties that will be overcome only with expanded knowledge of the mechanisms of genetic regulation. Attempts have been made to transplant the highly productive dairy animals of temperate zones to more tropical climates, where animal productivity is low and the need for dairy products great. These have failed; even when the cows themselves thrive, their milk production is dramatically reduced. But there is reason to suppose that genetic understanding, coupled with improved awareness of the environmental factors and the physiological response of milch cows to increased temperature and humidity will surmount the difficulty.

Like cattle breeders, swine breeders have been forced to react to increased consumer demands for unsaturated fats and avoidance of hard animal fat. Breeding for large deposits of fat beneath the skin and within the abdominal cavity is no longer profitable because a large portion of the market in which lard once brought high prices has been pre-empted by vegetable oils. Now, limited fat content, maximum muscle mass, and larger litter size are the important considerations, but crossbreeding has been less successful than commercial breeders had hoped. Again, one must turn to research for a solution.

The sea is one source of food that man has certainly not exploited wisely or fully. Today about 200 species of fish are used in the human diet. Their protein content is high and of excellent quality, their saturated fats meager. With few exceptions, including trout, salmon, and shrimp, fish are not harvested or farmed efficiently. Fishing is still a form of hunting economy. If fish farming were to be undertaken on an extensive scale worldwide, governments would be forced to reach clear and workable agreements governing fishing rights and territorial waters. And the scientific stage has not yet been set. Knowledge of the ecology of aquatic regions is inadequate to sustain much more extensive fish catches. Fish rearing, currently practiced for freshwater forms, is more art than science. The behavior patterns of fish species will have to be established, with knowledge of breeding grounds, reproductive potential, feeding habits, natural diseases, and predators. The females of many species produce millions of eggs. If adequate protection could be offered to fry and fingerlings, decidedly larger catches seem possible, based on current estimates of primary photosynthetic production at the sea surface. There is great potential for raising and stabilizing production by protecting the young of food species from hazards during this critical period.

But before such potential can be exploited, more thorough exploration of ecological factors must be accomplished. The consequences of introducing into the natural environment large numbers of certain kinds of animals with a high survival rate are not well known. There is a growing awareness that, even in the sea, the consequences of man-made changes in abundance can be far-reaching and not always to man's benefit.

There are other inadequately exploited opportunities. Shellfish could serve as an important source of dietary protein if full advantage were taken of available estuaries and shallow bays. Marine-biology stations have been studying the life cycles, physiology, and nutrition of such species for many years. Were legal obstacles removed and were there a genuine will to achieve these goals, a substantial industry could be established and improvement in human nutrition could result. There is reason to believe that several food species can be grown in inland impoundments. The per acre yield of catfish protein in Arkansas ponds is more than 10 times the per acre yield of chicken or beef protein. Current studies offer the prospect of success of shrimp culture in saltwater impoundments in areas such as the Louisiana bayou country. In each instance, long years of study of these species have prepared the way for an attractive future. But the task is by no means complete. Aquatic animals also are subject to diverse diseases; knowledge of those diseases is almost trivial, and knowledge of their control is even scarcer. Increased populations of edible pelagic fish must alter the ecology of a vast region, but with unforeseen consequences. Transplantation of major species from the Atlantic to the Pacific—e.g., shad and striped bass—has only occasionally been successful; each such trial must be followed closely for its secondary consequences.

One opportunity may already have been lost—the opportunity to harvest oceanic mammals, particularly some species of whales. These have already been overhunted. But it may yet be possible to manage the supply of the greatest of all animals in such a way as to utilize them as a steady supply of high-protein food.

Finally, it must be remarked that the marketing of fish offers special problems. Fish spoil readily; thus, processing plants must operate close to the sources and inspection procedures must be rigorous. A long-sought development now appears close to commercial realization. Fish-protein concentrate can be prepared from a variety of "trash" fish. An almost tasteless and odorless powder, it can serve as an important supplement to the diets of millions of malnourished individuals at very low cost. The alternative, enrichment of cereal flours with lysine, tryptophan, and methionine, may yet prove to be economically competitive, but both must be socially acceptable to the affected people. Meanwhile, the scientific bases for both endeavors have been well established.

We cannot close this subject without drawing attention to what should prove to be one of the major events in the history of human nutrition, already well under way—the "green revolution"—which consists in applying breeding and management to production of the basic grain crops in underdeveloped nations. The bleak image of lone farmers gathering meager crops is being supplanted in some areas by scenes of abundant harvest. Two to three years ago, for the first time since 1903, the Philippines produced more than enough rice for its own people, utilizing a new, short, stiff-strawed rice carefully engineered to thrive in Philippine paddies.

In the same year, India harvested a landmark crop of wheat, some of it the product of a sturdy short plant, bred for adaptation to the Indian farming milieu. Much of the harvest came from the north, from the fertile Punjab, but throughout the wheat-growing area, some fields were cultivated in the short new wheat, which yields 10 times or more the harvest of traditional varieties. In Pakistan and Turkey, farmers encountered similar success. In southern India, some farmers planted the new rice and were rewarded with abundant yields, as were Filipino farmers. To be sure, both crops were aided by a year of abundant rainfall. But the remarkable success of the new strains was self-evident, and the Indian Ministry of Agriculture declared that the nation had turned the corner to modern agriculture and predicted optimistically that within a decade India might be able to feed herself. A few hundred acres of new wheat were planted in 1964. By 1968, more than 20 million acres were under cultivation and plans called for 40 million acres in 1970. Withal, this represents only a fraction of India's farmland and is but a start.

The new wheats and rices emerged from painstaking crossbreeding with available strains to provide seeds that carry the most advantageous characteristics and with new mutants, each being examined for new useful properties. These new grains demand careful nurture. Designed to resist lodging, they also require and can make maximum use of fertilizers and water. They must be planted at the optimum moment and carefully tended. And, as in American agriculture, if the "revolution" is to be sustained, it will be essential that in these regions there be a continuing program of plant breeding to replace current strains as they fall victim to infection or pests.

The revolution began 30 years ago, when the Rockefeller Foundation sponsored a program to improve wheat and maize in Mexico. Through these efforts, the average yield per acre of wheat had been increased more than threefold by 1964. In 1959, the Rockefeller and Ford Foundations decided on a similar effort with rice, the staple food in the Orient. To this end, they established an International Rice Research Institute in the Philippines. The successful new rice varieties were bred there, drawing on the background of genetic information that had been accumulated in this country and Mexico. Success of the new varieties depends on packaging selected seed, fertilizer, and pesticides so that the farmer has these essential inputs at planting time. Perhaps the most significant aspect of this "green revolution" is that traditional farmers have been shown what can be done and are thus receptive to the further changes necessary to extend and stabilize these advances. There is no better illustration of the contribution of biological science to human welfare.

MAN AND HIS ENVIRONMENT

Science is applied to human affairs through an increasingly complex network of technologies. Each new technology finds acceptance if, for example, it solves a problem, eases a burden, enriches life, assures the food supply, or facilitates communication and transportation. But each such beneficial technology must be examined for its potential social cost. In this connection scientists must be particularly wary of threats to the public health, to the fertility of the soil, to the quality of air and water, and to the security of renewable resources.

Perhaps 500,000 distinct chemical entities and mixtures are in current use and hundreds more are added annually. Each must be considered for its effects on the biosphere, particularly on man himself—effects that may be acute, dramatic, and self-apparent or extremely slow, difficult to detect, and even indirect. For these reasons an increasing force of trained scientists is engaged in these activities. The level and pace of such activity are patently insufficient to the national need. Such agencies as the Food and Drug Administration and the Fish and Wildlife Service are seriously understaffed relative to national needs. Let us consider only a few pertinent problems typical of this large and disparate field of concern.

Water Supplies

Although it may not long be true, most American communities may still boast a supply of biologically safe water for domestic purposes. The character of the situation is such that potential hazards must be avoided from the beginning rather than removed after their introduction. The life scientist must be aware of these hazards, establish appropriate monitoring procedures, be aware of indications in the community of failure of controls, and establish reasonable standards. Avoidance of improper metals is now a long-established practice, as are a variety of procedures designed to minimize the presence of pathogenic bacteria.

That human disease can be transmitted through the water supply has been known in a qualitative way throughout history. Yet specific understanding of its role in disease transmission goes back only to the past century, with investigations of the spread of cholera and typhoid fever. Cooperative work by engineers, biologists, chemists, and physicians on the organisms responsible for such diseases and on development of methods for their control was so successful that these infectious diseases have been virtually eradicated from developed countries in which adequate treatment and sanitary control of water supplies are maintained.

Water can be freed of pathogenic microbial agents by (1) protection of water sources against initial contamination; (2) removal of organisms by filtration, adsorption, or similar physical means; and (3) chemical destruction of the organisms. All three approaches have been put under stress by increasing population density and demand for water. Because of these factors, coupled with a concurrent increase in sources of contamination—i.e., human and animal wastes—completely uncontaminated primary sources of water are becoming difficult to find.

As knowledge of the factors that affect the survival of waterborne pathogens and their sensitivity to various forms of water treatment increases, new methods of water management may well emerge. Research is needed to provide sanitary engineers with a rational rather than empirical basis for the design of facilities for collection, treatment, and distribution of water.

It has become apparent in recent years that traditional control measures are inadequate to prevent the spread of certain viral diseases that may be waterborne. Epidemics of infectious hepatitis have been known to be caused by water contamination. A most dramatic example was the 1956 epidemic in Delhi, India, where nearly 30,000 cases of hepatitis resulted from a temporarily contaminated water supply. There are now about 50 documented instances of similar, although much smaller, waterborne outbreaks of this disease, several in the United States. Because the hepatitis virus has only recently been identified and no animal other than man is known to be sensitive to this virus, thus precluding an animal model of the disease, there has until now been little opportunity to engage in the necessary studies.

Other viruses are assumed to be waterborne but have not been shown to transmit disease. With the possible exception of the polio virus, confirming epidemiological patterns have not yet been established. New techniques are required to gain definitive information about the health hazard these viruses present and the necessity for measures to remove them from communal water supplies.

Increasingly, water authorities must be aware of chemical contamination at the source, particularly by materials entering through groundwater or washed from the air by rain. Such materials do not announce themselves; one must be aware of the problem, perform appropriate analyses, establish standards, and, when feasible, institute procedures for removal of offending chemicals. Such actions may be taken on the general principle that no contaminants are acceptable, but there is greater conviction and urgency when the biological effects of a given contaminant are known. Most noteworthy, perhaps, are the various agricultural chemicals—insecticides, herbicides, and fertilizers. The soil burden of nondegradable insecticides is already such that they will be leached and present in communal waters for years to come. Current levels are such that no general risk is known to exist. Similar considerations apply to the herbicides. Although its teratogenic activity has resulted in a ban, 2,4,5-T concentrations in communal water supplies are trivial. The problems presented by fertilizers and nondegradable detergents are somewhat more serious. Nitrate from fertilizers, leached from manure piles or from sewage-disposal plants, occasions methemoglobinemia (the iron of hemoglobin is oxidized to the ferric state, Fe^{3+}, in which condition it is useless for internal oxygen transport). Significant levels of methemoglobin caused by such contamination have been detected in various populations. No known deaths have resulted, and, in almost all such cases, it has been possible to trace the contamination and act accordingly. These incidents serve to point up the need for intelligent, biologically sophisticated management of water supplies.

Air

The intensities of air pollutants in American cities today may be no greater than they were 25 years ago, except for a few areas such as Los Angeles or New York on bad days. But techniques for measuring the levels of chemicals or particulate matter in the air, then as now, were less than adequate. The most significant change in this quarter-century is that the American people are beginning to demand a higher level of quality in the environment. Levels of air pollution are difficult to measure, but new, sophisticated instrumentation is being developed. The body of knowledge concerning the biological effects of known pollutants is increasing but is unconvincing. Experience has shown that, when meteorological conditions heighten the concentration of atmospheric pollutants, individuals suffering from chronic lung disease and perhaps those with cardiac disorders may have serious reactions. Such episodes have no observable effect other than discomfort in the average, healthy member of the population, the individual of most concern to environmental biologists trying to evaluate the potential for serious consequences of long-term low levels of exposure. The most evident aspect of acute episodes of air pollution thus far is the increase in airborne sulfur dioxide and other sulfur-containing products from the combustion of coal and fuel oils. Normally less than 0.1 part per million (ppm), concentrations during prolonged inversions may rise to 0.5 ppm or more. Such concentrations are harmful to the most susceptible individuals and discomforting for the remainder. The fact that no evident illness is occasioned in healthy members of the population should not lead to a false sense of security. Effects not detectable by current procedures are not necessarily absent. In a few cases, toxic effects of pollutants have been documented. Beryllium, for example, is known to have produced serious disease downwind of beryllium-processing plants. Pollutants of well-established biological significance, e.g., carbon monoxide and lead, are being added to the atmosphere in immense quantities, primarily from automobile exhausts.

It is clear that all urban dwellers, because of the relatively high atmospheric concentrations of carbon monoxide, carry significant amounts of carboxyhemoglobin but not in sufficient quantity to limit physiological function. In episodes of striking increase, the consequence is an additional pumping burden on the heart, of no account in normal persons but perhaps sufficient to lead to serious crisis in those with incipient cardiac failure.

From epidemiological studies, the connection between cigarettes and lung cancer, heart disease, emphysema, and other diseases is now known. The carcinogenic action of cigarette smoke has been confirmed experimentally in animals, although specific chemical toxins have yet to be identified. Inhalation of smoke by rodents does not cause neoplasia; however, painting the tar of cigarette smoke on their skin is highly carcinogenic. There is a possibility that smoke contains agents that are not in themselves capable of producing cancer but that promote the growth of tumors by somehow interacting with otherwise innocuous doses of carcinogens. The causal relationship to cardiac disease is not understood. Nicotine does increase the oxygen requirement of the heart, while carbon dioxide from the smoke reduces the available oxygen supply, but these effects seem too small to account for known effects. Attempts to separate the effects of cigarette smoke from those of more general air pollution indicate a much higher correlation between smoking and disease than between community pollution and disease. It seems quite conceivable that a combination of cigarette smoking and general air pollution accounts for the higher statistical incidence of disease in the smoking population.

The necessity for monitoring the quality of air will not be lessened. New technologies will pose new hazards, and existing technologies will be used on larger scales. Thus, it is anticipated that combustion of fossil fuels for generation of electricity will quadruple by the end of the century, while that for transportation will double. Thus the potential gain from use of more efficient, less polluting automobiles may be totally offset by the increased level of use, a phenomenon demonstrated in the Los Angeles area.

Food and Drugs

Mounting concern with the effects of myriad chemicals in the environment has brought under closer scrutiny the agents that are deliberately added to food. Approximately 1,700 food additives are in use in the United States, and an equal number of additional materials go into animal feed and packaging materials. Each agent is subject to regulation by the Food and Drug Administration, which issues specific requirements to define closely the allowable concentrations of some agents and maintains a list of others that are "Generally Regarded As Safe" (the GRAS list). The safety of many of these materials that have enjoyed long tenure on the GRAS list is predicated on limited examinations performed years ago and on their long and seemingly innocuous usage by the public.

Safety is no longer easily assumed, nor is it a simple concept. When cyclamates—noncaloric sweeteners—were first introduced in the early

1950's, their use was limited to individuals who used them to sweeten coffee or tea. A decade later, the diet-soft-drink era came into being, followed by a mushrooming of the diet-food industry. The very fact that cyclamates were then consumed by many people in substantial amounts generated concern. Experiments voluntarily conducted by a pharmaceutical house revealed bladder tumors in a group of mice that had been fed vast doses of cyclamate over their entire life-spans (the equivalent of several thousand sweetened cups of coffee per day for man!). Although there was no evidence of similar effects in human beings despite the huge scale of the human "experiment," according to the Delaney amendment to the Pure Food and Drug Act of 1958, any agent in food that causes cancer in any species, regardless of dose, must be banned. Cyclamates, therefore, were ordered removed from ordinary foods.

But this points up major difficulties. Where is the rational limit? How shall one balance the beneficial effects of voluntary caloric restrictions and avoidance of obesity by millions of Americans against the very remote chance of tumors in a few? By any seemingly reasonable standard, cyclamates had been adequately screened for toxicity until an almost absurd experiment was undertaken. It is noteworthy that, among the group of animals at half the tumor-producing level in the diet, absolutely no lesions were encountered! Quite conceivably, an equivalently rigorous and extensive review of the GRAS list will yield some similar experiences. When there is available a substitute that, by the same yardsticks, is innocuous, the course is clear. But when there is no substitute? This dilemma—risk versus benefit, and not necessarily to the same individuals—characterizes most major decisions concerning the environment. The difficulty cannot readily be mitigated, but certainly each such decision should rest on thorough understanding of the biological implications.

It cannot be assumed that toxicological data are adequate for many of the familiar chemical entities in the environment, let alone the scores of new ones. It was only 25 years ago that investigators learned of the effect of nitrate on hemoglobin. Recognition that cadmium in low concentrations in water may have adverse physiological effects and that some water supplies occasionally carry appreciable quantities of this element is even more recent. A Public Health Service standard for an allowable limit of cadmium was not set until 1962.

The task of the toxicologist is complicated by the fact that what is usually required is an analysis not of the acute effects of large doses but of the effects of very small doses accumulated gradually, of variability of response within a large population, and of the effects of other environmental variables and of disease. If there is evidence that a toxic compound accumulates in the body and that no tolerance develops to it or that its effects are irreversible, that agent is more menacing than one that can be detoxified or readily excreted or whose effects are reversed when it is removed from the environment.

The interaction of chemicals is often difficult to determine. The interaction of trace amounts, difficult enough in themselves to detect, compounds the difficulty, but can be of critical importance. A few years ago, it was found that when malathion and EPN (ethyl nitrophenyl benzytriphosphonate), both organophosphate insecticides, are fed together to experimental animals, the toxic effects are considerably greater than the sum of the toxicities of both chemicals. In consequence, the Food and Drug Administration issued a regulation requiring that each new organophosphate be tested jointly with every organophosphate already approved before the new insecticide is cleared for sale. The pyramiding of tests that this would engender is apparent. Relief came when the basis of this hazardous interaction was elucidated, thus permitting use of simpler means for predicting dangerous combinations.

Both of these insecticides are toxic because they inhibit cholinesterase, an enzyme essential to normal functioning of the neuromuscular system. Alone, malathion is only mildly toxic to mammals because it is itself destroyed by another group of enzymes, the aliesterases, before it can render extensive damage to cholinesterase. The aliesterases, however, are inhibited by EPN, thus opening the way for the total toxic effect of the malathion that accumulates. Understanding of the ·underlying mechanisms in this situation has permitted direct measurement of the effects of new pesticides by testing them against appropriate enzyme systems, offering a rational approach to the design of new pesticides.

The need to overcome similar problems in testing procedures and to acquire the ability to predict adequately what will occur in given situations is evident. Toxicologists, aware of their own limitations and responsible for protecting the public health, would have to lean on crude and cumbersome procedures to avoid any uncertainty about the safety of products they stamp with approval. But even excessive caution cannot guarantee safety if the substance of fundamental biological knowledge is inadequate. Results of extensive testing on animals may not be applicable to man. Thalidomide only rarely deforms unborn rats though it consistently deforms human beings. Relatively early identification of the effects of this compound in man must be attributed to the fact that it results in a deformity that is so rarely seen ordinarily that the problem was readily evident. Phocomelia, an anomaly in which the limbs fail to develop, is so rare a congenital abnormality that it immediately aroused suspicions of some environmental agent when the deformity began to appear in a relatively small number of infants. However, an agent without effect on experimental animals, but which induces diabetes in man, for example, would probably go undetected for a long period.

Extrapolation of animal data to man must be done with the utmost care and caution. One study comparing the reactions of rats, dogs, and human beings to six standard test drugs revealed many similarities, but of 86 distinct recorded effects, 33 appeared only in man! In the final analysis, after careful and extensive animal experiments have been completed, controlled human trials are imperative to measure the full range of a drug's effects. But, at present, these should be undertaken only after extensive trials with animals, tissue preparations, and, when appropriate, enzyme systems.

In the end, whether our concern be with drugs, food, or the physical environment, the hard question is what the American public is willing to pay for. Monitoring the environment while insisting on the right to drive one's own car is a costly matter. The more rigorous the standards, the more costly it must be. Similarly, the only effective approach to a multitude of disorders to which man is subject is the development of new drugs, which, in our society, is largely the function of the pharmaceutical industry. If these are to be thoroughly tested for safety and efficacy before they are marketed, the public must be prepared to bear the costs, not only of the marketed drugs, but also of the studies that discard those chemical entities that prove either unsafe or inefficacious. The biological capability, thanks to years of fundamental research, is well established. Although much yet remains to be done, a national capability for maintaining the human environment is attainable—providing we continue to train the manpower and bear the costs. What other alternative is acceptable?

RENEWABLE RESOURCES

The biological and physical elements of the earth are vital to man. Soil, water, air, and populations of plants and animals can, under certain conditions, be used over and over again. These are man's renewable resources, and their sound management has become a prime concern to man, both for his well-being and, perhaps, for his survival on this planet. The greatest single threat to environmental resources and to man himself is his own "population explosion," with the concomitant rising pressure on food, land, and water needs. Only by understanding the function and interaction among biological and physical elements of the environment and applying that understanding to the management of resources can man control his numbers and keep his environment livable.

Although the major portion of man's food comes from only about 100 species of plants and animals, many thousands of species, including microorganisms, interact to provide the environment required by these major food sources. It has been estimated that at least 150,000 plant and animal species in the United States are involved in the collection and transfer of the sun's energy for the maintenance of life. In addition, some of these species are decomposers serving to break down waste products and dead organic material to make such essentials as carbon dioxide, nitrogen, and other elements available to plants for reuse and transmission to animals through the food chains of the biotic system. Beyond these material needs, living organisms are important in fulfilling esthetic and recreational needs.

Living systems have evolved for many millions of years to become a part of the environment as we know it today. Although civilization has developed throughout history at the expense of natural resources, population growth and technological achievements in the twentieth century have produced a disruptive assault on the environment on a greater scale than ever before. Contamination emanating from technological developments and urban concentrations has altered the chemical and physical characteristics of our seas, lakes, rivers, soils, and air. While simplified food chains have been exploited on some land to satisfy civilization's requirements for food and fiber, other vast land and aquatic areas have been developed for uses not associated with biological production. Economists project that within the next 20 years some 28 million acres (an area larger than Ohio) will be converted into urban areas and highways in the United States; four fifths of this land will come from croplands, pastures, and forests. Poor management in the past has resulted in loss of a third of the topsoil in the United States, with consequent lowered potential productivity.

It is difficult to return land to cultivation once it has been built upon. The quality of some of our surface waters and groundwaters can be restored, but, with the knowledge currently available to us, the results of pollution can be reversed only at great cost. For example, even if the introduction of fertilizing nutrients is terminated and if the waters of a historically heavily polluted lake can be completely exchanged over a period of time, the enormous amount of harmful matter bound in the bottom mud may continually replenish the pollutant materials.

How much more can we abuse our renewable resources—how much area can we remove from production, how many species can we destroy—before our resources will be unable to support man in an environment of acceptable quality? These crucial questions need answers now before renewable resources deteriorate irreversibly to an unacceptable level. We must maintain a continuing assessment of our renewable resources—land, water, air, and living things—because their status constantly changes. Only with such information can we find new and better ways to ensure their continuing availability.

Role of Science in the Management of Renewable Resources

The observations of early naturalists made important contributions to understanding of the environment. More recently, studies by pioneering systematists, geneticists, physiologists, evolutionists, and morphologists have provided much information of value to problem-solving ecology, although the significance of their contributions was not recognized for many years. During the past half-century or so, ecologists have searched for the principles underlying the interacting relationships of living things and their environments. Gradually they have come to recognize the complexity of these interrelationships and have categorized the influences on them as physical, biological, and, in some instances, social and cultural. Modern concepts of these dynamic arrangements recognize the constant interaction among all factors that make up the ecosystem.

Detailed studies of ecosystems or communities provide impressive demonstrations of mutual adaptation of species to one another and to their physical conditions. Host and parasite, prey and predator, and herbivore and plant are integrated in their life histories and requirements. These conditions can be understood in the light of modern evolutionary theory, which is based on genetic variability and natural selection and provides a satisfactory framework for understanding the diverse characteristics of the biological world. In this area, understanding and appreciation of population genetics is most critically needed. Understanding of the principles of natural selection is essential to the intelligent management of renewable resources, which almost always involves manipulation of populations by methods that depend heavily on selection of genetic traits governing such group properties as productivity, longevity, and reproduction rates. Ability to predict results will increase as more is learned about the mechanisms involved, both in the individual and in the interaction of populations.

Living organisms depend upon and are influenced by the physical and chemical elements of their environments. At the same time, they perform certain functions that are requisite to the structure and behavior of their physical environments—e.g., production of oxygen by plants through photosynthesis. Thus, understanding of the biosphere requires information about the physical nature of the environment (geology and soil science), the transport systems that move substances to and away from living things (meteorology and hydrology), the transformations that take place in the nonliving parts of the environment (physics and chemistry), and the means of modifying the environment (engineering, including weather modification).

Principles of Management

Rational plans for managing an environment either intensively (as in agriculture) or less intensively (as with wildlife) recognize that every area has a certain set of characteristics, that each living organism has a certain range of physical conditions that it can tolerate, and that for each physical condition there is some point or zone within the range that is near optimum. Organisms are aggregated into communities, the members of which are determined equally by their common ability to tolerate the physical conditions of the site and by their interactions with the other members of the community. The relationship is not passive, for the organisms in turn interact with and may change the site. Their tolerance levels are not necessarily identical, but they may overlap in the range of conditions present on a site. As conditions change, new forms, with tolerances that fall within the new ranges, may become a part of the community; some of those originally present may be eliminated. The less rigorous the conditions of the site the greater will be the variety of niches and inhabitants.

Two basic courses are open to us in using our surroundings: We can adapt our needs and demands to the capabilities of an area, or we can modify the area to change its capabilities. Urban and regional development, waste disposal without overloading the water or the air, and some recreational pursuits are examples of the former.

Environmental Management

AGRICULTURE

Agriculture has evolved beyond crop culture to become an environmental technology with emphasis on the management of land, water, air, and biological resources for the production of food and fiber and for the preservation of natural resources. The successful farm or ranch is, in fact, a well-regulated ecosystem in which renewable resources are effectively conserved. More than ever before in man's history, it is imperative to develop the technology by which agricultural practices can more effectively conserve our vast land, water, and biological resources.

Through sound management, agriculturists have been successful in making permanent use of renewable resources, especially land. In many places, the quality of the resources has been improved by careful use and management, with resulting increases in production and income. For example, in studies of individual farms in Illinois, yearly investments of about $35 per acre in conservation practices for soil and water returned about $41 per acre per year. Similarly, land that had yielded an average per acre of 15 bushels of corn, yielded 304 bushels per acre after six years of effective rotation and cropping practices. This kind of management makes possible the continuous and efficient use of the same natural resources year after year. In coming decades, with expanding world population, this aspect of conservation will become even more vital.

In sharp contrast is the unsound use of renewable resources that has led to disasters of the magnitude of the "Dust Bowl" of the 1930's. Before settlement in the 1870's and 1880's, the Great Plains had been protected against erosion during periods of drought by the natural cover of the short grasses. The first white settlers cultivated the land for wheat and in doing so destroyed the protective natural sod, exposing the bare soil to wind and other eroding forces until the soil structure was broken down. Thus, when severe droughts came in 1930 and 1931, soil conditions were ripe for devastation such as had never before occurred in the Great Plains areas. The Dust Bowl, involving 100 million acres, was a costly lesson to American agriculture; as a result of it, the Soil Conservation Service was formed in 1935 to devise and encourage sound land-management techniques. Only a small fraction of the Dust Bowl has been returned to production. Soil-conservation practices (contour farming, strip-cropping, rotation) illustrate effective use of applied ecology to maintain and even improve soil resources. Other ecological principles have been employed to increase crop and animal production but often have not been extended far enough to protect our renewable resources.

Water will always be a precious resource. In agriculture, much ground-water and surface water is lost or polluted by current practices. Manure, silts, and pesticides are some of the most serious pollutants. It is estimated that a fourth of all water stored for irrigation is lost by evaporation before use; yet water use in agriculture is increasing. Research has begun but more is needed to find ways to reduce evaporation of water in storage; some new chemical films offer considerable promise under special conditions.

Control of transpiration by plant hormones also offers a real opportunity to conserve water. This problem is well illustrated by the fact that, of the 500,000 gallons of water absorbed by an acre of corn in Illinois in one season, 498,750 gallons are lost to the atmosphere by transpiration.

FORESTRY

Forestry deals with the management of wooded lands for various goods and services. The term "wooded lands" is liberally construed to mean forest landscapes, including areas of alpine rockland, native grass, brush, and swamps. Such areas often influence management of adjacent lands. Lumber production is the principal objective of most large corporate ownerships, but water yield, watershed protection, recreation, grazing, and protection of wildlife and scenic values are explicitly recognized on many private lands and are primary aims in the management of most public holdings.

The obvious economic value of lumber has led to a tendency among many conservation writers—including some foresters—to equate forestry with timber production. A century and a half of historical development, as well as present-day practice over large areas, has emphasized game production, stream flora, steep-land protection, and nontimber products. This emphasis finds its modern expression in the "multiple use" doctrine, which Congress has now declared to be the guiding principle for some 180 million acres of national forest. It is likewise espoused in varying degree by many public and private forest landholders. For example, revenues from hunting-club leases approximately offset land taxes on some industrial forest holdings.

In forestry practice, biology is the major but by no means the exclusive scientific tool. The earth sciences (geology, physiography, hydrology, climatology, and soil science), engineering, and a large economic, social, and managerial component often dictate the framework for biological applications. Protection from accidental fires has been the *sine qua non* of forest management through much of North America and necessarily absorbs a substantial part of the resources and technical effort devoted to forest land. The protection, manipulation, and efficient use of vegetation are the dominant aim of most forestry activities. Hence, an understanding of the dynamics of this vegetation and its associated populations of animals interacting with the physical environment is the forester's primary tool.

The need for applications of biology to forest-land management are more readily appreciated in view of the very different levels of management currently practiced. The most extensive management for wood products is simply exploitation of useful trees, usually with protection against severe fire and pests, with the hope of natural renewal. The input of biological skill is minimal, and the results range from excellent, as in many of the eastern Canadian spruce fir pulpwood cuttings, to the destruction of the resource. As intensity of management increases, measures such as restricted harvesting, timing of operations, prescribed fire, and thinning are employed to favor reproduction and growth of desired species and reduce competition with less valuable species. Such measures may be insufficient to perpetuate recalcitrant species, such as the American bald cypress or the New Zealand podocarps, whose regeneration requirements are neither met nor understood. At the highest intensities of management, desired strains are planted or otherwise made dominant, and density and structure as well as composition of the forest are closely controlled. For this purpose, the environment is modified by reduction of competition and pests, and sometimes by soil treatments.

At the lowest intensity of management, reliance upon natural processes is complete. At intermediate intensities, great dependence is placed on understanding the requirements of individual species, their competitive positions, and the nature of successional trends that may be either reinforced or combated. This concern diminishes at the highest intensities as regeneration, composition, and density are brought under control, with marked reduction in age and species diversity. Attention then shifts to altering genotypes, additional manipulation of soil and plant features, and specific measures against injurious insects and diseases.

FISHERIES

True aquiculture, with complete control over all phases of a tended organism's life cycle, including a well-regulated harvest, has only regional importance (e.g., carp in the Far East and Israel, trout in North America and Europe). Fishery resources range from marine algae to whales and from the brook trout of alpine streams to benthic crustaceans at 200 fathoms in the sea. With few exceptions, fisheries are restricted to the lighted zone of the waters. Considering the gamut of aquatic plants and animals, the important species harvested are relatively few in number—about 200 among the over 20,000 kinds of fishes—and far fewer algae, mollusks, crustaceans, or aquatic mammals. Only 12 of these constitute 80 percent of the total catch.

Management for sustained yield is based on several factors, including (a) information about the stocks or populations and subpopulations that are often the effective breeding units (knowledge of age, growth, fecundity, longevity, and mortality due to natural causes and to exploitation); (b) information about the taxonomy, life histories, and behavior of the species under natural conditions and when confronted with capturing tools (included here are foods; food habits; sensory capacities; territorial or schooling behavior; knowledge of the action, including selectivity, of the capturing gear); and (c) information about the environment and the influences on the stocks of such variables as temperature, salinity, currents, and pollutants.

The deficiency in information needed for the adequate management of aquatic organisms can be ascribed to (a) lack of planning and failure of political boundaries to correspond to biological boundaries, (b) the short duration of studies in relation to the time span over which natural forces act and in which natural fluctuations take place, and (c) lack of funds, personnel, and interest.

The intensity of management measures applied to living aquatic resources decreases as the population dispersion and area occupied increase. Carp and trout, with their tolerance of confined freshwater areas, can be bred and tended intensively like domesticated animals with good control over their environment, while we can do little or nothing in the vast marine areas required by tuna or herring. With fishes of the latter type, management now depends upon prediction of population levels and controlled harvest. In the case of tuna in the Pacific, enough is now known about the relationship between tuna distribution and environmental conditions to permit satisfactory forecasts of distribution of tuna stocks several weeks in advance for the benefit of fishing fleets. Recent advances in tracing the life history of salmon at sea, coupled with detailed simulation models of the fishery, including the freshwater phase, provide an improved basis for prediction. Manipulation of spawning areas certainly provides opportunity for genuine management.

Management possibilities and the impact of man differ in the various regions of the hydrosphere. Fish and mammal stocks in the high seas can be managed only if characteristics of population and environment are known. Research on the high seas should be international in scope. Regional fisheries councils facilitate pelagic fisheries research. Agreements on apportionment of harvest through exclusive or joint exploitation are feasible. However, the common-property nature of high seas resources makes enforcement of harvest limitations difficult (e.g., only about 1,000 blue whales exist today). Furthermore, catch limits can be quickly filled with modern mechanized gear, and this leads to difficulties in keeping vessels and manpower profitably occupied, a problem encountered with the tuna stocks of the Western Pacific.

Offshore fish resources are most important in relation to bulk and dollar value. Such fishes as herrings, sardines, anchovies, and ground fishes (flat fishes) occur in abundance in various seas—the North Sea and the Caspian Sea for instance—and on the west coasts of certain continents, where currents and winds stimulate the upwelling and mixing of nutrients. Geographically, this region coincides with the continental shelf and overlying waters. Management in these areas, like that of the high seas stocks, must rely on regulation of gear and times of capture. More intensive methods of management are not presently feasible. There exist here common-property-resource problems that can be solved by bilateral agreements (e.g., Canada and the United States in the halibut fisheries). More and better agreements of this kind are needed; some may require new legal concepts because of the impending exploitation of this zone for other resources (minerals).

Near or inshore resources are often concentrated in shallow waters or near deltas and estuaries, or are associated with coral reefs. They are exploited by operators of small craft who, throughout history, have made up the bulk of the world's fishermen. Of the ocean environments, the inshore resources are most susceptible to overexploitation and to environmental deterioration caused by man. Along both coasts, estuaries are being filled with wastes at an alarming rate by industrial and housing developments. In addition, streams and rivers dump pollutants, collected from their drainage basins, into the estuaries. All this has already altered the ecology of these regions. Now many of these inshore resources will be subjected to further change by the addition of heated effluents from both nuclear and fossil-fuel power plants along our coasts. These plants require enormous quantities of water for cooling, and the low-grade waste heat carried by this water will also be enormous. If, for example, sufficient combined nuclear power and desalting plants were constructed on the West Coast to meet the needs for both fresh water and electric power, the rise in temperature of inshore waters might be as much as $4°$ F. Such a change in temperature would certainly alter the kinds, distribution, and abundance of animals inhabiting this area. We must be mindful of the opportunities to modify parts of this environment beneficially with this vast resource of low-grade heat, which could, with careless use, become a destructive pollutant. There is an attractive alternative, however—the use of this heat to maintain the environment of aquatic species that flourish in warmer waters, as has been done in England for cultivation of plaice.

The inshore marine environment, together with freshwaters that support the sport fisheries, suffer most from man's activities. Ecological imbalances resulting from events on the land are often difficult to correct, mainly because of traditional divisions in jurisdiction over and management of land and water. Authorities entrusted by society with the management of inshore waters have few or no organizational ties with those who determine land use, location and operation of industrial enterprises, and urban development.

Sport fishery in inland waters is strongly selective of predaceous fish (e.g., bass and pike) near the top of the food chain that constitute a small fraction of the total fish population available for harvest. Commercial fishermen stop fishing when it is no longer profitable. Anglers continue fishing for the large fish even though the numbers of fish decline and they have small chance of success. While the demand for large sport fish increases, the supply is limited. It may be preferable to modify the life habits of their predators, the anglers, so as to conserve the fish and their environment.

WILDLIFE

Wildlife may be defined as wild plants and animals in their natural environment (though often only animals are considered and here we consider mainly birds and mammals). The purpose of wildlife management is to maintain desired populations of wildlife. Wildlife management includes production and harvest of game species; maintenance of nongame species; and control of damage by wildlife to crops, forests, range, livestock, or human life. Management techniques have developed in a historical sequence that began with restrictions on time or methods of taking game, later included predator control and refuges, still later moved to artificial replenishment, and finally incorporated environmental manipulation.

Even though knowledge of habitat manipulation is considerable, we still rely on seasonal and bag-limit restrictions as the principal management measures for game species. Artificial game propagation attracts much public interest, but most wildlife biologists have come to regard this practice as better suited to intensively managed, private or commercial shooting preserves than to public hunting areas.

A great deal of the effort of fish and game agencies is directed toward gathering information on mortality, natality, and welfare factors that will be integrated to form the basis of the annual announcements concerning time and limits of harvest. Judgment gained from decades of trial and error still weighs heavily in the interpretation of field data; increasingly, however, sophisticated techniques are being employed. For example, in big-game management, many states conduct an annual survey of sex–age distribution in the population as well as estimating productivity from fawn–adult ratios and other population indices. Additional information on ovulation rate, placental scars, and weight and condition of carcasses is gained at hunter-checking stations. On many big-game ranges, annual surveys of forage are included as part of the information needed to establish the recommended harvest.

Habitat manipulation is potentially a far more responsive tool for managing areas intensively than is the regulation or restriction of harvest. Unfortunately, the manipulation of habitats is not feasible on some public and private lands, where other uses have high priority. The most successful widespread use of habitat manipulation came as a result of investigations in the fire ecology of the pinelands of the Southeast. Scientists have developed a high degree of skill in the use of fire in that region to manipulate forest communities for maximum wildlife production combined with timber or pulpwood production.

Among wild terrestrial vertebrates, particularly birds and mammals, much of the descriptive work at the species level has been accomplished, but many species occupying important niches over large areas are still little known. For example, until the appearance of a recent monograph, the mountain gorilla was largely a creature of mystery and misunderstanding. Similarly, the Wilson's snipe, an important migratory game bird in the United States, was largely unknown until a thorough field study of this species was completed recently. There are gaps in knowledge of some of the dominant members of the widespread communities in North America. For example, there is little knowledge of the actual effects of weather on deer, field voles, cottontail rabbits, and upland game birds, and of the physiological and behavioral adaptations of these species. The interactions among closely related species also need much more study. For instance, the effect of an expanding starling population upon other cavity-nesting species and the role of the starling as a vector of domestic-animal diseases should be studied more thoroughly.

The use of electronics, telemetry, and photography in remote sensing offers opportunities for real gains in dealing with wildlife problems. Research using some of these capabilities is under way, but progress is considerably short of what seems possible. The space program may launch a satellite with some components suitable for use by wildlife ecologists; however, many more applications are immediately feasible. For example, there are now microtransmitters with sufficient lifetime to permit following waterfowl or seabirds through an entire pattern of seasonal migration. With receivers that could be mounted in present satellite packages, continuous surveillance could be maintained on a sample of migrants fitted with microtransmitters. The same technique seems promising for marine mammals, large terrestrial predators, and wide-ranging ungulates.

Perhaps one of the greatest shortcomings in application of existing knowledge is reflected in the harvest of deer and other large ungulates. Satisfactory inventory techniques have been developed, but the public seems unconvinced of the high productivity of healthy deer in favorable habitats and fails to realize the resilience of a thriving deer population. Hunters, especially in the Northeast and Lake States, cling to their ideal of "bucks only" and frequently refuse to support a more flexible policy. The resulting underharvest of big-game herds has resulted in semipermanent damage to millions of acres of overutilized range.

Another area of confusion in applying research findings is in the control of pest animals. Numerous investigations have questioned the wisdom of pursuing traditional statewide predator control programs with little evaluation of either the need for the program or the effectiveness of the control effort.

Excessive populations of deer and elk are a nagging problem in national parks and on large military reservations, where hunting cannot be used to achieve population reduction. In these situations the use of chemosterilants offers promise of being a highly effectual technique. Considerable experimental research has already been done using these compounds on feral pigeons, gulls, and carnivores. This pattern of applied research should be extended to ungulates. Furthermore, increased effort in reproductive physiology would greatly enlarge our understanding of the effectiveness of antifertility compounds.

The effects of environmental pollution on wildlife is a subject of some importance. While the task of measuring direct effects of new spray materials is demanding, the subtle, pervasive phenomenon of bioaccumulation is of greater importance and is much more difficult to evaluate. The first step is to work out the pathways of pesticide-residue transfer and accumulation. The uptake, metabolism, and storage of pesticides are obvious objects for physiological studies that would support this effort. The ultimate fate of pesticide residues would be much better understood if concepts of major drainage basins as ecosystems were more clearly defined and described. The monitoring of pollution loads would be greatly expedited by advances in analyzing ecological systems.

RECREATION

Provision of adequate opportunities for outdoor recreation requires an understanding of the needs and desires of the potential participants, the kind and location of environment that will meet these needs, and the effects of use on these environments. Until we understand better why people seek outdoor recreation and the motivation that determines their recreation choices, and until there is general awareness of the deleterious effects of recreational activities upon the natural scene, much restorative effort will be of the stopgap variety. But treatment of the symptoms does not identify and eliminate the cause; thorough knowledge of the physical and biological components of the recreation environment is imperative. Equally necessary, however, is a deeper understanding of human behavior.

In many instances, biologists and other recreation-resource managers have not considered the visitor and the resource to be part of the same ecological situation. The use of the term "visitor," in this case, may be unfortunate. Nonetheless, at a time when perpetuation of the resource depends in part upon the visitor's understanding and cooperation, he and his fellow citizens, many of them urbanites, seem uninformed and careless about soils, plants, animals, and their interrelationships.

The recreation visitor and his activities influence not only the immediate site being occupied but also the adjacent areas that form the scenic backdrop. His presence may generate problems beyond those that already exist. Wilderness areas and national parks are good examples in which visitor-recreation problems extend beyond the immediate site being occupied. Although many such areas are more than several hundred thousand acres in size, the direct physical contact of visitors is concentrated on a very few acres. Within these small areas of intensive use, vegetation is trampled,

soils are compacted and eroded, and water supplies are subjected to pollution. Overuse and abuse—albeit unintentional—prevail. Further, these
sites of intensive use are often in aesthetically pleasing but fragile areas least
capable, biologically and physically, of withstanding great visitor pressure.
To a certain degree, the selection of such sites is a result of uninformed
management or poor planning of land use. There is some evidence, however, that these are the kinds of areas that many visitors—the recreation
public—prefer. To be sure, some persons wish to camp in relatively isolated sites in more stabilized vegetation systems regardless of the lack of
modern conveniences. However, most recreation campers prefer to con-
gregate in high-density campgrounds where electricity, sanitary facilities,
hot and cold water, cooking accommodations, and other refinements are
available, and where the vegetation is in a highly vulnerable, unstable
stage of development.

In periods of peak activity, present ability to handle the masses of out-
door-recreation enthusiasts is rapidly becoming quite inadequate. The
number of visitors to our national parks and national recreation areas begins
to pose a serious problem (Figure 32). The visitor load in these public
areas has increased nearly sixfold in 20 years and in 1968 was 151 million;
the number of parks has increased at a much slower pace. Indeed, the
number of units in the national park system, including national parks,
monuments, seashores, and historic sites, has increased only 50 percent in
this period. Potential sites for additional large-scale and magnificent recrea-

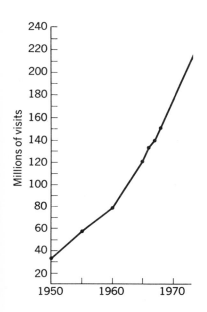

FIGURE 32 Total visits to national parks
and related areas, 1950–1968, and projected
visits to 1975. (From *Biology and the Future
of Man,* P. Handler, ed. Copyright © 1970
by Oxford University Press, Inc. Data from
Statistical Abstract of the United States, 1969,
Statistical Information Division, U.S. Department of Commerce.)

tion outlets are not unlimited. If visitor pressure on national forests, other wilderness areas, and state or local recreation facilities follows the general pattern experienced in the national parks, and if there are not substantial changes in the concepts of visitor management, we are clearly in danger of running out of space for certain types of recreational activities.

On wild lands managed for several purposes, the need for both more thorough ecological understanding of the landscape and greater insight into the physical requirements of an attractive landscape is coming into sharp focus. Much of the public seems more concerned about the "visual resource" than about the physical resource. The outdoor-oriented American public evidently does not wish to become reconciled to the fact that natural processes must sometimes be accompanied by temporary ugliness. Yet good silviculture may entail controlled burning to permit regeneration of more desirable trees, burning to maintain a plant community characteristic of a true prairie, reduction in an elk herd to forestall starvation of the animals and destruction of their range, or introduction of native predators to assist in the control of big game or other animal populations.

URBAN AND RURAL DEVELOPMENT

Man needs dwelling places, stores, industries, schools, museums, places of worship; he needs arteries for transport by rail or motor vehicle; he needs airports, canals, harbors, and dams. Once land is committed to these uses, the commitment is essentially irreversible. Man's activities in these places change raw materials and natural products into new forms, often resulting in waste products that must somehow be disposed of or recycled. When these waste products reach the air or water or land in forms or concentrations that are detrimental, they are "pollutants." Unacceptable means of waste disposal are the cause of one of the major impacts of man on his environment.

With increasing numbers of people, needs for food, fiber, industry, and transport—indeed for all kinds of goods and services—increase. Expansion of our cities converts more than a million acres of land a year to paved, biologically unproductive areas. At the same time that command of enormous amounts of energy for excavation, construction, and earthmoving gives ever greater freedom of choice in the location of cities and changes in the landscape, the changes are, all too often, unplanned and unthinking. The big changes—canals and dams, perhaps even interstate highways—are considered with some care, and the more obvious costs and benefits publicly weighed. The results are not always those biologically most desirable, but they are, for the most part, democratically acceptable. The more pervasive and uncontrollable changes result from incremental changes, as in creeping suburbia and filled-in wetlands. No single acre in these latter categories elicits much public defense, but the aggregate loss exceeds what we should be willing to accept.

In urban renewal or modification of existing metropolitan areas, the problem is to make the "best" use of the area. Zoning is useful to this end. Heavy, dirty industry can be positioned in relation to dwellings, open space, and other living parts of cities so that air pollutants are carried away, noise does not reach the dwellings, and offensive odors and the grime of industry are out of range of the senses of most inhabitants. Some trees, shrubs, and other plants will tolerate even existing congested and polluted conditions, and can thus be used for beautification. Transportation and communications systems can be planned so as to minimize conflicts among the diverse demands of metropolitan life.

The imminent location and construction of whole new cities affords both superb opportunities and difficult challenges. Before the turn of the century this nation will need to provide housing for an additional 100 million people, or a population equivalent to the sum of 500 Restons, 100 Columbias, 50 Atlantas, 5 Philadelphias, and 5 New Yorks. This new housing may either sprawl and congest the surroundings of existing cities or start afresh in entirely new locations. Much less concern will be necessary than heretofore with the needs for transportation and communications or nearness to primary resources. Locations can be based upon the amenities; water, raw material, and various modes of transportation can be brought to them as required.

Whether present cities are expanded or entire new ones built, it is imperative that their effects on the environment be considered. Paving of groundwater recharge areas, scalping of steep slopes, and placement of septic tanks in impermeable soil can all be avoided. Waste-treatment facilities can improve rather than damage their surroundings. Planners, architects, and engineers will be largely responsible for appropriate use of the environment. Development of understanding of land-use capabilities and a reciprocal interaction between the desired design and land-use-capability criteria will permit optimum use of the environment.

Local governmental and federal agencies should recognize a public right to live in an environment of acceptable quality. The true costs of any program in the management of renewable resources, be it in industry, agriculture, recreation, health, forestry, fisheries, or urban development, should be evaluated, and decisions should be made, upon the advice of groups of specialists, by representatives of society as a whole, seeking what is best for local, continental, and planetary ecosystems. Only by knowledge and understanding of the function and interaction of the biological and physical elements of the environment and by application of this knowledge and understanding in sound management programs can man expect to conserve the natural resources that are his great heritage.

INDUSTRIAL TECHNOLOGY

Biological science finds application in many aspects of the economy. There is no precise biology-related equivalent of the chemical or electronics industries, which bear one-to-one relationships with specific areas of science. As we have seen, agriculture, medicine, protection of the public health, and conservation of renewable resources all directly apply increments in biological understanding. Here we shall indicate briefly a few additional industries whose capabilities, scale, and quality rest on applied biology.

Pharmaceuticals

Reference has repeatedly been made to the role of the pharmaceutical industry; suffice it to note, then, the magnitude of this industrial endeavor. In 1967 the industry had gross sales of about $4.2 billion. Of this, 10.5 percent was allocated to its own research and development programs—about one fifth of the nation's entire biomedical research enterprise—in which just under 20,000 scientists and supporting personnel were employed. Table 5 summarizes the categories of drugs sold in 1967. The magnitude of the research task is shown by the number of animals required (Table 6).

By now, the general pattern of pharmaceutical research is somewhat standardized. Research directors, aware of the needs of human and veterinary medicine, monitor the output of worldwide fundamental biomedical research for clues to potential new drugs. Chemists then either purify some

TABLE 5   1967 U.S. Drug Shipments in Major Categories (In $ Millions)[a]

TOTAL HUMAN AND VETERINARY DRUGS	**4,143.0**
Drugs for treatment of cancer, endocrine and metabolic diseases	**439.6**
Hormones	413.5
Corticoids	141.7
Oral Contraceptives	103.5
Other	168.3
Other	26.1

TABLE 5—*continued*

Drugs acting on the central nervous system and sense organs	**1,130.1**
Internal Non-Narcotic Analgesics	448.4
Internal Narcotic Analgesics	40.3
Tranquilizers	321.4
Antidepressants	45.0
Central Nervous System Stimulants	80.5
Barbiturate Hypnotics	25.4
Nonbarbiturate Hypnotics	35.4
General and Local Anesthetics	47.4
Other	86.3
Drugs acting on the cardiovascular system	**204.0**
Drugs acting on the respiratory system	**366.6**
Drugs acting on digestive or genito-urinary system	**552.4**
Antacids	137.3
Antidiarrheal Drugs	41.4
Laxatives	79.0
Antiulcer Drugs	73.5
Motion-Sickness Remedies	24.6
Urinary Antibacterials and Antiseptics	28.7
Diuretics	94.8
Other	73.1
Dermatological preparations	**236.9**
Hemorrhoidal Drugs	27.6
Other Dermatological Drugs	209.3
Nutrients	**168.5**
Vitamins	**204.1**
Hematinics	58.5
Other	145.6
Drugs affecting parasitic and infective disease	**698.8**
Antibiotics	468.3
Sulfonamides	32.5
Antifungal agents	28.7
Antibacterials and antiseptics	133.1
Other	36.2
Pharmaceutical preparations for veterinary use	**140.4**
Pharmaceutical preparations not specified by kind	**1.6**

a Data from U.S. Department of Commerce, Bureau of the Census, *Current Industrial Reports,* "Pharmaceutical Preparations, Except Biologicals—1967," December 4, 1968.

TABLE 6   Animal Usage in the Pharmaceutical Industry in 1965 [a]

Mice	23,200,000
Rats	9,900,000
Hamsters	900,000
Guinea Pigs	350,000
Rabbits	250,000
Dogs	93,000
Monkeys	60,000
Cats	33,000

[a] Data from a survey made by the Institute of Laboratory Animal Resources, Division of Biology and Agriculture, National Research Council.

naturally occurring material or synthesize a desired chemical entity and a series of variations on this theme. The potential drug is then put through a "screen," an increasingly large and diverse battery of biochemical and physiological tests. If the material still appears to have the desired activity, it is tested in animal models of the human disorder, where such exist, and then screened for short-term and long-term toxicity in a variety of animal species. If all appears to be in order, permission is requested of the Food and Drug Administration to test the drug in man. After a tolerable dosage level is established in a few subjects, a much wider group of patients is treated with the drug for its specific use, while also being observed for any signs of toxicity. When a sufficient series has been tested, if the drug has proved efficacious for its intended purpose and if undesirable side reactions are minimal, permission is sought for free marketing. Finally, the Food and Drug Administration must balance hazard against benefit as it comes to a decision. *No* foreign compound is totally devoid of untoward effects. Indeed, were aspirin invented tomorrow, the Food and Drug Administration would have a difficult time in deciding whether to issue a license. If the benefits warrant and the hazards are tolerable—particularly if the drug offers decided advantages over existing drugs or is truly lifesaving—the Food and Drug Administration will issue the desired license, about 5 to 10 years and $5 million to $10 million after the start of the project.

The overall results are evident in the facts that no more than 10 percent of drug sales represent entities available before 1940, that mortality from all infectious diseases fell from 88,000 deaths in 1941 to 17,000 in 1961, that tuberculosis sanitariums are closed, tranquilizers have emptied thousands of sanitarium beds, and few of us any longer are asked to bear extreme pain, thanks to nonaddicting, powerful analgesics.

Food

Once harvested or slaughtered, the products of agricultural practice must be "brought to the table." This is accomplished by the many components of the food industry, employing 14 percent of the working population in an endeavor aggregating $90 billion in 1966. Through the combined efforts of applied biologists, chemists, and engineers, the housewife may now choose among 8,000 items in the supermarket. This team has solved the problems of raw storage; out-of-season processing; long-term storage; minimization of contamination by agricultural chemicals, bacteria, yeast, and molds; maintenance of moisture and nutritional content. It has upgraded the nutritional value of various native foods; monitored all stages of the preparative and marketing process; established the optimal conditions for transport and storage; devised such preservative mechanisms as ethylene oxide, sulfur dioxide, and nitrogen gas atmospheres, as well as procedures such as vacuum-, spray-, drum-, and freeze-drying, while monitoring such processes as fermentation of sauerkraut, cheese, or buttermilk and sterile filtration of beer, wine, and fruit juices. It has devised the wide variety of food additives now available and has developed specialized foods for diabetics, phenylketonurics, and galactosemics, as well as for those with heart or kidney disease or gastrointestinal limitations.

Pesticides

As noted earlier, the properties of DDT and 2,4-D inaugurated a new era in management of our living resources and gave rise to a new industry. Each touched off a wave of research that continues to the present, seeking newer compounds that are species-specific, safe, and degradable. For the moment, the use of such compounds is indispensable; until superior means and materials are found, these compounds are essential to the success of our agriculture, while assisting in maintenance of our woodlands and protection of our health. It is the scale of this use, rather than their intrinsic toxicity, that has properly generated public concern over the effects of these chemicals on the public health. In 1966, total production of all pesticides in the United States was 1,012,598,000 pounds.

The rapid increase in use occurred because new pesticides have been developed that control hitherto uncontrolled pests, and broader use of pesticides in large-scale agriculture has increased crop yields significantly. Current trends in crop production involving large acreages, greater use of fertilizers, and intensive mechanized cultivation and harvesting offer particularly favorable opportunities for insect pests and would result in large crop losses to these pests unless control measures were applied.

The increased number of new pesticides in part reflects a second generation of pesticides with more appropriate persistence for economic control of specific pests, more complete control of the pest, less hazard for the applicator, or less hazardous residues on the crop. An additional impetus to the development of new pesticides comes from the fact that many insect pests have developed resistance to the older pesticides. The development of pest resistance does not necessarily entail the development of more dangerous pesticides; the new agent need only be chemically different to overcome resistance. The continuing search for new, more nearly ideal pesticides requires the joint effort of research teams composed of organic chemists, biochemists, pharmacologists, physiologists, entomologists, and botanists. The effort is managed much like the development of new drugs, each chemical entity being tested in a "screen" of a variety of insects.

About 73 percent of the total insecticide usage is in agriculture, and about 25 percent is used in urban areas by homeowners, industry, the military, and municipal authorities. The remaining 2 percent is applied to forest lands, grassland pasture, and on salt and fresh water for mosquito control. Over 50 percent of the insecticide used in agriculture is applied to cotton acreage alone.

When insect-control measures are *not* used in agriculture, insect pests take 10 to 50 percent of the crop, depending on local conditions. Losses of this magnitude are not readily tolerated in the United States in the face of a rapidly increasing population and a concomitant decrease in agricultural acreage. In this sense, the use of insecticides might be deemed essential *at this time* for the production and protection of an adequate food supply and an adequate supply of staple fiber. While alternative methods of pest control are under investigation and development, they are not yet ready to displace completely the chemical pesticides, and it appears that a pesticide industry will be required for some years to come.

Pesticides have been tremendously effective, but individual pesticides, like sulfa drugs and antibiotics, tend to lose their effectiveness as species resistance to them develops. Hence, there will be a continuing search for new pesticides as long as pesticides are considered to be required for the economy or the public health. This search will require the continuing participation of able biologists. As with drugs, new pesticides, optimally, should be selectively toxic for specific pests, rather than broadly toxic against a wide variety of pests with serious side-effects on nonpest species. Broad-spectrum pesticides affect an essential enzyme or system common to a wide variety of pests. A selective pesticide, on the other hand, either should affect an essential enzyme or system peculiar to a particular pest or should be applied in such a way that only the particular pest gains access to it.

An interesting example of a selective pesticide is the rodenticide norbormide, which is highly toxic for rats, particularly for the Norway rat. By contrast, the acute oral toxicity of norbormide for other species is much lower, the lethal dose for a great variety of birds and mammals, per kilogram of body weight, being more than 100 times greater. The mechanism of the selective toxic action of norbormide for rats is not yet elucidated.

Achievement of target specificity requires a sophisticated knowledge of the anatomical, physiological, or biochemical peculiarities of the target pest as compared with other pests or vulnerable nonpests; a pesticide may then be developed that takes advantage of these peculiarities. This is obviously not easy to accomplish, and norbormide may prove to be unique for many years. An alternative is the introduction of a systemic pesticide into the host or preferred food of the target pest. Other pests or nonpests would not contact the pesticide unless they shared the same host or food supply. As an example, a suitable pesticide may be applied to the soil and imbibed by the root system of a plant on which the pest feeds. The pest feeding on the plant then receives a toxic dose. The application of attractants or repellents (for nontarget species) would increase the selectivity of the systemic pesticide. The use of systemic pesticides on plants used for food by humans or domestic animals poses an obvious residue problem.

There has been a strong public reaction against the continued use of pesticides on the grounds that such use poses a potential threat to the public health as well as being a hazard to wildlife. Careful investigations have so far failed to establish the magnitude of the threat to the public health; i.e., there are as yet few if any clear-cut instances of humans who have suffered injury clearly related to exposure to pesticides that have been used in the prescribed manner. Report No. 1379 of the 89th Congress (July 21, 1966)* concluded:

The testimony balanced the great benefits of disease control and food production against the risks of acute poisoning to applicators, occasional accidental food contamination, and disruption of fish and wildlife. . . . *The fact that no significant hazard has been detected to date does not constitute adequate proof that hazards will not be encountered in the future. No final answer is possible now, but we must proceed to get the answer.* (Italics ours)

Failure to establish such hazard does not mean that it does not exist. There are no living animals, including those in the Antarctic, that do not

* U.S. Congress. Senate. Committee on Government Operations. *Interagency Environmental Hazards Coordination, Pesticides and Public Policy* (Senate Report 1379). Report of the Subcommittee on Reorganization and International Organizations (pursuant to S.R. 27, 88th Cong., as amended and extended by S.R. 288), 89th Cong., 2d sess., Washington, D.C., U.S. Government Printing Office, 1966.

bear a body burden of some DDT. Large fish kills and severe effects on bird populations have been demonstrated. The large-scale use of these agents has been practiced for less than two decades, and use has increased annually until this year (1969). Whereas the anticholinesterase compounds, which have high acute toxicity (and hence are highly hazardous to the applicator), are readily and rapidly degraded in nature, the halogenated hydrocarbons are not. With time, their concentration in the soil and in drainage basins, lakes, ponds, and even the oceans must continue to increase, thereby assuring their buildup in plant and animal tissues. Over a sufficient time period, this is potentially disastrous. And should such a period pass without relief, the situation could not be reversed in less than a century. Because of the large economic benefit to the farmer, it is pointless to adjure him to be sparing; unless restrained by law, he will make his judgment in purely personal economic terms. But mankind badly needs the incremental food made possible by use of effective pesticides, and the enormous benefit to public health of greatly reducing the population of insects that are disease vectors is a self-evident boon to humanity. Thus it is imperative that alternative approaches to pest control be developed with all possible dispatch, while we learn to use available pesticides only where they are clearly necessary and desirable and to apply them in the minimal amounts adequate to the purpose.

A recent development in insect-pest control has been the possible use of juvenile hormone. This hormone, normally produced by insects and essential for their progress through the larval stages, must be absent from the insect eggs if the eggs are to undergo normal maturation. If juvenile hormone is applied to the eggs, it can either prevent hatching or result in the birth of immature and sterile offspring. There is evidence to suggest that juvenile hormone is much the same in different species of insects, and analogs have been prepared that are effective in killing many species of insects, both beneficial and destructive. There would, therefore, be great danger of upsetting the ecological balance if juvenile hormone were applied on a large scale.

What is needed, then, is development of chemical modifications of juvenile hormone that would act like juvenile hormone for specific pests but not for other insects. For example, a preparation from balsam fir, which appears to be such an analog, has been identified and is effective against a family of bugs that attack the cotton plant, but not against other species. If it proves possible to synthesize similar analogs specific for other pests, a new type of pesticide may emerge. If this happens, it will be extremely important to explore possible side-effects on other insect species and on warm-blooded animals before introduction of yet a new hazard into the biosphere.

We cannot rest with existing pesticides, both because of evolving pest resistance to specific compounds and because of the serious long-term threat posed by the halogenated hydrocarbons. While the search for new, reasonably safe pesticides continues, it is imperative that other avenues be explored. It is apparent that this exploration will be effective only if there is, simultaneously, ever-increasing understanding of the metabolism, physiology, and behavior of the unwanted organisms and of their roles in the precious ecosystems in which they and we dwell.

Fermentation Industry

Wine and leavened bread date back to antiquity, but the fermentation industry is a product of modern biological science. In sum, the disparate fermentation industry constitutes a major national resource. Each of the major companies in that industry retains a staff of microbiologists, biochemists, chemists, and engineers. Together, they are responsible for the continuing monitoring control of the fermentations with which they are concerned. The microbiologists constantly search, by the conventional techniques of bacterial genetics, for new strains of micro-organisms that will more efficiently or more rapidly conduct the fermentation in question. Rarely do these groups, as such, discover new fermentations yielding new products of value. Most have been encountered earlier in the course of systematic microbiology, and the industrial research team develops the procedures whereby a laboratory observation is scaled up to the requisite industrial magnitude. An important exception has been the systematic hunt for new antibiotics by the drug companies.

Bakers', food, and fodder yeasts were produced in excess of 180,000 tons in 1967. Alcohol fermentation amounting to 685 million gallons in 1945 has largely been replaced by a process starting with petroleum-cracking fractions, as has the fermentative production of acetone and butanol. But 5 billion pounds of raw grains were used to produce 110 million barrels of beer and ale, while 235 million gallons of wine and 185 million gallons of distilled spirits were also produced by fermentation. Other fermentation procedures produce lactic acid, vinegar, dextrans for drilling muds and as a plasma substitute, sorbose, and glutamic acid. Bacteria are grown as legume inoculants and as bioinsecticides. Molds and streptomyces are grown as a source of at least 30 distinct antibiotics in general use, as well as of citric acid and a variety of other organic acids. One mold is now used to make giberellin, which stimulates seed germination, improves growth of young trees, increases the flowering of plants, "sets" tomato fruit clusters, and breaks the dormancy of potatoes.

Allied to these processes is the use of molds, streptomyces, and bacteria in synthetic chemistry to accomplish specific reactions not readily feasible by chemical means. At least 25 such procedures are in current use in steroid synthesis in pharmaceutical laboratories. Related, also, is a relatively new industry, the manufacture of enzymes on a substantial scale. At least 20 enzymes are now articles of commerce. Thus, amylases from pancreas, barley malt, or fungi are used to de-size textiles, start brewing fermentation, precook baby foods, or cold-swell laundry starch. Papain from papaya, bromelin from pineapple, and subtilisin from *B. subtilis* are used as meat tenderizers, to stabilize chill-proof beer, and most recently as adjuncts to laundry detergents. Still other enzymes are used in candy manufacture, in clinical diagnostic procedures, to clarify wine and beer, to tan leather, and for debridement of wounds. The list of enzymes and their uses is growing, limited only by imagination.

No estimate of the magnitude of these diverse biological industries is available, but clearly in sum they represent several billion dollars of the gross national product. In every instance, biological understanding underlay the original industrial concept, guided the necessary research and development, gave direction to the industrial installation, and is required for continuing monitoring of the process.

Instrumentation

The scale and sophistication of modern biological research and its applications have necessitated birth of a new industry—the manufacture of biological instruments. The need for and use of most instruments usually arises in a research laboratory. But, thereafter, if it is to be more generally available, conveniently packaged, simple and reliable in performance, and readily serviced, its production must be taken over by a commercial manufacturer. Competition among manufacturers is the stimulus that has provided a stream of increasingly useful, sensitive, and reliable instruments, annual sales of which now approximate $1 billion. Appreciation of the diversity of such instruments may be gained from the data concerning use and need presented in Chapters 3 and 4.

When established, such instruments are modified to monitor and control industrial biological processes and for diagnostic and therapeutic use in hospital practice. An excellent example is the development of an automated apparatus that can accept an unmeasured small volume of blood, perform about 15 different analytical procedures thereon, calculate the results in conventional units, and record them on the patient's record. These data are decidedly more reliable than are individual determinations performed manually by technicians, and the cost is comparable to that of a single such manual procedure. In consequence, the physician's armamentarium is markedly expanded, at no additional cost to the patient or to society.

From this brief summary it will be evident that the skills and understanding of the modern biologist find their way into a remarkable variety of human endeavors, rendering life more secure, healthier, longer, more comfortable, and more pleasant, while giving employment to millions.

CHAPTER THREE

THE WORLD
OF BIOLOGICAL
RESEARCH

The life sciences embrace a great array of intellectual activity, a continuum extending from the search for the origin of life and the detailed structure of the macromolecules that make life possible to understanding of the total ecology of planet Earth. The millions of micro-organisms and plant and animal species interacting in the air, the soil, freshwater ponds and streams, and the oceans afford a never-ending variety of objects of fascinating inquiry. This endeavor has enhanced man's capacity to manage and protect his environment, to feed and clothe himself, and to prolong his comfortable and fruitful years. The inquiry itself is conducted in the laboratory, in research institutes and hospitals, in experimental tracts and ponds, by walks in the woods, by surveillance from the skies, from ships at sea, and on treks through the jungle, observing both undisturbed and managed nature. Those so engaged range from amateur nature lovers to directors of large institutes. They work in and out of institutions large and small; they work with private, state, and federal resources in institutions of higher learning, nonprofit research institutes, research hospitals, federal, state, and local laboratories, and in the organized multidisciplinary teams of industry.

In 1966 the National Register of Scientific Personnel identified approximately 84,000 individuals with diverse levels of training and educational backgrounds who classified themselves as working life scientists. The identification of these people was possible through the cooperation of the two major biological research societies, the Federation of American Societies for Experimental Biology and the American Institute of Biological Sciences.

The Federation issued questionnaires to approximately 24,300 people, the great majority of whom had earned doctoral degrees. Of these, some 20,100, or 83 percent, responded to the Register questionnaires. The American Institute of Biological Sciences contributed approximately 59,800 names, but the proportion of doctorate holders among this group is lower, and hence fewer of them meet the conditions for inclusion in our survey as individual life scientists. Approximately 40,000 people, or 67 percent, responded to the Institute's questionnaire and the proportion of doctorate holders represented by those respondents is higher than that of the original 59,800 individuals surveyed by that society. The overall response to the Register from the two societies was approximately 65 percent and should comprise most working biologists. From these numbers it can be estimated that 70 to 80 percent of doctoral-degree holders responded to the National Register in 1966. However, one can only guess what fraction of American biologists, with or without doctoral degrees, this represents.

It is estimated * that, in the aggregate, $2,264 million was invested in research in the life sciences in fiscal year 1967, of which 60.3 percent came from the federal government, 7.3 percent from the resources of nonprofit institutions, and 30.0 percent from industry. In its entirety, therefore, research in the life sciences has become one of the major pursuits of American society. This chapter is devoted to a description of some of the components of the life sciences research system, based largely on information gathered from responses to our two questionnaires (Appendixes A and B).

Detailed information on the gross parameters of the total system was revealed by the first of our two questionnaires: It contains 14,362 scientists, of whom 12,383 were investigators as here defined, viz., they devoted more than 20 percent of their time to research. In 1966 they published more than 24,000 original articles, 489 books, 1,100 reviews, and 7,500 in-house reports and other contributions. The universe revealed by the second questionnaire contains 1,256 academic departments with an aggregate continuing staff of 18,608 scientists, with available research funds (direct costs only) totaling $304 million, operating in 325 acres of laboratory space in which they directed the research and training of 23,287

---

* Basic Data Relating to the National Institutes of Health 1969, Associate Director for Program Planning and Evaluation and the Division of Research Grants, National Institutes of Health. U.S. Government Printing Office, Washington, D.C., 1969, p. 4.

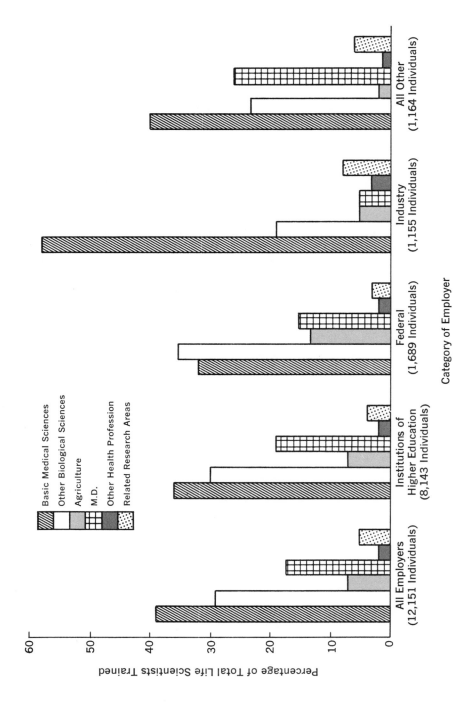

FIGURE 33 Type of employment of life scientists, by field of doctoral training. (Source: Survey of Individual Life Scientists, National Academy of Sciences Committee on Research in the Life Sciences.)

graduate students and 4,695 postdoctoral fellows and were assisted by 24,481 technicians, secretaries, and other personnel.

Of the 14,362 individuals who replied to the individual questionnaire, 3.4 percent were less than 30 years old and 6.3 percent were at least 60 years of age; 36.2 percent ranged from 30 to 39 years; 36.3 percent ranged from 40 to 49 years; and 17.8 percent were in the range 50 to 59 years. This distribution is fairly close to that of the scientific population at large. The average age of the group was 43.2 years, the median 41 to 42 years. Only 5.1 percent of the total population was female.

Every state of the Union was represented in the reporting of birthplaces. New York was represented by the largest number of scientists (1,989); Pennsylvania and Illinois followed with 880 and 855, respectively; and 631 were born in California; in all, 12,439 had been born in the United States, and 1,866 were foreign-born. All but 41 of the foreign-born regarded themselves as permanent residents of the United States at the time of the questionnaire. The foreign-born life scientists had come to our shores from 81 different nations. The major sources were Canada (292), Germany (236), England (162), Taiwan (142), India (97), Austria (89), Hungary (68), Poland (55), and Japan (50).

WHERE LIFE SCIENTISTS WORK

Two thirds of the 12,383 investigators were employed by institutions of higher learning; as shown in Table 7, 14 percent were employed by the federal government, 10 percent by industry, and the remaining 10 percent by a variety of nonprofit organizations—e.g., hospitals, clinics, museums, state and local governments—and a few are self-employed. In a general way, this pattern is relatively independent of the field in which these life scientists were trained (Figure 33). With the exception of horticulturists, those trained in the agricultural sciences are more likely to work for the federal government than those trained in any other scientific area. Of the 68 percent who were trained in the basic biological sciences, biochemists are by far the largest single group, constituting 15 percent of the total population of this study, with microbiologists and physiologists 8 percent and 7 percent of the total, respectively. Although, because of their numbers, these groups are predominant on the faculties of institutions of higher education, biochemists, microbiologists, and pharmacologists are also in great demand outside these institutions. Over 40 percent of those trained in these three disciplines operate in nonacademic environments, with all three unusually well represented in the laboratories of industry.

TABLE 7 Principal Employment of Life Scientists

FIELD OF TRAINING OF DOCTORAL DEGREE[a]	GRAND TOTAL	Institution of Higher Education	Non-academic Subtotal	Federal Government	ALL OTHER NONACADEMIC							
					Subtotal	Non-profit Organization	State and Local Government	Independent Hospital or Clinic	Federal Contract Research Center	Self-employed	Private Industry or Business	All Other
TOTAL RESPONDENTS	12,151	8,143	4,008	1,689	2,319	455	304	219	134	23	1,155	29
Agriculture Subtotal	855	559	296	200	96	6	19	1	4	—	65	1
Agronomy	347	218	129	93	36	3	4	—	1	—	28	—
Animal Husbandry	132	97	35	15	20	1	3	1	—	—	15	—
Fish and Wildlife	50	19	31	18	13	—	10	—	—	—	2	1
Forestry	88	54	34	33	1	1	—	—	—	—	—	—
Horticulture	158	131	27	14	13	1	1	—	3	—	8	—
Agriculture, Other	80	40	40	27	13	—	1	—	—	—	12	—
Biological Sciences Subtotal	8,269	5,524	2,745	1,129	1,616	335	177	93	106	6	885	14
Anatomy	196	176	20	6	14	5	4	—	—	1	3	1
Biochemistry	1,834	1,081	753	247	506	105	47	38	34	4	275	3
Biophysics	160	107	53	20	33	11	2	1	9	—	10	—
Cytology	109	84	25	5	20	12	2	2	1	—	3	—
Embryology	105	92	13	3	10	6	—	2	1	—	1	—
Microbiology	1,010	570	440	150	290	47	23	17	16	—	186	1
Pathology, Animal	77	52	25	7	18	3	—	5	2	—	8	—
Pharmacology	374	218	156	27	129	5	6	1	2	—	115	—
Physiology, Animal	805	594	211	91	120	18	12	11	8	—	71	—

Botany	365	285	**80**	43	**37**	7	6	2	9	—	13	—
Ecology and Hydrobiology	234	172	**62**	40	**22**	6	7	1	4	—	4	1
Entomology	415	271	**144**	98	**46**	14	13	—	—	—	18	1
Genetics	408	281	**127**	62	**65**	20	5	1	9	—	29	1
Nutrition	221	161	**60**	16	**44**	1	3	1	1	—	38	—
Paleontology and Systematics	72	50	**22**	13	**9**	5	2	—	1	—	—	1
Pathology, Plant	245	209	**136**	97	**39**	3	12	—	—	—	24	—
Physiology, Plant	353	236	**117**	68	**49**	11	4	1	2	—	30	1
Zoology	773	609	**164**	78	**86**	35	15	2	6	—	26	2
Bioscience, All Other	413	276	**137**	58	**79**	21	14	3	1	1	31	3
Health Professional Subtotal	**2,315**	1,616	**699**	288	**411**	76	87	113	**14**	**17**	**92**	**12**
M.D.	2,118	1,508	610	250	360	74	82	109	13	16	55	11
D.D.S.	65	45	20	14	6	—	1	—	1	1	2	1
D.V.M.	109	50	59	20	39	2	2	2	—	—	33	—
Other[b]	23	13	10	4	6	—	2	2	—	—	2	—
Related Areas Subtotal	**712**	444	**268**	72	**196**	38	21	12	**10**	—	**113**	**2**
Chemistry	442	252	190	41	149	34	13	7	5	—	90	—
Physical Sciences[c]	114	79	35	16	19	3	1	2	5	—	7	1
Psychology	105	75	30	13	17	—	5	2	—	—	9	1
All Other Fields[d]	51	38	13	2	11	1	2	1	—	—	7	—

[a] Respondents with Ph.D. or D.Sc. degrees are categorized by the field of training of their most recent degrees. Respondents having only health-professional degree(s) (M.D., D.D.S., D.O., D.P.H., D.Pharm., D.V.M., and equivalent foreign degrees) are categorized by type of most recent degree. Respondents having both health-professional degrees and Ph.D. or D.Sc. are categorized by the field of training of the most recent of the latter degrees.

[b] Includes D.O., D.P.H., D.Pharm., and other health-professional degrees not specified.

[c] Includes biometrics and biostatistics, computer science, earth sciences, engineering, mathematics, physics, and statistics.

[d] Includes anthropology, other social sciences, and related fields of training.

Source: Survey of Individual Life Scientists, National Academy of Sciences Committee on Research in the Life Sciences.

Of the 17 percent of our population who were originally trained as physicians, one third also obtained Ph.D. degrees. Seventy percent of the M.D.'s are on the faculties of universities, including virtually all the M.D.-Ph.D.'s; rather few research-performing M.D.'s are in industry, but there is unusually high representation in nonprofit institutions, particularly independent hospitals and clinics and public-health organizations. Those trained as physicians constituted 44 percent of the 3,170 reporting members of faculties of medical schools (and these schools corresponded to 39 percent of the total academic population); these were 87 percent of all reporting physicians. The remainder of the medical faculty was drawn largely from among those originally trained in the basic medical sciences; biochemists predominated in this last group (15 percent of the gross total), with major representation also from physiology, microbiology, and pharmacology.

Because of their relatively large total number, those trained in biochemistry are found throughout the system in substantial numbers. Of 1,834 trained biochemists reporting, 59 percent (1,069) were in institutions of higher learning, including 491 in medical schools, 225 on arts and sciences faculties, 126 in agricultural schools, and 37 in liberal arts colleges. Substantial numbers were also found elsewhere: 247 in the federal government, 275 in industry, and 231 in other nonacademic, nonprofit organizations. (The disciplinary designation, "biochemist," relates only to the field of original doctoral-level training, and not to the area of science in which the scientist is currently working.)

Of the life scientists in our sample employed by institutions of higher learning, slightly less than 5 percent were at liberal arts colleges. Undoubtedly, a much larger fraction of life scientists, particularly botanists and zoologists, are on the faculties of such institutions, but relatively few engage in research on a scale sufficient to have put them within the scope of this study.

The questionnaire addressed to department chairmen yielded an aggregate faculty for all responding departments of 17,172, of whom 3,852 were on faculties of arts and sciences, 3,907 on the faculties of agricultural schools, and 8,915 on the faculties of medical schools. Although the general employment patterns in the two questionnaire files are similar, the discrepancies are of some interest. Whereas 39 percent of all individual respondents were on the faculties of medical schools, 52 percent of the total departmental faculties reported were so employed. To place this in perspective, it should be noted that, of the 1,256 departments represented in the study, 267 are in agricultural schools, 246 in faculties of arts and sciences, and 694 in medical schools. Of the medical departments, 361 were departments of the preclinical and 333 of the clinical segments of medical schools. Undoubtedly, the returns from the chairmen's questionnaire should be taken as a more valid description of the distribution of the faculties of life scientists than that provided by the individual returns.

The 1,689 individual scientists who indicated that they are employed by the federal government appear to represent a large fraction of the senior life scientists in the federal establishment. The major employers of the 1,689 reporting life scientists within the federal establishment are the Departments of Agriculture (36 percent), Health, Education, and Welfare (27 percent), Defense (15 percent), and the Veterans Administration (11 percent). The patterns of employment of scientists in the various biological disciplines reflect the character of the agency missions rather closely. Thus, 84 percent of all those trained in agricultural sciences now in the federal establishment are employed by the Department of Agriculture; 53 percent of all federally employed M.D.'s actively engaged in research work for the Department of Health, Education, and Welfare; 29 percent of the M.D.'s work in the Veterans Administration; and 18 percent of the M.D.'s work in the Department of Defense.

The disciplinary employment patterns in other areas are repeated in the federal establishment: 32 percent of all federal life scientists were trained in the basic medical sciences, varying from 12 percent in the Department of Agriculture to 55 percent in the Department of Defense. Except for the physicians employed by the Department of Health, Education, and Welfare and the Veterans Administration and the agronomists employed by the Department of Agriculture, biochemists again constitute the largest single group of scientists in all federal agencies, ranging from 7 percent in the Department of Agriculture to 16 percent in the Department of Defense, 21 percent in the Department of Health, Education, and Welfare, and 26 percent in the Veterans Administration.

An additional 135 scientists were employed in federal contract research centers, which are managed by educational or other nonprofit organizations. State governments employed 229 life scientists (1.8 percent of the grand total), largely in hospitals or state health departments and their laboratories, and approximately half as many life scientists were found in municipally controlled institutions of the same character. A significant number, 462 scientists (3.7 percent of the total), were employed by nonprofit institutes, foundations, and privately controlled museums.

There are no reliable indicators to determine whether the 1,155 individual respondents who indicated that they are employed in industry constitute either a large or a true sample of the total number of senior life scientists employed in that sector of the economy. Seventy-six percent were employed by manufacturing industries; two thirds of these were in the pharmaceutical industry. Again, those trained in the basic medical sciences predominate: 262 biochemists were the largest group, followed

TABLE 8 Migration Patterns of Life Scientists[a]

PLACE OF BIRTH (CENSUS REGION)	CENSUS REGION IN WHICH CURRENTLY LOCATED								
	New England	Middle Atlantic	East North Central	West North Central	South Atlantic	East South Central	West South Central	Mountain	Pacific
New England	**210**	181	126	45	161	33	23	22	70
Middle Atlantic	280	**1,098**	374	133	536	70	77	73	286
East North Central	106	282	**733**	205	299	84	91	104	236
West North Central	60	134	260	**310**	198	49	108	83	197
South Atlantic	33	114	80	42	**362**	63	37	19	64
East South Central	6	22	49	16	100	**96**	51	15	15
West South Central	19	37	38	49	88	36	**200**	40	59
Mountain	19	62	89	54	59	16	43	**128**	143
Pacific	40	84	91	50	80	22	33	64	**431**
Non-U.S.A.	154	344	259	94	206	45	69	48	211

[a] Boldface figures are totals of life scientists who work within the census regions in which they were born.
Source: Survey of Individual Life Scientists, National Academy of Sciences Committee on Research in the Life Sciences.

by 178 microbiologists and 107 pharmacologists. The low representation of other disciplines among investigators in industry is somewhat disconcerting. For example, only two embryologists, three anatomists, four cell biologists, four ecologists, eight animal pathologists, 10 biophysicists, 13 botanists, and 25 zoologists reported that they were in the employ of some industrial establishment.

Finally, in this regard, it should be remarked that of the 12,151 life scientists responding, 442 had obtained Ph.D.'s in chemistry and 114 in other fields of the physical sciences (about one half in physics), while 105 individuals were originally educated as psychologists. (No questionnaires were sent to individual practicing research psychologists or to the chairmen of either psychiatry or psychology departments.) The employment distribution of these 662 converts to the life sciences among institutions of higher learning, the federal government, industry, and other organizations was much like that of the groups described earlier.

Mobility of Life Scientists

Geographic mobility, so prominently a characteristic of American society, is nowhere more evident than in the scientific community. As shown in Table 8, scientists born in each of the standard census regions can currently be found in each of the other census regions. Presumably, the direction of these migrations is dictated largely by increasing employment opportunities. This is particularly evident in the considerable migration from all other census regions to the Pacific Coast region and the South Atlantic region. Of at least equal interest, however, is the even greater tendency for relocation to regions likely to produce the least "cultural shock." Not only is there the expected tendency of a substantial fraction of all scientists in all census regions to remain within the states or census regions within which they were born, but the most frequent move from one region to another has been to an adjoining area where life patterns are similar—e.g., from the lower South to the upper South, or within the Midwest.

For the entire population of life scientists, the average length of employment in the current position was 9.6 years, with the median 6 to 7 years. Fifty-five percent of all respondents had held at least one previous position with a different employer, quite apart from any number of postdoctoral appointments. The average length of employment in that previous position was 4.7 years, and the median was 3 to 4 years. Although 90.5 percent of all such moves had been made after less than 10 years with the previous employer, employment translocation was reported by some scientists even after as long as 40 years with the initial employer.

The pattern of these moves is of interest in itself. Although institutions of higher learning were the principal source of those who entered the employ of the federal government, private industry, and other organizations, in a general way each employing entity in the system also tended to recruit from other institutions in the same category. For example, 36 percent of all those in private industry had been employed by a different corporation, and 19 percent of those now working for an independent hospital or clinic had previously worked for some other independent hospital or clinic.

Two thirds of those who had moved to an institution of higher learning had come from another such institution. Of the remainder, 13 percent had left the federal government, 5 percent private industry, 5 percent other nonprofit organizations, and 8 percent various other state and community institutions. Perhaps the major surprise in these data is the fact that, ignoring graduate and postdoctorate education, institutions of higher learning appeared to be a net importer of scientific employees. Whereas 1,750 individuals whose previous employers had been nonacademic institutions currently were employed by the universities, only 1,260 individuals currently employed by nonacademic institutions had previously been employed by universities or colleges.

Respondents to the questionnaire were not queried about their motivation in accepting offers of new positions. It may be assumed that these were responses to offers of higher pay, of opportunity to engage in independent research or research under more desirable conditions, or to locate in geographical areas attractive to the families of the scientists concerned.

Previous Education of Working Life Scientists

In the foregoing summary, the initial training of working life scientists was categorized in disciplinary terms that are familiar as the titles of academic departments and that are employed in most statistical collections. However, the reader who has considered earlier chapters will have recognized that these conventional subdisciplinary titles have, in considerable measure, lost their meaning and convey false distinctions. Whereas biochemists were formerly concerned largely with elucidation of metabolic maps, they may today be concerned with macromolecular structure, the chemistry of cell–cell recognition, or the phenomena responsible for atherosclerosis. Not so long ago, microbiologists were overwhelmingly concerned with the taxonomy of microbiological forms, yet today they may be concerned with genetic mechanisms or the nature of the immune response to invasion by some specific organism. Hematologists, who only yesterday were describing changes in the morphology of blood cells in leukemia as seen with a light microscope, are now intimately involved in understanding the manner in which nucleic acids control the differentiation process among white blood cell types. Physiologists, who formerly engaged in studies of the mechanics of muscular contraction or morphological changes induced by steroid hormones, are today inquiring into mechanisms of transmembranal transport or the molecular events by which steroid hormones affect protein biosynthesis in receptor cells. Botanists, once engaged in taxonomic studies or in gross plant physiology, are today concerned with the phenomena by which plants interact with other organisms and with their environment, the cardinal aspects of ecology, while zoologists may be concerned with all those aspects of the environment that have favored rapid proliferation of new species in one set of circumstances or remarkably prolonged survival, unchanged, of other species, studies that embrace all aspects of ecology, genetics, biochemistry, and physiology. Even more dramatic have been the changes in the character of research in clinical medicine, pathology, and pharmacology. Investigators in these areas have learned to use the most recent developments in understanding such phenomena as protein structure, enzyme kinetics, transmembranal transport, neural transmission, immunochemistry, viral reproduction, lipid metabolism, and behavioral genetics as they explore disease mechanisms in man or animals, design and test new drugs, or prepare a patient for organ transplantation. And their laboratories cannot be distinguished from those of other scientists so engaged.

Because of these rapidly evolving and profound trends, it appeared desirable to reconsider individual scientists, not under classical disciplinary labels, but in relation to the nature of the research conducted during their initial formal education in graduate school and in relation to the research in which they are currently engaged. That two individuals are studying cellular structure and function is more significant than that one considers himself a zoologist and the other a botanist. The plant pathologist may have more in common with an animal pathologist than with a plant taxonomist, and similar considerations are obvious for plant and animal physiologists, or for plant, animal, and microbial geneticists, for example.

Thus, we have found it useful to recategorize life sciences research into the following dozen classifications:

Behavioral biology	Genetics
Cell biology	Molecular biology and biochemistry
Developmental biology	Morphology
Disease mechanisms	Nutrition
Ecology	Pharmacology
Evolution and systematic biology	Physiology

TABLE 9 Comparison of Current Research Areas with Areas of Most Recent Ph.D. or D.Sc. Degree[a]

CURRENT RESEARCH AREA	GRAND TOTAL	Bioscience Subtotal	BIOSCIENCE RESEARCH AREA OF Ph.D. OR D.Sc. — Behavioral Biology	Cellular Biology	Developmental Biology	Disease Mechanisms	Ecology	Evolutionary and Systematic Biology	Genetics	Molecular Biology and Biochemistry	Morphology	Nutrition	Pharmacology	Physiology	Related Areas Subtotal	RELATED RESEARCH AREA OF Ph.D. OR D.Sc. — Chemistry	Physical Sciences[b]	Social Sciences[c]	Other
TOTAL RESPONDENTS	6,444	5,802	107	386	276	488	496	287	441	1,483	173	370	235	1,058	642	294	117	59	172
Bioscience Subtotal	5,979	5,665	107	382	273	475	484	285	430	1,439	167	356	229	1,038	314	201	37	28	48
Behavioral Biology	154	135	**76**	1	6	3	12	9	4	4	5	3	2	10	19	—	1	14	4
Cellular Biology	446	433	1	**235**	28	22	4	5	9	48	18	10	4	49	13	6	1	—	6
Developmental Biology	284	279	—	28	**186**	7	8	3	10	17	8	1	—	11	5	2	2	—	1
Disease Mechanisms	555	535	4	22	10	**378**	25	4	9	53	5	10	4	11	20	7	1	—	12
Ecology	529	512	9	4	3	3	**401**	30	6	7	9	8	—	32	17	1	3	4	9
Evolutionary and Systematic Biology	290	289	3	2	7	4	26	**219**	11	2	4	—	—	11	1	—	—	—	1
Genetics	417	409	2	7	6	7	1	4	**353**	7	4	2	—	16	8	1	2	4	1
Molecular Biology and Biochemistry	1,617	1,451	1	55	13	26	3	5	22	**1,196**	—	51	11	68	166	145	18	—	3
Morphology	119	115	1	5	3	1	3	3	—	3	**60**	1	1	34	4	3	1	—	—
Nutrition	287	278	1	—	—	4	1	—	2	19	1	**233**	3	14	9	7	—	—	2
Pharmacology	340	316	1	14	—	6	3	1	1	45	1	7	**195**	42	24	18	—	2	4
Physiology	941	913	8	16	14	8	25	4	9	38	12	31	7	**741**	28	11	8	4	5
Related Areas Subtotal	465	137	—	4	5	13	12	2	11	44	6	14	6	20	328	93	80	31	124
Chemistry	119	36	—	—	—	2	—	2	—	26	2	2	—	2	83	**79**	—	—	4
Physical Sciences[b]	106	23	—	2	—	—	—	—	2	14	1	1	1	2	83	2	**76**	—	5
Social Sciences[c]	36	3	—	—	—	—	2	—	—	—	—	—	—	1	33	3	—	**28**	2
Other	204	75	—	2	4	12	5	2	8	14	4	9	3	12	129	9	4	3	**113**

[a] Boldface figures are totals of life scientists who currently work in the same research area in which they carried out their doctoral research.

[b] Includes biometrics and biostatistics, computer science, earth sciences, engineering, mathematics, physics, and statistics.

[c] Includes anthropology, other social sciences, and related fields of training.

Source: Survey of Individual Life Scientists, National Academy of Sciences Committee on Research in the Life Sciences.

It will be evident that even these categories are somewhat arbitrary and are by no means mutually exclusive. They fail to make clear the fact that biochemistry, a research area itself, is also the common language and the tool for almost every other entry in the classification scheme. However, the questions being asked of nature by scientists within each category are sufficiently distinct to permit self-identification by our respondents, while providing a more revealing description of the life sciences endeavor than that offered by more traditional disciplinary titles.

Tables 9 and 10 summarize the current research areas of some of our respondents, comparing their current areas of involvement with the disciplines and research areas in which they had been trained as graduate students. As a consequence of an awkwardness in the design of the layout of the printed questionnaire, almost a quarter of all respondents failed to provide information concerning the research fields, as here categorized, in which they had been trained and in which they are currently engaged. However, as indicated in Appendix A, it appears fair to assume that the patterns revealed by those who did not overlook this question are representative of the total.

As indicated by the diagonal of Table 9, current research in any given area is conducted predominantly by individuals who were trained in that area, varying from 49 percent of those currently engaged in behavioral biology to 85 percent of those working in genetics. Equally impressive, however, is the degree of intellectual migration among research fields. Thus, 48 percent of all those trained in morphology are now engaged in some other area, as are 39 percent of those originally trained in cell biology, 33 percent of those trained in developmental biology, and 30 percent of those trained in physiology. Maximum field retention was found among those trained in pharmacology, ecology, genetics, and molecular biology and biochemistry. Perhaps the most striking fact shown by the table is that every possible crossover was reported. Noteworthy, too, are the fields that, on balance, have either attracted more investigators than they have lost, or vice versa. The "gainers" include molecular biology and biochemistry, behavioral biology, cellular biology, disease mechanisms, ecology, and pharmacology. The most significant "losers," in absolute numbers rather than percentages, were genetics, morphology, nutrition, and physiology, with developmental biology and systematic biology remaining approximately in balance.

Many biologists currently consider that there has been a rapid growth in the opportunities for fruitful studies in behavioral and developmental biology and in ecology. But these data indicate that, although there has been some modest influx into these fields, it is not yet particularly striking,

TABLE 10 Comparison of Current Research Area with Disciplines in Which Life Scientists Were Trained

FIELD OF TRAINING OF DOCTORAL DEGREE [a]	GRAND TOTAL	Bioscience Subtotal	Behavioral Biology	Cellular Biology	Developmental Biology	Disease Mechanisms	Ecology	Evolutionary and Systematic Biology	Genetics	Molecular Biology and Biochemistry	Morphology	Nutrition	Pharmacology	Physiology	Related Areas Subtotal	Chemistry	Physical Sciences [b]	Social Sciences [c]	Other
			BIOSCIENCE RESEARCH AREA												RELATED RESEARCH AREA				
TOTAL RESPONDENTS	8,005	7,459	178	585	331	973	557	301	464	1,853	172	313	468	1,264	546	126	117	45	258
Agriculture Subtotal	576	453	5	7	13	4	97	1	114	19	3	70	1	119	123	10	58	5	50
Agronomy	227	161	2	—	5	1	28	—	59	6	—	18	—	42	66	3	40	1	22
Animal Husbandry	93	88	1	2	2	—	—	—	24	8	—	32	—	19	5	—	—	—	5
Fish and Wildlife	36	36	1	—	2	—	32	—	—	—	—	—	—	1	—	—	—	—	—
Forestry	69	46	1	—	4	1	26	—	7	—	—	1	—	6	23	1	9	4	9
Horticulture	98	91	—	3	—	1	1	—	22	3	1	15	—	45	7	—	1	—	6
Agriculture, Other	53	31	—	2	—	1	10	1	2	2	2	4	1	6	22	6	8	—	8
Biological Sciences Subtotal	5,549	5,330	123	437	272	577	433	296	299	1,420	118	215	325	815	219	48	28	6	137
Anatomy	130	126	5	25	20	7	1	3	1	1	43	2	1	17	4	—	—	—	4
Biochemistry	1,271	1,205	5	32	17	58	4	2	5	925	2	67	53	35	66	38	4	1	23
Biophysics	115	107	2	22	1	5	—	—	2	53	1	—	2	19	8	1	4	—	3
Cytology	80	79	1	46	7	3	—	2	6	9	2	1	2	—	1	—	—	1	—
Embryology	80	79	—	11	57	—	—	—	—	4	1	7	—	1	1	—	1	—	—
Microbiolog…	656	613	4	95	16	202	27	9	14	173	1	7	8	57	43	5	6	—	32

Physiology, Animal	535	531	7	38	14	11	2	2	1	51	3	3	50	349	8	6	—	1	5
Botany	230	225	1	15	25	10	36	69	14	13	22	1	1	20	5	6	2	—	4
Ecology and Hydrobiology	159	155	5	2	1	1	122	10	2	1	1	3	1	9	4	1	2	—	2
Entomology	288	270	45	4	15	12	84	47	2	18	1	1	4	33	18	4	1	—	16
Genetics	287	282	2	8	10	4	2	11	217	23	3	—	—	4	5	5	1	—	3
Nutrition	153	148	—	2	1	3	—	45	—	14	—	1	2	11	5	1	2	—	4
Paleontology and Systematics	50	48	—	—	—	—	1	—	—	—	—	115	—	—	2	—	1	—	1
Pathology, Plant	227	213	6	8	9	117	24	2	6	17	2	—	1	21	14	—	1	—	14
Physiology, Plant	241	235	2	29	34	4	8	2	3	37	—	2	2	107	6	1	4	—	1
Zoology	512	503	31	53	39	32	110	84	18	18	26	7	10	80	9	—	—	1	8
Bioscience, All Other	258	240	4	32	4	76	10	6	6	39	8	2	11	41	18	1	1	2	15
Health-Professional Subtotal	**1,391**	**1,322**	**21**	**130**	**40**	**383**	**18**	**3**	**38**	**209**	**47**	**19**	**116**	**298**	**69**	**6**	**8**	**8**	**47**
M.D.	1,270	1,217	20	122	38	333	12	3	36	207	41	15	101	289	53	6	8	7	32
D.D.S.	35	28	—	5	2	10	1	—	1	1	3	3	3	2	7	—	1	1	6
D.V.M.	74	67	1	3	—	39	3	—	—	1	2	10	5	7	—	—	—	7
Other[d]	12	10	1	—	—	1	2	—	1	—	1	2	2	5	2	—	—	—	2
Related Areas Subtotal	**489**	**354**	**29**	**11**	**6**	**9**	**9**	**1**	**13**	**205**	**4**	**9**	**26**	**32**	**115**	**62**	**23**	**26**	**24**
Chemistry	305	238	2	5	1	7	—	1	1	190	—	7	19	6	67	61	—	1	5
Physical Sciences[b]	77	53	3	4	1	1	6	1	8	15	2	1	11	24	—	22	22	2
Psychology	73	49	24	1	—	—	—	—	2	—	2	—	6	14	24	1	—	20	3
All Other Fields	34	14	—	1	1	3	2	2	—	—	1	1	1	1	20	14	1	5	14

[a] Respondents with a Ph.D. or D.Sc. degree are categorized by the field of training of their most recent degree. Respondents having only a health-professional degree(s) (M.D., D.D.S., D.O., D.P.H., D.Pharm., D.V.M., and equivalent foreign degrees) are categorized by type of most recent degree. Respondents having both a health-professional degree and a Ph.D. or D.Sc. degree are categorized by the field of training of the most recent non-health-professional degree.

[b] Includes biometrics and biostatistics, computer science, earth sciences, engineering, mathematics, physics, and statistics.

[c] Includes anthropology, other social sciences, and other related fields of training.

[d] Includes D.O., D.P.H., D.Pharm., and other health-professional degrees not specified.

Source: Survey of Individual Life Scientists, National Academy of Sciences Committee on Research in the Life Sciences.

TABLE 11 Projected Research Areas of Some Life Scientists

CURRENT RESEARCH AREA	TOTAL	Behavioral Biology	Cellular Biology	Developmental Biology	Disease Mechanisms	Ecology	Evolutionary and Systematic Biology	Genetics	Molecular Biology and Biochemistry	Morphology	Nutrition	Pharmacology	Physiology	Related Research Areas
TOTAL RESPONDENTS	741	37	86	91	110	64	33	18	138	10	16	20	56	62
Bioscience Subtotal	705	36	86	91	107	59	32	18	126	10	15	19	53	53
Behavioral Biology	24	—	1	1	3	7	1	—	2	—	1	1	6	2
Cellular Biology	93	1	13	18	14	1	—	3	47	—	—	1	2	6
Developmental Biology	44	2	8	—	6	3	2	2	11	—	—	1	4	—
Disease Mechanisms	57	1	1	6	—	10	7	—	11	2	3	3	6	4
Ecology	43	11	1	2	6	—	11	—	—	1	1	—	3	8
Evolutionary and Systematic Biology	27	2	2	1	2	7	—	2	2	4	—	—	3	2
Genetics	34	1	4	6	3	5	1	—	5	—	1	—	3	5
Molecular Biology and Biochemistry	150	1	33	39	33	3	—	8	—	1	5	6	9	6
Morphology	35	2	9	5	6	1	2	1	1	—	—	—	5	3
Nutrition	31	1	2	2	5	6	—	—	5	—	—	1	5	6
Pharmacology	42	7	2	2	9	1	—	—	12	—	—	—	7	2
Physiology	125	7	11	16	20	15	2	2	30	2	4	7	—	9
Related Areas Subtotal	36	1	—	—	3	5	1	—	12	—	1	1	3	9
Chemistry	10	—	—	—	1	1	—	—	7	—	—	—	1	1
Physical Sciences[a]	13	1	—	—	1	3	1	—	2	—	1	—	2	3
Social Sciences[b]	5	1	—	—	1	1	—	—	—	—	—	—	1	2
Other	8	—	—	—	—	—	—	—	3	—	—	1	1	3

(Columns Behavioral Biology through Related Research Areas fall under the heading PROJECTED RESEARCH AREA.)

[a] Includes biometrics and biostatistics, computer science, earth sciences, engineering, mathematics, physics, and statistics.
[b] Includes anthropology, other social sciences, and related fields of training.
Source: Survey of Individual Life Scientists, National Academy of Sciences Committee on Research in the Life Sciences.

although graduate enrollments have been affected in the predicted directions. Moreover, the changes are generally immediately lateral in the sense that most of those who have changed research areas have moved into areas in which they can apply the skills and insights of their primary training. This is most certainly the case for the 184 of 287 individuals who left molecular biology and biochemistry to enter upon studies in cellular biology, disease mechanisms, pharmacology, or physiology, as it must also be true for the 317 individuals who left physiology to enter other biological categories.

Only 741 scientists were sufficiently certain of their plans to change research areas in the future to so indicate. And again, the planned changes were, in the main, relatively conservative (Table 11) and into closely related areas, e.g., molecular biology to genetics, genetics to molecular biology, physiology to pharmacology, botany to ecology. Molecular biology will be the chief gainer (19 percent of all who plan to change), largely from cellular biology and physiology. However, it will lose a slightly larger number (20 percent), mainly to cell biology, developmental biology, and disease mechanisms. Disease mechanisms attracts the second largest group (15 percent), largely from among those now engaged in cellular biology, biochemistry, and physiology, while developmental biology also seems attractive to those in the same group of research areas (12 percent). The survey revealed a particularly interesting trend. Some ecologists indicated plans to enter behavioral biology, while a significant number of physiologists and students of disease were seriously considering switching to ecology.

Moreover, the perhaps not unexpected conservative migratory pattern is again evident from the responses of life scientists who intended to change the biological material with which they were working. In a general way, those now seriously contemplating such a change are, in the main, thinking of switching either to the next higher or the next lower level of biological organization, e.g., from broken cell preparations to cells or tissue culture or to molecular systems; or from intact organs to either intact organisms or cellular preparations.

Table 12 relates research areas to the principal employers of the 8,139 individuals for whom such information is available. Of this subset, institutions of higher learning employed 68 percent, the federal government 14 percent, industry 9 percent, and all other nonprofit organizations, hospitals, etc., 9 percent. Noteworthy are the high levels of employment by the federal government of those studying ecology and disease mechanisms; the government shows much less interest in developmental biology, morphology, and pharmacology. Private business employs an unusually high fraction of all nutritionists and pharmacologists, but appears to have little interest in ecology, systematic biology, or morphology.

TABLE 12 Distribution of Investigators in Various Research Areas by Principal Employer

CURRENT RESEARCH AREA	GRAND TOTAL	Institution of Higher Education	Non-academic Subtotal	Federal Government	ALL OTHER NONACADEMIC							
					Subtotal	Non-profit Organization	State and Local Government	Independent Hospital or Clinic	Federal Contract Research Center	Self-employed	Private Industry or Business	All Other
TOTAL RESPONDENTS	8,139	5,476	2,663	1,149	1,514	315	200	133	81	8	761	16
Bioscience Subtotal	7,582	5,210	2,372	1,039	1,333	301	175	122	80	8	633	14
Behavioral Biology	181	127	54	29	25	6	13	—	—	—	6	—
Cellular Biology	593	447	146	57	89	29	8	11	16	1	23	1
Developmental Biology	342	280	62	19	43	13	6	7	6	—	11	—
Disease Mechanisms	987	594	393	193	200	35	34	27	12	3	84	5
Ecology	569	404	165	111	54	7	24	2	6	—	12	3
Evolutionary and Systematic Biology	309	230	79	41	38	28	5	1	—	1	2	1
Genetics	470	316	154	81	73	19	9	3	6	—	35	1
Molecular Biology and Biochemistry	1,883	1,250	633	248	385	116	41	34	25	1	168	—
Morphology	173	143	30	14	16	5	3	5	1	—	2	—
Nutrition	317	218	99	37	62	7	7	2	1	—	45	—
Pharmacology	476	246	230	40	190	14	5	2	—	—	168	1
Physiology	1,282	955	327	169	158	22	20	28	7	2	77	2
Related Areas Subtotal	557	266	291	110	181	14	25	11	1	—	128	2
Chemistry	128	46	82	17	65	3	4	4	—	—	54	—
Physical Sciences [a]	120	63	57	36	21	2	1	—	—	—	17	1
Social Sciences [b]	45	33	12	6	6	—	2	2	—	—	2	—
Other	264	124	140	51	89	9	18	5	1	—	55	1

a Includes biometrics and biostatistics, computer science, earth sciences, engineering, mathematics, physics, and statistics.
b Includes anthropology, other social sciences, and related fields of training.
Source: Survey of Individual Life Scientists, National Academy of Sciences Committee on Research in the Life Sciences.

A small insight into the changing dynamics of the life sciences is provided by observation of the fraction of the total population within each research area under 34 years of age. This fraction is remarkably close to 21 percent for virtually all research areas, with a few interesting exceptions. Only 11 percent of those engaged in the study of disease mechanisms are within this age group, presumably reflecting the long period of residency training for physicians. In contrast, 23 percent of those in developmental biology and 28 percent of those in molecular biology and biochemistry were under the age of 34 at the time of this survey, indicating that in the recent past these two fields, as compared with the other research areas, have become increasingly attractive to young scientists. Only 18 percent of all those attracted into the life sciences from the physical sciences were within this age group, indicating that there has been no dramatic upsurge of interest in the life sciences among young chemists or physicists.

The reverse situation is in accord with the same suggestions. For the entire population, 18 percent were 50 years of age or older, but only 12 percent of those in molecular biology and biochemistry fell within that age range, in contrast with 25–28 percent in the areas of disease mechanisms, evolutionary and systematic biology, morphology, and nutrition.

Of some interest are the attributes of the group of investigators originally trained only as M.D.'s or in the other health professions. They are older, with only 15 percent under 34 years of age, but 42 percent within the age span 40–49. Logically, disease mechanisms constitute their principal single interest (27 percent of the total), but they are also represented in every other research area with the exception of systematic biology, major interests being physiology (22 percent), molecular biology and biochemistry (15 percent), cellular biology (9 percent), and pharmacology (8 percent).

The 456 women showed only a few distinct tendencies to differ from the distribution of the men. Women tended to avoid physiology, ecology, and systematic and behavioral biology, and 28 percent of all female respondents work in molecular biology and biochemistry.

Postdoctoral Training

Prior to World War II, postdoctoral research training experience was a privilege granted very few young scientists. Fellowships were scarce, and only the most highly talented could aspire to such opportunity. Since available research grants were decidedly limited in size, few senior academic investigators commanded the means to support eligible new M.D.'s or Ph.D.'s desirous of embarking upon the apprentice training characteristic of the postdoctoral experience. That situation no longer obtains. Post-doctoral experience has become almost the norm rather than the exception, and we are entirely convinced that this is in the national interest.

However, the situation has given rise to concern among those less closely associated with research in these disciplines. For example, agencies that provide support for postdoctoral training are uncertain of its value. Educational institutions in which postdoctoral fellows abound are uncertain of their institutional responsibility for this enterprise. Institutions that, perhaps until 1969, have had difficulty in recruiting sufficient staff to meet teaching obligations—largely the four-year colleges and junior colleges, but also a significant number of medical schools, as well as industry and some federal laboratories—have complained that the postdoctoral system is a holdup in the pipeline that, in the steady state, keeps a substantial number of bright young investigators out of the regular job market. We appreciate these problems, but consider that the benefits of postdoctoral education far outweigh these transient difficulties. Let us consider here the postdoctoral training experience of our responding population of life scientists. In the following chapter there is a summary of the numbers and activities of postdoctoral fellows in training in 1967–1968, as well as an analysis of the contribution of postdoctoral education to the operation of the entire endeavor.

Of the 12,151 investigators in the study, 5,041 had had at least one postdoctoral appointment, including 1,402 M.D.'s who had had postdoctoral experience in which research was their major responsibility. Three fourths of those who had had postdoctoral experience are now in academic life. Indeed, 45 percent of the 8,143 scientists now employed by universities had enjoyed postdoctoral experience, compared with 21 percent of the scientists in industry and 31 percent of those in the federal establishment. Taken across all disciplines, postdoctoral experience somewhat enhances the opportunity for employment in the federal government and markedly enhances the opportunity for employment in the universities. It is our impression that in universities with major commitments to graduate education and research, measured in supporting dollars and number of graduate students, faculty appointments for individuals who have not had postdoctoral experience are probably rare indeed. According to a National Academy of Sciences study of postdoctorals,* 74 percent of all new appointees to the rank of instructor or assistant professor in 21 departments of biological sciences in 10 "leading" institutions either came from other university faculties or had just held postdoctoral appointments.

* The Invisible University: Postdoctoral Education in the United States, Report of a Study Conducted under the Auspices of the National Research Council, National Academy of Sciences, Washington, D.C., 1969.

However, the trend to postdoctoral education is not universal across all biological fields. For example, of the 855 individuals with graduate training in agricultural fields, only 35 had had postdoctoral appointments. In contrast, postdoctoral training was commonplace among M.D.'s since it has become the conventional medium for obtaining research training among this group.

As shown in Table 13, postdoctoral training was less frequent among botanists (29 percent) than among biochemists (53 percent), with the other disciplines ranging in between. Postdoctoral training was frequently taken in fields other than those in which scholars had their initial doctoral experience. Thus, of the zoologists and botanists who did take postdoctoral training, less than half did so in zoology and botany departments. Again, the biochemists appear as the other extreme. Not only did a larger fraction of biochemists than other life scientists take postdoctoral training, but a decidedly larger fraction remained within biochemistry for their postdoctoral experiences. Since an additional 540 individuals who had taken their original graduate education in fields other than biochemistry sought postdoctoral training in biochemistry, postdoctoral education is a major aspect of life in biochemistry departments. Large numbers of those trained in biochemistry in graduate school later work in other disciplinary areas, while many individuals enrich their original disciplinary education by a one- or two-year postdoctoral experience in biochemistry and then, when they become independent investigators, return to their original disciplines and research areas or enter yet other research areas.

These data uphold one of the primary arguments in support of the trend toward postdoctoral experience as a normal component of the education of those who later will espouse careers in which research is a major activity, *viz.,* that this constitutes a unique opportunity to broaden one's horizons, learn new techniques, and become familiar with the style of other sub-disciplines, while profiting by the examples of different master scientists. The overall situation is reflected in the totals of Table 13. Of 5,765 Ph.D.'s in this file, 2,395 undertook postdoctoral experience, of whom 1,463, or 61 percent, extended their experience in the same disciplines in which they had studied in graduate school. But the impression that postdoctoral experience is a continuation of graduate education in 61 percent of all cases is misleading, since it is weighted by the fact that more than half of all of those who did experience this continuation were biochemists. If the biochemists are excluded, only 50 percent of the remaining scientists who undertook postdoctoral training did so in their graduate disciplines. Moreover, such an experience is but rarely a mere continuation of graduate education. This is borne out by the following consideration: In a subfile of 3,234 postdoctoral fellows, only 14 percent had taken postdoctoral edu-

TABLE 13   Postdoctoral Experience of Scientists in a Limited Group of Biological Disciplines

FIELD OF GRADUATE STUDY	NUMBER OF Ph.D.'s	POSTDOCTORAL EXPERIENCE		POSTDOCTORAL EXPERIENCE IN FIELD OF GRADUATE STUDY	
		Number	Percent	Number	Percent
ALL BIOLOGY	**5,765**	**2,395**	*41*	**1,463**	*61*
Anatomy	196	74	*37*	30	*41*
Biochemistry	1,834	968	*53*	752	*77*
Botany	365	108	*29*	40	*37*
Genetics	408	157	*38*	116	*73*
Microbiology	1,010	359	*35*	198	*55*
Pharmacology	374	131	*35*	84	*64*
Physiology	805	329	*40*	167	*50*
Zoology	773	269	*35*	76	*28*

Source: Survey of Individual Life Scientists, National Academy of Sciences Committee on Research in the Life Sciences.

cation in the same university in which they had obtained their doctoral degrees, and only 6 percent in the same departments that had awarded their doctoral degrees. This migratory pattern is particularly evident among the M.D. population. However, about one third of all Ph.D.'s in agriculture and forestry who undertook their postdoctoral training—a rather small group—did so in their original universities and, indeed, in the departments that had awarded their degrees. The rather small proportion of students who remained in the same department for postdoctoral study was almost twice as great in public universities as in private universities.

In sum, it is clear that the norm for postdoctoral experience, by a wide measure, consists of apprenticeship to a different set of investigators in an environment different from that in which graduate education has been completed. Further, in the experience of our panelists, the current internal heterogeneity of the classical disciplines assured that even the postdoctoral trainee who remains within his original discipline is likely to engage in a problem remote from his graduate research experience. The biochemist who studied intermediary metabolism may later become preoccupied with the mechanism of enzyme action; the physiologist who traced neural pathways as a graduate student may focus upon ion transport across the nerve membrane during his postdoctoral years. The botanist who was concerned with nutritional requirements for plant growth may later become involved in the ecology of a cornfield, while the entomologist concerned with patterns of insect distribution may switch to a study of insect sex attractants. Intellectual inbreeding is rare in the life sciences community, and the postdoctoral experience is among the chief means of assuring the hybrid vigor of the entire enterprise.

A few notes comparing the bioscience subculture with the subcultures of the physical and social sciences may be warranted. The data in support of the following statements are derived largely from the recent National Research Council study of postdoctoral education, *The Invisible University.**

In the nation's leading academic institutions, postdoctoral experience has become the expected prelude to faculty appointment. In recent times, 70 to 80 percent of all initial faculty appointments at such institutions in physics, in chemistry, in biology departments of faculties of arts and sciences, and in the preclinical departments of medical schools have been made to individuals with postdoctoral experience either at the same or at some other institution. In contrast, initial faculty appointments in the social sciences, the humanities, and engineering relatively rarely require postdoctoral experience. The play of the academic marketplace is such that the frequency of postdoctoral experience among initial appointees to the faculty decreases with the general academic status of the institution. Postdoctoral experience is less frequent among the faculties of "developing" universities, is rare for scientists who are appointed to the faculties of liberal arts colleges, and is even less common among those who enter industry.

The converse is equally evident; 30 to 40 percent of all relatively young faculty at all universities who have not had postdoctoral experience feel this lack in their current professional lives. In all branches of natural science, promotion up the academic ladder occurs somewhat less rapidly for those who have not had postdoctoral experience, although this may reflect similar appraisal of human potential by the committees who select postdoctoral-fellowship recipients and those who recommend academic promotions, rather than the intellectual rewards of postdoctoral study. These trends are undoubtedly enhanced by the advice given to aspiring scientists by their mentors in graduate school, who strongly urge students in the natural sciences to undertake postdoctoral experience if they aspire to academic careers but rarely do so when this is not the case. In general, such mentors recommend a postdoctoral experience of about two years, with a specific senior scientist in a field somewhat different from that in

* *The Invisible University: Postdoctoral Education in the United States,* Report of a Study Conducted under the Auspices of the National Research Council, National Academy of Sciences, Washington, D.C., 1969.

which the student's dissertation research was conducted, thereby broadening his understanding of his discipline. When queried, postdoctoral students advance the same general purpose as their reason for undertaking postdoctoral study, but place more emphasis than do their graduate mentors upon the acquisition of additional research techniques.

Attempts by statistical means to assess the influence on subsequent scientific productivity of postdoctoral training are not revealing. Differences among those who took postdoctoral training immediately after graduate school, those who deferred such training for several years, and those who had no such training are trivial when measured by counting numbers of scientific publications, reviews, books written, and similar measures. What cannot be assessed by this means is the quality of the work or its significance to the field. One indicator has been reported in *The Invisible University**: the fact that papers published by those who have had postdoctoral experience are cited about twice as frequently in the Citation Index † as are papers by those who have not had such experience. Statistically, frequency of citation of a paper is some measure of its significance or fundamentality. It is our contention that, in all scientific fields, scientific boldness—willingness to venture beyond the frontier or to undertake large and challenging problems—is established relatively early. Certainly, if this is not encouraged in graduate school or in the immediate postdoctoral years, it is rarely evident in subsequent careers. But statistical assessment of this all-important quality is not readily feasible; hence, the enhanced opportunity to develop such habits of mind is another argument that we would advance in support of a year or two of postdoctoral study, preferably not in the same institution or with the same mentor that provided the graduate experience.

Data purporting to compare the consequences of graduate or postdoctoral study in the 10 or 20 leading academic institutions with those in other institutions are probably not completely valid. The selection process that operates at the level of admission to graduate school and then to postdoctoral study in the most productive academic laboratories already serves as a screen almost sufficient to assure the ultimate outcome. It is not readily possible to distinguish between the consequences of differences in the quality of the educational experiences in such institutions and the consequences of the quality of the initial human input. Certainly it must be undeniable that those most highly qualified will benefit most from a stimu-

* *The Invisible University: Postdoctoral Education in the United States,* Report of a Study Conducted under the Auspices of the National Research Council, National Academy of Sciences, Washington, D.C., 1969.
† *Science Citation Index; An International Interdisciplinary Index to the Literature of Science.* (Published by Institute for Scientific Information, Philadelphia.)

lating environment in which science is being conducted at its outermost
frontiers.

EDUCATIONAL LIMITATIONS

An attempt was made to estimate the extent to which working life scientists
sense deficits in the educational preparation for their careers. Respondents
to the questionnaire were asked to state whether their current research
programs are significantly limited by their own educational preparation in
chemistry, mathematics, physics, electronics, statistics, other areas of
the life sciences, or the use of computers. In all, 4,396 scientists, 30.6
percent of the entire responding population, indicated that full development of their current research effort is indeed very seriously hindered by
insufficient personal training in one or more of these disciplines. Lack of
knowledge of chemistry was most frequently felt to be limiting (1,766
individuals), followed by computer science (1,569), mathematics (1,427),
statistics (1,136), other biological sciences (1,085), and electronics
(983), with only 498 life scientists acutely aware of insufficient personal
training in physics.
    Scientists in academic institutions were not distinguished from those
working in nonacademic institutions with respect to this pattern of perceived inadequacies, although 38 percent of academic personnel were
aware of some such limitation, and only 30 percent of nonacademic scientists were. In both groups, those in the middle of the age range (35–50
years) were about 30 percent more likely to be aware of such deficits than
were younger or older investigators. Again, however, age was essentially
without influence on the pattern of perceived disciplinary insufficiency; the
rank order of disciplines cited above for the entire population was characteristic of the youngest, oldest, and midrange investigators alike.

WITH WHAT MATERIALS DO LIFE SCIENTISTS WORK?

The panorama of the biological universe offers such remarkable and diverse
organisms, ecological situations, environmental responses, and unanswered
questions at levels varying from the molecular to the cosmic that it is not
surprising that research biologists employ an almost equally disparate and
diverse variety of approaches to the questions they put to nature. In
Table 14 is displayed a representation of primary research materials and

TABLE 14   The Research Materials of Life Scientists

PRIMARY RESEARCH MATERIALS	TOTAL	Behavioral Biology	Cellular Biology	Developmental Biology	Disease Mechanisms	Ecology	Evolutionary and Systematic Biology	Genetics	Molecular Biology and Biochemistry	Morphology	Nutrition	Pharmacology	Physiology	Related Research Areas
TOTAL	7,881	176	581	330	949	558	299	457	1,845	168	303	452	1,246	517
Mathematical Models	166	2	6	2	9	8	6	18	27	—	3	6	44	35
Atomic and Molecular Models	46	—	3	1	—	2	—	—	26	1	—	3	—	10
Design/Fabrication of Apparatus	68	2	1	2	7	1	—	2	15	2	—	2	12	22
Development of Analytical Procedures or Methodology	456	9	13	13	72	15	1	11	149	4	16	34	51	68
Molecular Systems	788	1	33	23	34	—	1	20	612	1	3	11	21	28
Cell Fractions	1,025	3	192	45	89	2	5	25	524	10	17	33	74	6
Disassociated Cells	223	—	50	16	21	1	7	7	77	2	7	11	24	6
Cell Cultures	473	1	130	29	118	11	5	41	87	4	2	6	31	8
Tissue/Organ Systems	977	11	84	73	104	1	7	9	140	92	26	122	291	17
Artificial Organs	27	—	—	—	7	—	2	—	—	—	—	3	12	3
Whole Organisms	2,388	107	47	101	373	161	138	188	132	35	195	203	594	114
Populations of Organisms	681	31	13	11	60	237	85	114	25	2	133	5	40	45
Ecosystem Studies	144	2	2	2	8	104	2	—	2	—	1	—	3	18
Comparative Studies in Single Phylum/Division	110	3	2	6	9	5	41	5	4	4	6	1	16	8
Comparative Studies in 2 or More Phyla/Divisions	26	2	1	—	1	1	3	2	4	2	1	3	5	1
None of the Above	283	2	4	6	37	9	2	15	21	9	13	9	28	128

Source: Survey of Individual Life Scientists, National Academy of Sciences Committee on Research in the Life Sciences.

246

the extent to which these are utilized by those who work in various biological research areas.

It may come as a surprise to some that mathematical models are utilized by representatives of almost every research area, most frequently by those engaged in the study of physiology, molecular biology and biochemistry, genetics, or biophysics and, increasingly, in studies of ecology. Molecular models are to be found in virtually every biochemical laboratory, and the refined, precise models now available have become an extremely important tool for those seeking to relate molecular structure to biological function. Indeed, 46 individuals stated that such models constitute their primary materials.

It was somewhat surprising to find 6 percent of the entire surveyed population engaged primarily in the development of analytical procedures of various types. Study of molecular systems, utilizing highly purified materials of natural origin, engaged 10 percent of the total population, including one third of the biochemists. A somewhat greater proportion of life scientists were studying the behavior of subcellular organelles, isolated or *in situ*. Such materials are utilized by scientists, except the ecologists, in all research areas and, as one might expect, are a principal preoccupation of cell biologists and biochemists. A small proportion (3 percent) of our population, most notably the cell biologists, were learning to use disassociated preparations of living cells, from either plant or animal sources, as primary tools in their studies. Tissue culture was twice as popular and was utilized by at least some scientists, including behavioral biologists, in every research area, while intact tissues and organs claimed the attention of 12 percent of the total population, involving all research categories except ecology—most notably morphologists, pharmacologists, physiologists, and developmental biologists.

Intact individual organisms were the test objects of one third of all life scientists in the study, notably the behavioral biologists and those studying disease mechanisms, ecology, systematic biology, genetics, nutrition, pharmacology, and physiology. Decidedly smaller numbers of scientists addressed themselves to entire populations of organisms or to ecosystems.

Of interest is the fact that the pattern of use of materials by those with original training in the health professions cannot be distinguished from that of the remainder of the population; their primary research materials simply reflect the pattern of all others in the research areas in which they now engage. Accordingly, their major research materials are whole organisms (32 percent), tissue and organ systems (23 percent), subcellular fractions (13 percent), cell cultures (8 percent), and molecular systems (9 percent).

Within each research area a few individuals are engaged in comparative studies either within a single phylum or plant division or across several phyla or plant divisions. Although students of evolution and systematic biology were the most numerous such group, these were only 44 of the 123 individuals so engaged.

WITH WHAT SPECIES DO LIFE SCIENTISTS WORK?

The diversity of living nature never fails to astonish. The workings of evolution have resulted in millions of distinct species of living forms, unicellular, plant, and animal, all located in the thin web of life, which is a film on the surface of our planet. These are the objects of study for life scientists. But which species should one study? The answer depends upon the question that has been raised. Some species are of interest because they are the basis of our agricultural economy. Some make the world more beautiful and exciting; some cause disease of man, plant, or animal. Sometimes even the most obscure species provide excellent models for study of complex biological phenomena. And surely a proper object for study by man is man himself! Thus there are valid reasons for the study of a great variety of species.

Some species are of interest because they are intermediate links in a food chain, because they survive under what appear to be improbable conditions, or because they represent evolutionary extremes. Still others are of interest because they offer unique opportunities to study phenomena of general importance but difficult to analyze or observe in more common species. For example, the nerve net of the crab is of interest as a prototype of the more complex nervous system of the mammal; the response of certain insects to sugars can serve as a model for some aspects of the physiological bases of behavior; the "alarm reaction" of the clam is highly instructive with respect to certain reflex activities; the photosynthetic properties of the chromatophores of purple bacteria and of certain algae are more readily studied than is photosynthesis in a higher plant; regulation of the genome of a bacterium serves as a model for the process of differentiation in a higher organism; and the giant axon of the squid is the favorite test object of numerous neurophysiologists. Nutritionists long since seized on the omnivorous white rat as a model for human nutritional requirements, but primates may be more instructive with respect to human behavior or reaction to disease. The pig offers a surface area and mass somewhat comparable to that of man, and thus should serve as a model for human response to radiation. Comparison of the properties of hemoglobins from a wide variety of species elucidates those properties of the hemoglobin molecule that are imperative to its physiological function, and frog muscle has taught us much of what we understand of muscle physiology and its molecular aspects. The list is well-nigh endless.

And so it is that life scientists continue to study or exploit the properties of a great diversity of organisms. In a highly compressed form, this is displayed in Table 15. Each of the respondents to the questionnaire was given a choice of 58 genera, phyla, or larger divisions of the plant, animal, and microbial kingdoms and was asked to indicate no more than two that most closely described the objects of his study. Hence, the number of specific responses exceeded the number of respondents. But hundreds of investigators indicated that necessarily and properly they should indicate more than two such entries.

Perhaps the aggregated totals are of greatest interest: 21 percent of all scientists dealt with one or another micro-organism, 15 percent with plant forms, and 54 percent with animal forms. None of the categories of living forms was totally ignored by the current activities of life scientists but, clearly, some are more attractive than others. Viruses and bacteria are the concern of scientists in each research area, particularly those who study disease mechanisms, cell biology, and molecular biology and biochemistry. Lower plants engage the attention of all but the nutritionists and pharmacologists, while higher plants attract the attention of all but the pharmacologists. Invertebrates are of great interest to the ecologists and the systematists as well as to the behavioral biologists, who see in them models for the behavior of more advanced forms. Surprisingly little attention is being given to the species of fish that dominate our commercial harvests, whereas other fish, amphibia, reptiles, and birds are receiving greater attention. Of the mammalia, man and the common laboratory rodents are the most frequent study objects. The great utility of the latter is indicated by the fact that, whereas ecologists and systematic biologists pay them scant heed and only 6 percent of all geneticists make use of their particular attributes, these species are utilized by 12 percent of the behavioral biologists and 37 percent of the pharmacologists. Domestic mammals, i.e., cats and dogs, are particularly useful to the physiologists, pharmacologists, nutritionists, and morphologists and are used to some degree by almost all other groups.

Although 5 percent of all behavioral biologists and 4 percent of the morphologists report that they work with small primates, primates are little used by workers in other scientific areas. However, there is reason to think that this reflects not the utility of these species, but the great costs involved in

TABLE 15 Biological Materials Studied by Life Scientists

ORGANISMS	TOTAL	Behavioral Biology	Cellular Biology	Developmental Biology	Disease Mechanisms	Ecology	Evolutionary and Systematic Biology	Genetics	Molecular Biology and Biochemistry	Morphology	Nutrition	Pharmacology	Physiology	Related Research Areas
TOTAL RESPONDENTS	7,634	169	566	322	937	545	295	438	1,783	166	290	441	1,208	474
Micro-organisms Subtotal	1,651	8	174	35	332	79	20	59	661	11	42	33	110	87
Virus[a]	698	3	69	16	158	19	4	40	267	4	27	18	41	32
Bacteria	758	4	62	11	142	33	8	14	361	2	13	9	54	45
Actinomycetes, Mycoplasma, etc.[b]	60	—	6	4	9	4	2	—	18	2	1	1	10	3
Plankton	28	1	1	—	—	18	3	—	—	—	—	—	2	3
Protozoa	107	—	36	4	23	5	3	5	15	3	1	5	3	4
Plants Subtotal	1,260	13	59	74	91	172	102	189	155	28	46	2	215	114
Lower Plants[c]	343	6	31	21	76	30	28	26	60	11	1	—	39	14
Higher Plants Subtotal	884	7	27	52	11	135	72	162	89	17	44	1	172	95
Seed Plants Horticultural and Field Crops	266	3	4	9	8	15	3	50	28	2	25	1	83	35
Other[d]	618	4	23	43	3	120	69	112	61	15	19	—	80	60
Three or More Divisions	33	—	1	1	4	7	2	1	6	—	1	1	4	5
Animals Subtotal	4,510	148	327	213	510	292	170	190	873	124	199	401	873	190
Invertebrata Subtotal	614	51	29	37	43	134	85	56	41	11	3	9	92	23
Parasitic Worms[e]	81	1	1	5	28	14	5	—	4	3	1	2	11	5

250

Organism													
Insecta	335	40	11	12	9	76	52	54	22	5	1	1	14	1
Arachnida	58	6	5	3	2	15	9	2	3	1	2	2	9	15
Other[f]	47	2	5	12	1	1	5	—	6	1	—	2	11	1
Vertebrata Subtotal	3,709	95	282	170	444	114	82	132	783	109	194	385	762	157
Pisces, Commercial	11	—	—	—	—	1	1	1	4	2	2	1	2	1
Pisces, Other	108	4	4	6	—	29	25	1	14	5	2	4	19	1
Amphibia and Reptilia	138	4	13	26	3	9	16	1	16	5	2	5	41	—
Aves, Domestic	118	4	4	30	9	—	1	10	20	—	19	6	16	—
Aves, Other	173	10	9	11	10	37	17	11	14	5	18	6	21	4
Mammalia Subtotal	3,062	69	243	92	418	31	21	104	692	91	149	355	649	148
Common Laboratory Rodents	1,136	20	139	52	129	—	1	26	376	39	68	149	163	24
Domestic Mammals[g]	333	6	9	4	28	1	—	14	40	11	21	41	148	9
Wild Mammals[g]	13	—	1	1	—	4	2	—	—	1	—	—	4	—
Small Primates	68	9	2	1	9	—	2	—	8	7	3	6	18	3
Large Primates (Excluding Man)	8	—	1	—	—	—	—	—	2	—	—	—	4	1
Man	859	24	49	21	209	8	4	48	126	23	28	48	175	96
Other Mammals	595	10	42	13	43	18	12	15	140	10	29	111	137	15
Vertebrata, Other	99	4	9	5	4	7	2	4	23	6	4	13	14	4
Three or More Phyla	187	2	16	6	23	44	3	2	49	4	2	7	19	10
Other Phylum or Division	8	—	—	—	—	1	2	—	1	1	1	—	1	1
No Organism Used	205	—	6	—	4	1	1	—	93	2	2	5	9	82

[a] Includes viruses, bacteriophages, animal viruses, and plant viruses. [b] Includes actinomycetes, mycoplasma, and other bacteria-like organisms. [c] Includes algae, fungi, and nonvascular green plants other than algae. [d] Includes vascular nonflowering plants and seed plants; forest species. [e] Includes platyhelminthes and nematoda. [f] Includes porifera, coelenterata, rotifera, bryozoa, annelida, echinodermata, and tunicata. [g] Includes carnivores, and ungulates.
Source: Survey of Individual Life Scientists, National Academy of Sciences Committee on Research in the Life Sciences.

their acquisition and maintenance, which have inhibited, if not prohibited, their utilization for a variety of studies in which they could be extraordinarily useful.

In contrast, millions of species currently go unstudied, and many others are under scrutiny by only one or two investigators. When, from time to time, such an investigator directs attention to some unique or remarkable attribute of a seemingly esoteric species, it can rapidly claim the attention of many other scientists, an incident that has recurred many times in the past. Thus, the bacterium *Escherichia coli* has become the most thoroughly studied of all cells, while both neurophysiologists and molecular biologists have recently seized upon the tiny marine organism *Aplysia* because of its easily studied giant nerve cells. In any case, the diversity of species under study demands an equal diversity of laboratory accommodations for their culture or maintenance. This may engender substantial expenditures and contribute much to the cost of scientific investigation, particularly in extreme instances. Elaborate facilities are required for the conduct of research employing cells in culture. Inadequate accommodations, overcrowding, or infestation can render a colony of dogs or rodents useless to the investigator and give rise to misleading data. Humane considerations demand that larger domestic mammals—cats, dogs, and primates—be housed in decent quarters, be wellnourished, and be subjected to the minimum of trauma commensurate with the purposes of study. This in turn creates further serious financial requirements, which should be borne by some institutional mechanism and not met by taking funds from personal research grants made to individual investigators. Certain plants and animals require carefully controlled environments; a continuing supply of virus may require a colony of host animals, a large-scale fermentor, or a large tissue-culture facility. Most importantly, all these demand substantial expenditures merely to assure a supply of the biological entity to be studied before the research proper can be undertaken.

What Facilities and Tools Do Life Scientists Use?

The classic image of the biologist is an aging gentleman, wrapped in a dirty laboratory apron, in a musty laboratory surrounded by museum jars, an ancient, battered microscope, staining jars for microscope slides, and perhaps an unwashed dissecting table. If that image ever corresponded to reality, it no longer does. As the questions we ask of nature become more sophisticated and the information we seek becomes more remote from that which we can acquire with our naked senses, the requirements for the conduct of research in the life sciences become more complex. Today, in order to achieve his ends, the investigator may have to travel thousands of miles from his home base, armed with telemetering equipment, tape recorders, or remote sensors. He may require a floating laboratory, a deep-submersible vessel, a reconnaissance plane, or even a satellite equipped with infrared sensors. He may utilize the gadgetry of modern biochemistry— ultracentrifuges, equipment for optically following the course of kinetic processes on the scale of milliseconds or of molecular-relaxation times (10^{-9} sec), for the quantitation of visible or ultraviolet light or radioactivity. His laboratory may be what amounts to a small electronics plant equipped with the complex electronic apparatus needed for the study of neurophysiology, and his experiment may be guided by an on-line computer. Increasingly, the tools of any biological subdiscipline tend to become the tools in many other areas of biology. As we have noted repeatedly, this is particularly true of the tools of the biochemist, which have become the tools of all biologists.

Specialized Biological Research Facilities

Table 16 summarizes the replies from respondents whose completed questionnaires usefully indicated their utilization of specialized research facilities. The spectrum of such activity is broad indeed. For example, we were surprised at the high rate of utilization of controlled field areas, which seemingly are employed by participants in each of the research areas. Computer centers are available to and utilized by a strikingly high fraction of all life scientists, and general animal care facilities appear to be utilized by almost half the scientists covered by our survey. Indeed, it is difficult to correlate specific types of facilities with specific research areas. Notable exceptions include the 87 percent of all systematists and 44 percent of ecologists who utilized taxonomic research collections, the 51 percent of cell biologists who employed cell- or tissue-culture facilities, and the 76 percent of all pharmacologists who made use of general animal care facilities. The existence of the specialized facilities listed here was known to the Survey Committee, but the extent of use was not anticipated.

Rarely can the cost of acquisition and maintenance of such facilities be justified by the research program of a single investigator; hence, no small or medium-sized institution can hope to have a complete selection of these opportunities for conduct of research. This has the effect of either limiting the capabilities of the staff of such institutions or so affecting their recruitment patterns that, at each institution, there are clusters of investigators whose research requires easy access to the same major research facility.

TABLE 16 Utilization of Specialized Biological Research Facilities

SPECIALIZED FACILITY	RESEARCH AREA													
	TOTAL	Behavioral Biology	Cellular Biology	Developmental Biology	Disease Mechanisms	Ecology	Evolutionary and Systematic Biology	Genetics	Molecular Biology and Biochemistry	Morphology	Nutrition	Pharmacology	Physiology	Related Research Areas
TOTAL NUMBER OF DIFFERENT RESPONDENTS[a]	7,019	169	513	297	910	549	301	417	1,407	144	280	435	1,157	440
Field Areas	2,035	91	47	82	213	476	202	208	106	39	98	19	267	187
Zoo and Aquarium	304	12	24	21	32	55	32	6	38	11	5	8	49	11
Taxonomic Research Collection	914	46	24	36	71	240	262	38	49	32	5	5	64	42
Organism-Identification Service	674	38	37	28	141	133	49	30	74	10	14	13	60	47
Tropical Terrestrial Station	124	8	1	1	9	32	29	12	6	1	4	2	14	5
Tropical Marine Station	82	5	4	3	1	24	16	—	11	2	1	2	11	2
	200	9	38	33	12	58	25	3	47	10	2	5	60	7

Facility	Total		11	7	17	4	—	1	14	1	2	12	30	7
Programmed Climate-Controlled Rooms	**677**	24	36	53	69	94	20	50	86	7	19	9	160	50
Computer Center	**2,483**	72	99	66	262	247	46	226	412	26	147	170	484	226
Primate Center	**259**	11	21	3	51	4	2	8	46	9	3	33	52	16
Other Specialized Animal Colony	**1,044**	44	83	45	190	55	17	71	165	19	67	88	154	46
Germ-Free Facility	**242**	2	32	17	76	7	3	8	39	6	13	10	19	10
Animal-Surgery Facility	**1,645**	38	129	58	285	12	7	23	248	50	65	219	456	55
Animal Quarantine Facility	**718**	20	54	21	184	15	3	11	83	18	15	118	149	27
General Animal Care Facility	**3,113**	63	272	119	546	58	23	82	723	72	157	331	567	100
Cell and Tissue Culture Facility	**1,325**	11	264	98	293	23	17	75	340	22	14	39	86	43
High-Intensity Radioactive Sources	**570**	4	70	29	99	37	—	45	126	11	16	25	74	34
Center for Production of Biological Materials	**449**	7	23	10	79	18	4	27	173	3	13	22	35	35
Clinical Research Ward	**905**	20	59	25	279	7	4	32	150	15	26	78	175	35
Greenhouse	**1,307**	40	55	85	129	194	80	182	155	22	44	10	207	104
Ships Longer than 18 ft	**165**	8	14	9	5	59	19	1	16	3	1	3	22	5
Electrically Shielded Room	**287**	20	10	5	16	6	1	4	21	7	2	59	123	13
Instrument Design and Fabrication Facility	**1,430**	34	111	38	126	74	10	31	385	19	21	122	370	89

[a] Number of life scientists reporting the use of one or more specialized facilities.
Source: Survey of Individual Life Scientists, National Academy of Sciences Committee on Research in the Life Sciences.

For smaller institutions, this fact, in turn, may well prevent the assembly of a staff broadly representative of biology.

Major Instruments

Table 17 displays the utilization of major instruments by life scientists during 1966–1967. Like Table 16, this table is limited to those respondents whose replies to the questionnaire were found adequate to the purpose. And, as in Table 16, what is impressive is the extent of use of the wide variety of instruments listed and the relative amount of use without regard to specific research areas, again with a few notable exceptions. This table well illustrates how the tools developed for biochemical studies have become the tools of biology in general; this is evident in the use pattern of centrifuges, gas chromatographs, amino acid analyzers, scintillation counters, infrared and ultraviolet spectrophotometers, as well as electrophoresis apparatus. These common tools of the biochemical laboratory are now the common tools of the biological laboratory. Specialized uses of instruments will, however, be found in the table. For example, large-scale fermentors are used largely by biochemists; multichannel recorders are required by physiologists and pharmacologists; small special computers by physiologists. Biochemists are pioneering in the use of ultrasonic probes, and electron paramagnetic resonance and nuclear magnetic resonance spectrometers, as well as instruments for measuring circular dichroism. The physiologists are the major users of infrared carbon dioxide analyzers, and the clinicians interested in disease mechanisms utilize complex electronic systems for monitoring human physiology, while systematists use telemetry and sensitive tape recorders.

The utilization of the electron microscope is particularly revealing. This instrument, slowly introduced into biological laboratories in the years following World War II, is now used by investigators in every research area. In absolute numbers, those interested in molecular biology and biochemistry, cellular biology, disease mechanisms, and physiology are the principal users. But 48 percent of all those studying morphology and 44 percent of those studying cellular biology made use of this instrument. The great expense of acquisition and maintenance of these instruments prevents the figures for utilization from approximating 100 percent of those in both of the latter research areas.

One should not leave the subject of instruments without a tribute to the instrument-manufacturing industry. This highly competitive industry has frequently been a jump ahead of most life scientists. In general, instrument manufacturers have recognized needs and potential uses before the scientific community has. Yet, as each instrument has become available—e.g., ultraviolet spectrophotometers, electrophoresis apparatus, scintillation counters, electron microscopes, and multichannel recorders—not long thereafter the scientists involved have wondered how they had ever made progress before these commercial instruments became available. As the markets grow, the instruments become more refined, more reliable, and more versatile, thereby enormously enhancing the reliability, sophistication, and ease of performance of biological research. The availability of such instruments has been made possible by the very scale of federal support of the life sciences. By creating a sufficient market, the manufacturer has, in turn, been able to achieve economies of large-scale production, keeping the unit cost and sales price down. (It is ironic that, although the electron microscope was developed by an American firm, and this country is the major market for this instrument, no American manufacturer now supplies it.)

Nor should we fail to acknowledge our debt to our brethren in physics, chemistry, and engineering. From them came the electron microscope, spectrophotometers, the electron paramagnetic and nuclear magnetic resonance spectrometers, ultrasonic gear, the great variety of oscilloscopes, x-ray crystallographic analysis systems, the laser, telemetry, and a host of other devices. To their designers and developers, the biological community extends its gratitude.

THE RESEARCH GROUP

Research in the life sciences is "small science"; only rarely is it organized around some very large and expensive piece of apparatus or facility. Whereas much research in other areas of science revolves about large accelerators, research vessels, telescopes, balloon-launching facilities, rocket facilities, or large magnets, for example, there are few parallels in the life sciences. Occasional exceptions include relatively elaborate hyperbaric facilities, primate colonies, colonies of germ-free animals, phytotrons or biotrons, biosatellites, museums, or marine-biology stations. But these are the exceptions rather than the rule, and even in these instances, the facilities in question are actually utilized by numbers of small research groups, each pursuing its own questions in its own way, while taking advantage of the availability of the facilities. In very few instances have the various groups that, collectively, used such a facility comprised a coordinated whole with common goals and objectives. The functional unit of research in the life sciences, therefore, usually consists of a principal investigator and the

TABLE 17 Utilization of Instruments by Life Scientists

| | BIOSCIENCE RESEARCH AREAS | | | | | | | | | | | | | |
MAJOR INSTRUMENTS	TOTAL	Behavioral Biology	Cellular Biology	Developmental Biology	Disease Mechanisms	Ecology	Evolutionary and Systematic Biology	Genetics	Molecular Biology and Biochemistry	Morphology	Nutrition	Pharmacology	Physiology	Related Research Areas
DIFFERENT RESPONDENTS [a]	6,702	128	567	284	866	298	147	280	1,828	130	239	416	1,122	397
Acoustic														
Acoustic-Analysis Equipment	110	21	6	2	7	12	6	3	3	2	1	6	36	5
Sonar	53	4	3	1	5	16	3	—	7	2	1	1	7	3
Ultrasonic Probes and Sensoring System	486	8	45	18	59	4	1	9	215	3	12	14	70	28
Centrifuges														
Analytical Ultracentrifuge	1,351	9	95	46	163	20	4	25	704	12	34	73	102	64
Preparative Ultracentrifuge	2,555	16	270	103	382	26	6	56	1,245	15	46	131	198	61
Refrigerated Centrifuge	3,978	29	396	166	612	65	23	121	1,608	29	131	219	450	129
Chromatography														
Amino Acid Analyzer	961	11	35	24	95	9	3	33	508	3	69	30	83	58
Gas Chromatograph	1,438	24	63	31	140	45	8	31	472	7	115	125	244	133
Programmed Gradient Pump	272	5	22	5	26	2	—	2	163	1	6	11	22	7
Counters														
Automatic Particle Counter	784	10	93	23	114	26	1	28	255	9	30	44	108	43
Scintillation Counter	2,591	21	228	92	269	43	4	49	1,107	16	84	194	392	92
Whole-Body Counter	141	—	8	6	24	6	1	4	17	3	22	8	31	11
Microscopy														
Electron Microscope	1,320	11	248	84	213	26	14	38	397	62	15	39	132	41
Electron Probe for Microscopy	30	—	3	1	5	1	1	—	10	3	1	1	2	2
Fluorescence Microscope	985	8	170	74	309	36	2	28	162	36	10	31	85	34
Metallograph	16	1	1	—	4	—	—	—	3	2	—	—	2	3
Microtome-Cryostat	821	11	154	78	180	14	7	21	111	54	18	39	116	18
Phase-Contrast Microscope	2,487	36	386	173	388	130	89	156	580	75	36	68	277	93
Spectrometers														
Electron Paramagnetic												7	14	13

	Total[a]													
Resonance Spectrometer	313	2	7	2	10	4	1	7	179	1	8	26	19	47
Spectrophotometers, Polarimeters, Fluorimeters														
Circular Dichroism Analyzer	62	1	1	—	2	1	—	—	46	—	2	2	1	6
Infrared Spectrophotometer	959	9	38	14	70	28	5	14	430	3	38	80	118	112
Microspectrophotometer	345	1	55	22	39	5	2	11	101	9	15	16	55	14
Spectrofluorimeter	891	14	48	16	96	7	2	15	354	6	38	144	120	31
Spectropolarimeter	208	3	8	2	14	1	2	2	147	1	—	6	5	19
Ultraviolet Spectrophotometer	3,230	29	269	105	329	42	11	80	1,552	13	90	209	340	161
X-Ray														
X-Ray Crystallographic Analysis System	133	—	12	3	6	10	—	—	45	4	4	4	9	36
X-Ray Diagnostic System	484	7	27	14	140	5	3	12	46	16	9	35	144	26
X-Ray Source	480	4	82	29	77	13	19	59	68	15	8	12	70	24
Miscellaneous														
Apparatus for Measuring Fast Chemical Reactions	156	2	9	3	14	2	—	2	75	—	11	7	19	12
Artificial Kidney	129	1	11	3	36	—	—	1	15	2	2	10	42	6
Cine and Time–Motion Analysis Equipment	305	17	57	23	45	13	4	7	28	7	1	7	83	13
Closed-Circuit TV	387	18	24	11	50	11	5	11	43	15	6	58	112	23
Electrophoresis Apparatus (Various Types)	2,380	13	199	95	398	32	23	88	1,051	13	70	92	221	85
Intensive-Care Patient-Monitoring System	215	6	9	6	67	2	—	4	18	2	1	20	76	4
Infrared CO_2 Analyzer	360	1	15	13	22	25	1	4	32	2	11	34	193	7
Laser System	66	1	13	1	6	1	1	1	21	1	1	6	8	5
Large-Scale Fermenter	268	—	25	6	16	3	3	5	153	1	8	5	23	20
Light-Scattering Photometer	163	—	14	5	19	5	—	4	72	—	3	4	21	15
Microcalorimeter	53	1	1	—	5	17	—	—	14	—	4	1	7	3
Multichannel Oscilloscope	777	29	39	21	58	12	3	5	88	8	6	118	361	29
Multichannel Recorder	1,289	40	48	29	109	49	5	15	167	10	39	205	501	72
Osmometers	502	2	51	12	63	7	3	6	89	3	15	48	187	16
Small Specialized Computer System (CAT/LINC, etc.)	250	9	2	3	15	4	2	6	53	4	3	30	102	17
Stimulus Programming and Operant Conditioning Equipment	130	19	2	3	3	2	1	—	5	1	1	40	42	11
Telemetering System	194	11	4	2	14	27	2	3	13	1	5	25	71	16

[a] Number of life scientists reporting use of one or more specialized instruments.
Source: Survey of Individual Life Scientists, National Academy of Sciences Committee on Research in the Life Sciences.

postdoctoral fellows, graduate students, and technicians who work with him. According to data collected by the Study of Postdoctoral Education of the National Academy of Sciences,* the mean such research group, in addition to the faculty member, is 6.1 members in academic biology departments, 7.6 in biochemistry departments, 5.3 in physiology departments, and 4.0 in clinical specialties. These may be compared with 5.8 members in physics and 8.3 in chemistry. When, however, research groups without postdoctorals are considered, these units are distinctly smaller, receding to 4.6, 3.9, and 4.0 in biology, biochemistry, and physiology, respectively, and 3.2 and 5.2 in physics and chemistry.

This scale of operation was borne out by reports from the individual investigators surveyed in the study. For all principal investigators, the mean was 6.5 persons per research group, in addition to the principal investigator himself, ranging from 4.4 for investigators engaged in studies of systematic biology to 8.0 for those studying disease mechanisms. Perhaps surprisingly, the sizes of groups were much the same in academic and nonacademic laboratories. Approximately equal numbers of co-investigators and professional staff are found in both classes of laboratories. The graduate students, who vary in academic laboratories from 1.5 to 4.0 students per group (the extremes being represented by morphology and behavioral biology, respectively), with an overall average for all biological disciplines of 2.2 students per group, are replaced in nonacademic laboratories by technicians and other supporting staff.

Thus, in general, the typical academic laboratory contains a principal investigator, a co-investigator, and one other scientist with a doctoral degree who may be a visiting scientist, postdoctoral fellow, or continuing research associate, two technicians, and two or three graduate students. Federal laboratories may have one or two postdoctorals in place of the graduate students, while industrial laboratories utilize additional technicians. The routine tasks of the laboratory are generally performed by the technicians, while the graduate students and postdoctoral fellows serve as junior co-investigators and colleagues for the principal investigator. In our view, such a research group does indeed constitute something close to optimal for the conduct of "small science," particularly in the life sciences. Graduate students and postdoctorals are spared some of the drudgery of routine analyses after they have learned to perform such analyses and understand their limitations, and the total group combines a mixture of experience, expertise, ideas from other disciplines, and youthful enthusiasm. We can only conclude that, however haphazard the various mechanisms by which such an enterprise is funded, the average working unit is sufficiently large

---

* *The Invisible University: Postdoctoral Education in the United States,* Report of a Study Conducted under the Auspices of the National Research Council, National Academy of Sciences, Washington, D.C., 1969.

to attain an intellectual critical mass and to sustain the pace of exciting investigation while training the novice investigator for his future career.

Although this report gives emphasis to the research and education endeavor of the universities, it remains possible for dedicated scholars to pursue meaningful research in the biology departments of the independent four-year colleges. Biology is still mainly "small science," and research in many subdisciplines can be conducted with relatively modest support. When access to major equipment is required, this is frequently arranged with the faculty of a nearby university or undertaken during the summer at some properly equipped institution. These efforts constitute a significant part of the total life sciences research endeavor.

There are, however, important exceptions to this "small science" pattern. Decidedly larger aggregates of scientists, focused on a single goal, have been brought together to design a biological experiment for a space probe or to study the ecology of a major biome. The integrated approach to environmental research, stimulated by the International Biological Program, promises to open new levels of understanding of the functioning, resilience, and critical sensitivities of man-dominated ecosystems. In this program, teams of ecologists, social scientists, and physical scientists— as many as 150 individuals—cooperate in the analysis of entire ecosystems, such as the Western grasslands, the Eastern deciduous forests, or the Southwestern desert. Their data are compiled, coordinated, and utilized to construct mathematical models of these large systems, one day to be integrated with models of the atmospheres of the same regions. These systems involve so many components and multiple interactions that realistic abstractions or simplifications must be designed for simulation on large digital computers. The model is a combination of mathematical expressions and statistical probability distributions representing the processes and interactions of the system, as from soil to plant or plant to animal, and the impact of temperature on energy flow. A properly designed model can be used to suggest the potentially most fruitful field experiments from among the multitude that might be conducted, to identify gaps in existing knowledge through deficiencies in model performance, and to suggest optimal courses of action in managing real-world ecosystems. In the medical schools, large groups with representatives from several clinical or preclinical departments coalesce to collaborate on some aspect of cardiovascular, neurological, or neoplastic disease. These groups can number from 20 to 200 scientists and may well serve as forerunners of an era of "big biology."

WHAT DO LIFE SCIENTISTS DO?

The average life scientist employed in an institution of higher learning devotes about half his time to research, 10 to 20 percent to administration,

TABLE 18 Percentage Distribution of Work Time of Some Life Scientists

ORIGINAL FIELD OF TRAINING	ACADEMIC SCIENTISTS					NONACADEMIC SCIENTISTS				
	Research	Administration	Instruction	Patient Care	Other	Research	Administration	Instruction	Patient Care	Other
Ph.D. in										
An Agricultural Science	59	9	28	—	4	72	19	2	—	7
Anatomy	49	8	41	<1	2	79	13	7	—	1
Biochemistry	65	9	23	<1	2	78	16	2	<1	4
Microbiology	55	10	32	<1	3	70	22	2	<1	5
Pharmacology	60	8	27	<1	4	70	24	1	—	5
Physiology	55	9	34	—	2	73	18	3	<1	5
Botany	41	11	45	—	3	77	11	3	<1	5
Entomology	58	9	28	<1	4	72	18	2	—	9
Zoology	44	12	41	<1	3	64	21	5	<1	9
M.D.	52	13	19	11	4	56	16	9	14	5
Other Health Professionals	50	17	27	3	6	66	16	3	4	6

Source: Survey of Individual Life Scientists, National Academy of Sciences Committee on Research in the Life Sciences.

a fourth to a third of his time to instruction, and the balance to assorted other responsibilities. The actual distribution, of course, varies with the type of institution and the specific disciplinary field and according to whether he has clinical responsibility. This pattern is clearly in contrast with that of life scientists employed by nonacademic institutions, for whom research is, to an even greater extent, their dominant responsibility, demanding about 70 percent of their effort, while the remainder of their time is largely devoted to administrative responsibilities. Surprisingly, nonacademic scientists report that they engage in instruction that varies in percentages of their time from 0 to 10 percent—about 3 percent for the entire group but 8.5 percent for physicians. The physicians also give a sixth of their time to clinical care and hence can devote only about half their time to research. Some pertinent data in this regard are summarized in Table 18.

The same set of respondents, 6,125 scientists in academic institutions and 3,054 scientists in nonacademic institutions, were also queried with respect to whether the research in which they were engaged was basic, clinical, or applied. It was made clear that these designations were not necessarily mutually exclusive and, indeed, that an individual could check more than one of these categories if he felt that this was appropriate, particularly if he was engaged in more than one research project. Some of the resultant data are shown in Table 19. It is not surprising that scientists outside the academic world engage in applied and clinical research. But it may be surprising that 22 percent of all life scientists in institutions of higher learning indicated that their research is applied in some degree. By their own judgment, 76 percent of academically employed physicians indicate that they are engaged in basic research, and only 12 percent state that the research that they are doing is "applied" in some fashion. Quite logically, entomologists and the faculty of agriculture schools consider that a large fraction of their research is directed toward application. Conversely, while it was to be anticipated that 48 percent of all life scientists employed outside the academic world engage in applied research, the fact that 79 percent of all such scientists consider that they are engaged in some fundamental research was somewhat surprising. It indicates that the prejudices of many young scientists against careers outside the academic setting, for lack of opportunity to engage in basic research, may well be ill founded.

In any case, the reader will recognize that there is no meaningful close definition of the terms "basic" and "applied" in these regards and that these indications by our respondents reflect their motivation in addressing specific problems and not the character of the work. By this measure, one investigator studying sodium transport in human erythrocytes may classify it as "basic" research; another may consider the same study "clinical," only because human cells are employed for the purpose; and a third may view

TABLE 19   Types of Research Conducted by Some Life Scientists[a]

DISCIPLINE OF ORIGINAL TRAINING	ACADEMIC			NONACADEMIC		
	Basic	Clinical	Applied	Basic	Clinical	Applied
ALL BIOLOGISTS	**90**	**18**	**22**	**79**	**19**	**48**
Ph.D. in						
Agronomy	71	1	91	72	1	85
Fish and Wildlife	54	8	54	80	—	90
Forestry	75	—	78	71	—	90
Anatomy	99	6	1	100	—	11
Biochemistry	99	7	5	87	14	36
Microbiology	94	6	23	72	15	62
Animal Pathology	100	10	—	—	—	100
Pharmacology	97	5	9	82	15	68
Physiology	97	5	14	86	15	47
Botany	99	1	8	89	2	48
Entomology	91	<1	54	77	1	73
Ecology	96	1	24	80	2	66
Plant Pathology	90	3	71	86	—	80
Zoology	99	<1	7	86	3	36
M.D. Only	**76**	**65**	**12**	**68**	**59**	**22**

[a] All numbers represent the percentage of responding scientists in each class of employing institutions who indicated their work was, in any degree, basic, applied, or clinical, and horizontal columns are not additive since a respondent may indicate more than one type of research. The figures do not reflect the proportions of their effort.
Source: Survey of Individual Life Scientists, National Academy of Sciences Committee on Research in the Life Sciences.

it as applied, since he hopes to develop a new drug. Taking into consideration these broad caveats, the data of Table 19 provide a useful description of the world of biological research.

FINANCIAL SUPPORT OF RESEARCH IN THE LIFE SCIENCES

Research in the life sciences is a substantial national enterprise in which the United States invested $2,264 million in fiscal year 1967*; of this, 30 percent was provided by industry, 4.1 percent by foundations and other private granting agencies, 1.2 percent by academic institutions from their own resources, 0.3 percent by local and state governments, and 60.3 percent

* *Basic Data Relating to the National Institutes of Health 1969,* Associate Director for Program Planning and Evaluation and the Division of Research Grants, National Institutes of Health. U.S. Government Printing Office, 1969, p. 4.

by the federal government, principal patron of the endeavor. Table 20 summarizes federal expenditures for life science research in fiscal year 1968. Research supported by industry was largely conducted in-house. In all, biomedical research conducted within federal laboratories required the expenditure of approximately $435 million. In part because of the proprietary nature of industrial biomedical research, and largely because the "principal investigator" in industrial and federal laboratories functions with a large supporting organization for whose expenditures he is not responsible, it was patently impossible to obtain, by questionnaire, meaningful data concerning research expenditures from individual scientists in these two sectors. Our data, therefore, are restricted to information provided by individual life scientists employed by academic institutions and by academic department chairmen. Only the former are considered in this chapter; the latter are discussed in the succeeding chapter. The collected data, summarized in Tables 21, 22, and 23, indicate that in fiscal year 1967 the 4,046 responding academic life scientists, each of whom was principal investigator of one or more research grants or contracts, had available to them, collectively, $162,883,000 in support of the direct costs of research. The growth of this system is indicated by the fact that, in the previous year, the same investigators had available $134,726,000 and, in the prior year, $115,319,000. It is most unfortunate that we have no data for the same group in fiscal years 1969 or 1970, and, hence, no realistic data base with which to examine the consequences of the alterations in federal funding of science that have occurred since our questionnaires were distributed.

It will be seen that, using our categorizations of the life sciences, molecular biology and biochemistry commanded one fourth of all reported support, a substantial fraction of which went to individuals with appointments in clinical departments. Following, in rank order, were physiology (17 percent) and disease mechanisms (14 percent). Only 1 percent of the total support went to scientists who stated that they were studying morphological problems and 2 percent, each, to those engaged in behavioral biology and in the study of systematic biology and evolution, with other research areas distributed in between.

The magnitude of support reported for the research area of disease mechanisms is disturbing in that, proportionally, it is very significantly under-represented. While the relative support per research area for all other areas may be considered a reasonably fair indication of the fraction of total national support that they command, this is surely not the case for disease mechanisms, presumably due to the disproportionately low response to our questionnaire by clinical investigators. Thus, it is highly doubtful that the support of research directly concerned with disease mechanisms

TABLE 20 Federal Obligations for Research in Life Sciences, by Agency and Discipline—Fiscal Year 1968 (In Thousands of Dollars)

AGENCY AND SUBDIVISION	ALL RESEARCH IN THE LIFE SCIENCES			BASIC RESEARCH IN THE LIFE SCIENCES			APPLIED RESEARCH IN THE LIFE SCIENCES		
	Total	Biological	Clinical Medical	Total	Biological	Clinical Medical	Total	Biological	Clinical Medical
TOTAL ALL AGENCIES	**1,537,362**	**635,976**	**901,386**	**638,265**	**429,378**	**208,887**	**899,097**	**206,598**	**692,499**
Departments									
Department of Agriculture Total	**149,768**	**129,466**	**20,302**	**69,347**	**56,411**	**12,936**	**80,421**	**73,055**	**7,366**
Agricultural Research Service	88,547	70,793	17,754	45,083	33,115	11,968	43,464	37,678	5,786
Cooperative State Research Service	38,715	36,167	2,548	14,405	13,437	968	24,310	22,730	1,580
Forest Service	22,506	22,506	—	9,859	9,859	—	12,647	12,647	—
Department of Defense Total	**105,114**	**56,374**	**48,740**	**31,124**	**13,416**	**17,708**	**73,990**	**42,958**	**31,032**
Department of the Army	52,401	20,591	31,810	19,903	6,856	13,047	32,498	13,735	18,763
Department of the Navy	12,117	4,872	7,245	8,487	3,826	4,661	3,630	1,046	2,584
Department of the Air Force	31,484	22,384	9,100	2,734	2,734	—	28,750	19,650	9,100
Defense Agencies	8,072	7,823	249	—	—	—	8,072	7,823	249
Departmentwide Funds	1,040	704	336	—	—	—	1,040	704	336
Department of Health, Education, and Welfare Total	**950,989**	**159,250**	**791,739**	**316,251**	**148,346**	**167,905**	**634,738**	**10,904**	**623,834**
Consumer Protection and Environmental Health Service	53,159	17,121	36,038	18,876	9,230	9,646	34,283	7,891	26,392
Health Services and Mental Health Administration	99,248	13,101	86,147	12,215	10,388	1,827	87,033	2,713	84,320
	779,706	128,728	650,978	285,160	128,728	156,432	494,546	—	494,546

Agency									
Bureau of Land Management	691	691	—	5,333	5,333	—	15,914	15,914	—
Bureau of Reclamation	16	16	—	15	15	—	691	676	15
Bureau of Sport Fisheries and Wildlife	11,882	11,882	—	4,081	4,081	—	7,801	7,801	—
Federal Water Pollution Control Administration	6,828	6,828	—	366	366	—	6,462	6,462	—
Geological Survey	420	420	—	420	420	—	—	—	—
National Park Service	401	401	—	401	401	—	—	—	—
Office of Water Resources Research	1,962	1,962	—	392	392	—	1,570	1,570	—
Department of Justice, Bureau of Narcotics and Dangerous Drugs	356	356	—	—	—	—	356	356	—
Department of State Total	3,895	3,419	476	—	—	—	3,419	3,419	476
Agency for International Development	3,879	3,419	460	—	—	—	3,419	3,419	460
Peace Corps	16	—	16	—	—	—	—	—	16
Department of Transportation Total	1,481	1,247	234	—	—	—	1,247	1,182	55
Federal Aviation Administration	1,299	1,182	117	—	—	—	1,182	1,182	—
Federal Highway Administration	172	55	117	—	—	—	55	55	—
United States Coast Guard	10	10	—	—	—	—	10	10	—
Atomic Energy Commission	79,363	72,072	7,291	69,675	66,384	3,291	9,688	5,688	4,000
National Aeronautics and Space Administration	98,313	91,713	6,600	74,560	71,141	3,419	23,753	20,572	3,181
National Science Foundation	57,204	57,204	—	55,501	55,501	—	1,703	1,703	—
Office of Economic Opportunity	311	—	311	311	—	311	—	—	—
Office of Science and Technology	271	155	116	100	57	43	171	98	73
Smithsonian Institution	5,272	5,272	—	5,272	5,272	—	—	—	—
Tennessee Valley Authority	1,621	1,621	—	—	—	—	1,621	1,621	—
Veterans Administration	39,957	14,380	25,577	5,115	1,841	3,274	34,842	12,539	22,303

Source: National Science Foundation Federal Funds for Research, Development, and Other Scientific Activities, Volume XVIII, U.S. Government Printing Office, Washington, D.C., 1969.

TABLE 21 Financial Support of Academic Research in the Life Sciences (In Millions of Dollars)

SUPPORTING ORGANIZATION	TOTAL	RESEARCH AREA												
		Behavioral Biology	Cellular Biology	Developmental Biology	Disease Mechanisms	Ecology	Evolutionary and Systematic Biology	Genetics	Molecular Biology and Biochemistry	Morphology	Nutrition	Pharmacology	Physiology	Related Research Areas
TOTAL, ALL SOURCES	162.9	4.1	12.3	7.8	22.2	8.5	3.0	10.1	40.3	2.4	5.6	9.7	27.2	9.9
Own Institution	25.1	0.36	1.4	2.0	3.0	2.7	0.54	3.3	4.2	0.39	2.0	0.98	3.8	1.5
Federal Government Subtotal	120.4	3.2	9.8	4.7	16.1	4.9	2.2	6.0	32.2	1.9	2.8	7.8	21.1	7.7
Department of Agriculture	5.2	0.11	0.13	0.08	0.71	0.68	0.07	0.64	0.78	—	0.72	<0.01	0.89	0.38
Atomic Energy Commission	7.3	0.15	0.31	0.25	0.51	0.94	0.01	0.86	2.21	0.04	1.0	0.08	1.4	0.41
Department of Defense [a]	6.2	0.33	0.26	0.02	1.1	0.19	0.05	0.07	0.61	0.03	0.07	0.89	1.2	1.4
Department of Health, Education, and Welfare Subtotal	85.2	2.1	7.7	3.3	12.7	1.4	0.78	0.5	24.1	1.6	1.5	6.7	15.1	4.7
National Institutes of Health (NIH)	75.5	1.7	7.3	2.9	11.2	0.66	0.74	3.2	20.6	1.5	1.3	6.5	14.0	3.7
Public Health Service (other than NIH)	6.3	0.23	0.29	0.27	1.2	0.71	0.04	0.23	1.2	0.0	0.07	0.14	0.9	0.84
Other Department of Health, Education, and Welfare	3.4	0.15	0.14	0.13	0.31	0.03	—	0.06	2.2	—	0.14	0.02	0.16	0.1
Department of the Interior	1.2	0.01	—	0.02	0.5	0.57	0.06	0.07	0.02	0.03	0.04	—	0.16	0.13
National Aeronautics and Space Administration	2.2	0.08	0.12	0.01	0.33	0.03	0.04	0.1	0.23	—	0.11	0.05	0.91	0.15
National Science Foundation	11.6	0.35	1.2	1.0	0.32	0.93	1.2	0.7	4.1	0.2	0.05	0.07	1.2	0.3
Other Federal	1.6	0.07	0.02	0.01	0.36	0.16	0.01	0.08	0.23	0.01	0.14	0.01	0.26	0.21
State and Municipal Agencies	2.4	0.18	0.15	0.03	0.37	0.33	0.09	0.2	0.36	0.01	0.17	0.02	0.33	0.2
Industry	2.9	0.13	0.06	0.09	0.48	0.26	0.01	0.2	0.38	0.02	0.38	0.39	0.39	0.07
Private Foundations	7.4	0.16	0.55	0.8	1.4	0.26	0.03	0.12	2.2	0.1	0.24	0.3	1.0	0.24
Voluntary Societies	2.3	0.01	0.33	0.13	0.51	<0.01	0.02	0.03	0.69	<0.01	0.07	0.11	0.33	0.07
Other [b]	1.4	0.02	0.05	0.02	0.24	.01	0.03	0.26	0.18	0.01	0.03	0.12	0.26	0.13

[a] Includes Air Force, Army, Navy, and other Defense agencies.
[b] Includes all other institutions and other sources of support amounting to 5 percent or more of the total support.
Source: Survey of Individual Life Scientists, National Academy of Sciences Committee on Research in the Life Sciences.

TABLE 22 Numbers of Research Grants and Contracts Awarded to 4,046 Academic Life Scientists

SUPPORTING ORGANIZATION	TOTAL	Behavioral Biology	Cellular Biology	Developmental Biology	Disease Mechanisms	Ecology	Evolutionary and Systematic Biology	Genetics	Molecular Biology and Biochemistry	Morphology	Nutrition	Pharmacology	Physiology	Related Research Areas
TOTAL, ALL GRANTS	5,479	192	386	273	649	386	160	286	1,393	84	251	274	925	220
Federal Government Subtotal	4,097	120	317	194	438	279	146	212	1,108	70	149	184	707	173
Department of Agriculture	392	12	13	9	36	59	9	41	45	—	64	—	66	38
Atomic Energy Commission	132	2	11	8	6	17	2	14	38	1	2	3	21	7
Department of Defense [a]	181	9	9	3	35	16	4	1	36	2	4	14	36	12
Department of Health, Education, and Welfare Subtotal	2,435	59	220	105	330	32	27	107	758	52	58	157	451	79
National Institutes of Health (NIH)	2,202	48	212	91	288	22	24	95	700	49	51	150	406	66
Public Health Service (other than NIH)	185	10	5	11	33	10	3	10	40	3	6	6	37	11
Other Department of Health, Education, and Welfare	48	1	3	3	9	—	—	2	18	—	1	1	8	2
Department of the Interior	82	1	—	1	4	48	5	—	2	1	1	—	11	8
National Aeronautics and Space Administration	56	5	2	1	1	1	3	2	12	—	3	3	21	2
National Science Foundation	711	25	60	65	19	88	94	42	196	10	4	7	82	19
Other Federal	108	7	2	2	7	18	2	5	21	4	13	—	19	8
State and Municipal Agencies	192	11	7	5	16	30	4	22	40	3	10	2	31	11
Industry	525	43	3	28	91	54	1	27	54	3	71	54	76	20
Private Foundations	449	14	31	33	70	21	5	21	133	7	17	21	66	10
Voluntary Societies	188	2	26	12	28	2	4	4	52	1	4	12	37	4
Other [b]	28	2	2	1	6	—	—	—	6	—	—	1	8	2

[a] Includes Air Force, Army, Navy, and other Defense agencies.
[b] Includes all other institutions and other sources of support amounting to 5 percent or more of the total support.
Source: Survey of Individual Life Scientists, National Academy of Sciences Committee on Research in the Life Sciences.

269

TABLE 23 Average Size of Research Grant (Direct Costs) in Thousands of Dollars

SUPPORTING ORGANIZATION	TOTAL	Behavioral Biology	Cellular Biology	Developmental Biology	Disease Mechanisms	Ecology	Evolutionary and Systematic Biology	Genetics	Molecular Biology and Biochemistry	Morphology	Nutrition	Pharmacology	Physiology	Related Research Areas
AVERAGE, ALL GRANTS	20	16	23	19	23	11	14	18	22	20	12	28	21	33
Federal Government, Average	24	22	25	21	30	13	14	22	24	22	16	39	25	40
Department of Agriculture	11	8	8	8	15	9	6	13	14	—	10	—	11	9
Atomic Energy Commission	30	67	18	29	62	31	6	43	26	14	43	26	24	27
Department of Defense [a]	24	20	16	8	28	12	11	67	16	14	17	52	29	113
Department of Health, Education, and Welfare, Average	29	31	29	28	33	27	26	26	28	26	23	39	29	54
National Institutes of Health (NIH)	29	30	28	28	33	27	27	26	27	27	23	40	30	54
Public Health Service (other than NIH)	24	23	53	25	31	28	14	23	22	14	7	22	21	58
Other Department of Health, Education, and Welfare	44	150	33	20	22	—	—	26	65	—	136	10	20	20
Department of the Interior	12	9	—	20	11	11	11	—	10	31	40	—	15	13
National Aeronautics and Space Administration	26	16	27	4	33	6	12	50	18	—	20	12	36	75
National Science Foundation	14	13	17	14	14	10	12	14	18	12	11	10	11	14
Other Federal	9	9	10	6	5	7	5	7	11	2	11	—	13	22
State and Municipal Agencies	10	9	15	6	17	8	24	9	9	2	11	7	10	14
Industry	4	3	4	3	5	3	2	5	4	7	3	6	3	3
Private Foundations	13	8	15	22	14	7	3	5	13	13	11	11	13	17
Voluntary Societies	10	3	11	10	13	1	5	6	11	4	12	8	7	18
Other [b]	8	2	8	1	4	—	—	11	11	—	—	—	13	9

RESEARCH AREA

[a] Includes Air Force, Army, Navy, and other Defense agencies.
[b] Includes all other institutions and other sources of support amounting to 5 percent or more of the total support.
Source: Survey of Individual Life Scientists, National Academy of Sciences Committee on Research in the Life Sciences.

by the National Institutes of Health is only 15 percent of its extramural research program, since half of its total extramural research support is granted to clinical investigators.

Caution is necessary in interpreting these data, however, because of the failure of the questionnaire to be sufficiently precise in guiding the respondents. Although "disease," broadly taken, is the concern of clinicians and pathologists, there are no aspects of the study of disease, other than access to human patients, that are unique to their endeavors. In addressing himself to cardiac disease, the clinician may actually function as a physiologist who studies vector cardiography or analyzes the composition of blood obtained by catheterization of one of the cardiac chambers; or he may be concerned with the etiology and pathogenesis of atherosclerosis and so utilize the techniques and understanding of the biochemist or nutritionist. Concerned with a hereditary disorder, he may consider himself a human geneticist; if studying changes in the architectonics of the brain, he may view himself as a morphologist or even a student of evolution. If engaged in elucidation of the causative agent of an infectious disease, he may function, variously, as a cell biologist or a biochemist, while, if he is testing a drug in the hope of finding a successful therapeutic procedure, he is, at least for the time being, a pharmacologist. Accordingly, it is entirely possible that students of disease, its etiology, pathogenesis, incidence, or therapy, may well have indicated that their current research area lies in some category other than "disease mechanisms," thus unintentionally distorting the interpretation that might be applied to these data.

The pattern of support from the National Science Foundation contrasts with that from the National Institutes of Health. Both supported molecular biology and biochemistry more heavily than any other category, but, whereas the National Institutes of Health also contributed in a large way to the study of physiology and disease mechanisms, the National Science Foundation was clearly the principal supporter of systematic biology. The Atomic Energy Commission and the Department of the Interior, while contributing only 4 percent and 1 percent, respectively, to the total support of these life sciences, were particularly concerned with ecology. The principal thrust of support by the National Aeronautics and Space Administration, which contributed only 1 percent of the reported federal total, was in physiology, while only the Department of Agriculture and diverse industrial contributors allocated as much as one seventh of their research funds to studies involving nutrition.

Of interest is the fact that, whereas the voluntary societies were organized to combat the dread diseases, only 22 percent of their funds went to scientists who classified their own research as bearing directly on disease mechanisms, whereas one third of their support went to investigators in molecular biology and biochemistry, and one seventh each to studies of physiology and cellular biology. Clearly, the administrators of these societies were sufficiently understanding of the problems involved in treating and preventing these diseases to recognize the need for relevant basic research.

Table 21 indicates clearly that indeed the federal government is the principal patron of these areas of scientific endeavor. Three fourths of all funds in direct support of research derived from the federal government, while one sixth of such funds was provided out of the academic institutions' own resources. The low figures quoted for support by state and municipal agencies refer to direct granting activity, but the state budgets for the public universities contributed in major degree to the 16 percent of all directly research-supporting funds that are stated to have come from the institutions' own resources.

Particularly disappointing is the low order of contribution to research support provided by industry, private foundations, voluntary societies, and individual contributors shown in Table 21. This is the consequence not so much of a low frequency of granting activity as it is of the relatively small awards actually made by these sources, as shown in Tables 22 and 23. Thus, the average grant from industry was only $4,000, that from the voluntary societies, $10,000, and that from private foundations, $13,000. These figures are in contrast to grants from the National Science Foundation ($14,000), the National Institutes of Health ($30,000), and the federal average of $25,000.

Of some interest is the pattern of support by discipline. Typical grants in nutrition, ecology, and systematic biology are of the order of $15,000 per year, whereas grants to investigators in most of the other research areas were about twice as large.

Utilization of Research Grants

Typically, a research grant is utilized to provide consumable supplies, major and minor equipment, salaries of technicians and clerical staff, travel and publication costs, stipends for graduate students, postdoctoral fellows, and visiting investigators, as well as a variable fraction of the salary of the principal investigator not to exceed that fraction of his annual effort invested in the research project in question. Uniquely, research grants to clinical investigators may require expenditures in support of the basic costs of maintaining patients in hospitals; other grants may provide for unusual purposes such as ship time, international travel either to meetings or for work in the field, and, increasingly frequently, computer time. The relative

TABLE 24   Utilization of Funds from an Average Two-Year Research Grant in the Life Sciences—National Science Foundation—1968

	DOLLARS	PERCENTAGE
GRAND TOTAL	**39,961**	*100.0*
Total Direct Costs	**32,184**	*80.5*
Salaries + Wages, Subtotal	19,699	*49.3*
Principal + Coinvestigators	3,901	*9.8*
Research Associates	3,529	*8.8*
Graduate Students	3,264	*8.2*
Other Professional Personnel	3,551	*8.9*
Technical Personnel	3,573	*8.9*
Clerical Personnel	588	*1.5*
Fringe Benefits on All Above	1,293	*3.2*
Equipment	4,123	*10.3*
Supplies	4,374	*11.0*
Travel, Subtotal	1,346	*3.4*
Within the United States	978	*2.5*
International	368	*0.9*
Publication	505	*1.3*
Computer Time	152	*0.4*
Other	1,986	*5.0*
Total Indirect Costs	**7,777**	*19.5*

distribution of expenditures among these various areas from research grants in support of research in the life sciences was not ascertained by the present study. However, data describing the general patterns of funding by the National Science Foundation are summarized in Table 24.

Research Support as a Function of the Investigator's Age

In a general way, increasing research support comes to the academic investigator as he gains seniority in the system. As shown in Figure 34, this is clearly true for investigators supported by the National Institutes of Health

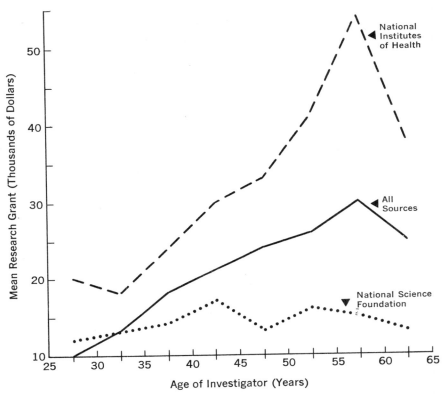

FIGURE 34   Research support of life scientists as a function of their age. (Source: Survey of Individual Life Scientists, National Academy of Sciences Committee on Research in the Life Sciences.)

and most other sources. The figures shown for "all sources" represent the simple arithmetic means for all grants from all sources. Because of the relatively large number of small grants from the National Science Foundation, industry, foundations, and voluntary societies, the mean grant size for all sources is decidedly less than that shown for the National Institutes of Health. Nevertheless, the trend is quite apparent: individual research support attains a maximum at 50 to 60 years of age and declines thereafter. This phenomenon is scarcely visible for the National Science Foundation, largely because this beleaguered agency strives to stretch its available resources as far as it can to support all qualified applicant investigators whose proposals fall within its purview, thus markedly reducing the amount of money available per applicant investigator.

RESEARCH INSTITUTES

The preceding survey of the major parameters of the world of biological research fails to convey the myriad arrangements for both research and education in biology. It ignores the dozens of small research institutes in which excellent investigators quietly pursue their research, occasionally with profound impact on the conceptual development of biology. The Cold Spring Harbor Laboratory for Quantitative Biology has had a brilliant record of achievement, and its summer courses have trained virtually all those who have led the modern development of virus and bacterial genetics, a major segment of molecular biology. Developmental biology and some aspects of neurophysiology have received great stimulus from the research and education programs of marine-biology stations such as that at Woods Hole, Massachusetts. Much of the current understanding of neurochemistry and the physiology of the brain has been obtained at small research institutes under private or state auspices, while ecology has grown at a multitude of field stations remote from their parent institutions.

NATURAL HISTORY MUSEUMS

Natural history museums, with their combinations of scientists, research collections, and field stations are unique non-degree-granting academic institutions for research and graduate training. Quite apart from its role in public education through exhibits, a natural history museum contributes to the acquisition of scientific knowledge in two principal ways.

1. Its staff of scientists may engage in original research in systematic biology, evolutionary biology, ecology, geophysics, astrophysics, oceanography, and many other fields of science, depending upon their academic training and scientific interests. While many museum scientists depend on specialized collections in conducting their investigations, an increasing number engage in field and laboratory experimental studies of living organisms, or of ecological problems in natural settings. Their collections provide the basis for taxonomic-classification services necessary to many other scientists and also provide a base line for ecological studies.

2. The combination of resident scientists, research collections, and field research facilities provides intellectually attractive settings for visiting scientists. The number of graduate students who receive part or all of their graduate training in natural history museums is impressive and increasing.

Natural history museums, as both forums and research settings for systematists, ecologists, and environmental scientists, are becoming increasingly important as a national scientific resource, despite a long history of public neglect.

BIOLOGICAL DISCIPLINES

For brevity and conciseness, we found it useful to structure all the life sciences into a dozen research areas. But this should not conceal the rich and diverse infrastructure of the life sciences. As we have seen, classical disciplinary labels have lost their meaning, but one could readily describe a hundred or more subdisciplines based on the work of groups of like-minded scientists who have blended the approaches of several older disciplines in attacks on some specific subsets of biological problems. A few examples are cited in the following paragraph.

Photobiologists, well versed in optics and the physics of light, are variously concerned with the mechanism of vision, the events in photosynthesis, the emission of light by bacterial and animal forms (the biological purpose of light emission by all but fireflies being not at all evident), and the photo-inactivation of enzymes and viruses. Neuroscientists bring the skills of electrophysiology, cellular biology, molecular biology, and communications theory to bear on studies of information processing in the nervous system. Oncologists, focusing on the essential nature of the transformation of normal cells into malignant ones, are similarly a group apart, borrowing from every major discipline that may be of help, while vascular physiologists necessarily borrow from hydrodynamics and studies of urban traffic flow as they study the operation of a capillary bed or a major blood vessel. Physical anthropology is a subdiscipline that contributes to the total endeavor while it provides a bridge from the biological to the social sciences. It is the study of the bodily manifestations of human variation—in particular, the description of human body size, shape, and function in the light of man's history—and the role of heredity, environment, and culture in bringing about man's present diversity. The biological anthropologist aims to understand human physical variation and to apply his knowledge for human betterment through medicine and engineering.

As concern with the environment grows, an increasing number of physicians and biologists of many backgrounds have generated the area of research and practice called "environmental health," the concern of one of the panels of this survey. More sophisticated understanding of this field should permit society to enjoy the fruits of an advancing technology, a superior living environment, and freedom to develop a society with fewer restraints and tensions. Past effort is minuscule compared with the magnitude of the problem. Since the problems increase with increasing population density and developing technology, efforts at controlling the environment, and thus the health of the population, must keep pace. Indeed, in a very real sense, students of environmental health serve technology by providing the knowledge permitting its benefits to be enjoyed without adventitious adverse effects on the health of man and, more broadly, on the environment of man. Thus, support of an adequate level of competence in environmental health is indispensable to a society that elects to make optimal use of the fruits of technology. Accordingly, the environmental-health resources of the nation must first be expanded to catch up with the problems now with us and thereafter be developed, along with technological development, to provide an adequate preventive program. Current support of research in environmental health probably lies between $30 million and $50 million per year; support for training for both research and practice is between $9 million and $18 million per year and is known to support (in 1969) 974 candidates for the master's degree, 981 candidates for the Ph.D., and 148 postdoctoral fellows.

A broad federal policy is needed, with a long-range plan of attack upon the whole problem of environmental deterioration and with better identification of the separate missions and responsibilities of the several federal departments and agencies. Only with such a policy will it be possible to develop in an orderly way the required training programs to supply the personnel needed for both research and practice, both within and outside the government, necessary to build a strong foundation for effective control programs against environmental-health hazards, a foundation that must rest on the entire current understanding of the life sciences.

Thus, the world of research in the life sciences is marvelously diverse. Tens of thousands of scientists in a thousand institutions contribute to its progress. They migrate between institutions, between classes of institutions, and between subfields of biology. They are quick to seize upon any new instruments or techniques, without regard to whether these are initially devised for use in the physical sciences or for some other research area in the life sciences. Biochemistry has become the language of biology, providing the bridge to the physical sciences, but it has yet to be applied to the farthest reaches of organismal biology. The federal government is the principal sponsor of the entire endeavor and, for the indefinite future, only the federal government can sponsor an effort of this magnitude. Its success will affect all aspects of our lives, and its conduct has become one of the central purposes of our civilization.

CHAPTER FOUR

THE ACADEMIC
ENDEAVOR IN THE
LIFE SCIENCES

The present studies, which revealed a wealth of detail descriptive of the life sciences endeavor in academic institutions, are insufficient to characterize research in the life sciences in private industry, nonprofit research institutes, or government laboratories. This is a consequence of a number of circumstances: the greater ease of identification of academic institutions, their departments, and individual academic investigators; the diversity of organizational forms in both industry and government, which make it difficult to identify the equivalent of an academic "principal investigator"; the difficulty of locating industrial laboratories that employ significant numbers of research-performing biological scientists; and the fact that most previous statistical compilations have focused on academia.

Our studies have also provided a more complete picture of life in schools of arts and sciences and agriculture and a clearer picture of the preclinical component of medical schools than of clinical departments. Thus, our pair of questionnaires located approximately equal numbers of preclinical and clinical academic faculty, yet the latter are known to comprise 69 percent of total medical faculty. The very nature of our questionnaires, with their emphasis on elements of graduate education, undoubtedly discouraged responses from numbers of clinical investigators and chairmen of clinical departments. And the task of completing the chairmen's questionnaire might well have appalled the chairmen of some very large clinical departments. In consequence, the funds in support of clinical research reported to us cannot be reconciled with the much larger amounts known to have been committed by federal agencies, particularly the National Institutes of Health. Withal, it has been possible to construct a quantitative description of the academic research endeavor in the life sciences that is adequate to our task.

A total of 1,256 academic departmental chairmen provided our information; their data are summarized in Tables 25 and 26. Of these, 246 chaired departments in colleges of arts and sciences (including engineering and graduate studies), 267 chaired departments in colleges of agriculture (including forestry), and 694 chaired departments in colleges of medicine, approximately equally divided between clinical and preclinical departments. Decidedly smaller groups from schools of dentistry, pharmacy, public health, and veterinary medicine were treated together as "other health-professional." One third of all departments were in private universities; two thirds were in state universities, and only 26 departments were in municipal institutions. Collectively, these departments reported 17,172 faculty members, of whom three fourths were devoting 20 percent or more of their effort to research, and 1,436 were continuing senior research associates. They employed just under 11,000 technicians and animal care personnel, 512 business and laboratory managers, 6,416 supporting personnel, and 6,700 clerical and secretarial staff. Of the 5,223 postdoctoral appointees, 1,945 were post-M.D. and 3,278 post-Ph.D. In addition, there were stated to be another 1,546 M.D.'s engaged in residency training that included a significant research component. Of all postdoctoral appointees, 2,013 were foreign nationals, only 309 of whom had received their doctoral degrees in the United States.

These departments, containing 23,287 graduate students, of whom 15,755 were adjudged to be "potential Ph.D. candidates," collectively awarded 2,332 Ph.D.'s in academic year 1966–1967. In addition, 2,138 medical students, not enrolled in programs leading to the Ph.D. degree, were engaged in research among these departments.

Collectively, these 17,172 individuals utilized 13,423,000 net square feet of useable research space, and their efforts were supported by $304 million (direct costs) in research grants, in addition to funds provided from institutional resources, training grants, and fellowships.

ACADEMIC DEPARTMENTS

Research in the life sciences is conducted in three major organizational entities of universities—colleges of arts and sciences, colleges of agriculture,

TABLE 25 The Academic Universe of Life Sciences Research

TYPE OF SCHOOL	DEPARTMENTS		FACULTY		POST-DOCTORALS		GRADUATE STUDENTS		Ph.D.'s IN 1967		SUPPORT PERSONNEL		FEDERAL RESEARCH SUPPORT		NON-FEDERAL RESEARCH SUPPORT		SPACE 1,000 ft²	
	No.	%	No.	%	No.	%	No.	%	No.	%	No.	%	$M	%	$M	%		%
TOTAL, ALL SCHOOLS	1,256	100	17,172	100	5,223	100	23,287	100	2,332	100	24,481	100	253.1	100	50.9	100	13,423	100
Arts and Sciences [a]	246	20	3,852	23	1,168	22	10,743	46	946	41	4,469	19	60.6	24	7.7	15	4,201	31
Agriculture [b]	267	21	3,907	23	355	7	6,373	27	771	33	4,270	17	23.5	9	12.8	25	3,374	25
Medical	694	56	8,915	52	3,498	67	5,658	25	555	24	15,062	61	163.3	65	29.5	58	5,561	41
Preclinical	361	29	3,777	22	1,466	28	4,348	19	523	22	6,936	28	81.6	32	9.3	18	3,434	26
Clinical	333	27	5,138	30	2,032	39	1,310	6	32	2	8,126	33	81.8	33	20.3	40	2,127	15
Other Health Professional [c]	49	4	498	3	202	4	513	2	60	3	680	3	5.6	3	0.82	1	286	2
TOTAL, PRIVATE	401	33	6,032	36	2,799	54	5,191	22	616	27	9,915	41	124.7	49	20.9	41	4,317	32
Arts and Sciences [a]	70	6	1,189	7	622	12	3,067	13	367	16	1,891	8	27.3	11	3.4	7	1,457	11
Agriculture [b]	—	—	—	—	—	—	—	—	—	—	—	—	—	—	—	—	—	—
Medical	322	26	1,812	28	2,146	41	2,057	9	230	10	7,885	32	96.3	38	17.3	34	2,803	21
Preclinical	162	13	1,812	11	775	15	1,586	7	215	9	3,203	13	43.8	17	4.6	9	1,547	12
Clinical	160	13	2,931	17	1,371	26	471	2	15	1	4,682	19	52.5	21	12.7	25	1,256	9
Other Health Professional [c]	9	1	100	1	31	1	67	<1	19	1	139	1	1.2	<1	0.18	<1	57	<1
TOTAL, PUBLIC	855	68	11,140	65	2,424	46	18,096	78	1,716	74	14,566	59	128.4	51	29.9	58	9,106	67
Arts and Sciences [a]	176	14	2,663	16	546	10	7,676	33	579	25	2,578	11	33.4	13	4.3	8	2,744	20
Agriculture [b]	267	21	3,907	23	355	7	6,373	27	771	33	4,270	17	23.5	9	12.8	25	3,374	25
Medical	372	30	4,172	24	1,352	26	3,601	16	325	14	7,177	29	67.0	27	12.2	24	2,758	20
Preclinical	199	16	1,965	11	691	13	2,762	12	308	13	3,733	15	37.7	15	4.7	9	1,887	14
Clinical	173	14	2,207	13	661	13	839	4	17	1	3,444	14	29.3	12	7.5	15	872	6
Other Health Professional [c]	40	3	398	2	171	3	446	2	41	2	541	2	4.5	2	0.64	1	229	2

[a] Includes schools of engineering and schools of graduate studies.
[b] Includes schools of forestry.
[c] Includes schools of dentistry, pharmacy, public health, and veterinary medicine.
Source: Survey of Academic Life Science Departments, National Academy of Sciences Committee on Research in the Life Sciences.

TABLE 26 The Academic Universe of Life Sciences Research

CLASS OF DEPARTMENT	DEPARTMENTS		FACULTY		POST-DOCTORALS		GRADUATE STUDENTS		Ph.D.'s IN 1967		SUPPORT PERSONNEL		FEDERAL RESEARCH SUPPORT		NON-FEDERAL RESEARCH SUPPORT		SPACE	
	No.	%	No.	%	No.	%	No.	%	No.	%	No.	%	$M	%	$M	%	1,000 ft²	%
TOTAL, ALL CLASSES	**1,256**	*100*	**17,172**	*100*	**5,223**	*100*	**23,287**	*100*	**2,332**	*100*	**24,481**	*100*	**253.1**	*100*	**50.9**	*100*	**13,423**	*100*
Biological Sciences Subtotal	**917**	*73*	**11,960**	*70*	**3,166**	*61*	**21,860**	*94*	**2,397**	*99*	**16,208**	*66*	**170.2**	*67*	**30.5**	*60*	**11,251**	*84*
Animal Husbandry	90	*7*	1,193	*7*	137	*3*	1,271	*5*	177	*8*	1,727	*7*	7.2	*3*	4.3	*8*	976	*7*
Agronomy and Forestry	96	*8*	1,652	*10*	78	*1*	2,319	*10*	230	*10*	1,557	*6*	5.4	*2*	5.0	*10*	1,163	*9*
Anatomy	65	*5*	690	*4*	194	*4*	670	*3*	75	*3*	910	*4*	9.4	*4*	0.90	*2*	523	*4*
Biochemistry and Nutrition	107	*9*	1,122	*7*	684	*13*	2,621	*11*	288	*12*	1,937	*8*	29.3	*12*	4.4	*9*	1,362	*10*
Biology and Ecology	111	*9*	1,853	*11*	654	*13*	4,589	*20*	397	*17*	2,367	*10*	29.6	*12*	3.5	*7*	1,962	*15*
Biophysics and Biomedical Engineering	16	*1*	205	*1*	105	*2*	338	*1*	48	*2*	597	*2*	7.6	*3*	0.42	*1*	291	*2*
Botany	52	*4*	821	*5*	97	*2*	1,694	*7*	165	*7*	542	*2*	6.4	*3*	1.4	*3*	701	*5*
Genetics	10	*1*	90	*1*	70	*1*	185	*1*	32	*1*	215	*1*	3.7	*1*	0.12	*<1*	189	*1*
Microbiology	87	*7*	792	*5*	276	*5*	1,781	*8*	188	*8*	1,473	*6*	19.3	*8*	2.3	*5*	878	*7*
Pathology	65	*5*	899	*5*	272	*5*	460	*2*	111	*5*	1,673	*7*	12.7	*5*	1.9	*4*	678	*5*
Pharmacology [a]	58	*5*	510	*3*	195	*4*	527	*2*	94	*4*	758	*3*	10.6	*4*	1.5	*3*	452	*3*
Physiology	75	*6*	746	*4*	252	*5*	1,113	*5*	108	*5*	1,112	*5*	13.9	*5*	2.0	*4*	700	*5*
Zoology and Entomology	85	*7*	1,387	*8*	152	*3*	4,292	*18*	385	*17*	1,340	*5*	15.2	*6*	2.7	*5*	1,375	*10*
Clinical Medical Sciences [b]	**339**	*27*	**5,212**	*30*	**2,057**	*39*	**1,427**	*6*	**34**	*1*	**8,273**	*34*	**82.9**	*33*	**20.3**	*40*	**2,172**	*16*

[a] Includes a college of pharmacy.
[b] Includes a department of oral biology.
Source: Survey of Academic Life Science Departments, National Academy of Sciences Committee on Research in the Life Sciences.

and medical schools. Departmental titles are remarkably varied in all three organizational components. Indeed, the 1,256 responses to this questionnaire revealed 195 distinct departmental titles; returns from individual biologists in their questionnaires indicated about 300 others! For our purposes, these were reclassified as Animal Husbandry, as Agronomy and Forestry, as one of 11 other scientific disciplines, or as clinical medical sciences. Although some of these compressions were rather arbitrary, there is little likelihood that they gave bias to summary data.

A major general trend is discernible in the academic colleges of arts and sciences. The unity of biology has compelled consolidation into single biology departments in the private universities, but, as yet, this trend is not nearly as pronounced in the public universities. Reports were obtained from 54 departments of biology in the arts and sciences faculties of private schools, and from 57 departments so titled in public universities. However, the private schools reported only four departments of botany and five departments of zoology, as compared with 48 and 80, respectively, in the public schools. The chief beneficiaries of this consolidation are the students, both undergraduate and graduate, since consolidated departments facilitate consolidated course planning and a unified presentation of current understanding of living systems. The relative sluggishness of public universities in this regard is not a reflection on the sophistication of their faculties but, rather, is the consequence of their teaching responsibilities to their very large student bodies. This is further reflected in the mean faculty sizes— 17 for the 54 biology departments in private schools and 16 in both the botany and zoology departments of the public schools. Were the public schools to consolidate these departments, their faculties would be twice the mean size of the consolidated departments of private schools. Moreover, on the campuses of the 57 public universities that do have consolidated biology departments, the mean faculty size is only 13, so that, as in the private sector, consolidation has occurred when the faculty group is not too large.

The other major departmental titles are those common to the preclinical component of medical schools: anatomy, biochemistry, microbiology, pathology, pharmacology, and physiology, to which were added, for separate analysis, the 16 departments of biophysics and the 10 departments of genetics that were reported. However, included within these categories in our tabulated data are numerous departments with similar titles that are not based in medical schools. Thus, of 107 reporting biochemistry departments, 24 are components of colleges of arts and sciences and 18 of agriculture schools. Nineteen departments of microbiology were components of arts and sciences faculties, while five were components of agriculture schools. One department of pathology was stated to be a component of an arts and sciences college, while 17 were actually departments of plant pathology.

Approximately half of all solicited department chairmen responded to the questionnaire. Test examinations revealed no particular bias; the percentage returned from public and private schools, preclinical medical school departments, and departments in the colleges of agriculture and arts and sciences were all much the same. A somewhat lower rate of return, however, was experienced from clinical departments, as indicated earlier. A distinct effort was made, by letter and telephone, to assure response from the largest known departments in each category; the failure rate in these instances was extremely low. Thus, the sample available to us represents a very large fraction of academia and, except for the clinical departments, in the main, the missing departments are likely to be departments with relatively small on-going programs of research and graduate education. Table 1, Appendix B, contains a summary of departmental returns.

Tables 25 and 26 summarize major aspects of responding departments according to their places in the organization of the university and by discipline, respectively.

The Life Sciences Faculty

In a general way, one fourth of the reported life sciences faculty functions in the arts and sciences schools, another fourth in the agriculture schools, and approximately half (52 percent) in the medical schools, of which the preclinical component represents 22 percent and the clinical component 30 percent. An additional 3 percent were on the faculties of a variety of such other health-professional schools as dentistry, veterinary medicine, and public health. The agriculture schools are, without exception, in state institutions or in institutions under combined state and private auspices. Thirty-six percent of all reported academic life scientists were in private institutions, where the arts and sciences group was about one third the size of the medical faculty. Two thirds of all reported faculty were in publicly sponsored institutions, which include, in addition to the agricultural faculty, the arts and sciences faculty (16 percent), the preclinical faculty (11 percent), and the clinical faculty (13 percent).

The title "instructor" has essentially fallen into disuse except in clinical departments. In the total system there were approximately equal numbers of professors, associate professors, and assistant professors; full professors were the largest single component of the arts and sciences and agricultural faculties, while assistant professors were the most numerous group among medical faculties.

UNFILLED FACULTY POSITIONS

Seven percent of all budgeted positions were unfilled in 1967, the year for which these data were available, varying from 5 percent in the agricultural schools to 8 percent in the medical schools. Roughly one sixth of these unfilled positions were at the rank of full professor, one third at the rank of associate professor, and one half at the rank of assistant professor. That general pattern obtained in both public and private schools but was more evident in the latter, where 10 percent of all budgeted preclinical positions, half of them at the rank of assistant professor, were unfilled in that year. It is our impression that, with the opening of several new medical schools, this circumstance has been further exacerbated. This is borne out by data collected by the American Medical Association*: In academic year 1968–1969, 579 preclinical and 1,112 clinical faculty positions were budgeted but unfilled. Of all unfilled positions reported to us, 16 percent were in the colleges of agriculture, 23 percent in the arts and sciences faculties, and 60 percent in the medical schools, distributed evenly between preclinical and clinical departments. Thus, the overall distribution of unfilled positions is much like that of the existing reporting faculties.

Each department chairman also indicated the extent to which he expected the departmental faculty would grow in the next four years. This overall growth pattern was not significantly different from that of the existing faculties. Each segment of the system anticipated growth by 20–30 percent, averaging 27 percent within that period. The most optimistic group was the health-professional, other than medical, schools, which anticipated 35 percent growth. The medical schools anticipated faculty growth of about 29 percent, arts and sciences departments 27 percent, and the agricultural schools about 21 percent, predominantly in the lower echelons of the academic hierarchy. Similarly, when examined by disciplinary departments, each discipline anticipated 20–30 percent growth in the subsequent four years.

In all, 1,257 budgeted but unfilled positions were reported by department chairmen for fiscal year 1967. The overall increment to which these department chairmen looked forward would have required the addition of yet

* *Medical Education in the United States 1968–1969.* "Medical School Faculties," p. 1477. Reprinted from *J.A.M.A.* 210(8):1455–1587, 1969.

another 4,941 budgeted positions for a total of 6,200 new faculty, i.e., slightly more than one third of the already existing faculty, in addition to positions created by retirement and death. About 2,000 of these additions would necessarily be from among those with medical degrees. These numbers may be related to the annual output of the overall system. If it is assumed that all appointments for these positions would be filled by individuals with postdoctoral experience, and that the mean postdoctoral experience time is two years, it follows that these 6,200 new faculty positions must be filled from a total through-put of about 4,000 postdoctoral physicians and 6,600 post-Ph.D's known to come through the system in the same period of four years (one half current postdoctoral population $\times$ 4). Thus, the total operation of the system produces a surplus of about 60 percent more postdoctorals than there could be new budgeted academic positions to be filled in this system during the same period. The surplus, about 2,000 M.D.'s and 2,300 Ph.D.'s, with an average of two years of postdoctoral training each, plus all the Ph.D.'s who do not take postdoctoral training, will become available over this four-year period to fill positions in government, industry, and a variety of nonprofit research-performing establishments. This pattern is not markedly different from that which was noted earlier; about two thirds of all responding individuals who had had postdoctoral training remained within the academic world.

Half of the four-year projection period has now elapsed, and it is uncertain what fraction of this anticipated growth has been realized. The gross preclinical faculty of the nation grew from 6,004 to 7,098 during this period, while the clinical faculty increased from 13,292 to 15,916, but much of this was due to the opening of new medical schools. Meanwhile, the fraction of all budgeted positions that were unfilled remained essentially constant. It is our impression, based on quite inadequate documentation, that the anticipated growth rate in our reporting departments has not occurred. The decline in the growth of federal expenditures for research has curtailed the growth of a system that, as we shall see, is substantially dependent upon federal funding. The operation of the selective service system has not affected medical school enrollments, but it has begun to limit the contributions of the medical school preclinical departments to graduate education. The growth of graduate enrollments in those two years was about as rapid as anticipated, but was probably significantly affected by the draft in the Fall of 1969. This diminution in the growth of graduate education affects the teaching responsibilities of three fourths of the academic life scientists in our population, viz., all but the clinical departments. Accordingly, failure of the combined faculties to increase quite as rapidly as had been anticipated has been mitigated, in part, by the unfortunate circumstances that have deflected significant numbers of bright young men from graduate education in the life sciences. It is, as yet, too early to establish the extent to which the changing mood of the country with respect to the conduct of science will affect graduate enrollments in the life sciences.

Graduate Education in the Life Sciences

Department chairmen reported the presence in their departments of a total of 23,287 bona fide graduate students, of whom 15,755 were stated to be Ph.D. candidates. Unfortunately, the definitions provided were an inadequate guide and it is not clear how to interpret the discrepancy of 7,500 students. Department chairmen interpreted "Ph.D. candidate" variously. In some instances, "Ph.D. candidate" was taken as a student who had completed all requirements but the dissertation; in others, only those students who had passed preliminary examinations for the Ph.D.; in still other instances, the term was used simply to exclude those graduate students known to have enrolled for training leading only to the master's degree. Further, a significant number of departments contain graduate students who are simply extending their educations and aspire to no advanced degree. Under these circumstances, the figure of 15,755 must be viewed as a minimum of bona fide Ph.D. candidates and the true value must remain unestablished. Accordingly, in the discussion that follows, the total number of graduate students in residence will be taken as the base, though this must exceed the true number of Ph.D. aspirants.

Of all graduate students, 27 percent were in agricultural schools, 46 percent were in graduate schools of arts and sciences, and 25 percent were in medical schools. The 6 percent of graduate students in clinical departments was a surprise to the authors of this report, and one may suspect that, in some instances, medical students engaged in extracurricular research in clinical-science departments may have been recorded as graduate students. The other health-professional schools combined accounted for 2 percent of all graduate students. As expected, more than three fourths of all graduate students were in universities under public auspices. All agricultural graduate students are so located, as are almost three fourths of the students in graduate colleges of arts and sciences, two thirds of those in the medical schools, and essentially all those in other health-professional schools. When the graduate student body is examined by disciplines, 20 percent were enrolled in departments of biology and 18 percent in departments of zoology, followed in rank order by biochemistry, the plant sciences of the agricultural schools, microbiology, botany, the clinical medical sciences, and physiology.

A decade ago, support of graduate students as research assistants by stipends defrayed from research grants made to individual faculty members was the norm, rivaled only by university teaching assistantships. This mode of financing has receded considerably in importance; only 13 percent of all the students in the system examined here were so paid (Table 27). In contrast, 38 percent of students were supported with institutional or other nonfederal funds, largely teaching assistantships, while 42 percent received stipends originating in one of several federal programs designed to support research training. The major single federal source was the disciplinary training-grants programs of the National Institutes of Health, largely those of the National Institute of General Medical Sciences. National Science Foundation training and institutional grants combined supported only 3 percent of all graduate students in this system. Direct competitive national fellowships were provided to 8 percent of the students, and National Defense Education Act Awards to but 4 percent.

The distribution of support of disciplines differs significantly from the overall pattern in a number of instances. For example, only 15 percent of all students in agronomy and forestry and 20 percent of the students in botany received any form of federal support, whereas more than half of the latter had university assistantships. This pattern was only slightly different from that in zoology, in which 29 percent of students had federal support and 40 percent held teaching assistantships. The biology departments of private universities resemble the preclinical departments more than they do their own counterparts in public universities, in the sense that 50 percent of their students had federal support from some source. Departmental training grants supported one fourth of their students, and their graduate students were most successful of all in competing for fellowships on the national scene, some 15 percent being so supported.

In greatest contrast to the classical biology departments is the pattern of graduate-student support in biochemistry. Biochemistry graduate students enjoyed 20 percent of all support from the National Institutes of Health training-grant system, which thus supported 30 percent of all biochemistry graduate students; nevertheless, because of the relative lack of teaching assistantships in this discipline, 19 percent of all biochemistry graduate students were still supported from faculty research grants. National Institutes of Health training grants loomed much larger in the support of students in pharmacology, genetics, anatomy, physiology, and microbiology (providing 62, 50, 42, 39, and 39 percent of their support, respectively) and nearly two thirds of all students in clinical science departments were similarly supported.

Sixty-one percent of all graduate students in the system were supported on a year-round basis. Eighty-one percent of those in the preclinical de-

TABLE 27 Sources of Support of Graduate Students

PRIMARY SOURCE OF SUPPORT	GRAND TOTAL ALL DISCIPLINES	Biological Sciences Subtotal	Animal Husbandry	Agronomy and Forestry	Anatomy	Biochemistry and Nutrition	Biology and Ecology	Biophysics and Bioengineering	Botany	Genetics	Microbiology	Pathology	Pharmacology [a]	Physiology	Zoology and Entomology	Clinical Medical Sciences [b]
TOTAL Ph.D. CANDIDATES	13,897	13,764	673	1,210	466	1,920	2,582	309	1,018	172	1,294	351	441	763	2,565	133
Total with Federal Training Support	**5,803**	**5,692**	**180**	**178**	**306**	**1,001**	**1,116**	**204**	**200**	**105**	**760**	**96**	**336**	**475**	**735**	**111**
Competitive Fellowships	1,073	1,071	16	28	38	181	307	54	21	11	113	8	26	71	197	2
National Defense Education Act Awards	609	607	37	66	9	66	132	13	67	—	27	22	14	29	125	2
Department Training Grants, Total	3,255	3,170	96	65	217	602	506	125	35	87	525	48	275	322	267	85
National Institutes of Health	2,857	2,778	62	15	195	572	445	105	19	86	507	28	272	298	174	79
National Science Foundation	145	145	6	22	11	15	26	3	8	1	6	5	3	4	38	6
Other	253	247	28	28	11	15	35	17	8	—	12	15	3	20	55	6
Institutional Grants, Total	866	844	31	19	42	152	171	12	77	7	95	18	21	53	146	22
National Institutes of Health	415	397	27	8	38	94	36	4	10	1	62	8	16	36	57	18
National Science Foundation	240	236	2	9	1	30	65	2	28	3	23	9	2	7	55	4
National Aeronautics and Space Administration	211	211	2	2	3	28	70	6	39	3	10	1	3	10	34	—
Total with Nonfederal Support	**5,216**	**5,207**	**404**	**636**	**141**	**478**	**908**	**46**	**611**	**37**	**284**	**168**	**65**	**143**	**1,286**	**9**
University Assistantship	3,849	3,848	292	396	101	325	678	32	528	29	191	125	25	96	1,030	1
Fellowship, Local Sources	445	440	17	38	27	40	139	3	29	1	33	9	14	19	71	5
Fellowship, National (Nonfederal)	119	119	6	5	5	40	20	—	3	—	9	1	4	1	25	—
Department Training Grant	81	81	2	21	—	12	19	1	1	—	2	4	1	3	16	—
Institutional Grant (Foundation)	127	127	6	16	5	11	19	3	11	4	13	4	4	12	19	—
Industry	261	260	41	57	—	24	8	5	17	—	10	5	11	4	78	1
Foreign Student, Own Country	334	332	40	103	3	26	25	3	22	3	26	20	6	8	47	2
Research Assistantship from Faculty Grant [c]	1,799	1,797	59	225	12	364	262	50	167	27	171	39	33	77	311	2
Other (Including Self-supporting)	1,079	1,068	30	171	7	77	296	9	40	3	79	48	7	68	233	11

DEPARTMENTAL DISCIPLINE

[a] Includes a college of pharmacy.
[b] Includes a department of oral biology.
[c] Largely from federal sources.
Source: Survey of Academic Life Science Departments, National Academy of Sciences Committee on Research in the Life Sciences.

partments of medical schools, but only 50 percent of those in the colleges of arts and sciences, were so supported. A fourth of the latter received support for eight to ten months and another fourth for some lesser fraction of the year. The pattern in the schools of agriculture lies between these two extremes. For graduate students who work in clinical departments, short-term support appears to be the norm: 68 percent of all graduate students working in clinical departments received support for less than seven months of the year. It seems likely that many of these have home bases in other departments during the rest of the year.

CAPACITY OF THE CURRENT GRADUATE EDUCATION SYSTEM

An attempt was made to compare the attributes of departments that have conferred Ph.D. degrees ("performer departments") with those that now have graduate students whom they consider to be Ph.D. candidates but have never awarded this degree previously ("promiser departments"). These two classes were examined relative to a variety of criteria: size of faculty, number of postdoctoral fellows, amount of space, and volume of federal research support.

The promiser departments were not significantly smaller than the per-former departments in their faculty size, available physical plant, or scientific facilities. They had been decidedly less successful in attracting postdoctoral fellows and federal research funds as well as graduate students. In a general way, it is probably safe to conclude that the mean quality of the faculty of the promiser departments—as it might be judged by their peers—is less impressive than the quality of the faculties of the performer departments and that this is the primary reason for the relatively smaller attraction to graduate students, postdoctoral fellows, and federal research-supporting agencies. This should be read as a fact of history, not as criticism. The "performers" were off to an earlier start, and, in general, have been more generously supported by their parent universities. As indicated by a recent report of the National Science Board,* the chief mechanism for upgrading the total performance of a research-performing graduate department is to offer competitive salaries for the faculty.

The intentions of the "promisers" are evident in the fact that 126 of these 146 departments (86 percent), scattered among all disciplines in the

---

* *Toward a Public Policy for Graduate Education in the Sciences, Report of the National Science Board,* National Science Board–National Science Foundation, U.S. Government Printing Office, Washington, D.C., 1969, p. 328–298.

Graduate Education: Parameters for Public Policy, Report Prepared for the National Science Board, National Science Board–National Science Foundation, U.S. Government Printing Office, Washington, D.C., 1969, p. 331–173.

study, indicated that they could significantly expand their graduate student bodies with the current space and faculty available to them, whereas 70 percent of the performer departments similarly indicated that their capacity was less than saturated. This varied from 56 percent of all zoology departments to 84 percent of the agricultural sciences departments. In all, the performer departments indicated their ability to accept another 3,255 graduate students, while the promiser departments considered that they had room for another 774 students; i.e., the total system could accommodate a 27 percent increase in graduate students, with no requirement for additional faculty or for increased space other than that which was available, under construction, or in advanced planning in the summer of 1967. Again, the pattern varied by discipline: The classical departments on campus (biology, zoology, botany) appeared to be closest to saturation, with room for a mean increment of only about 15 percent in their graduate enrollments. In general, the other disciplinary departments could accommodate increments that were about twice as large, expressed as percentages of their current student enrollments.

Space and faculty were the chief limitations to further enrollments in those departments that indicated that they were already saturated; 74 departments were limited only by space, 14 only by faculty size, and 88 departments required both space and faculty if graduate student enrollments were to be augmented.

These data must be regarded in the light of anticipated graduate enrollments over an extended time scale. Historically, graduate enrollments in the United States have doubled each decade since the Civil War. This trend has accelerated somewhat since World War II; indeed, enrollments almost trebled in the last decade. Although the Vietnam war may cause a temporary deflection in this growth curve, the overall trend should continue until about 1980 if demographic projections are correct, unless the federal government deliberately fails to support such growth, a trend that is evident in the budget proposed for fiscal year 1971. Moreover, data from the Office of Education indicate that graduate enrollments in the life sciences have recently been increasing more rapidly than have those in the physical sciences, while the proportion of all graduate students enrolled in the natural sciences has decreased somewhat. Accordingly, were it not for the Vietnam episode, one would have anticipated that the entire capacity of the national system for graduate education in the life sciences would be saturated within two years. That day has been postponed by the Vietnam war and continuation of the draft, but it cannot be more than four years off unless either a substantial construction program is soon inaugurated or students are discouraged by lack of stipends or potential jobs.

It is uncertain whether the market for those with advanced degrees is similarly reaching saturation. Industry requires considerably more life scientists than it can find; if the 16 new medical schools now under development, which are urgently required to meet the national demands for medical care, are to be staffed, if undergraduate students are to have the kind of educations for which they are clamoring, if the students in the junior colleges are to have instruction from suitably trained scientists, the graduate enterprise in the life sciences must be augmented in substantial fashion to meet the national need. It is clear that practically all those receiving Ph.D.'s in 1968 and 1969 have found positions in which they can utilize their education. But, at this date, there is growing apprehension that there will be insufficient opportunities for the graduate class of 1970, despite the urgency of national needs.

STUDENT STIPENDS

Department chairmen reporting graduate students with 11 to 12 months of financial support—69 percent of our file—were queried concerning the extent to which the stipend levels set by the national competitive fellowships offered by the National Institutes of Health and the National Science Foundation have established a norm for graduate students' stipends. From their reports, we find that 40 percent of all such graduate students were receiving stipends of the magnitude stipulated by the federal agencies in their competitive fellowship programs, 13 percent received stipends that were lower in varying degree, and 47 percent received stipends that exceeded these national norms. In those departments in which National Institutes of Health training programs or national fellowships constituted a major input to graduate stipends, this tended to maintain the national norm; accordingly, in the preclinical departments of private medical schools, four times as many students received the basic federal stipend as received increments beyond that level. One third of the students in the graduate colleges of arts and sciences in the same institutions received stipends greater than those of the national fellowships. In the public universities, 42 percent of the graduate students in the arts and sciences faculties received "excessive" stipends while, in the agriculture schools, 72 percent of all the graduate students received stipends greater than the national norms. No information was requested concerning the specific sources of funds utilized to pay these higher stipends.

In our view, no national purpose is served by interdepartmental competition for bright graduate students based upon funds available to pay stipends. Nor should increments in stipend levels be paid for a reasonable amount of teaching. We believe that a modest amount of undergraduate teaching should be an intrinsic aspect of the experience of all graduate students, except for those who obviously have no talent for such teaching, regardless of the source of funds for their stipends. *We believe, further, that it is in the national interest that there be a standard stipend level for essentially all graduate students, albeit with two provisos. (a) It appears reasonable to adjust the basic stipend to reflect the cost of living in the region of each university. (b) Increased stipends are appropriate when it is deemed necessary and desirable to attract students into an underpopulated research area that should be expanded in the national interest,* e.g., research in problems germane to monitoring the effects of environmental pollutants on the public health and in some aspects of agriculture research.

We are persuaded of the great virtue of training grants of the type awarded by the National Institutes of Health. These grants provide both graduate and postdoctoral stipends as well as funds for consumable supplies, communally used equipment, visiting lecturers, and, when appropriate, for additional faculty positions. They are concerned, therefore, not only with the quantitative scale but also with the quality of graduate education and have markedly upgraded such education in most departments that have known such support. We regret that equivalent programs are not available in support of other life science departments, indeed of academic departments generally. In our view, the total support system should so evolve that virtually all federal stipends for graduate students would be provided by this mechanism. As the federal government assumes its responsibility for graduate education, it seems pointless to enlarge the fellowship system of one-to-one correspondence between federal agencies and individual graduate students who must, nevertheless, win admission to graduate study in the departments of their choice. Moreover, the National Institutes of Health experience demonstrates that it is possible for an external jury of peers to evaluate periodically the quality and appropriate scale of graduate education in a given department without generating undue rancor and in a far more sophisticated manner than can either the inadequately informed graduating senior or undergraduate adviser in some remote institution or the local graduate dean.

One hurdle for this evolutionary development is the support of students who for purely personal, perhaps geographic, reasons choose to undertake study in graduate departments that fail to qualify for such training grants. This contingency could be met by an appropriately designed limited fellowship program or simply by stipends defrayed from faculty research grants. Where the latter do not exist, graduate education is scarcely of a quality to recommend it to prospective students.

Postdoctoral Fellows

In all, 5,223 postdoctoral appointees were located in the system (Tables 28 and 29). Of these, 22 percent were in colleges of arts and sciences, 7 percent in agricultural schools, 28 percent in preclinical departments of medical schools, 39 percent in their clinical departments, and 4 percent in other health-professional schools; 1,945 of the total held medical degrees. Seventy-six percent of all postdoctoral appointees received support from federal sources.

As we have already noted, the role of the preclinical departments of medical schools, in postdoctoral as well as in graduate education, continues to expand. This is corroborated by data from other sources: In academic year 1962–1963, there were 1,061 postdoctoral fellows, both post-Ph.D. and post-M.D., in preclinical departments in all medical schools; by 1967–1968 this group had increased to 1,383. During the same interval, the graduate student population of the departments increased from 4,105 to 7,421, and in the following year, it increased by yet another 570 students. Table 28 reveals the disproportionate roles of the private universities, particularly their medical schools, in postdoctoral education.

FINANCING POSTDOCTORAL EDUCATION

As we have seen, two thirds of all postdoctoral fellows are in departments of medical schools, and 58 percent of these are, in turn, in clinical departments. It may safely be assumed that all postdoctoral fellows find financial support, the going rate in 1970 for an initial fellowship being approximately $6,500 a year plus $500 for each dependent. Of those reported to us, three fourths were supported with funds provided through federal agencies. Somewhat unexpectedly, perhaps, postdoctoral trainees in the arts and sciences colleges were most frequently supported in this way (85 percent) and those in clinical departments least frequently (69 percent). Direct fellowships from federal agencies were received by 23 percent of the 5,223 postdoctoral appointees in the system, while 13 percent were the recipients of competitive fellowships from other sources. Again, reckoned on a percentage rather than an absolute basis, it was postdoctoral appointees in the arts and sciences faculties who most frequently obtained federal fellowship support. Thus, fellowships from the Public Health Service were distributed in approximately equal numbers to postdoctoral appointees in the arts and sciences, preclinical, and clinical faculties, whereas National Science Foundation support was offered three times as frequently to fellows in the arts and sciences faculties as to both Ph.D.'s and M.D.'s studying with medical

TABLE 28 Distribution of Postdoctoral Support by Type of School

PRIMARY SOURCE OF SUPPORT	GRAND TOTAL	ALL SCHOOLS					PRIVATE SCHOOLS					PUBLIC SCHOOLS					
		Agriculture[a]	Arts and Sciences[b]	Preclinical Medical	Clinical Medical	Other Health Professional[c]	Subtotal	Arts and Sciences[b]	Preclinical Medical	Clinical Medical	Other Health Professional[c]	Subtotal	Agricultural[a]	Arts and Sciences[b]	Preclinical Medical	Clinical Medical	Other Health Professional[c]
TOTAL, ALL POSTDOCTORALS	5,223	355	1,168	1,466	2,032	202	2,799	622	775	1,371	31	2,424	355	546	691	661	171
Total with Federal Support	3,972	282	989	1,161	1,409	131	2,079	532	604	916	27	1,893	282	457	557	493	104
Fellowships, Subtotal	1,181	102	373	330	339	37	566	189	161	214	2	615	102	184	169	125	35
U.S. Public Health	926	48	274	283	291	30	466	143	135	186	2	460	48	131	148	105	28
National Science Foundation	124	34	68	21	1	—	43	29	14	—	—	81	34	39	7	1	—
Other	131	20	31	26	47	7	57	17	12	28	—	74	20	14	14	19	7
Training Grants	1,509	16	174	413	848	58	909	112	234	547	16	600	16	62	179	301	42
Institutional Grants	128	46	20	20	39	3	42	9	9	24	—	86	46	11	11	15	3
Research Funds	1,154	118	422	398	183	33	562	222	200	131	9	592	118	200	198	52	21
Total with Nonfederal Support	927	65	156	256	389	61	467	87	141	235	4	460	65	69	115	154	57
Fellowships, Subtotal	682	43	118	191	273	57	320	62	101	154	3	362	43	56	90	119	54
State	137	18	24	17	51	27	11	2	1	7	1	126	18	22	16	44	26
Industrial	44	5	5	15	12	7	17	2	7	8	—	27	5	3	8	4	7
Foreign	153	12	37	57	28	19	78	20	34	22	2	75	12	17	23	6	17
Other	348	8	52	102	182	4	214	38	59	117	—	134	8	14	43	65	4
Training Grants	27	2	5	11	9	—	16	5	6	5	—	11	2	—	5	4	—
Institutional Grants	73	4	7	14	47	1	56	5	9	42	—	17	4	2	5	5	1
Research Funds	145	16	26	40	60	3	75	15	25	34	1	70	16	11	15	26	2
Unspecified	324	8	23	49	234	10	253	3	30	220	—	71	8	20	19	14	10

[a] Includes schools of forestry.
[b] Includes engineering schools and schools of graduate studies.
[c] Includes schools of dentistry, pharmacy, public health, and veterinary medicine.
Source: Survey of Academic Life Science Departments, National Academy of Sciences Committee on Research in the Life Sciences.

TABLE 29 Sources of Support of Postdoctoral Fellows by Class of Department

PRIMARY SOURCE OF SUPPORT	GRAND TOTAL ALL DISCIPLINES	Biological Sciences Subtotal	Animal Husbandry	Agronomy and Forestry	Anatomy	Biochemistry and Nutrition	Biology and Ecology	Biophysics and Bioengineering	Botany	Genetics	Microbiology	Pathology	Pharmacology [a]	Physiology	Zoology and Entomology	Clinical Medical Sciences [b]
TOTAL, ALL POSTDOCTORALS	5,223	3,166	137	78	194	684	654	105	97	70	276	272	195	252	152	2,057
Total with Federal Support	3,972	2,548	104	44	139	591	539	90	78	53	228	213	150	187	132	1,424
Fellowships, Subtotal	1,181	837	31	16	49	187	209	28	31	19	70	47	26	80	44	344
U.S. Public Health	926	631	15	2	45	152	157	24	10	16	59	35	22	64	30	295
National Science Foundation	124	123	8	7	3	17	37	2	15	1	7	7	—	7	12	1
Other	131	83	8	7	1	18	15	2	6	2	4	5	4	9	2	48
Training Grants	1,509	653	40	—	48	90	91	28	5	16	72	124	59	60	20	856
Institutional Grants	128	88	15	5	1	14	11	—	2	—	8	2	2	7	21	40
Research Funds	1,154	970	18	23	41	300	228	34	40	18	78	40	63	40	47	184
Total with Nonfederal Support	927	528	30	32	44	86	97	9	18	14	45	43	42	54	14	399
Fellowships, Subtotal	682	399	22	18	37	56	77	9	15	11	31	37	31	47	8	283
State	137	86	13	7	4	2	13	1	8	—	5	14	4	14	1	51
Industrial	44	26	1	2	—	2	1	3	—	—	6	4	7	2	1	18
Foreign	153	122	4	7	13	12	26	3	3	6	11	9	7	18	3	31
Other	348	165	4	2	20	40	37	5	4	5	9	10	13	13	3	183
Training Grants	27	18	2	—	5	4	4	—	—	—	—	—	2	—	1	9
Institutional Grants	73	26	4	4	2	4	1	—	—	2	1	1	1	4	2	47
Research Funds	145	85	2	10	—	22	15	—	3	1	13	5	8	3	3	60
Unspecified	324	90	3	2	11	7	18	6	1	3	3	16	3	11	6	234

[a] Includes a college of pharmacy.
[b] Includes a department of oral biology.
Source: Survey of Academic Life Science Departments, National Academy of Sciences Committee on Research in the Life Sciences.

school faculties. Only a single National Science Foundation postdoctoral fellowship was reported by clinical department chairmen. This was compensated by the greater success of postdoctoral appointees in the clinical sciences, and to a lesser degree in the preclinical sciences, in finding other forms of fellowship support.

However, the dominance of federal funds in supporting postdoctoral appointees rested not so much on the direct fellowship programs as on training grants and utilization of research funds, more than half of all postdoctoral appointees in all groups being so supported. Training grants made to individual departments or to multidisciplinary programs contributed most heavily to the support of postdoctoral appointees in the clinical sciences, somewhat less to those in the preclinical sciences, and to a much lesser extent to those in the arts and sciences. A third of all postdoctorals in the arts and sciences and in agriculture were supported by stipends made from the research grants of their mentors, as were a fourth of all postdoctoral appointees in the preclinical sciences; less than a tenth of those in the clinical sciences were so supported. Adequate data are not available to describe the specific manner in which foreign postdoctorals are supported. Since, however, they are ineligible under the terms of federal fellowship and training programs, it may be assumed that, in the main, they are supported by stipends from research grants, nonfederal sources, and, occasionally, funds from their home countries.

FOREIGN POSTDOCTORALS

Figures 35 and 36 show the distribution of U.S. and foreign national postdoctoral fellows revealed by the departmental questionnaires. Thirty-eight percent of all postdoctoral fellows in the system were not of American origin, of whom about one sixth had obtained their doctoral degrees within the United States. About 30 percent of all postdoctoral fellows in the clinical departments were of foreign origin; virtually all of these may be presumed to have medical degrees. At the other extreme, 62 percent of all postdoctorals in the agriculture and forestry schools were foreign nationals, while 42 percent of the postdoctorals in the preclinical sciences departments and 41 percent in the life sciences departments of the colleges of arts and sciences in the universities had come from overseas.

In our view, these figures are not seriously excessive. Like their American equivalents, foreign postdoctoral fellows contributed significantly to research progress. Already rather well trained, they are capable of conducting semi-independent work in projects under the direction of faculty supervisors. They contribute stimulating new ideas not only for their own

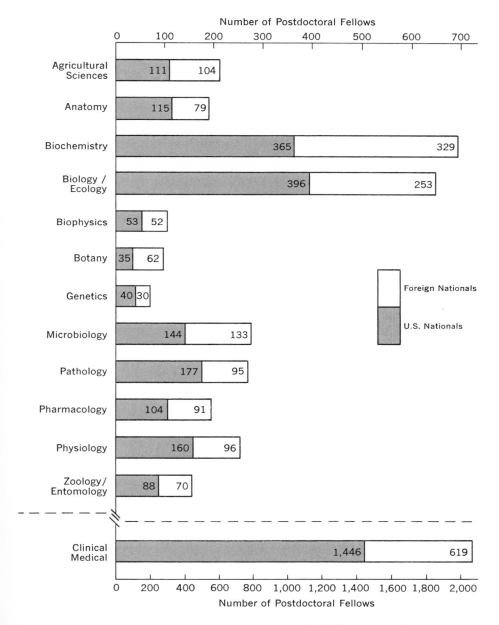

FIGURE 35   Distribution of postdoctoral fellows by discipline and national origin. Source: Survey of Academic Life Science Departments, National Academy of Sciences Committee on Research in the Life Sciences.

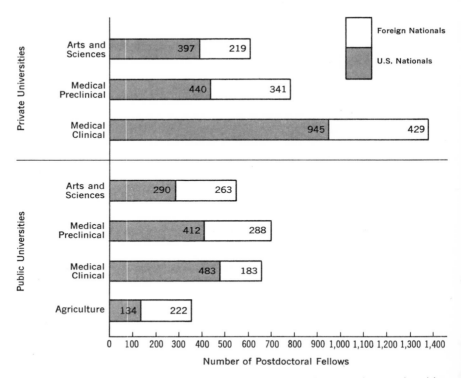

FIGURE 36 Distribution of postdoctoral fellows in public and private universities, by type of school and national origin. Source: Survey of Academic Life Science Departments, National Academy of Sciences Committee on Research in the Life Sciences.

immediate research but to the entire research group with which they are intimately connected. Frequently, they bring with them laboratory skills acquired in their laboratories of origin and divergent points of view learned in those laboratories. With their participation, the work goes more rapidly and progresses more satisfactorily, occasionally in unforeseen directions because of the differing viewpoints of the foreign fellows. They also contribute significantly to the education of American graduate students in the same environment. In view of the fact that 12 percent of all working U.S. life scientists in our sample are of foreign birth, including many of our most distinguished scientists, to return significant numbers of trained scientists to those countries seems only just. When postdoctorals return to their countries of origin, not only do they take with them an enhanced capability for research and education in some discipline, but they also go as ambassadors of good will who have had an intensive experience in American life under favorable circumstances.

For those postdoctoral fellows who return to the developed nations— e.g., western Europe, Australia, Japan—there need be no concern about the relevance of their postdoctoral experience to life in their home countries. However, postdoctoral fellows educated in the United States may return to developing nations only to find no market for the sophisticated science in which they have become skilled. This is most frequently true for postdoctoral fellows from India, Southeast Asia, and various African and Latin American countries. It is heartening to note that a significant fraction of postdoctoral fellows in the agricultural schools and in clinical departments are from the developing nations. Their experience here will make them more valuable citizens of their homelands, and their newly acquired skills can immediately and profitably be put to work.

Although we cannot find it in ourselves to deny experience in enzyme kinetics, neurophysiology, or the more recondite aspects of the physiological bases of behavior to any qualified would-be postdoctoral student, we are painfully aware of the fact that significant numbers of such individuals have returned to their homelands in developing countries only to become bitter and frustrated upon finding no opportunity to take advantage of their highly individualized educations. Numerous such individuals naturally attempt to remain in or return to the United States, in conflict with our national policy. We have no useful recommendations to make in this regard since we feel it unjust to deny further education in science to such qualified individuals, to condemn them to less intellectually rewarding careers than they might otherwise know, or to insist that only applied science and developmental research be conducted in developing nations and studied by their citizens while in our country. But the situation could be ameliorated, if only in some part, if before each such postdoctoral fellow is admitted to an American laboratory, the problem of his future is explored in advance, so that the circumstances are understood by his sponsors at home, by himself, and by the Americans who will receive him.

FOREIGN INTERNS AND RESIDENTS

The situation is even more complicated with respect to M.D.'s seeking advanced clinical training or clinical research experience. There can be no doubt that those who take advanced training as fellows in the clinical departments of American medical schools or who take internship or residency training in American hospitals and then return to their native countries will bring with them training and skills that will permit them to make valuable contributions to their compatriots. And there can be no doubt, also, that a large fraction of foreign M.D.'s who spend a year or two as postdoctoral fellows engaged in research in either clinical or science departments of American universities contribute to the programs of those departments and laboratories in much the same manner as do American-born postdoctoral fellows. For these reasons we think it appropriate that this nation welcome such individuals to our shores.

Moreover, foreign graduates have become essential to the continued operation of a large number of hospitals. Foreign graduates now constitute 32.7 percent of all interns and residents in American hospitals. For those concerned not only with academic training but also with the delivery of health care to the American public to deplore this fact, when 11,000 other opportunities for interns and residents, one quarter of the national total, remain untaken, is unreasonable. The 14,500 foreign graduates filling these positions today make a large contribution to American life, even though most arrive less skilled and less knowledgeable than are American medical graduates at the same level of training.

We are comfortable with the fact that this large number of foreign graduates is receiving advanced medical training in the United States, since, when they return to their home countries, they can make large contributions to life there after several years during which they had made extremely useful contributions to the delivery of medical care in the United States. Indeed, it is entirely evident that, at this time, this group of foreign nationals is indispensable to our system. However, we do require assurance that a large percentage of them will indeed return home in due course.

Without them, 45 percent of all budgeted internships and residencies in American hospitals would go vacant, and this country would be the worse for such a disaster. Even now, 15 percent of the existing openings cannot be filled. We also recognize that, with a profound shortage of physicians to provide for the growing medical needs of the United States, one might well be tempted to welcome into our society immigrant physicians, particularly those who have received advanced training in the United States, physicians from advanced nations, and physicians from developing nations with a surplus, if such there be. The bulk of the foreign postdoctoral physician group, however, comes from developing nations that suffer an acute lack of physicians for their own medical care programs. We can best reconcile the current situation with the general moral posture of the United States by urging that, after completion of their training, foreign physicians return to their native lands.

These considerations are highly germane to the present study. Foreign graduates today fill more than 30 percent of all internships and residencies, and were we, by statute, to forbid their entry, it would be imperative that the supply of physicians from American medical schools be increased as rapidly as possible to compensate for this deficit. The capitalization cost of such a venture is a major consideration (30 new medical schools at $50 million each for a total of $1.5 billion), and it would be essentially impossible to meet the requirement for new faculty, a requirement that is barely being met at the current rate of growth of the medical education system. Additional approaches to provision of sufficient manpower to assure adequate levels of medical care—basic changes in medical school curricula, expansion of medical student bodies, more efficient use of para-medical personnel, and so on—are outside the scope of this report but warrant early and serious consideration.

Laboratory Space

The aggregate net usable laboratory space reported to us by department chairmen was 13,423,000 ft²; 31 percent of the space was allocated to departments in colleges of arts and sciences, 25 percent to agriculture and forestry schools, and 41 percent to medical schools (26 percent to pre-clinical and 15 percent to clinical departments). Two thirds of all this space was in universities under public auspices and one third in private universities. If the agriculture and forestry schools, common only to the public universities, are omitted from these totals, 43 percent of the labora-tory space was found in private institutions and 57 percent in the public institutions. This distribution is not seriously disproportionate, since, as shown in Table 30, space per faculty member is of the same order in both classes of institution.

A total of 9,211 individual scientists provided information concerning the working laboratory space available to them, exclusive of office space or common service areas utilized by laboratory groups other than their own. A third of the total reported that they had less than 500 ft² for their own work, a quarter had 500 to 750 ft², 13 percent 750 to 1,000 ft², 10.7 per-cent 1,000 to 1,250 ft², and 5.4 percent 1,250 to 1,500 ft². Three percent reported quarters varying from 1,500 to 2,500 ft² each, and 4.8 percent of all scientists had laboratory space in excess of 2,500 ft².

Instances of both small and large laboratories were found in all dis-ciplines. Table 31 shows the mean net square footage available to the various classes of disciplinary departments for full-time faculty members. The unusually high figures quoted for genetics departments should be

TABLE 30 Research Laboratory Space by Type of School

TYPE OF SCHOOL	RESEARCH LABORATORY SPACE (FT²)					
	All Schools		Public Schools		Private Schools	
	Per Department	Per Faculty Member	Per Department	Per Faculty Member	Per Department	Per Faculty Member
TOTAL, ALL TYPES	**11,482**	**781**	**11,382**	**820**	**11,699**	**715**
Agriculture [a]	13,283	863	13,283	863	—	—
Arts and Sciences [b]	18,425	1,090	16,834	1,030	22,415	1,225
Preclinical Medical	9,839	909	9,726	960	9,980	835
Clinical Medical	7,259	413	5,736	395	8,907	428
Other Health-Professional [c]	6,355	574	6,189	575	7,125	570

[a] Includes schools of forestry.
[b] Includes engineering schools and schools of graduate studies.
[c] Includes schools of dentistry, pharmacy, public health, and veterinary medicine.
Source: Survey of Academic Life Science Departments, National Academy of Sciences Committee on Research in the Life Sciences.

viewed in light of the fact that there are only 10 genetics departments in the sample.

Contrary to the expectations of some, members of arts and sciences faculties enjoy some of the largest of the individual laboratories, and clinical scientists, by and large, operate in minimal laboratory space, with other disciplines ranged between. The mean laboratory space available to individual members of the faculties of public universities, which is somewhat larger than that of their counterparts in private universities, is largely accounted for by the space available to their agricultural and preclinical scientists. Available figures concerning space per department or per individual scientist in given disciplines are not genuine reflections of the space utilization appropriate to these disciplines, but rather are the consequence of historical trends, opportunities for construction, and related factors. It will be recognized, however, that working space in general is fairly tight. The mean academic research group consists of a professor and 7.5 other individuals. Since the mean space available per full-time faculty member is 781 ft² for the population as a whole, this provides only 90 ft² per working body. Since space distributions are surely uneven, as reported above, it is evident that many laboratories have decidedly less space per working individual—a deplorable situation when one considers that modern planning practice assumes the desirability of 150–200 ft² per body.

TABLE 31   Research Laboratory Space by Class of Department

CLASS OF DEPARTMENT	RESEARCH LABORATORY SPACE (FT²)	
	Per Department	Per Faculty Member
ALL DEPARTMENTS	**11,482**	**781**
Agricultural Sciences	13,283	863
Anatomy	8,864	757
Biochemistry and Nutrition	13,352	1,215
Biology and Ecology	19,235	1,058
Biophysics and Biomedical Engineering	18,187	1,419
Botany	13,745	853
Genetics	21,000	2,100
Microbiology	10,452	1,108
Pathology	10,593	754
Pharmacology [a]	7,793	886
Physiology	9,722	938
Zoology and Entomology	18,092	991
Clinical Sciences [b]	7,264	416

[a] Includes a college of pharmacy.
[b] Includes a department of oral biology.
Source: Survey of Academic Life Science Departments, National Academy of Sciences Committee on Research in the Life Sciences.

The Tools of Biological Research

SPECIALIZED RESEARCH FACILITIES

The impressions gained in considering the facilities and instruments used by individual scientists are fortified by examination of those available to entire departments in the biological sciences (Table 32). Biology, botany, and ecology departments are expected to have access to and utilize field areas, but it is a surprise that 27 percent of all pathology departments, 40 percent of genetics departments, and 12 percent of biochemistry departments also indicate some such usage. Similarly, one would hardly expect 7 percent of biophysics departments to utilize an organism-identification service or 14 percent of biochemistry departments to make use of a marine biological station. High-pressure chambers, thought to be the special province of some physiologists and clinical scientists, found use among disciplinary departments of almost every category, but not among geneticists or in agricultural schools. To the authors of this report, the extent of use of programmed climate-controlled rooms across the entire spectrum of disciplines came as a considerable surprise, and we were not aware that pri-

TABLE 32   Specialized Facilities Available to Disciplinary Departments

SPECIALIZED FACILITY	GRAND TOTAL ALL DISCIPLINES	Animal Husbandry	Agronomy and Forestry	Anatomy	Biochemistry and Nutrition	Biology and Ecology	Biophysics and Engineering	Botany	Genetics	Microbiology	Pathology	Pharmacology [a]	Physiology	Zoology and Entomology	Clinical Medical Sciences [b]
TOTAL DISCIPLINARY DEPARMENTS [c]	1,200	88	93	64	99	107	15	51	10	83	63	57	75	84	311
Field Areas	432	51	89	8	12	77	2	47	4	13	17	3	11	77	21
Zoo and Aquarium	136	2	2	14	7	44	1	3	—	6	7	—	9	27	14
Taxonomic Research Collection	272	3	37	5	7	67	1	46	1	11	11	—	2	73	8
Organism-Identification Service	220	18	14	4	9	28	—	23	1	20	15	—	1	40	47
Tropical Terrestrial Station	67	—	2	2	5	13	—	11	—	2	2	—	2	20	7
Tropical Marine Station	52	—	—	2	3	18	2	3	—	7	—	2	4	11	4
Marine Station Other Than Tropical	136	1	—	10	14	40	1	15	—	5	5	—	6	31	5
	64	1	3	3	7	9	1	5	—	3	3	3	9	7	13

							3	1	—	5	7	6	13	3	42
Programmed Climate-Controlled Rooms (Phytotron, Biotron, Etc.)	291	14	45	6	19	47	3	36	2	3	19	3	14	45	35
Computer Center	997	71	84	44	89	96	14	49	7	64	51	48	62	80	238
Primate Center	210	7	—	25	16	21	4	1	1	11	7	10	17	13	77
Other Specialized Animal Colony	335	46	2	27	23	34	5	5	4	26	12	9	18	35	89
Germ-Free Animal Facility	113	74	1	8	8	10	—	5	—	19	8	3	3	6	31
Animal Surgery Facility	641	49	1	52	47	41	8	2	2	43	35	37	60	27	237
Animal Quarantine Facility	394	34	4	27	27	27	5	4	2	38	20	28	35	18	131
General Animal Care Facility	799	70	4	56	75	75	8	1	4	69	37	49	61	41	245
Cell and Tissue Culture Facility	512	31	22	37	34	49	8	5	6	67	41	11	15	29	139
High-Intensity Radioactive Sources	485	26	38	21	40	53	9	23	3	35	25	15	18	42	130
Center for Large-Scale Production of Biological Materials	93	13	10	4	11	11	—	30	1	13	4	—	3	10	9
Clinical Research Ward	324	8	—	13	33	8	4	4	2	12	19	24	18	1	182
Greenhouse	366	4	84	4	28	81	5	48	4	11	19	8	2	57	11
Ships Longer than 18 ft	59	—	1	1	6	22	—	8	—	2	—	—	2	11	6
Electrically Shielded Room	229	1	3	16	7	28	5	1	—	2	2	20	40	14	90
Instrument Design and/or Fabrication Facility	470	16	22	24	37	47	10	18	1	26	16	33	54	28	138

[a] Includes a college of pharmacy.
[b] Includes a department of oral biology.
[c] Number of departments reporting availability of one or more specialized facilities.

Source: Survey of Academic Life Science Departments, National Academy of Sciences Committee on Research in the Life Sciences.

mate centers, animal surgery facilities, and tissue-culture facilities were so heavily utilized. The data indicate that clinical research wards are not quite the unique province of clinical medical scientists, while at the same time indicating that only 59 percent of all departments of clinical medical science consider that they even have access to clinical research wards.

UTILIZATION OF RESEARCH INSTRUMENTS

The unity of the life sciences, the pervasiveness not only of concepts but of research tools, is again evident in the instruments available to the purportedly disciplinary departments (Table 33). Most particularly, as we saw earlier, the tools of biochemistry are employed in research in all areas of the life sciences. Ultracentrifuges, amino acid analyzers, gas chromatographs, scintillation counters, ultraviolet spectrophotometers, and electrophoresis apparatus find intensive use across all the biological disciplines, as do electron and phase-contrast microscopes. Indeed, some of the data are particularly surprising, such as those indicating that the 38 clinical science departments utilize or have access to mass spectrometers, as do 43 departments of biology, 21 departments of botany, and 32 departments of zoology. Although intensive-care patient-monitoring systems remain in the particular purview of clinical science departments, large-scale fermentors are available across numerous disciplines, as are multichannel recorders, which have usually been thought of as especially associated with physiologists and pharmacologists. In short, we see once again that biologists, whatever the discipline of their training, are relatively quick to seize upon any useful tools that will assist them in their continual inquiries of nature.

MEDICAL SCHOOLS AS RESEARCH AND EDUCATIONAL ENTERPRISES

In academic year 1968–1969, 85 four-year medical schools and six approved two-year schools providing preclinical training were in operation. Eight additional schools that had not yet been certified for approval had opened their doors, and 17 four-year schools were in various stages of development, from being about to accept their first classes to advanced planning on the drawing boards. Enrolled in all four medical classes were 35,833 students, and, at the end of the year, 8,059 students were awarded medical degrees. Total enrollment had grown by 6,200 medical students in the previous decade. However, as shown in Table 34, these medical students constitute less than half of the total educational responsibilities of these institutions.

The growth of graduate education in preclinical departments is shown in Table 35. Patently, an appropriate fraction of the attention of the medical faculties must be addressed to each category of student as well as to research and to the responsibility for medical care essential to operation of such institutions. The growth of the total student body has been paralleled by growth of the faculty. From academic 1962–1963 to 1968–1969 the total full-time faculty of these institutions grew from 13,681 to 23,014, which was accomplished by increasing the basic sciences faculty from 4,716 to 7,098 while the clinical faculty grew from 8,965 to 15,916. The latter figures may be misleading in some part as they may reflect the conversion from part-time to full-time clinical faculty status as much as they do an absolute increase in the number of individual teaching faculty members.

The total number of basic sciences faculty personnel bears almost a one-to-one relationship to the total number of graduate students in the system, while the ratio of arts and sciences faculty personnel to students is close to one-to-three. This should not suggest an excessive faculty-staffing pattern in the medical schools since the basic sciences faculties have primary responsibility for the instruction of medical students as well as for the conduct of research and graduate education. Despite this growth rate, or perhaps because of it, throughout the decade of the 1960's, there has always been a substantial number of budgeted but unfilled faculty positions within the medical schools. For the academic year 1968–1969, 1,112 clinical and 579 preclinical positions were budgeted but vacant; these were distributed among all disciplines, and the unsatisfied demands for pathologists, anatomists, and biochemists were most acute. Overall, 6.8 percent of the budgeted places on medical school faculties were unoccupied during that year; this fractional level of vacancies had persisted for a decade.

In contrast to graduate students and postdoctoral fellows, virtually all of whom receive financial support from some source, by and large it is still necessary for the families of medical students to provide the funds necessary for their maintenance and tuition during their medical educations. Internships and residencies are increasingly better paid than they have been historically, so the sense of exploitation of this group has been mitigated in small degree. But it remains true that, with rare exceptions, only the children of the upper economic segments of American society can afford attendance at medical school. To be sure, in academic year 1968–1969, 35 percent of all students received some scholarship help, and 56 percent were the beneficiaries of loans. But the average loan was $924 and the

TABLE 33 Major Research Instruments Available to Disciplinary Departments

MAJOR INSTRUMENT	GRAND TOTAL ALL DISCIPLINES	DEPARTMENTAL DISCIPLINE													
		Animal Husbandry	Agronomy and Forestry	Anatomy	Biochemistry and Nutrition	Biology and Ecology	Biophysics and Engineering	Botany	Genetics	Microbiology	Pathology	Pharmacology [a]	Physiology	Zoology and Entomology	Clinical Medical Sciences [b]
TOTAL DISCIPLINARY DEPARTMENTS [c]	**1,192**	**85**	**86**	**65**	**106**	**107**	**16**	**49**	**10**	**84**	**64**	**57**	**74**	**83**	**306**
Acoustic															
Acoustic-Analysis Equipment	**104**	2	2	3	2	15	3	—	—	1	—	2	4	23	47
Sonar	**30**	2	1	2	—	9	—	—	—	1	—	—	—	5	10
Ultrasonic Probes and Sensoring System	**150**	8	8	2	11	17	2	6	—	13	3	2	10	7	61
Centrifuges															
Analytical Ultracentrifuge	**541**	31	28	23	81	51	10	21	7	54	33	14	30	41	117
Preparative Ultracentrifuge	**703**	42	25	41	97	68	13	29	8	77	48	35	47	47	126
Refrigerated Centrifuge	**918**	67	55	46	100	90	13	41	8	83	55	53	59	69	179
Chromatography															
Amino Acid Analyzer	**534**	49	32	18	81	50	6	23	5	44	34	18	22	41	111
Gas Chromatograph	**749**	66	78	24	91	72	7	32	3	45	41	38	37	64	151
Programmed Gradient Pump	**165**	4	6	6	33	14	5	8	2	20	13	5	3	11	35
Counters															
Automatic Particle Counter	**398**	22	27	19	33	50	10	22	4	32	22	20	20	30	87
Scintillation Counter	**845**	47	57	35	97	83	12	39	8	69	44	45	57	60	192
Whole-Body Counter	**163**	19	5	15	9	12	5	1	—	2	9	6	11	7	62
X-Ray															
X-Ray Crystallographic Analysis System	**154**	6	27	6	18	28	9	6	—	7	5	3	3	9	27
X-Ray Diagnostic System	**308**	19	6	13	7	11	7	2	—	7	20	10	19	8	179
X-Ray Source	**385**	27	12	21	15	50	11	19	4	22	22	10	14	35	123
Microscopy															
Electron Microscope	**813**	58	56	59	65	78	12	38	3	72	55	21	39	68	189
Electron Probe for Microscopy	**96**	10	8	2	8	12	2	4	1	4	7	2	1	9	26
[illegible]	[illegible]	[illegible]	[illegible]	[illegible]	[illegible]	[illegible]	[illegible]	19	1	75	49	16	14	29	127

The following table lists instruments (rows) with a total column followed by breakdown columns. Values of "—" indicate none reported.

Instrument	Total[c]														
Microtome-Cryostat	494	32	11	52	21	57	7	29	—	34	49	15	19	47	121
Phase-Contrast Microscope	791	47	39	61	53	95	12	46	7	83	54	21	40	74	159
Spectrometers															
Electron Paramagnetic Resonance Spectrometer	145	5	8	5	29	28	5	8	—	7	6	7	5	13	19
Mass Spectrometer	304	23	32	11	39	43	5	21	2	19	19	10	10	32	38
Nuclear Magnetic Resonance Spectrometer	208	2	18	4	52	39	6	14	—	18	6	11	6	14	18
Spectrophotometers, Polarimeters, Fluorimeters															
Circular Dichroism Analyzer	71	2	3	2	24	10	5	—	—	4	2	3	4	5	7
Infrared Spectrophotometer	492	23	40	14	83	54	9	22	1	34	32	29	25	43	83
Microspectrophotometer	219	7	10	17	21	31	8	7	—	15	17	5	5	21	55
Spectrofluorimeter	365	21	12	11	66	32	8	8	2	26	19	40	25	17	78
Spectropolarimeter	126	4	4	5	34	18	5	2	1	7	10	6	5	9	16
Ultraviolet Spectrophotometer	680	33	36	27	99	67	12	28	6	71	38	47	38	51	127
Miscellaneous															
Apparatus for Measuring Fast Chemical Reactions	123	5	8	2	22	19	3	3	1	2	7	5	10	9	27
Artificial Kidney	158	3	—	4	6	7	3	—	—	3	8	4	6	—	114
Cine and Time–Motion Analysis Equipment	246	14	5	24	7	27	6	7	—	16	17	6	10	21	97
Closed-Circuit TV	403	9	9	31	19	40	7	12	1	22	14	30	37	28	139
Electrophoresis Apparatus (Various Types)	780	58	41	38	98	82	11	34	8	77	51	36	34	64	148
Intensive-Care Patient-Monitoring System	166	1	—	1	7	3	2	18	—	3	4	6	3	—	137
Infrared CO_2 Analyzer	217	2	37	6	5	18	3	3	1	3	4	11	37	9	64
Laser System	95	2	2	5	5	20	4	8	—	33	4	1	6	6	34
Large-Scale Fermenter	110	4	2	1	21	18	3	5	2	11	5	2	1	3	8
Light-Scattering Photometer	116	4	4	4	28	21	2	8	1	3	1	8	4	4	15
Microcalorimeter	69	2	6	2	8	17	3	5	—	7	6	1	2	18	5
Multichannel Oscilloscope	465	14	10	33	17	63	11	1	1	14	6	37	63	41	158
Multichannel Recorder	563	25	25	28	23	66	12	10	2	3	17	40	63	42	189
Osmometers	303	7	7	17	26	32	2	7	—	10	12	19	51	25	92
Small Specialized Computer System (CAT/LINC, etc.)	258	5	5	13	18	28	8	7	4	10	12	18	26	15	89
Stimulus Programming and Operant Conditioning Equipment	104	3	1	10	2	16	3	—	—	1	1	11	15	6	35
Telemetering System	118	8	4	5	4	17	1	2	1	1	2	4	16	14	39

[a] Includes a college of pharmacy.
[b] Includes a department of oral biology.
[c] Number of departments reporting availability of one or more Major Research Instruments.
Source: Survey of Academic Life Science Departments, National Academy of Sciences Committee on Research in the Life Sciences.

TABLE 34   Composition of the Student Body of 85 Four-Year Medical Schools in the United States

TOTAL	**91,046**
Medical Students	35,833
Interns and Residents	23,462
Graduate Students	9,743
Postdoctoral Fellows	6,166
Other	15,842

Source: Data from "Medical Education in the United States (1968–1969)." *J.A.M.A.* 210(8):1478, November 1969.

TABLE 35   Growth of the Graduate Academic Endeavor in Preclinical Departments of 85 Four-Year Medical Schools in the United States

CLASS OF DEPARTMENT	1962–1963			1968–1969		
	Graduate Students	Ph.D.'s Awarded	Post-doctorals	Graduate Students	Ph.D.'s Awarded	Post-doctorals
TOTAL	**4,105**	**382**	**1,061**	**7,892**	**934**	**1,200**
Anatomy	531	49	112	958	118	125
Biochemistry	1,158	125	267	2,029	246	356
Microbiology	680	65	162	1,374	143	173
Pathology	—	—	—	473	29	133
Pharmacology	499	50	135	940	131	140
Physiology	688	57	152	1,268	171	176
Other[a]	549	36	233	850	96	97

[a] Includes genetics, pathology (1962–1963), and other programs not listed above.
Source: 1962–1963 data from "Medical Education in the United States (1967–1968)." *J.A.M.A.* 206(9):2012, November 1968; 1968–1969 data from "Medical Education in the United States (1968–1969)." *J.A.M.A.* 210(8):1479, November 1969.

average scholarship $860, while annual tuition varied from $500 to $2,600, quite apart from the cost of living. It is anticipated that, by academic year 1972–1973, medical schools will admit 11,650 students annually; unless important new initiatives are taken, the financial hardships they and their families will be asked to endure clearly will increase with the passage of time. *We find little merit in this circumstance, and recommend that federal funds on a decidedly larger scale be made available for the personal assistance of medical students.*

Meanwhile, the medical schools themselves face increasingly difficult times. American society now looks to the medical schools for the management of undergraduate medical education, advanced clinical training, graduate education, and the provision of acute medical care for a large fraction of the American citizenry, while also educating large numbers of paramedical personnel—nurses, technicians, physical therapists, dieticians, medical librarians, x-ray technicians, and others. Public funding of these institutions has not kept pace with the increased expectations of either the public or those responsible for management of the medical schools and the universities of which they constitute a part.

We recommend that, at an early date, block funding in support of medical schools be provided from an appropriate federal program. The general approach suggested by the Carnegie Report on Higher Education,* in which it is proposed that such funding be based on a capitation formula, seems entirely appropriate, and we recommend it earnestly.

At the same time, the medical schools, collectively, have become a major component of the national research endeavor in the life sciences. In 1967–1968, the aggregate expenditure within the medical schools for direct costs of research was $473 million, of which $389 million came from federal sources, very largely the National Institutes of Health, and $83.6 million from all other sources (state and local appropriations, endowed income, foundations, societies to combat dread diseases, and others). It is noteworthy that the total expenditure of federal research funds by medical school departments revealed by our questionnaire to department chairmen was about one half of that reported by medical school deans for their aggregate totals. In addition, a substantial fraction of all clinical research is conducted in hospitals not under the management of medical schools, as well as in federal and other nonprofit laboratories. Nevertheless, it is clear that the overwhelming bulk of all patient-oriented clinical research does occur under the auspices of the medical schools.

At the same time, research in the preclinical departments of medical schools is at the leading edge of many scientific frontiers. Indeed, a medical school, in its entirety, constitutes a rather unique system in which scientists in the preclinical departments can remain *au courant* with the advancing edges of the biological, physical, and social sciences, translate these for their clinical colleagues, and collaborate with the latter in both clinical research and relevant fundamental research that are also part of the education of graduate, medical, and postdoctoral students. When the process is successful, the findings can be both tested and then immediately put to

* *Quality and Equality: New Levels of Federal Responsibility for Higher Education,* A Special Report and Recommendations by the Carnegie Commission on Higher Education, McGraw-Hill Book Company, Hightstown, New Jersey, p. 33–36. (Copyright © 1968 by Carnegie Foundation for the Advancement of Teaching.)

TABLE 36   Medical Students in Research

TYPE OF STUDENT	GRAND TOTAL ALL DISCIPLINES	DEPARTMENTAL DISCIPLINE													
		Animal Husbandry	Agronomy and Forestry	Anatomy	Biochemistry and Nutrition	Biology and Ecology	Biophysics and Engineering	Botany	Genetics	Microbiology	Pathology	Pharmacology [a]	Physiology	Zoology and Entomology	Clinical Medical Sciences [b]
TOTAL	2,154	30	48	165	213	17	31	6	8	110	149	121	253	5	998
Students not Seeking Ph.D. Degrees	1,875	4	30	139	143	11	24	6	6	92	140	105	197	2	976
Students Seeking Ph.D. Degrees	279	26	18	26	70	6	7	—	2	18	9	16	56	3	22

[a] Includes a college of pharmacy.
[b] Includes a department of oral biology.
Source: Survey of Academic Life Science Departments, National Academy of Sciences Committee on Research in the Life Sciences.

312

work for the benefit of patients in the same institution. The closest analogy in the academic world is the college of agriculture, in which there are basic scientists, applied scientists, students, and postdoctoral fellows, as well as experimental fields and herds, and direct contact with the network of county farm agents who translate recent applicable research findings into agricultural practice in the surrounding area.

The engineering school cannot behave quite analogously except by virtue of the consulting activities of the engineering faculty, which provide contacts with industrial organizations in position to utilize and exploit recent research findings. It is, perhaps, unfortunate that there are few equivalent built-in mechanisms within the university for assuring that the research endeavors of academic physicists, chemists, sociologists, economists, or attorneys can equally rapidly be tested and put into practice for the benefit of society.

Medical Students in Research

Increasingly, medical students are caught up in the research enterprise. Some desire careers as academic clinicians, a smaller group seeks careers on the faculties of preclinical departments, and some simply taste research and go on to careers in private practice. For all, experience in research will enhance their capabilities as physicians. Research is an exercise in biological problem solving; so too is clinical diagnosis. And the physician will be more proficient at his art if he fully appreciates the caveats to be associated with claims made in the medical literature, which he must continue to study throughout his professional career. There can be no better means for learning how extraordinarily difficult it is to "prove" something in the laboratory or clinic than trying one's own hand at it.

Some medical students work at research only during the summers, some work at odd hours throughout the academic year, and others drop out of regular medical school curricula for a year, most frequently between the second and third year of the curriculum, to engage in research full time. The smallest group undertakes to complete all the requirements for both a Ph.D. and an M.D., usually in one of the preclinical scientific disciplines.

The magnitude of this endeavor, as reported by department chairmen, is shown in Table 36. The 2,154 medical students in the sample were working in 413 individual departments, and it may be assumed that both numbers represent about half the actual totals. Almost half the students undertake their research experience in clinical departments. The others are distributed over the preclinical science departments, with a scattering elsewhere on the university campus. Physiology departments attract a larger number of students to research than do any other of the preclinical disciplines, but a fourth of all medical students who undertake to obtain Ph.D. degrees are enrolled in biochemistry departments.

A few institutions have established curricula providing an intensive combined experience leading jointly to the M.D. and Ph.D. In the long run, M.D.-Ph.D. graduates may well prove to be the most valuable citizens of the medical school faculty community. When such a graduate later manages a joint appointment in both a clinical and preclinical department, he will serve as the translator of science to his clinical colleagues and bring back to the science departments the problems of clinical medicine in terms that may make for successful scientific approaches to understanding of disease. Students so aspiring should be strongly encouraged but should also be clearly apprised of the difficulties of the careers they have embarked upon. It is asking much to expect anyone, over an entire career, to remain a vigorous, effective, and knowledgeable clinician while also keeping abreast of developments in one of the basic biological disciplines.

AGRICULTURAL SCHOOLS AS RESEARCH AND EDUCATIONAL ENTERPRISES

There are 68 four-year land-grant colleges or schools of agriculture in the United States. Enrolled in these schools in 1968–1969 were 50,717 undergraduates and 15,734 graduate students. Undergraduate enrollments have been increasing steadily in these professional schools during the past four years (Table 37), while graduate enrollments have increased more slowly during the same period. Even so, about 27 percent of all graduate students

TABLE 37   Undergraduate and Graduate Enrollments from 1965 to 1968 in Agriculture and Related Curricula

YEAR	TOTAL, ALL STUDENTS	UNDERGRADUATE STUDENTS	GRADUATE STUDENTS
1965	56,339	41,757	14,582
1966	59,296	44,621	14,675
1967	63,917	47,704	16,213
1968	66,451	50,717	15,734

Source: Data from *Proceedings of The National Association of State Universities and Land-Grant Colleges,* 82nd Annual Convention, edited by C. K. Arnold. National Association of State Universities and Land-Grant Colleges, Washington, D.C., 1968.

in the life sciences in the United States are studying in agricultural schools.

Agricultural colleges at land-grant institutions perform three functions: resident teaching, research, and extension teaching. The latter two functions account for about two thirds of faculty time and often do not involve students. For this reason, the faculty-to-student ratio is generally quite high; the ratio of faculty to graduate students is about 1:1, in contrast to a 1:3 ratio in arts and sciences colleges.

Research dominates the budgets of agricultural schools, which are a major component of the U.S. research effort in the life sciences. About $213 million was spent in fiscal year 1968 on research in these schools; the sources of these funds were state ($133 million), federal ($73 million), and private ($7 million).

Research is the foundation for the dramatic advance in technology that has characterized American agriculture. Research has provided hundreds of improved varieties and strains of domesticated plants and animals, has developed methods of protecting man and his plants and animals from the ravages of pests, and has provided a basis for the protection and wise management of our natural resources.

The research program includes both basic and applied research. Many scientists in agricultural schools are making exciting contributions to the development of the life sciences and adding to our understanding of the principal manifestations of life. At the same time, agricultural scientists are seeking answers to some of the major problems facing mankind. An important stimulus to the life scientist in agricultural schools is the fact that he knows that his research results can immediately be translated into action by the extension program and into use by growers and the public. Certainly one of the important reasons for the continuing success of the agricultural colleges has been the well-functioning interrelationships of basic and applied research, extension, and teaching. The need for research expertise is steadily increasing for those students who plan to teach or direct extension programs as well as those interested mainly in research or teaching in agricultural institutions. Of the undergraduate students in agricultural schools, 3 percent (or more than 1,000) are from foreign countries, attracted by the recognized success of American agriculture.

Productive agricultural programs are the foundation for the economies of all nations and especially for those of the developing nations. The developing nations are in critical need of specialists to help them increase food production while protecting their natural resources, both to feed their own populations and to generate capital for other economic development. Working in international agriculture gives faculties another valuable dimension of experience and also fosters contacts between foreign and U.S.

students. The major problem in educating foreign graduate students relates to making their experience relevant to conditions they face at home. Learning to use elaborate equipment in the United States does not prepare them to deal successfully with problems at home, where equipment and facilities are often limited. Ideally, these foreign students should do their thesis research in their homelands to avoid this danger; this is being arranged by some agricultural schools.

The relative and absolute numbers of postdoctorates in agricultural schools are small compared to those in arts and sciences and medical schools. This reflects both the great demand for trained life scientists in agriculture, with vigorous recruitment of young investigators directly into their first positions, and the lack of available postdoctoral fellowships for those engaged in this area of applied research.

Finally, we should note that agricultural schools have a large concentration of life scientists with special interests in environmental biology. A special effort is required to intensify the research and teaching efforts of these scientists and to engage them in coordinated fashion with the endeavors of those in other units of the university and in other institutions concerned with the physical, biological, and social environment.

FINANCING ACADEMIC RESEARCH IN THE LIFE SCIENCES

The department chairmen reported a total of $304,000,000 in support of the direct costs of research in 1966–1967, of which $253,000,000 was derived from federal sources. These figures should be related to the reported total of $2,264,000,000 * for biomedical research from all sources in all categories of performing institutions, both direct and indirect costs, in that year, and $1,459,000,000 * of federal funds for this purpose. Again, it would appear that our questionnaire responses revealed approximately one half of the total academic universe from which they were derived.

Of the reported total, 22 percent went to support research in the colleges of arts and sciences, 12 percent to the agricultural schools, and 63 percent to the medical schools, divided approximately evenly between preclinical and clinical science (Table 38). Approximately half of all these funds (47 percent) went to the private universities, and half (52 percent) to the

---

* Basic Data Relating to the National Institutes of Health 1969, Associate Director for Program Planning and the Division of Research Grants, National Institutes of Health, U.S. Government Printing Office, Washington, D.C., 1969, p. 51.

TABLE 38   Distribution of Research Funds from All Sources

TYPE OF SCHOOL	TOTAL		FEDERAL		NONFEDERAL	
	$M	%	$M	%	$M	%
TOTAL, ALL SCHOOLS	**304.0**	100	**253.1**	100	**50.9**	100
Agriculture[a]	**36.3**	12	23.5	9	12.8	25
Arts and Sciences[b]	**68.4**	22	60.6	24	7.7	15
Medical	**192.9**	63	163.3	65	29.5	58
Preclinical	**90.8**	30	81.6	32	9.3	18
Clinical	**102.1**	33	81.8	33	20.2	40
Other Health-Professional[c]	**6.5**	2	5.6	2	0.9	1
TOTAL, PRIVATE	**145.6**	47	**124.7**	49	**20.9**	41
Arts and Sciences[b]	**30.7**	10	27.3	11	3.4	7
Medical	**113.6**	37	96.3	38	17.3	34
Preclinical	**48.4**	16	43.8	17	4.6	9
Clinical	**65.2**	21	52.5	21	12.7	25
Other Health-Professional[c]	**1.4**	—	1.2	—	0.2	—
TOTAL, PUBLIC	**158.3**	52	**128.4**	51	**29.9**	58
Agriculture[a]	**36.3**	12	23.5	9	12.8	25
Arts and Sciences[b]	**37.7**	12	33.4	13	4.3	8
Medical	**79.3**	26	67.0	27	12.2	24
Preclinical	**42.4**	14	37.7	15	4.7	9
Clinical	**36.8**	12	29.3	12	7.5	15
Other Health-Professional[c]	**5.1**	2	4.5	2	0.6	1

[a] Includes schools of forestry.
[b] Includes schools of engineering and schools of graduate studies.
[c] Includes schools of dentistry, pharmacy, public health, and veterinary medicine.
Source: Survey of Academic Life Science Departments, National Academy of Sciences Committee on Research in the Life Sciences.

public universities. This division was even closer when only federal research funds are considered.

Support of clinical research exceeded that of preclinical research in the private schools (21 percent and 16 percent, respectively, of the overall total), whereas this relationship was inverted in the public medical schools

TABLE 39   Distribution of Research Funds by Class of Department

CLASS OF DEPARTMENT	TOTAL		FEDERAL		NONFEDERAL	
	$M	%	$M	%	$M	%
GRAND TOTAL	**304.0**	*100*	**253.1**	*100*	**50.9**	*100*
Agricultural Sciences	**21.9**	*7*	12.6	*5*	9.3	*18*
Anatomy	**10.3**	*3*	9.4	*4*	0.9	*2*
Biochemistry and Nutrition	**33.7**	*11*	29.3	*12*	4.4	*9*
Biology and Ecology	**33.1**	*11*	29.6	*12*	3.5	*7*
Biophysics and Biomedical Engineering	**8.0**	*3*	7.6	*3*	0.4	*1*
Botany	**7.8**	*3*	6.4	*3*	1.4	*3*
Genetics	**3.8**	*1*	3.7	*1*	0.1	*<1*
Microbiology	**21.6**	*7*	19.3	*8*	2.3	*5*
Pathology	**14.6**	*5*	12.7	*5*	1.9	*4*
Pharmacology[a]	**12.1**	*4*	10.6	*4*	1.5	*3*
Physiology	**15.9**	*5*	13.9	*5*	2.0	*4*
Zoology and Entomology	**17.9**	*6*	15.2	*6*	2.7	*5*
Clinical Sciences[b]	**103.3**	*34*	82.9	*33*	20.3	*40*

[a] Includes a college of pharmacy.
[b] Includes a department of oral biology.
Source: Survey of Academic Life Science Departments, National Academy of Sciences Committee on Research in the Life Sciences.

(12 and 14 percent). Moreover, the support of research in the colleges of arts and sciences was also divided approximately equally between private and public schools, despite the wide disparity in the numbers of faculty and students involved.

Table 39 summarizes the distribution of research funds among disciplinary departments. The clinical sciences received slightly more than one third of the total; roughly equal amounts went to departments of biochemistry and of biology/ecology, with other disciplinary departments receiving lesser amounts.

The data reported here were gathered from both individual scientists and department chairmen in the last fiscal year (1967) during which appropriations for these purposes received a significant increment. Since then, the overall federal appropriation has remained essentially constant, while the number of mature scientists, postdoctoral fellows, graduate students, graduate departments, medical schools, and even universities has continued to increase and the dollar has depreciated steadily. Assuming that a rational growth rate would be about 12 percent per year in absolute dollars, by fiscal year 1971 this appropriation should be greater by 56 percent than that for fiscal year 1967. Clearly, this will not be the case. The consequence, in our collective experience, has been a significant decline in the mean size of individual grants and a lengthening list of competent scientists whose research is limping along for lack of supplies, equipment, and assistants. The absolute extent of this problem has not been documented; its effects are only now becoming evident on every campus and in every medical school.

Failure to exploit the capabilities and talents of the scientific personnel on whose educations the nation has spent great sums while the major problems of human biology in health and disease remain unsolved, the human brain is still mysterious, the principles of ecology—imperative to management of our renewable resources—are still unrevealed, the essentials of the process of differentiation in the development of an organism are still obscure, the menace of pollution is inadequately evaluated, and overpopulation and mass starvation are in the offing is, in our view, woefully false economy. At a time when understanding of the living state may provide enormous benefits to the average citizen, public funds in support of the efforts to achieve that understanding are actually being reduced. And the succeeding generation will be penalized.

It may reasonably be assumed that the overall pattern of utilization of these funds conforms to the gross pattern of National Science Foundation grants, described earlier. Whereas in the early history of federal research-grant programs investigators had great latitude in the disposition of the funds made available to them, as among the various categories of expenditure, investigators are increasingly expected to manage research funds within the various budget categories negotiated at the time of award. We have no fault to find with this practice as long as agency administrators remain flexible in granting permission to transfer funds between categories when this can be justified by the course of research in progress, unanticipated needs, and related factors. One category of expenditure, faculty salaries, is a matter of special concern and will be discussed in some detail.

Faculty Salaries

Essentially similar pictures emerged from the data supplied by individual investigators and department chairmen. Let us consider first the information from investigators. Institutional budgets supplied 68.6 percent of the salaries of all the academic life scientists in our sample; 18.8 percent of all salaries came from research grants and contracts, 4.2 percent from federal training grants, 3.7 percent from advanced fellowships, and 2.7 percent from federal institutional grants. These data and those considered below include all reported salary payments, whether annual, nine-month, or summer salaries. However, these data were collected before the growth of both the Health Sciences Advancement Awards program of the National Institutes of Health and the Institutional Science Development Award program of the National Science Foundation. Accordingly, contributions from federal institutional grants should now loom somewhat larger as a fraction of the total. It is too early to appraise the impact of the recent decision to reduce the institutional-grants program of the National Institutes of Health by 50 percent, but this may be presumed to be of yet greater effect and, indeed, may well be disastrous in some schools.

Scientists in the agricultural schools are positioned most securely, since institutional sources provided 88.2 percent of all such salaries, the only discrepancy being in the field of fish and wildlife, where research grants and contracts provided 25 percent of all salaries. As seen in Table 40, the M.D. population was least secure, as only half of the M.D. salaries are provided from institutional sources. The preclinical faculties and those in colleges of arts and sciences were in intermediate positions, the seriousness of which varied among the disciplines. It should be understood that the values shown do not represent any single scientist but rather the total contribution of each of the sources shown to the total salaries of all the scientists in each pool. Actually, of 7,621 faculty members concerning whom complete salary breakdowns were made available, 50 percent (3,799) derived their entire salaries from nonfederal sources, 13 percent (994, of whom one fourth were M.D.'s) were entirely supported from federal sources, and 37 percent reported that their salaries included both federal and nonfederal monies. Of this total sample, 50 percent (3,817) of all individuals received part or all of their salaries from federal sources, and 42 percent received salaries from more than one source, usually distributed between the institutional budget and some federal source. Of those whose entire incomes derived from federal funds, 58 percent were supported by research grants or contracts, 23 percent by training grants, 10 percent by institutional grants, and 19 percent through one of several advanced fellowship programs, e.g., senior postdoctoral fellowships from several agencies and the Career Awards of the National Institutes of Health.

Table 41 indicates the percentage of salary from institutional budgets paid to Ph.D.'s, D.Sc.'s, and M.D.'s working in the various research areas defined by this study. Patently, the institutional contribution to the salaries of M.D.'s is decidedly less than that of the Ph.D.'s or D.Sc.'s, although the

TABLE 40   Percentage of Salary Derived from Various Sources for Some Academic Life Scientists

DOCTORAL TRAINING	INSTITU-TIONAL BUDGET	RE-SEARCH GRANT	TRAIN-ING GRANT	INSTITU-TIONAL GRANT	FELLOW-SHIP
ALL BIOLOGISTS	**68.5**	**18.8**	**4.1**	**2.7**	**3.7**
Ph.D. or D.Sc. in:					
An Agricultural Science	88.2	8.0	0.1	2.6	—
Anatomy	81.1	12.6	3.4	0.8	1.2
Biochemistry	57.7	27.1	4.3	3.2	6.6
Microbiology	70.7	18.1	4.9	2.4	3.3
Pharmacology	62.1	18.0	9.8	5.8	3.2
Physiology	69.2	18.6	3.7	3.4	3.6
Botany	84.2	11.3	1.2	1.3	1.3
Entomology and Hydrobiology	88.8	8.1	0.4	1.6	—
Zoology	78.3	14.0	2.7	1.9	2.6
Other[a]	62.6	26.7	3.4	4.1	2.5
M.D. Degree Only	50.8	24.0	7.9	3.1	7.1
Other Health-Professional Degree	69.0	15.7	3.0	4.0	3.0

[a] Not originally in a life science, e.g., physics, chemistry, or sociology.
Source: Survey of Individual Life Scientists, National Academy of Sciences Committee on Research in the Life Sciences.

salary scale of the M.D.'s may be somewhat greater. It is also apparent that, as a fraction of total salaries, molecular biology, cellular biology, and pharmacology are, of all research areas, least well supported from institutional resources. The differences between institutional payments and total salaries are very largely made up by payments from various federal sources. Since molecular biology and biochemistry is the most populous research category, federal contributions in support of this field by salary payments are greater than those of any other research area within the study. Further, of 696 investigators receiving their total salaries from federal sources, 40 percent were in the field of molecular biology and biochemistry, followed by physiology with 16 percent of the total. Moreover, investigators in molecular biology and biochemistry and in physiology constituted 23 percent and 18 percent, respectively, of all those whose incomes derived from a mixture of institutional and federal sources. We were surprised to learn that investigators in the area of disease mechanisms constituted only 9 percent of this total, despite the large number of M.D.'s in this

TABLE 41   Average Percentage of Salary Derived from Institutional Budgets of Ph.D.'s or D.Sc.'s and M.D.'s in Various Research Areas

RESEARCH AREA	Ph.D.'S OR D.Sc.'S	M.D.'S [a]
ALL AREAS	**72.6**	**54.2**
Behavioral Biology	75.8	69.4
Cellular Biology	68.3	44.1
Developmental Biology	77.1	63.5
Disease Mechanisms	73.7	58.7
Ecology	81.8	86.3
Evolutionary and Systematic Biology	86.2	79.5
Genetics	78.0	50.5
Molecular Biology and Biochemistry	59.5	41.6
Morphology	88.0	70.4
Nutrition	81.4	58.2
Pharmacology	66.1	57.1
Physiology	75.5	51.8
Other Related Areas	76.1	69.4

[a] Includes some M.D.'s who also have a Ph.D. or D.Sc. degree.
Source: Survey of Individual Life Scientists, National Academy of Sciences Committee on Research in the Life Sciences.

field and the inadequacy of their support from institutional funds. This appears to reflect the fact, noted earlier, that the M.D.'s in our population very largely consider themselves to be engaged in fundamental research, particularly in molecular biology, physiology, and cellular biology.

The impressions gained from the responses of individual scientists are well borne out by those from department chairmen. Table 42 indicates the average percentage of salaries paid from funds derived from an institution's own resources for the full-time faculty engaged in research, according to the class of college and university. In accord with previous data, it is the clinical departments that, necessarily, are forced to seek non-institutional funds for faculty support. This table vividly shows the great dependence of the medical schools, particularly the private medical schools, on federal funding for payment of faculty salaries. Thus the public universities, which provide 84 percent of salaries in the faculties of arts and sciences and agriculture, provide only 77 percent of salaries for their pre-clinical departments and 62 percent for their clinical faculties. In the private schools, in which the faculties of arts and sciences received 76 per-

TABLE 42   Average Percentage of Faculty Salaries Derived from Institutional Funds

TYPE OF SCHOOL	PRIVATE UNIVERSITIES	PUBLIC UNIVERSITIES
ALL TYPES	**55**	**78**
Agricultural School[a]	—	84
Arts and Sciences[b]	76	84
Medical	50	71
Preclinical	57	77
Clinical	43	62
Other Health Professional[c]	73	84

[a] Includes schools of forestry.
[b] Includes schools of engineering and schools of graduate studies.
[c] Includes schools of dentistry, pharmacy, public health, and veterinary medicine.
Source: Survey of Academic Life Science Departments, National Academy of Sciences Committee on Research in the Life Sciences.

cent of their support from institutional sources, the preclinical departments received only 57 percent and the clinical departments only 43 percent of faculty salaries from internal sources. In each case, the balance came very largely from the federal government.

The extent of this dependence is revealed further by the fact that, among the private medical schools, two preclinical departments and 12 clinical departments reported that they had absolutely no institutional funds for faculty salary payments; 36 preclinical departments and 67 clinical departments managed to find 60 to 100 percent of the funds required to pay faculty salaries (Table 43). In the main, this was accomplished by use of federal funds appropriated in support of biomedical research.

This bleak picture of the great dependence of medical schools, particularly private medical schools, on federal research funding for payment of faculty salaries has been amply corroborated in an independent study reported by the American Medical Association * utilizing data derived from reports from all functioning medical schools. In 1961–1962, federal research funds were utilized to defray part or all of the salaries of 31 percent of the individuals on then-existing medical faculties. By 1968–1969, the number receiving such subsidies had grown to 49 percent of the entire national medical faculty. Moreover, the pattern of contribution of federal funding shown in Table 44 was much the same as that in clinical and preclinical departments reporting in our survey. The fraction of the faculty

* *Medical Education in the United States 1968–1969.* "Medical School Faculties," p. 1477. Reprinted from *J.A.M.A.* 210(8):1455–1587, 1969.

TABLE 43   Departments Receiving 60 to 100 Percent of the Salaries of Full-time Faculty from Federal Sources

TYPE OF SCHOOL	ALL SCHOOLS		PUBLIC SCHOOLS		PRIVATE SCHOOLS	
	Number	Percentage	Number	Percentage	Number	Percentage
ALL TYPES	**194**	*17.8*	**83**	*7.6*	**111**	*10.2*
Agriculture [a]	**22**	*2.0*	22	*2.0*	—	—
Arts and Sciences [b]	**14**	*1.3*	6	*0.6*	8	*0.7*
Preclinical Medical	**52**	*4.8*	16	*1.5*	36	*3.3*
Clinical Medical	**103**	*9.5*	36	*3.3*	67	*6.2*
Other Health-Professional [c]	**3**	*0.3*	3	*0.3*	0	*0.0*

[a] Includes schools of forestry.
[b] Includes engineering schools and schools of graduate studies.
[c] Includes schools of dentistry, pharmacy, public health, and veterinary medicine.
Source: Survey of Academic Life Science Departments, National Academy of Sciences Committee on Research in the Life Sciences.

federally supported, in whole or in part, ranged from 38 percent of pathologists to 58 percent of pharmacologists and from 19 percent of all academic anesthesiologists to 58 percent of all academic neurologists. The requisite funds were found from the National Institutes of Health institutional grants to each medical school and from advanced fellowships, but most frequently from grants for research or research training.

These salary payments were drawn from $154.3 million in teaching and training grants and $389.6 million in research grants from the federal government to the medical schools, as well as from $83.7 million in research funds from other sources. The absolute magnitude of these payments for medical school faculty salaries from federal funds is not known. If we assume a mean faculty salary for all three ranks of $16,000 per annum, total payments for faculty salaries from federal funds appropriated in support of research and research training must have been well in excess of $80 million per annum, a figure that should be compared with such items of nationally aggregated medical school income in 1967–1968 as: tuition and fees, $48 million; indirect costs on all federal grants and contracts, $74 million; endowment income, $30 million; and state appropriations, $143 million. Since the faculty so supported performs in the manner normal to such faculties and distributes its time much as do those whose salaries derive entirely from institutional sources, federal support of medical schools through the agencies of research and training grants patently has become a very significant part of the financial structure of the national medical edu-

TABLE 44 Medical Faculty Who Received Complete or Partial Salary Payments from Federal Research and Training Grants, 89 U.S. Medical Schools

BASIC SCIENCE DEPARTMENTS	PERCENTAGE OF SALARY FROM FEDERAL SOURCES BY RANK												FACULTY SURVEYED[a]		
	Professor			Associate Professor			Assistant Professor			All Ranks			Total	With Federal Support	
	100	50–99	1–49	100	50–99	1–49	100	50–99	1–49	100	50–99	1–49		No.	%
TOTAL FACULTY	364	544	1,435	695	785	1,357	1,267	1,103	1,533	2,326	2,432	4,325	18,518	9,083	49
Total Preclinical	166	181	594	269	199	495	391	307	473	826	687	1,562	6,094	3,075	50
Anatomy	22	33	90	21	22	88	39	41	88	82	96	266	934	444	48
Biochemistry	34	27	94	49	28	79	76	47	82	159	102	255	957	516	54
Microbiology	21	13	87	41	22	62	40	52	60	102	87	209	775	398	51
Pathology	12	36	103	37	34	99	66	43	88	115	113	290	1,322	518	38
Pharmacology	21	22	85	28	30	72	76	34	45	125	86	202	722	413	58
Physiology	26	27	107	54	39	75	44	64	78	124	130	260	918	514	56
Other	30	23	28	39	24	20	50	26	32	119	73	80	466	272	58
Total Clinical	198	363	841	426	586	862	876	796	1,060	1,500	1,745	2,763	12,424	6,008	48

[a] Records the total medical school life scientists responding to the survey, the total receiving federal support of salary payments, and the percentages of the total respondents receiving federal support.
Source: Data from "Medical Education in the United States 1968–1969." *J.A.M.A.* 210(8):1578, November 1969.

cational system. When it is recognized that more than half of this form of support goes to the one fourth of all medical schools that are under private auspices, the very high degree of dependence of the private medical schools on research funding for their very existence becomes startlingly and painfully evident.

The equitable geographic distribution of federal funds in support of research and graduate and postdoctoral education is a matter of perennial public and congressional concern. An attempt to assess this situation is summarized in Tables 45 through 48, which compare the distribution of the American population, by major census regions, with the distribution of federal research support, life sciences faculties, Ph.D. production, and postdoctoral appointees among the same census regions. Research dollars are somewhat more generously available to the Northeast, Middle Atlantic, and Pacific Coast states than their fractions of the national population would suggest; this correlates best with the distribution of postdoctoral fellows. It does not correlate as well either with the gross distribution of faculty or with Ph.D. production, to which the Midwest makes disproportionate contributions. In general, these distributions are very markedly affected by the geographical locations of perhaps the 20 major research-performing universities, but the disparities appear to offer little cause for concern. Figure 37 summarizes these data.

TABLE 45   Federal Research Dollars Reported by Some Life Sciences Departments in Academic Year 1966–1967 (In Millions of Dollars)

CENSUS REGION	Agricultural[a] Public		Arts and Sciences[b] Public		Private		Preclinical Medical Public		Private		Clinical Medical Public		Private		Total All Types[c]		U.S. Population, 1967[d] (In Thousands)	
	No.	%	No.	%	No.	%	No.	%	No.	%	No.	%	No.	%	No.	%	No.	%
TOTAL U.S.A.	23.6	100	33.4	100	27.2	100	38.2	100	43.4	100	29.3	100	52.5	100	253.1	100	197,863	100
New England	1.3	5.7	1.4	4.1	11.1	40.7	0.7	1.8	4.7	10.7	<0.5	0.2	9.6	18.2	28.9	11.4	11,321	5.7
Middle Atlantic	1.9	7.9	3.9	11.7	5.6	20.5	2.7	7.1	17.5	40.3	1.6	5.6	18.1	34.5	53.5	21.1	36,968	18.7
East North Central	5.3	22.5	7.3	21.9	4.4	16.3	6.4	16.8	4.1	9.4	8.7	29.8	4.1	7.8	40.7	16.1	39,123	19.8
West North Central	3.4	14.3	2.5	7.5	<0.5	1.4	6.0	15.8	2.4	5.5	3.3	11.2	2.3	4.4	20.7	8.2	15,961	8.1
South Atlantic	3.5	14.9	2.3	6.9	2.8	10.4	5.1	13.5	6.6	15.2	4.5	15.3	10.8	20.5	35.8	14.2	29,480	14.9
East South Central	1.1	4.8	0.5	1.6	<0.5	0.1	3.5	9.2	1.7	4.0	<0.5	1.4	1.0	1.9	8.4	3.3	12,970	6.6
West South Central	0.9	4.0	3.4	10.2	<0.5	1.5	1.9	4.9	2.9	6.8	3.2	10.9	2.1	4.0	15.5	6.1	18,993	9.6
Mountain	1.9	5.4	2.1	6.3	<0.5	<0.1	3.9	10.3	—	—	2.4	8.3	—	—	10.5	4.2	7,796	3.9
Pacific	4.8	20.5	9.9	29.7	2.5	9.0	7.9	20.6	3.5	8.2	5.1	17.4	4.5	8.6	39.1	15.5	25,249	12.8

TYPE OF SCHOOL

[a] Includes schools of forestry.
[b] Includes schools of engineering and schools of graduate studies.
[c] Includes health-professional schools other than schools of medicine.
[d] Source: *Statistical Abstract of the United States, 1968*, Department of Commerce, Bureau of the Census; *Current Population Reports*, Series P-25, U.S. Government Printing Office, Washington, D.C., 1968, p. 25.
Source: Survey of Academic Life Science Departments, National Academy of Sciences Committee on Research in the Life Sciences.

TABLE 46 Full-Time Faculty in Some Academic Life Sciences Departments, Academic Year 1966–1967

CENSUS REGION	Agricultural[a] Public		Arts and Sciences[b] Public		Private		Preclinical Medical Public		Private		Clinical Medical Public		Private		Total All Types[c]		U.S. Population, 1967[d] (In Thousands)	
	No.	%	No.	%	No.	%	No.	%	No.	%	No.	%	No.	%	No.	%	No.	%
TOTAL U.S.A.	3,907	100	2,663	100	1,189	100	1,965	100	1,812	100	2,207	100	2,931	100	17,172	100	197,863	100
New England	227	5.8	97	3.6	339	28.5	16	0.8	197	10.9	36	1.6	403	13.7	1,327	7.2	11,321	5.7
Middle Atlantic	387	9.9	223	8.4	346	29.1	149	7.6	702	38.7	209	9.5	1,104	37.7	3,278	19.1	36,968	18.7
East North Central	559	14.3	711	26.7	157	13.2	432	22.0	201	11.1	611	27.7	289	9.9	2,977	17.3	39,123	19.8
West North Central	640	16.4	318	11.9	13	1.1	233	11.8	125	6.9	198	9.0	135	4.6	1,777	10.4	15,961	8.1
South Atlantic	668	17.1	269	10.1	157	13.2	309	15.7	291	16.0	388	17.6	662	22.6	2,764	16.1	29,480	14.9
East South Central	250	6.4	105	3.9	12	1.0	213	10.8	88	4.9	146	6.6	54	1.8	886	5.2	12,970	6.6
West South Central	280	7.2	254	9.5	46	3.9	164	8.3	116	6.4	221	10.0	134	4.6	1,251	7.2	18,993	9.6
Mountain	333	8.5	303	11.4	31	2.6	130	6.6	—		181	8.2	—		1,016	5.9	7,796	3.9
Pacific	563	14.4	383	14.4	88	7.4	319	16.2	92	5.1	217	9.8	150	5.1	1,896	11.0	25,249	12.8

TYPE OF SCHOOL

[a] Includes schools of forestry.
[b] Includes schools of engineering and schools of graduate studies.
[c] Includes health-professional schools other than schools of medicine.
[d] Source: *Statistical Abstract of the United States, 1968*. Department of Commerce, Bureau of the Census; *Current Population Reports*, Series P-25, U.S. Government Printing Office, Washington, D.C., 1968, p. 25.
Source: Survey of Academic Life Science Departments, National Academy of Sciences Committee on Research in the Life Sciences.

TABLE 47   Postdoctoral Appointees in Some Academic Life Sciences Departments, Academic Year 1966–1967

CENSUS REGION	Agricultural[a] Public		Arts and Sciences[b] Public		Arts and Sciences[b] Private		Preclinical Medical Public		Preclinical Medical Private		Clinical Medical Public		Clinical Medical Private		Total All Types[c]		U.S. Population, 1967[d] (In Thousands)	
	No.	%	No.	%	No.	%	No.	%	No.	%	No.	%	No.	%	No.	%	No.	%
TOTAL U.S.A.	**360**	*100*	**544**	*100*	**577**	*100*	**691**	*100*	**776**	*100*	**668**	*100*	**1,317**	*100*	**5,134**	*100*	**197,863**	*100*
New England	24	*6.7*	10	*1.8*	274	*47.5*	4	*0.6*	132	*17.0*	7	*1.0*	305	*23.2*	**762**	*14.8*	11,321	*5.7*
Middle Atlantic	51	*14.2*	47	*8.6*	64	*11.1*	53	*7.7*	302	*38.9*	45	*6.7*	415	*31.5*	**1,034**	*20.1*	36,968	*18.7*
East North Central	120	*33.3*	131	*24.1*	82	*14.2*	127	*18.4*	52	*6.7*	207	*31.0*	94	*7.1*	**822**	*16.0*	39,123	*19.8*
West North Central	30	*8.3*	45	*8.3*	9	*1.6*	101	*14.6*	49	*6.3*	98	*14.7*	33	*2.5*	**422**	*8.2*	15,961	*8.1*
South Atlantic	22	*6.1*	28	*5.1*	45	*7.8*	84	*12.2*	113	*14.6*	66	*9.9*	335	*25.4*	**699**	*13.6*	29,480	*14.9*
East South Central	8	*2.5*	15	*2.8*	—	—	48	*6.9*	22	*2.8*	18	*2.7*	16	*1.2*	**129**	*2.5*	12,970	*6.6*
West South Central	4	*1.1*	48	*8.8*	2	*0.3*	17	*2.5*	42	*5.4*	66	*9.9*	52	*3.9*	**239**	*4.7*	18,993	*9.6*
Mountain	8	*2.2*	23	*4.2*	—	—	43	*6.2*	—	—	61	*9.1*	—	—	**155**	*3.0*	7,796	*3.9*
Pacific	93	*25.8*	197	*36.2*	101	*17.5*	214	*31.0*	64	*8.2*	100	*15.0*	67	*5.1*	**872**	*17.0*	25,249	*12.8*

TYPE OF SCHOOL

[a] Includes schools of forestry.
[b] Includes schools of engineering and schools of graduate studies.
[c] Includes health-professional schools other than schools of medicine.
[d] Source: *Statistical Abstract of the United States, 1968*, Department of Commerce, Bureau of the Census; *Current Population Reports*, Series P-25, U.S. Government Printing Office, Washington, D.C., 1968, p. 25.
Source: Survey of Academic Life Science Departments, National Academy of Sciences Committee on Research in the Life Sciences.

TABLE 48   Ph.D. Degrees Awarded by Some Life Sciences Departments in Academic Year 1966–1967

CENSUS REGION	Agricultural[a] Public		Arts and Sciences[b] Public		Arts and Sciences[b] Private		Preclinical Medical Public		Preclinical Medical Private		Total All Types[c]		U.S. Population, 1967[d] (In Thousands)	
	No.	%	No.	%	No.	%	No.	%	No.	%	No.	%	No.	%
TOTAL U.S.A.	771	100	579	100	367	100	308	100	215	100	2,332	100	197,863	100
New England	19	2.5	10	1.7	85	23.2	1	0.3	16	7.4	138	5.9	11,321	5.7
Middle Atlantic	74	9.6	47	8.1	130	35.4	24	7.8	67	31.2	368	15.8	36,968	18.7
East North Central	188	24.4	163	28.2	44	12.0	71	23.1	38	17.7	516	22.1	39,123	19.8
West North Central	141	18.3	90	15.5	2	1.0	52	16.9	9	4.2	312	13.4	15,961	8.1
South Atlantic	112	14.5	48	8.3	63	17.2	39	12.7	48	22.3	318	13.6	29,480	14.9
East South Central	24	3.1	12	2.1	2	1.0	27	8.8	6	2.8	72	3.1	12,970	6.6
West South Central	36	4.7	74	12.8	7	1.9	28	9.1	20	9.3	167	7.2	18,993	9.6
Mountain	43	5.6	39	6.7	—	—	9	2.9	—	—	102	4.4	7,796	3.9
Pacific	134	17.4	96	16.6	34	9.3	57	18.5	11	5.1	339	14.5	25,249	12.8

[a] Includes schools of forestry.
[b] Includes schools of engineering and schools of graduate studies.
[c] Includes health-professional schools other than schools of medicine.
[d] Source: Statistical Abstract of the United States, 1968, Department of Commerce, Bureau of the Census; Current Population Reports, Series P-25, U.S. Government Printing Office, Washington, D.C., 1968, p. 25.
Source: Survey of Academic Life Science Departments, National Academy of Sciences Committee on Research in the Life Sciences.

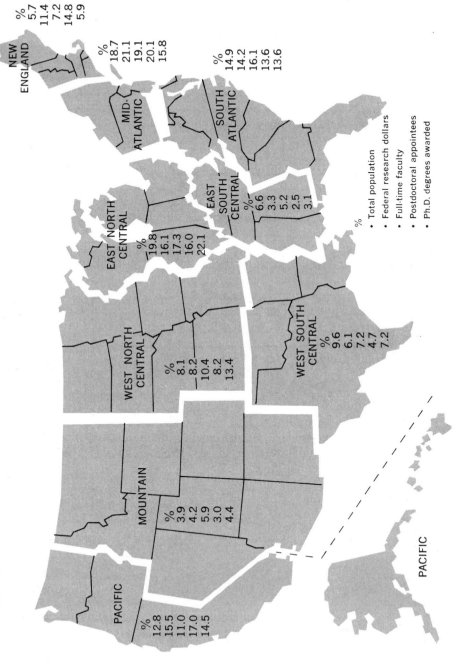

FIGURE 37 Geographic distribution of total population, federal research dollars, full-time faculty, postdoctoral appointees, and Ph.D. degrees awarded in academic year 1966–1967 in the biological sciences. Source: Survey of Academic Life Science Departments, National Academy of Sciences Committee on Research in the Life Sciences.

NEW
ENGLAND

%
5.7
11.4
7.2
14.8
5.9

MID-
ATLANTIC

%
18.7
21.1
19.1
20.1
15.8

SOUTH
ATLANTIC

%
14.9
14.2
16.1
13.6
13.6

EAST NORTH
CENTRAL

%
19.8
16.1
17.3
16.0
22.1

EAST
SOUTH-
CENTRAL

%
6.6
3.3
5.2
2.5
3.1

WEST NORTH
CENTRAL

%
8.1
8.2
10.4
8.2
13.4

WEST SOUTH
CENTRAL

%
9.6
6.1
7.2
4.7
7.2

MOUNTAIN

%
3.9
4.2
5.9
3.0
4.4

PACIFIC

%
12.8
15.5
11.0
17.0
14.5

PACIFIC

· Total population
· Federal research dollars
· Full-time faculty
· Postdoctoral appointees
· Ph.D. degrees awarded

%

CHAPTER FIVE

REQUIREMENTS FOR THE FUTURE OF THE ACADEMIC ENDEAVOR IN THE LIFE SCIENCES

The substance of the case for societal support of the research endeavor in the life sciences is presented throughout this report. No natural constraints to this endeavor are evident other than the supply of qualified investigators. If the people of the United States, and indeed all mankind, are to be adequately nourished in the future, if we are to pursue with utmost vigor the attempt to understand the nature of life, the nature of man, the diseases to which he is subject, and the environment in which he dwells so that, one day, this knowledge may be utilized to alleviate the human condition, to minimize suffering, and to extend the useful and enjoyable life-span, then the research endeavor in the life sciences must be supported in a manner and on a scale commensurate with our national aspirations in these regards.

Both individual investigators and department chairmen were queried with respect to current constraints to research progress. The responses obtained from 10,083 individual scientists engaged in all research areas and employed in all classes of academic and nonacademic institutions were almost monotonously uniform. These are summarized in Tables 49 and 50. Funds, space, and research staff were considered to be serious problems by more than half the scientists who responded.

TABLE 49 Factors Seriously Limiting the Research Programs of Individual Life Scientists

LIMITING FACTOR	ACADEMIC SCIENTISTS		NONACADEMIC SCIENTISTS	
	Number	%	Number	%
GRAND TOTAL [a]	**7,007**	*100*	**3,076**	*100*
Space	**3,529**	*50*	**1,277**	*42*
Budget for:	**5,399**	*77*	**2,187**	*71*
Supplies	1,779	*25*	371	*12*
Equipment	2,806	*40*	803	*26*
Professional Staff	1,554	*22*	957	*31*
Technicians	3,353	*48*	1,454	*47*
Clerical–Administrative	1,169	*17*	404	*13*
Student Stipends	1,742	*25*	196	*6*
Postdoctoral Stipends	1,621	*23*	339	*11*
Computer Time	245	*3*	100	*3*
Travel	1,329	*19*	661	*21*
Insufficient Research Time due to:	**3,429**	*49*	**1,118**	*36*
Heavy Teaching Responsibilities	1,851	*26*	75	*2*
Service	848	*12*	371	*12*
Administration	1,913	*27*	867	*28*
Unfilled Funded Positions for:	**1,446**	*21*	**725**	*24*
Professional Staff	759	*11*	430	*14*
Technicians	832	*12*	432	*14*
Clerical–Administrative	149	*2*	62	*2*

[a] Life scientists reporting one or more limiting factors.
Source: Survey of Individual Life Scientists, National Academy of Sciences Committee on Research in the Life Sciences.

INDIVIDUAL SCIENTISTS

Academic Scientists

Half the scientists considered that the space available to them was seriously limiting. A fifth to a quarter were hampered for lack of access to one or another specialized research facility. Three fourths of all scientists declared that they had insufficient funds, with pharmacologists least needful in this regard and those studying morphology most severely constrained. Although the lack of funds resulted in different difficulties for different individuals

TABLE 50   Additional Factors Seriously Limiting the Research Programs of Individual Life Scientists

LIMITING FACTOR	ACADEMIC SCIENTISTS		NONACADEMIC SCIENTISTS	
	Number	%	Number	%
GRAND TOTAL[a]	**7,007**	*100*	**3,076**	*100*
Constraints on Nature of Research Problem due to:	**1,835**	*26*	**1,054**	*34*
Conditions of Employment	539	*8*	643	*21*
Source of Research Funds	1,384	*20*	516	*17*
Other	236	*3*	109	*4*
Inadequacy of Personal Training in:	**3,120**	*45*	**1,238**	*40*
Chemistry	1,303	*19*	449	*15*
Statistics	776	*11*	350	*11*
Mathematics	1,056	*15*	356	*12*
Computer Use	1,121	*16*	434	*14*
Electronics	707	*10*	261	*8*
Physics	360	*5*	129	*4*
Other Biological Sciences	696	*10*	376	*12*

[a] Life scientists reporting one or more limiting factors.
Source: Survey of Individual Life Scientists, National Academy of Sciences Committee on Research in the Life Sciences.

in the population, the pattern, by research areas, showed few significant variations. Twenty to thirty percent were limited by available consumable supplies; about a third required specialized equipment, a requirement stated most frequently by biochemists and least frequently by systematic biologists. Almost every individual in the study would expand his research group somewhat if he could: one fourth indicated a desire for additional professional staff, a desire least often asserted by the developmental biologists and most often by those studying disease mechanisms; half felt an urgent need for additional technicians, the most numerous being the nutritionists. About a sixth of these scientists have insufficient clerical and administrative assistance while a fourth desire additional support for students; those studying pharmacology (13 percent) and disease mechanisms (19 percent) felt this need least severely, whereas those engaged in studies of ecology (40 percent) and systematic biology considered it considerably more urgent (35 percent). A fifth of the group expressed a need for additional postdoctoral fellows, a need apparently least frequent among ecologists (14 percent) and most urgent among the molecular biologists and biochemists (30 percent), reflecting the style of research in these disciplines and the existing distribution of postdoctoral fellows.

Approximately one fifth of the life scientists polled considered themselves hampered by insufficient funds in support of travel, most particularly the systematic biologists, reflecting their requirement for far-flung field studies.

Insufficiency of funds was most common among the faculty of agricultural schools (86 percent) and less urgent in the medical schools (71 percent), with those in the graduate colleges of arts and sciences in between. Those working in agricultural schools also felt most pressed for lack of supporting technical help, which was in better supply in the laboratories of the medical schools. Funds to support and, presumably, to increase the number of graduate students were in great demand in the colleges of arts and sciences and in the agricultural schools but less pressing in the medical schools, where National Institutes of Health training grants had alleviated the pressure in considerable degree. The events of Fiscal Year 1970 and 1971 may very well alter that situation dramatically, and for the worse.

These statements, descriptive of circumstances during the summer of 1967, should be read in the light of the subsequent serious deterioration in the federal funding of all research, particularly that in the life sciences. If three fourths of all life scientists considered their research limited by lack of funds at the beginning of fiscal year 1968, at this time the problem must be both well-nigh universal and considerably more urgent and constraining. Our best guess concerning the magnitude of the deficit between current support and a level that would be commensurate with current capability, opportunities, and needs is 20–25 percent.

Research is an all-engrossing, compelling aspect of a scientist's life, and it is hardly unexpected that this research-performing population feels itself pressed for lack of time to pursue research wherever it leads. This problem, common to all research areas and to all schools within the universities is, however, most pressing in those groups with the largest teaching responsibilities (biologists in the colleges of arts and sciences) or that engage in the delivery of patient care.

Nonacademic Scientists

Nor does it appear that the research needs of scientists employed by nonacademic organizations are significantly better met than are those of their academic colleagues. Forty-two percent of all such scientists indicated that they have insufficient research space, and 71 percent reported insufficient funds. The primary difficulty occasioned by insufficient funds is an insufficiency in the supply of supporting help, particularly of professional associates. As one might expect, the demand among this group for funds for student fellowships or postdoctoral appointments is relatively minor. They too, although in slightly lesser degree, wish that there were more time available to pursue their research, but it is usually administrative or service duties rather than instruction that claim their time.

Personal Constraints on Research

All respondents were asked whether the conditions of their employment or the sources of the funds that support their research in some manner imposed very serious constraints on the nature of the research problems on which they are engaged. The question might have been stated, "If you were employed elsewhere with complete freedom to choose your own research problem, or if the funds that support your research had no strings attached, would you be engaged in a research problem other than that which now claims your attention?" Of the entire group under study, 2,889 scientists (29 percent) consider themselves to be constrained by the conditions of their employment and 1,900 (19 percent) appear to have tailored their research, in some degree, to meet the requirements of a funding organization. Members of academic faculties, in all colleges, feel relatively unconstrained by such considerations, whereas a fifth of all of those in nonacademic institutions appeared to consider that the conditions of their employment adversely affect their choice of research problems. This is somewhat surprising in view of the very large fraction of life scientists in such organizations who also stated that they are engaged in the performance of fundamental research. This aspect of life was twice as troublesome to federal employees (17 percent) as it was to academic employees, while it gave concern to 35 percent of all employees of private industry. At the same time, almost a fifth of all academic life scientists are disturbed by the constraints implicit in the sources of the funds that support their research, most notably those studying nutrition (28 percent), disease mechanisms (24 percent), and ecology (23 percent). No major deviations from this pattern were apparent among the faculties of the various colleges of the universities.

In view of the multiple and diverse opportunities for employment and the remarkable diversity of sources of research funds, the relatively high degree of direction seemingly given to the research endeavor in all sectors by funding agencies, which overrides the personal research preferences of so many investigators, particularly those in academic life, came as a surprise. The constraints imposed by insufficiency of space, funds, and other requirements can be alleviated by an expansion of the total funding of the research enterprise. It is not at all clear that the reported constraints on the *choice* of research problems either can or should also be mitigated in this way.

Over the long term, a totally *laissez faire* system of choice of research area might well serve society adequately. But it is a responsibility of government to assure that, at all times, the most pressing problems of society— e.g., food supply, population control, quality of the environment, alleviation of disease, national security, habitability of the cities—are receiving vigorous attention from the research-performing community while also assuring a fundamental research effort sufficient to long-term goals. The primary leverage available to the government to assure a satisfactory balance of this effort is a combination of mission-oriented laboratories and the pattern of funding of academic research. It is regrettable that there are scientists engaged in mission-oriented research who would prefer to be engaged in fundamental research of their own choosing, but it would be a grave error to reduce the level of mission-oriented research on that account. Indeed, we draw hope from the observation that increasing numbers of talented young investigators seek means of serving the nation by addressing their research to significant scientific aspects of the great variety of societal problems.

It appeared to be of interest to ascertain whether the factors seriously limiting research productivity vary with the age of the investigators. Only one such correlate was found among academic life scientists: the feeling that there simply is not enough time to pursue research as vigorously as one would like increases with the passage of the years! Thus 30 percent of investigators under 30, 41 percent of those in the age range 30 to 39, 55 percent of those in the age range 40 to 49, and 56 percent of those in the age range 50 to 59 are disturbed about the lack of time for research. The fraction of time devoted to teaching does not change markedly with years of academic service, remaining constant at about 22–24 percent for all academic ranks in medical schools and decreasing from 35 percent for assistant professors to 30 percent for full professors on the rest of the campus. Accordingly, it is increasing administrative duties, which rise steadily from 7 percent of time for those under 30 to 39 percent of time for those in the 50-to-59-year-old bracket, that is the major encroachment on the opportunity for research. The democratic advantages of administration by committee are not without penalty!

Similar considerations dominate the pattern of responses from life scientists employed by nonacademic institutions. The time for research is eroded by increasing administrative duties with the acquisition of seniority, but there are no other correlates of age with the general pattern of factors that limit the research productivity of nonacademic life scientists. In view

TABLE 51 Facilities Cited as First-Priority Requests of Life Scientists[a]

SPECIALIZED FACILITY	TOTAL	Behavioral Biology	Cellular Biology	Developmental Biology	Disease Mechanisms	Ecology	Evolutionary and Systematic Biology	Genetics	Molecular Biology and Biochemistry	Morphology	Nutrition	Pharmacology	Physiology	Related Research Areas
TOTAL RESPONDENTS	3,537	99	267	152	418	324	135	191	765	73	152	187	621	153
Field Areas	42	5	—	3	5	7	6	2	2	2	1	1	6	2
Zoo and Aquarium	45	3	2	3	1	12	3	4	7	3	—	1	8	3
Taxonomic Research Collection	42	1	1	3	7	12	6	4	—	3	—	—	4	—
Organism-Identification Service	101	1	8	3	10	26	9	1	12	1	7	3	11	6
Tropical Terrestrial Station	74	2	2	4	8	12	23	—	3	5	3	1	8	2
Tropical Marine Station	41	1	4	5	3	6	4	—	5	1	1	2	7	2
Marine Station, Other	47	—	8	2	3	9	2	2	9	2	2	1	8	1
High-Altitude Laboratory	45	1	1	1	4	5	2	2	4	4	—	3	18	1
Low-Pressure Chambers	38	—	2	3	2	3	1	—	2	3	2	3	19	—
High-Pressure Chambers	41	1	2	2	6	2	—	—	5	1	1	4	16	1
Programmed Climate-Controlled Rooms	**599**	**21**	23	**26**	**54**	**95**	**31**	**76**	**51**	**11**	**43**	11	**126**	**31**
Computer Center	85	3	3	—	9	12	3	2	12	1	3	13	18	6
Primate Center	216	10	16	12	35	2	3	6	30	8	8	**25**	53	8
Other Specialized Animal Colony	118	4	16	9	17	10	5	1	19	3	8	9	14	3
Germ-Free Facility	202	4	**38**	7	**64**	5	2	10	23	4	**17**	3	17	8
Animal-Surgery Facility	66	1	7	2	4	1	—	2	7	1	8	7	25	1
Animal-Quarantine Facility	74	2	5	2	16	2	—	6	9	—	3	9	15	5
General Animal Care Facility	107	3	6	7	10	9	2	7	13	2	10	12	23	3
Cell and Tissue Culture Facility	219	—	28	**21**	22	6	3	18	**70**	**8**	4	6	29	4
High-Intensity Radioactive Sources	54	1	3	5	7	2	1	2	11	1	3	3	8	7
Center for Production of Biological Materials	**499**	11	**42**	8	49	12	1	12	**321**	3	4	11	17	8
Clinical Research Ward	144	5	6	5	39	1	—	3	19	2	8	19	26	11
Greenhouse	97	1	2	6	2	19	14	9	18	1	6	1	10	8
Ships Longer than 18 ft	30	2	—	1	1	13	7	—	1	1	—	—	1	3
Electrically Shielded Room	55	3	3	—	2	—	—	1	9	—	—	8	28	1
Instrument Design/Fabrication Facility	456	13	39	12	38	41	7	23	103	5	10	31	106	28

[a] Boldface indicates two most frequent first-choice requests by those engaged in each research area.
Source: Survey of Individual Life Scientists, National Academy of Sciences Committee on Research in the Life Sciences.

of our own ages, the authors of this report will eschew comment concerning inherent personal limitations as a concomitant of the advancing years!

Specialized Facilities

Each scientist was also asked what specialized facilities and instruments, currently unavailable to him, he requires and would use were they made available. The general pattern of response of those engaged in various research areas with respect to the potential utility of both specialized facilities and instruments was rather like the general pattern of the current distribution and availability of facilities and instruments—*viz.,* those in most common use are in greatest demand. To sharpen the question, however, each investigator was asked to indicate his first, second, and third priorities for acquisition of these research tools.

The most seriously unfilled requirements for specialized facilities, as indicated by first-priority choices (Table 51), are programmed climate-controlled rooms, centers for the production of biological materials, and facilities for instrument design and fabrication, named by 18, 14, and 13 percent, respectively, of the total responding population. The second tier of requests for facilities was comprised of primate centers, germ-free facilities, and facilities for growth of cells and tissues in culture, each of which was the first priority of about 6 percent of respondents.

A few special requirements are noteworthy. There was little demand for tropical biology stations other than those already available, except for 23 systematic biologists who required access to a tropical terrestrial station; only 41 scientists, of all categories, expressed need for a tropical marine station, unavailable to them at present, as a first-priority request. Climate-controlled rooms were in demand by scientists from all research areas; unexpectedly, this was most frequent among the geneticists. Germ-free facilities were most desired by scientists studying disease mechanisms and cell biology, while the molecular biologists and biochemists expressed a most acute need for centers for production of biological materials (43 percent of their first choices).

Instrumentation

Responses to the question, "Which major instruments, currently unavailable to you, would you use if they were available?", indicate that there is a considerable backlog of unmet demand for a wide variety of biological research equipment. Again to sharpen the question, each was asked to state his first priority; 4,045 scientists provided such information, a fourth of whom were molecular biologists and biochemists (Table 52).

For some instruments, the response was almost independent of research area. Lack of electron microscopes conditioned the first-priority requests of investigators in every research area (9 percent of the total), the highest percentage frequencies being found among systematic biologists and scientists studying disease mechanisms, developmental and cell biologists, and morphologists (27, 18, 15, 11, and 14 percent, respectively), and with biochemists as the largest single source of such requests. The second most frequently cited instrument was an amino acid analyzer, requested by 317 individuals (8 percent of all respondents), a third of whom were molecular biologists, the remainder being distributed across all research areas. Small specialized computers were in third place on this list, requested by 275 individuals (7 percent of the population) among whom were 84 physiologists. Other instruments, for which there was only somewhat less demand, included analytical ultracentrifuges, mass spectrometers, gas chromatographs, and telemetering systems.

DEPARTMENT CHAIRMEN

Specialized Facilities

The priorities of these selections, presumably indicative of genuine limitations on current research, were borne out by the equivalent first-priority selections made by academic department chairmen, who were asked to indicate the prime unmet requirements of their departments for access to specialized facilities. Of the 725 department chairmen who provided such information (Table 53), 103 indicated programmed climate-controlled rooms; 93, instrument design and fabrication facilities; 70, facilities for growth of cells and tissues in culture; and 61, access to a center for large-scale production of biological materials (14, 13, 10, and 8 percent of all chairmen, respectively). Agricultural scientists, ecologists, zoologists, and botanists were most numerous among those chairmen seeking climate-controlled rooms; clinical departments, biochemists, physiologists, and ecologists most frequently sought instrument fabrication facilities; and the biochemists and microbiologists together accounted for 61 percent of the requests for centers for large-scale production of biological materials. For obvious reasons, clinical science chairmen almost uniquely indicated a serious lack of closed clinical research wards (40 departments); it was also the clinical departments that felt the most pressing need for access to primate centers (21 departments). This need for clinical research wards was expressed even prior to the recent closing of 20 such wards for budgetary reasons.

Instrumentation

Eight hundred twenty-five department chairmen reported the first-priority needs of their departments for instruments (Table 54). The pattern was strikingly like that of personal-use priorities of individual scientists; the most frequently requested instruments were electron microscopes (83 departments), amino acid analyzers (78 departments), small specialized computers (64 departments), and analytical ultracentrifuges (48 departments). But, in keeping with all previous indications, no remarkable use pattern emerged. Biochemistry departments constituted only 14 percent of the demand for amino acid analyzers, and there were only three departments of anatomy and four departments of pathology among the 83 departments seeking electron microscopes! Although there were nine biochemistry departments among the 28 departments that hoped to find funds to acquire mass spectrometers, it is more significant that the other 19 departments so requesting were scattered among eight other disciplines.

It is difficult to know to what extent these priority statements by department chairmen and investigators represent significant restrictions on the quantity or quality of on-going research and to what extent they are mere "wish lists." Where the indicated instrument or facility is truly unavailable to the investigators concerned, there can be no doubt that it might well constitute an absolute restriction on research productivity, a barrier to the logical pursuit of a research program in the light of information already obtained. In our experience, such instruments as electron microscopes, amino acid analyzers, analytical ultracentrifuges, and mass spectrometers are programmed weeks or months in advance, and it is difficult if not impossible either to borrow them or to have runs made on request. And there can be no substitute for a controlled-climate chamber when it is required.

Few working life scientists have not occasionally encountered the need for an instrument that is not available through commercial channels, but only the more fortunate have access to adequately equipped instrument fabrication facilities managed and operated by skillful, imaginative machinists, knowledgeable in the fields of electronics and instrumentation generally. The modern biochemist, molecular biologist, microbiologist, and geneticist frequently require bacteria, other micro-organisms, or animal cells in 25-pound to 200-pound lots, unavailable through any commercial source; hence the understandable requests for facilities for large-scale

TABLE 52 Major Instruments Cited as First-Priority Requests of Life Scientists[a]

MAJOR INSTRUMENTS	TOTAL	Behavioral Biology	Cellular Biology	Developmental Biology	Disease Mechanisms	Ecology	Evolutionary and Systematic Biology	Genetics	Molecular Biology and Biochemistry	Morphology	Nutrition	Pharmacology	Physiology	Related Research Areas
TOTAL RESPONDENTS	**4,045**	**105**	**320**	**183**	**436**	**275**	**95**	**176**	**1,105**	**66**	**154**	**228**	**690**	**212**
Acoustic														
Acoustic-Analysis Equipment	23	5	1	—	—	6	3	1	—	—	—	—	5	2
Sonar	8	—	—	—	—	5	1	—	—	—	—	1	—	1
Ultrasonic Probes and Sensoring System	35	3	2	1	4	3	—	5	3	—	3	—	7	4
Centrifuges														
Analytical Ultracentrifuge	210	—	19	10	41	3	2	7	89	—	4	8	21	6
Preparative Ultracentrifuge	110	—	10	8	15	2	—	3	41	1	6	7	12	5
Refrigerated Centrifuge	78	—	6	6	4	5	1	2	13	2	10	2	24	3
Chromatography														
Amino Acid Analyzer	**317**	3	**22**	7	30	22	5	**22**	**106**	2	**30**	11	**45**	**12**
Gas Chromatograph	143	5	8	5	9	16	2	3	38	—	7	10	31	9
Programmed Gradient Pump	101	—	8	3	10	1	—	4	66	—	2	2	4	1
Counters														
Automatic Particle Counter	81	—	14	3	14	9	—	1	17	—	3	4	8	8
Scintillation Counter	129	2	15	11	8	9	—	5	38	1	7	11	22	—
Whole-Body Counter	47	2	—	—	12	4	—	3	2	—	9	4	10	1
Microscopy														
Electron Microscope	**359**	9	**41**	**34**	**57**	10	**23**	10	68	**13**	7	22	**51**	**14**
Electron Probe for Microscopy	33	—	12	2	2	1	3	1	2	3	1	—	5	1
Fluorescence Microscope	84	—	11	5	16	6	4	7	8	4	—	6	14	3
Metallograph	1	—	—	—	—	—	—	—	—	1	—	—	—	—
Microtome-Cryostat	54	2	7	7	8	2	4	3	4	4	—	4	8	1
Phase-Contrast Microscope	108	3	15	9	14	10	**12**	**15**	10	**5**	—	2	8	5
Spectrometers														

342

Equipment	Total													
Resonance Spectrometer	64	—	2	—	5	—	—	3	34	—	4	1	5	10
Spectrophotometers, Polarimeters, Fluorimeters														
Circular Dichroism Analyzer	50	—	—	—	—	—	—	3	45	—	1	—	1	4
Infrared Spectrophotometer	65	1	2	3	13	3	—	9	21	—	1	1	12	5
Microspectrophotometer	75	2	14	9	7	1	—	2	21	2	2	2	6	—
Spectrofluorimeter	70	3	5	2	6	1	—	—	30	1	5	4	6	5
Spectropolarimeter	42	—	1	1	2	2	—	—	33	—	3	3	—	4
Ultraviolet Spectrophotometer	33	—	5	2	2	2	—	2	11	—	3	—	5	1
X-Ray														
X-Ray Crystallographic Analysis System	30	—	4	—	3	1	—	—	13	2	1	—	1	5
X-Ray Diagnostic System	9	3	—	—	1	—	—	1	—	—	1	2	3	1
X-Ray Source	44	3	7	1	6	—	6	6	6	1	—	3	3	2
Miscellaneous														
Apparatus for Measuring Fast Chemical Reactions	110	—	6	4	2	—	—	3	72	—	3	6	9	5
Artificial Kidney	14	—	2	—	2	—	—	—	1	1	1	2	4	1
Cine and Time-Motion Analysis Equipment	117	**10**	18	6	18	10	7	4	6	6	1	3	23	5
Closed-Circuit TV	64	**9**	4	3	6	4	2	1	3	4	2	5	17	4
Electrophoresis Apparatus (Various Types)	81	2	7	4	15	3	6	12	19	—	4	—	8	1
Intensive-Care Patient-Monitoring System	54	2	1	2	11	—	—	3	6	—	2	6	18	3
Infrared CO_2 Analyzer	114	2	4	6	5	**33**	—	3	4	2	5	4	**40**	6
Laser System	44	—	6	7	4	1	—	2	11	1	2	—	8	2
Large-Scale Fermenter	127	—	13	3	12	4	—	5	84	—	—	—	5	1
Light-Scattering Photometer	18	—	—	—	1	1	—	—	13	—	1	1	4	2
Microcalorimeter	30	2	2	2	1	**16**	—	—	3	—	1	2	15	2
Multichannel Oscilloscope	32	7	—	1	1	1	—	1	1	—	2	2	27	1
Multichannel Recorder	106	6	3	3	14	16	2	7	15	2	3	5	16	3
Osmometers	51	—	5	—	6	2	—	—	15	3	2	2	16	—
Small Specialized Computer System (CAT/LINC, etc.)	275	11	10	3	26	19	1	13	41	1	5	32	84	**29**
Stimulus Programming and Operant Conditioning Equipment	17	4	1	1	—	—	—	1	—	—	1	2	5	2
Telemetering System	202	7	2	7	18	**40**	9	3	3	3	7	35	56	12

[a] Boldface indicates two most frequent first-choice requests by those engaged in each research area.
Source: Survey of Individual Life Scientists, National Academy of Sciences Committee on Research in the Life Sciences.

TABLE 53 Facilities Cited as First-Priority Needs of Academic Departments by Their Chairmen[a]

SPECIALIZED FACILITY	GRAND TOTAL ALL DISCIPLINES	DEPARTMENTAL DISCIPLINE													
		Animal Husbandry	Agronomy and Forestry	Anatomy	Biochemistry and Nutrition	Biology and Ecology	Biophysics and Engineering	Botany	Genetics	Microbiology	Pathology	Pharmacology[b]	Physiology	Zoology and Entomology	Clinical Medical Sciences[c]
TOTAL DEPARTMENTS	**725**	**64**	**52**	**34**	**63**	**85**	**8**	**31**	**3**	**57**	**27**	**30**	**43**	**71**	**157**
Field Areas	15	—	1	1	—	3	—	2	—	—	1	—	—	4	4
Zoo and Aquarium	3	—	—	1	—	1	—	—	—	—	—	—	—	—	1
Taxonomic Research Collection	4	1	1	—	—	3	—	—	—	—	—	—	—	—	—
Organism-Identification Service	8	1	2	—	—	3	—	—	—	1	1	1	—	—	1
Tropical Terrestrial Station	6	—	—	—	1	1	—	3	—	—	1	—	—	1	—
Tropical Marine Station	3	—	—	—	—	2	—	—	—	—	1	—	—	—	—
Marine Station Other than Tropical	16	—	1	1	1	7	1	2	—	1	1	—	1	2	1
High-Altitude Laboratory	3	1	—	—	—	—	—	—	—	—	—	—	1	1	1
Low-Pressure Chambers	6	—	—	1	—	—	—	—	—	—	—	—	3	1	1
High-Pressure Chambers	10	—	—	1	—	1	—	—	—	—	—	1	—	—	7
Programmed Climate-Controlled Rooms (Phytotron, Biotron, etc.)	103	**21**	**25**	1	4	**19**	—	**7**	—	5	4	3	4	**8**	2
Computer Center	17	1	—	4	2	1	—	—	1	—	1	—	3	1	9
Primate Center	45	1	2	2	3	2	—	—	—	2	1	3	**7**	2	**21**
Other Specialized Animal Colony	37	5	2	4	2	2	—	—	1	4	2	2	4	**6**	4
Germ-Free Animal Facility	41	**11**	—	—	2	3	1	—	1	**12**	**6**	—	—	1	3
Animal Surgery Facility	21	**6**	—	4	1	3	1	—	—	1	1	1	1	2	**7**
Animal Quarantine Facility	26	3	—	1	3	4	1	—	1	5	1	1	1	2	3
General Animal Care Facility	42	3	1	**8**	4	8	3	1	—	2	3	4	4	**8**	7
Cell and Tissue-Culture Facility	70	5	6	1	4	8	3	5	—	2	3	—	1	5	**17**
High-Intensity Radioactive Sources	6	—	1	—	—	1	—	—	—	—	—	—	—	1	3
Center for Large-Scale Production of Biological Materials	61	3	—	1	**25**	4	1	1	—	**12**	2	4	1	4	3
Clinical Research Ward	48	2	—	—	2	2	—	2	—	—	—	2	—	—	**40**
Greenhouse	18	—	**6**	—	1	4	—	—	—	—	—	—	—	5	—
Ships Longer than 18 ft	6	—	—	1	—	1	1	—	—	—	—	—	—	4	—
Electrically Shielded Room	17	—	—	1	—	—	—	2	—	—	—	4	4	2	4
Instrument Design or Fabrication Facility	93	1	5	4	**15**	**8**	—	6	—	9	2	3	9	**12**	19

[a] Boldface denotes two most frequent requests of each class of department.   [b] Includes a college of pharmacy.   [c] Includes a department of oral biology.
Source: Survey of Academic Life Science Departments, National Academy of Sciences Committee on Research in the Life Sciences.

preparation of biological materials, now available to only a handful of institutions. Were it possible to make available these indicated choices of department chairmen and of individual scientists, both the quality and tempo of research in the life sciences would most certainly be augmented.

Improvement and Expansion of the Academic Research Endeavor

Another set of indicators of what may be required to assure the quality and magnitude of the research endeavor was obtained from department chairmen by asking whether, and for what purposes, their departments *very seriously* required additional funds to finance their *current* research endeavors. Once again, the question was sharpened by a request to indicate first-priority needs. One thousand thirty-three department chairmen responded to this set of questions.

As can be seen in Table 55, funds for graduate student stipends and for additional faculty salaries were most numerous among such responses (25 and 29 percent, respectively). The stipends were the first-priority request of almost half of all departments of biochemistry, physiology, and agronomy and forestry; the need for faculty salaries was about equally urgent among all disciplines, except for clinical departments, 51 percent of which indicated this as their first priority. Funds for instruments were the most acute need of 10 percent of all departments, including six of the 15 reporting biophysics departments. Funds to support the research of junior faculty were listed as the first priority of only 9 percent of all chairmen; strangely, this was the first priority of the chairmen of 15 of 98 general-biology departments but of only 1 of 73 chairmen of zoology departments. When the chairmen were asked to indicate their second and third choices as well, this item was selected by an aggregate of 30 percent of departments.

These data must again be regarded with a caveat; they were obtained before the impact of the current plateau in federal funding, with all its implications, and it is difficult to distinguish clearly between acute need and general "wish list."

Some 205 departments indicated that they had no need of additional funds to improve their *current* research endeavors, and it was of interest to examine the pattern of the considered needs of this limited group for *expansion* of the departmental research endeavor. These departments were well distributed among all the disciplines, colleges, and both classes of universities. Once again, it was salaries for additional faculty that loomed as the prime requirement, cited by 43 percent of this group of departments. Again this stipulation was mentioned most frequently by the private medical schools, in both their clinical and preclinical departments, by an inter-

TABLE 54 Instruments Cited as First-Priority Needs of Academic Departments by Their Chairmen

MAJOR INSTRUMENT	GRAND TOTAL ALL DISCIPLINES	Animal Husbandry	Agronomy and Forestry	Anatomy	Biochemistry and Nutrition	Biology and Ecology	Biophysics and Engineering	Botany	Genetics	Microbiology	Pathology	Pharmacology[a]	Physiology	Zoology and Entomology	Clinical Medical Sciences[b]
TOTAL NUMBER OF DIFFERENT DEPARTMENTS[c]	825	63	53	42	89	85	12	39	6	62	39	43	55	55	182
Acoustic															
Acoustic-Analysis Equipment	3	—	—	—	—	1	—	—	—	—	—	—	—	1	1
Sonar	1	—	—	—	—	—	—	—	—	—	—	—	1	—	—
Ultrasonic Probes and Sensoring System	8	2	—	1	—	1	—	—	—	—	—	—	1	—	3
Centrifuges															
Analytical Ultracentrifuge	48	1	1	—	9	8	—	4	1	9	3	2	1	4	5
Preparative Ultracentrifuge	23	3	1	—	1	6	—	—	1	1	1	5	—	2	2
Refrigerated Centrifuge	11	1	3	1	1	—	—	—	—	1	1	—	—	1	3
Chromatography															
Amino Acid Analyzer	78	10	8	2	11	7	1	3	1	10	5	1	5	2	12
Gas Chromatograph	24	1	3	1	—	—	—	2	—	2	1	—	1	3	10
Programmed Gradient Pump	11	1	—	1	3	—	—	—	—	2	1	—	1	1	1
Counters															
Automatic Particle Counter	21	—	1	2	2	3	—	—	—	4	—	—	3	1	5
Scintillation Counter	23	—	1	3	1	1	—	—	—	1	1	4	2	5	4
Whole-Body Counter	6	—	—	—	—	—	1	—	—	—	—	—	—	—	5
X-Ray															
X-Ray Crystallographic Analysis System	8	—	—	1	2	1	1	—	—	—	—	—	—	—	3
X-Ray Diagnostic System	4	—	—	—	—	—	—	—	—	—	—	—	—	1	3
X-Ray Source	4	1	1	—	1	1	—	—	—	—	—	—	—	—	—
Microscopy															
[instrument label cut off at page bottom]	83	3	8	3	6	11	1	6	—	8	4	7	11	4	11

Equipment															
Microtome-Cryostat	11	2	—	1	—	1	1	—	1	—	1	—	—	—	1
Phase-Contrast Microscope	8	3	2	1	—	—	3	—	1	1	1	—	—	1	1
Spectrometers															
Electron Paramagnetic Resonance Spectrometer	4	—	—	3	—	—	1	—	—	—	3	—	—	—	1
Mass Spectrometer	28	1	5	9	1	1	1	—	1	3	—	3	—	3	4
Nuclear Magnetic Resonance Spectrometer	12	1	—	6	1	—	1	—	1	1	—	1	—	1	—
Spectrophotometers, Polarimeters, Fluorimeters															
Circular Dichroism Analyzer	7	—	—	5	—	—	1	1	1	—	—	1	—	—	1
Infrared Spectrophotometer	22	1	6	4	1	1	2	2	2	1	1	2	1	3	1
Microspectrophotometer	15	—	—	—	2	3	—	2	—	—	—	2	1	2	4
Spectrofluorimeter	18	—	1	5	1	—	1	—	—	5	—	2	2	—	5
Spectropolarimeter	6	—	—	4	2	—	2	—	4	1	—	—	—	—	—
Ultraviolet Spectrophotometer	15	2	—	—	3	—	3	1	—	—	1	—	—	2	3
Miscellaneous															
Apparatus for Measuring Fast Chemical Reactions	15	2	—	4	1	—	1	1	1	1	—	3	—	—	2
Artificial Kidney	3	1	—	—	—	3	—	1	3	1	—	—	1	—	3
Cine and Time-Motion Analysis Equipment	25	1	3	—	6	2	1	1	1	1	—	—	4	1	9
Closed-Circuit TV	27	1	2	—	4	—	1	1	2	1	—	—	2	2	10
Electrophoresis Apparatus (Various Types)	7	1	1	—	1	1	1	1	1	—	—	—	1	1	1
Intensive-Care Patient Monitoring System	21	2	—	1	2	—	—	1	1	—	—	1	—	—	18
Infrared CO_2 Analyzer	20	3	4	—	1	—	5	—	1	—	1	1	1	1	3
Laser System	9	1	—	1	1	1	—	1	1	—	—	—	1	—	3
Large-Scale Fermenter	26	2	—	8	6	—	1	1	9	—	1	—	1	1	3
Light-Scattering Photometer	—	—	—	—	—	—	—	—	—	—	—	—	—	—	—
Microcalorimeter	4	—	1	—	—	1	1	—	1	—	—	1	—	—	—
Multichannel Oscilloscope	10	2	2	—	2	—	—	—	1	—	—	1	—	1	3
Multichannel Recorder	14	3	2	—	—	—	—	1	1	—	2	—	—	1	4
Osmometers	4	—	—	—	—	—	1	—	1	—	2	—	1	—	1
Small Specialized Computer System (CAT/LINC, etc.)	64	2	2	3	5	2	2	1	1	2	2	4	17	7	15
Stimulus Programming and Operant Conditioning Equipment	7	2	2	—	—	—	5	—	1	—	—	5	2	—	2
Telemetering System	38	6	2	—	3	1	1	1	1	—	—	1	2	3	17

[a] Includes a college of pharmacy.
[b] Includes a department of oral biology.
[c] Boldface denotes the two most frequent requests by each class of department.
Source: Survey of Academic Life Science Departments, National Academy of Sciences Committee on Research in the Life Sciences.

TABLE 55 First-Priority Needs for Funds To Improve Current Research Endeavor of Academic Departments

SELECTED FACTORS TOTAL NUMBER OF DIFFERENT DEPARTMENTS REQUIRING FUNDS FOR:	GRAND TOTAL ALL DISCIPLINES	Animal Husbandry	Agronomy and Forestry	Anatomy	Biochemistry and Nutrition	Biology and Ecology	Biophysics and Engineering	Botany	Genetics	Microbiology	Pathology	Pharmacology [a]	Physiology	Zoology and Entomology	Clinical Medical Sciences [b]
	1,033	83	72	59	96	98	15	47	7	71	55	40	58	73	259
Predoctoral Stipends	255	24	32	22	46	17	3	11	2	19	6	11	27	21	14
Postdoctoral Stipends	69	5	5	3	7	7	—	—	—	5	2	1	4	5	25
Salary, Additional Faculty	295	19	7	17	11	18	5	11	2	18	17	12	9	17	132
Salary, Additional Nonfaculty	122	17	13	2	4	14	1	10	1	6	11	4	3	11	25
Specialized Research Facilities	64	7	5	—	2	7	—	5	—	5	4	2	4	7	16
Major Instrument	105	6	4	5	12	17	6	6	—	10	8	5	4	4	18
Consumable Supplies and Equipment	23	2	1	3	3	3	—	1	—	1	—	—	1	7	1
Travel and Publishing Costs	7	—	1	1	—	—	—	—	—	1	1	—	—	—	4
Research Support for Junior Faculty	93	3	4	6	11	15	—	3	2	7	6	5	6	1	24

DEPARTMENTAL DISCIPLINE

[a] Includes a college of pharmacy.
[b] Includes a department of oral biology.
Source: Survey of Academic Life Science Departments, National Academy of Sciences Committee on Research in the Life Sciences.

mediate number of public medical schools, and least frequently by chairmen of arts and sciences faculties. Predoctoral stipends were again the second most frequent problem if expansion of the research program was to be undertaken by this group.

The same query, concerning requirements for *expansion* of their research endeavors, was answered by a total of 1,052 department chairmen (Table 56); faculty salaries were cited by 42 percent and predoctoral stipends by 22 percent. The other seven possible uses of additional funds were cited by 1 to 9 percent of chairmen, the variation among disciplinary departments being unremarkable. The faculty salary problem loomed as the major need to 52 percent of chairmen in private universities and to 38 percent in public universities, as expected from their funding problems. This item would surely be even more pressing today.

In short, by each form of this examination, if the academic research endeavor in the life sciences is to be improved or expanded, additional funds must be provided to assure payment of faculty salaries and predoctoral stipends. Until these urgencies have been met, all other considerations appear to be secondary.

Expansion of the Graduate Education Endeavor

The system was tested in yet another manner; the department chairmen were asked to state whether they could accept additional Ph.D. candidates at present; if not, whether their inability was due to lack of space or faculty. Further, they were asked whether they planned to accept additional numbers of graduate students over the next four years and whether this would require additional space or faculty. Seven hundred seventy-six of 779 chairmen with Ph.D. programs responded to these questions, of whom 565 (73 percent) indicated that their departments were below graduate-training capacity at that time. These 565 departments, rather uniformly distributed among all types of colleges and universities, had a current enrollment of 9,397 students. Their chairmen reported an unused capacity of 4,000 students, three fourths of which was in public universities and one fourth in private universities. The total graduate enrollment in all departments was reported as 14,764.

Of those unable to expand their graduate enrollments, roughly one third had insufficient space, one third had both insufficient space and insufficient faculties, and the remainder were unable to expand for a variety of reasons. Unexpectedly, only 2 percent cannot expand merely for lack of faculty.

In sum, both private and public universities anticipated 50 to 60 percent

TABLE 56 First-Priority Needs for Funds To Expand Research Endeavor of Academic Departments

SELECTED FACTORS	GRAND TOTAL ALL DISCIPLINES	Animal Husbandry	Agronomy and Forestry	Anatomy	Biochemistry and Nutrition	Biology and Ecology	Biophysics and Engineering	Botany	Genetics	Microbiology	Pathology	Pharmacology[a]	Physiology	Zoology and Entomology	Clinical Medical Sciences[b]
TOTAL NUMBER OF DIFFERENT DEPARTMENTS REQUIRING FUNDS FOR:	1,052	80	76	60	98	100	16	47	8	76	56	43	65	75	252
Predoctoral Stipends	230	21	25	21	38	20	4	12	3	19	3	10	22	21	11
Postdoctoral Stipends	93	7	3	4	9	5	—	3	—	6	8	8	4	6	30
Salary, Additional Faculty	442	26	28	26	29	35	7	12	3	35	30	16	27	27	141
Salary, Additional Nonfaculty	69	9	11	—	3	11	—	4	2	1	2	1	3	7	14
Specialized Research Facilities	70	9	4	3	3	8	2	5	—	3	5	2	2	2	17
Major Instrument	72	3	4	3	9	12	2	6	—	5	5	1	3	7	16
Consumable Supplies and Equipment	27	3	1	1	2	5	—	4	—	2	1	1	2	1	4
Travel and Publishing Costs	4	—	—	—	—	—	1	—	—	1	—	1	—	—	1
Research Support for Junior Faculty	45	2	—	2	5	4	—	1	—	4	2	3	2	2	18

DEPARTMENTAL DISCIPLINE

[a] Includes a college of pharmacy.
[b] Includes a department of oral biology.
Source: Survey of Academic Life Science Departments, National Academy of Sciences Committee on Research in the Life Sciences.

increases in graduate student enrollments between fiscal year 1968 and fiscal year 1972. It is unlikely that the increment will be realized. From the aggregated responses of the department chairmen, an increment of about 20 percent in the total life sciences faculty of all universities appeared to be required to meet the demands of such increased enrollments. Figure 38 summarizes these needs by department title. In absolute terms, the largest faculty increases were specified by chairmen of departments of biology, biochemistry, and zoology; the percentage increment required, per department, varied from 8 percent in the agricultural sciences to 31 percent in pharmacology. Clinical department chairmen found expansion of the pre-Ph.D. population no problem, although they would take quite a different view toward the requirements for expansion of the medical student bodies.

Seventy-six percent of all department chairmen also indicated that, to meet the projected fall 1971 enrollment, their departments would require an increment in available research space. However, two thirds of that increment, nationally, was already either under construction or in an advanced planning phase. The public universities were in a somewhat better position than the private universities in this regard. The total estimated increment in space required by the aggregate of all reporting departments was 3,752,000 square feet, of which 2,537,000 square feet was either under construction or in a very advanced planning phase. There remained, therefore, approximately 1.2 million square feet of laboratory space, ostensibly required by fiscal year 1972, which, in June of 1967, was not even in the planning phase.

Assuming that our total sample embraced 60 percent of all eligible research-performing departments in the life sciences, and neglecting the clinical departments, it would appear that approximately 2 million square feet of space was required to meet the total anticipated increment in graduate student enrollment, which was, in 1967, not even in the planning phase. Assuming further that net usable space is approximately two thirds of each building, the gross square feet required would be 3 million, which, at $50 per square foot, would require $150 million in construction costs. The lag in the growth of graduate enrollments permits this time scale to be extended by about two years, after which either there will be serious overcrowding or qualified applicants to graduate school will have to be denied entrance.

NATIONAL CONSIDERATIONS

The data summarized above were collected from individual scientists and departments, all expressive of the needs perceived for furtherance of their

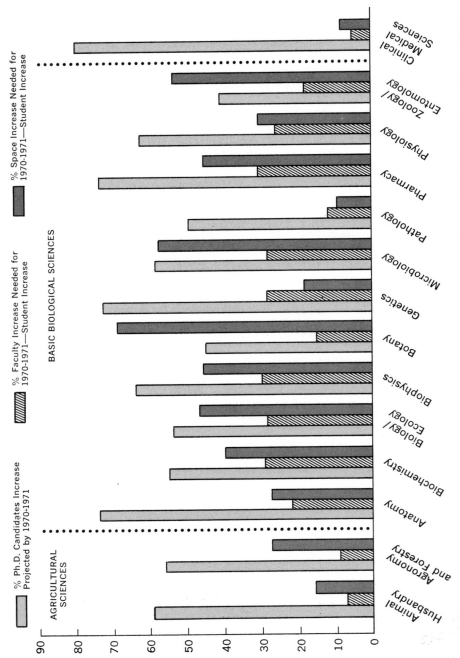

FIGURE 38 Projected increase in Ph.D. candidates, faculty, and space for academic year 1970–1971. (Source: Survey of Academic Life Science Departments, National Academy of Sciences Committee on Research in the Life Sciences.)

goals—research and education in the life sciences. But it is insufficient to the national purpose to use these data as the exclusive basis for assessment of national needs or even for the development of biological disciplines.

Thus, whereas very few scientists and fewer department chairmen cited a tropical station as a first-priority request, the future of tropical agriculture demands that a tropical terrestrial or terrestrial–marine station be available to qualified agricultural scientists and ecologists. No one scientist may feel that he can properly request a deep-submersible vehicle, yet this advancing technology can open a new scientific frontier, providing access to deep marine forms whose existence was unknown. Ecologists will require tracts of hundreds of acres of land near universities, near cities, and near industrial areas to be retained as ecological base lines or to be polluted and restored experimentally. As noted in Chapter 7, most biologists are satisfied with their access to computer services. But when academic computer centers convert to operation on a use-time-charge basis, large numbers of biologists will find themselves short of the necessary funds to defray these costs. And as larger ecosystems are studied, permitting mathematical modeling, as human pedigrees are collected, as human biology is more systematically examined using all the potentially linked data banks available, as attempts at computer diagnosis of human ailments become more sophisticated, as biologists join with engineers and sociologists in planning new communities, as cell biologists acquire more understanding of the multitude of controls that regulate the extraordinarily complex chemical life of a single cell, as communications theory, neuroanatomy, and physiology are applied in creating conceptual models of brain functioning, the requirements of biologists for access to the largest, fastest computers must necessarily increase, perhaps by an order of magnitude.

Moreover, these responses from working life scientists fail to indicate the unique and special requirements of specific federal or local mission agencies. These may include vessels and experimental hatcheries for the Bureau of Fisheries as it attempts to develop "fish factories" analogous to modern "chicken factories," control of an estuary or bay for experimental aquiculture by the Department of Agriculture, test facilities for the Atomic Energy Commission to assess the reality of "thermal pollution," a large policy-oriented environmental studies institute sponsored by the National Science Foundation, the Department of the Interior, or any new agency concerned with environmental quality, the totality of biological sophistication and technology required for an effective program in the National Institute for Environmental Health, satellites for monitoring gross ecological effects by remote sensors, a manned orbiting laboratory to establish the influence of lack of gravity on embryonic development, adult physiology, and circadian rhythms by the National Aeronautics and Space Administration, adequate facilities for a sophisticated upgrading of the Bureau of Sport Fisheries and Wildlife, facilities for study and control of the process of eutrophication by the Federal Water Pollution Control Administration, or a test facility wherein the Atomic Energy Commission could monitor the biological consequences of its capability to use nuclear explosives to generate harbors. And surely some agency must explore means of coping with the mounds of chicken manure accumulating near every major "chicken factory," as well as developing techniques for dealing with the ever-growing liquid- and solid-waste-disposal problems of our society. The Public Health Service will require more elaborate facilities for operating screens for prospective antitumor drugs and for quick identification of the endless procession of new viruses that spread through our population, while encouraging the development of bioengineering both in-house and in the universities. And the Food and Drug Administration must acquire the competence and facilities for the chronic, low-dose examination of thousands of new organic compounds introduced into American society annually. The list is long, the needs are many, and the ultimate payoff in human welfare is beyond assessment.

We have not completely discussed here all the problems of support for academic biology. We have spoken of the requirements for a vigorous research endeavor through the project-grant system. Equally important are the needs for support of the institutions in which this endeavor occurs, for support of the educational efforts of the life science departments, and for construction of new facilities. Our recommendations in these regards are offered in the Summary that begins this report.

Museums

One group of institutions, museums, was inadequately represented both in our sampling procedure and in the design of the questionnaires. Their systematic collections of plants and animals are the only permanent record of the earth's biota, and the specialized libraries attached to these collections are the written record of the earth's natural history. The major systematics facilities are few in number; some are free-standing institutions, others are located at private or state universities. Systematic biology forms the basic framework and reference system for the study of all organisms. Many major systematics collections were begun and supported by wealthy private individuals. As patronage passed to governments and foundations, these collections were left bereft; they remained housed in nineteenth century buildings and became increasingly crowded as the collections grew. The National Science Foundation has become the only significant source of funds for this enterprise, but the level of funding is scarcely commensurate with the need.

These collections are the laboratories of the systematist, and the earth's surface is his field station. The huge number of plants and animals render the task confronting the taxonomists and systematists overwhelming. This task would have been urgent even if man had not begun to change the earth's surface in such a way as to make the job of recording the biota and understanding the evolutionary web of life absolutely desperate. In too many areas of the globe, collections must be made within a generation, or it will be too late. In other areas, the materials already in the collections are the only testimony concerning the natural history of the region. If collections are not made now or in the near future, if the great systematics collections already in existence are to continue to deteriorate, then future generations of biologists will be denied the opportunity of understanding the biological evolution of planet Earth. Although the natural history of the Earth will never return to its state before the existence of man, an understanding of evolution, of the natural distributions of plants and animals, and of their interrelationships are essential to man's well-being.

As society becomes more complicated, the demand for knowledge concerning systematic biology becomes more acute. Yet systematic collections, although growing in size, have been deteriorating in quality for many decades. Constant use in the museums and herbaria plus a continuous exchange of materials for study by investigators throughout the world have eroded collections. Great collectors and philanthropists have presented to this nation some of the greatest records of natural history, which have in recent times been treated with singular indifference. If we cannot retain undisturbed the various natural ecosystems of the world, at least we should preserve the records of their plant and animal communities.

The financial needs of the systematics collections are relatively small compared to the sums currently spent for facilities in other branches of science. Yet the systematics collections consist of fragile, perishable materials that cannot be replaced if damaged and that deteriorate when not cared for.

A small questionnaire was sent to the directors of several major collections who were asked three questions: What are the needs of your institution for new buildings in order to house the collections for the next 10 years? What annual increase in support for curatorial work is required by your institution? What annual increase in support do you anticipate for research involving fieldwork to collect and for work with the collections? The replies from 25 institutions were as follows:

TOTAL NEED FOR NEW BUILDINGS	INCREASED SUPPORT ON ANNUAL BASIS		
	Curatorial	Research	Other
$60 million	$3 million	$4 million	$3 million

If one were to attempt to satisfy the needs of all major institutions not included in this survey, the total need might double. Universities, colleges, and other independent museums that possess smaller systematics collections are nevertheless important for education of both students and the public. Hence, new buildings for all systematics facilities in the nation could cost $120 million over perhaps the next 10 years. But many of these collections are in buildings that have not been overhauled in decades and may be almost a century old. The major systematic biological collections are national assets and should be treated as such; many of them desperately need help now.

In Conclusion

The total cost of all these programs becomes a significant fraction of the gross national product. Some years ago, a distinguished committee that reviewed the programs and the billion-dollar operation of the National Institutes of Health * concluded that for no equivalent sum did the American people receive greater value. In the short interval since, biological science has made the equivalent of a quantum jump in understanding and experimental sophistication. Today, the life sciences are poised to explore the most arcane mysteries of life and, with the aid of only a moderate rate of increase in public support, they can attempt to provide a foundation for the measures required to counteract some of the oldest enemies of man— the diseases to which he is subject—to protect the quality of the environment and the habitability of planet Earth, to assist in limiting the burden of an excessive human population, and to assure an adequate food supply for all.

No guarantees can be offered in good conscience. But should the effort fail, it should not be for lack of trying. Today, as yesterday, it is difficult to imagine more noble goals or more appropriate use of public funds. For much of the endeavor the pace of progress will be determined by the generosity of public support. But as, increasingly, the life sciences become "big science," then, as in the physical sciences, the magnitude of public funding will determine not merely the pace of progress but whether, indeed, there is to be progress.

---

* *Biomedical Science and Its Administration,* A study of the National Institutes of Health. A report of the NIH Study Committee to the President, February 1965. U.S. Government Printing Office, Washington, D.C.

CHAPTER SIX

EDUCATION
IN
BIOLOGY

Education in the life sciences is seen as an enormous undertaking when its full national scope is considered. The pyramid of numbers begins with unreckoned millions of elementary school children whose first exposure to things scientific comes from experience with living things. It continues through the approximately 2.5 million high school students enrolled in formal biology courses each year, and perhaps 200,000 who receive some formal exposure to the life sciences in college. The next level comprises the 25,000 students who, each year, complete baccalaureate training with concentration in one of the life sciences. At the apex of the pyramid are the approximately 3,500 new Ph.D.'s, 8,000 M.D.'s, and several thousand other life science professionals (dentists, foresters, and others) per year who are the next generation of practitioners and research workers. Our present concern rests primarily with education in biology rather than that for the biology-based professions.

Eighty percent of Americans who graduate from high school take their *only* formal science courses in biology. At the college level, the life sciences attract a high proportion of those who enroll in single science courses while majoring in nonscientific disciplines. Thus, biology fulfills a unique role in providing large numbers of our citizens with their only view of science and its impact upon the problems of our society. This opportunity also entails an obligation of professional biologists to provide effective training for the teachers who are entrusted with this task.

The heterogeneous academic system that provides training for the research workers and teachers who make up the community of more than 80,000 professional biologists in the United States is difficult to describe. As we noted earlier, the life sciences are uniquely diverse: Biologists belong to more different professional societies and read more different journals than do other scientists; the life sciences are taught in departments ranging in degree of specialization from biology to forest pathology. While this rich diversity may be useful to society, it adds up to a system that escapes easy characterization.

One difficulty is measurement of the extent and efficiency of research training. Almost all the 3,500 Ph.D.'s earned annually in the life sciences are awarded by 145 institutions; but a much larger number train students at the master's and baccalaureate levels, and these play a critical role in forming the intellectual backgrounds and experimental habits of many potential research workers. We cannot be certain of how many of the 25,000 college majors in the life sciences graduated annually enter upon Ph.D. programs, but it is probably in the neighborhood of 7,500. Only about half of these complete the highest degree and enter the national pool of potential research workers. The attrition is difficult to measure, but, even among the holders of highly prized Woodrow Wilson fellowships in 10 distinguished graduate schools in the United States, only about half of those who began doctoral studies in the biological sciences between 1958 and 1960 had earned the Ph.D. degree by June 1966. Figures for the other sciences are comparably low. A subsequent, less well-measured and less understood attrition is that which causes fully trained biologists to fail to function as independent investigators.

One might conclude from these facts that graduate education is an inefficient process, and, in a sense, this may be true. Inefficiency is inevitable in a system in which even some able students find along the way that they lack the interest or drive to carry through a successful program of independent research, or that their strongest motivations lie elsewhere. A rigorous selection procedure at the beginning of graduate training, which would guarantee a 100-percent yield of research scientists, would probably also eliminate many of those who later become the most productive creative scientists.

Several features of research training in the life sciences present a pattern quite different from that found in other disciplines. Because of the heterogeneity of biology, graduate training is often offered in small, relatively specialized departments. Most departments granting the Ph.D. in the basic medical sciences have only six to eight faculty members; this is often true also of departments in schools of liberal arts or agriculture in which the life sciences are relatively fragmented—into departments of genetics, microbiology, wildlife management, and so on. The type of training received by doctoral students in such departments is inevitably strikingly different from that provided in coherent departments of biology. The tendency to merge relatively specialized departments into departments of biology has grown during the last decade, as discussed in Chapter 4. The fraction of students trained in these more inclusive departments, however, has not increased, owing to the dramatic increase in Ph.D. programs in the basic medical sciences during the same period.

Perhaps in part because of the "specialness" of many graduate programs, and perhaps also because of the diversity of demands placed on the investigator by a multiplicity of techniques and experimental systems, it has become traditional over the past decade for experimental biologists to follow their Ph.D. programs with postdoctoral training. In 1966–1967, 5,223 biologists were postdoctoral fellows in the departments surveyed by this committee. Over three fourths of these were supported by national postdoctoral fellowship programs, primarily from the National Science Foundation and the National Institutes of Health. This pattern contrasts with that in mathematics and the social sciences, in which a much smaller percentage of Ph.D.'s undertake postdoctoral appointments.

The number of persons undertaking and completing advanced training in the life sciences has been increasing dramatically. The number of Ph.D.'s awarded annually in biology more than doubled in the decade from 1955 to 1965. Annual output had increased at about 7 percent per year for several decades but rose sharply after 1964; the increment in 1969 over 1968 was about 12 percent. To contend with the rise in demand for trained biologists, the fellowship and traineeship programs of the National Science Foundation and the National Institutes of Health supported growing numbers of students until fiscal year 1969. These programs should continue to grow as the demands of the educational system as well as those of the pure and applied research establishments grow to keep pace with a growing and increasingly complex society. Moreover, quickening excitement in biological research itself—occasioned by a decade of especially dramatic progress—produced an unusual shift in the career plans of gifted young people. A recent survey by the American Council on Education showed that whereas, in 1961, 3.2 percent of all freshmen entering Stanford University intended to major in the biological sciences, that figure had doubled in 1966—a period during which similar figures for the physical sciences had fallen by half. Thus the requirements for support of research training in the life sciences are generated not only by the demands of society, but also by the wants of intelligent students who sense that biology is ready for new discoveries and that the life sciences seem both most relevant to human problems and most distant from military use or the furtherance of polluting forms of technology.

An important activity of the biological community is neglected if one considers only its contribution to the store of knowledge. It must also provide, directly and through the training of primary and secondary school teachers, for the instruction of large numbers of future citizens who, though not scientists themselves, will be asked to make decisions concerning public issues affecting or affected by science.

ELEMENTARY AND SECONDARY EDUCATION

Education in the life sciences at these levels is apt to involve exposure to "science" or "general science" in the elementary or junior high school grades and to a course in biology in high school. The overwhelming majority of secondary school students study biology in the tenth grade; a small but increasing fraction are exposed to an advanced-level course later. Two aspects of the secondary school experience are of special interest: it provides the only formal exposure to a science for many of the large number of citizens who will not become professional scientists; and it must supply the background and the motivating force for those students who will undertake work in the life sciences at the university level.

The elementary level is nearly *terra incognita* in understanding of how learning about science occurs. Among the few scientists now beginning to work seriously in this area, the general feeling is that methods of science teaching need drastic reorientation—in particular, toward taking advantage of the child's natural tendency to explore and to make use of materials. These considerations lead naturally into an experimental approach in which the child is left to reach independent conclusions, not to work toward a set of results that have been given textbook justifications in advance. Some experienced scientists who have worked with elementary school children state that this new approach demands a new philosophy of educating the teacher. There is little correlation between the success of teachers in communicating science in this manner and the teacher's own science training. Unfortunately, this approach is also expensive. Even the simplest collection of experimental materials—caterpillars, flowerpots, and aquaria—cost more than books.

The secondary school curriculum has been the more frequent target of reform by professional life scientists. As a part of the assault on science education in general, biologists examined the secondary school curriculum and found it wanting. Compared to the excitement and stir in "modern biology," the average high school course was strongly oriented toward systematics and comparative morphology. The new insights of molecular biology and genetics were missing; so were up-to-date treatments of the biology of populations, of animal behavior, and of physiology.

Secondary school curriculum revision in the life sciences has been principally effected by one group, the Biological Sciences Curriculum Study—an organization staffed largely by professional biologists, with its policies established by a steering committee drawn from the scientific community. The Study was organized in 1958 by the American Institute of Biological Sciences and has been financed by the National Science Foundation since 1961. Using writing teams drawn from both secondary school and university faculties, the Study has completed three major secondary-level textbooks. All are comprehensive and reasonably modern in that they have increased emphasis on chemical biology, genetics, and other topics, but they differ in approach. The "blue" version is more biochemically oriented, the "yellow" uses development and genetics as a unifying focus, and the "green" version stresses environmental biology. With their supplementary volumes (teachers' guides, laboratory exercises, options for advanced students, film clips, and the like) these books comprise an imposing shelf. They have been produced at a cost of over $10 million, much or all of which will be recovered by the government in royalty payments from users.

What results has this curriculum reform produced? As with all such efforts, evaluation is difficult. About 1.5 million of these books had been sold through 1967; apparently, however, less than 20 percent of the students in secondary school biology used the Study materials in 1967. Surveys currently in progress tend to show that colleges receiving students who used this material are pleased with their performance, but this is difficult to separate from the general improvement in secondary education and from increasingly selective admissions policies at most colleges. Critical appraisal of the Study materials has been generally favorable, although biologists have criticized them for being too molecular, for treating complex modern topics in a way that provides verbal facility without real understanding, for slighting development and "organismic biology," and for a variety of more minor sins. These judgments notwithstanding, the Study materials represent a major increment in quality over what had gone before.

For a substantial percentage of high school students, the biology course is not a foundation for other studies; it is a terminal course. Some biologists have asked whether "professionally oriented" curricula like the Biological Sciences Curriculum Study are optimally designed to meet the needs of these students as well as the college-preparatory group. To meet this need, future consideration might be given to a course that might be described as "humanistic biology." It could be called "human biology"—as indeed it is in some nascent university curricula—except that, at the secondary level, by previous usage "human biology" is equated with human physiology. Human physiology should be a proper part of the secondary school course, but only a part; more attention should be paid to man's place in the living world. This would include a recounting of the evolution of the physical world and its biological inhabitants in such a way as to give an appreciation of their dimensions in space and in time; attention should be given to the present position of man in the biosphere. Current concern for environmental quality suggests that the approach should be essentially ecological. The complex interdependencies of living organisms lead to the basic rule of practical ecology that "we can never do merely one thing to an ecosystem," that the more man manages the natural environment, the more he generates the necessity for yet more management. By historical examples, the future citizen needs to be shown that knowledge is a prerequisite to intelligent interference in the scheme of nature, and that there are practical limits to what man can do. Problems of food production, pesticides, radiation, pollution, conservation, and population all have their place in secondary school biology courses and will find an appreciative, understanding audience. These considerations are crucial if we want a citizenry equipped to make intelligent decisions about the variety of questions facing society that have biological roots. The nature of our environment is determined in large part by decisions about landscape management that are made at *local* levels—by zoning boards, county supervisors, etc. These decisions are thus especially sensitive to the wishes of immediately concerned local constituencies, and it is in the nation's interest that these decision-makers have the greatest possible awareness of the scientific issues and complexities underlying their decisions. It seems likely that a course so structured could be maximally useful to future citizens, while including a sufficient presentation of cellular and genetic biology to afford a tempting glimpse of the elegance and intellectual attraction of current frontiers of biological progress.

Perhaps the greatest problem confronting the adoption of this and other experimental courses in the high schools is the education of a sufficient number of teachers with the training, insight, and enthusiasm to teach them effectively. The difficulties are grounded not so much in intellectual considerations as in the sociology of science. Biology teachers must be trained in colleges, but the curricula of colleges and universities are often structured almost entirely to meet the needs of college teachers, research biologists, or future physicians. With rare exceptions, college biology curricula are neither broad nor humanistic. High school teachers, so trained, tend to structure their high school courses in the same way. The resultant instruction may be well suited to proselytizing students for careers in biological research, but it is ill suited to the education of the citizen. With the best of motives and largely ignorant of what they do, university faculties deflect would-be high school teachers from preparing themselves for educating the citizen. Since the number of high school biology teachers is several times larger than the number of Ph.D. researchers, this is a massive and indefensible deflection.

Various remedial actions can be suggested, depending on local circumstances. One general proposal—that universities create curricula in human biology—should be seriously considered. Such a program could well include instruction in the specialties of human physiology, physical anthropology, human engineering, human genetics, ecology (both general and human), and population studies. Whatever the specialties of its members, the faculty should be recruited on the basis of interest in the scientific aspects of the relation of man to man and man to his environment. The education of teachers of biology in high schools and junior colleges would be a central concern of a program of this kind.

Many secondary school teachers are simply unprepared for new curricular materials. In some areas of the United States, only three semester units of college biology is considered adequate preparation to teach on the secondary level. Over 50 percent of the nation's high school teachers in the life sciences have had less than an undergraduate college minor in biology. Clearly, retraining and updating of teachers, as well as methods of recruiting better trained ones, are a large part of the secondary education problem; such methods are discussed in a later section.

New curricula developed by the professionals in a particular discipline often carry the taint of paternalism; occasionally, they have been resented by local officials or teachers. An additional pressure confronts biology: The programs of the Biological Sciences Curriculum Study have been attacked because their treatment of evolution offended some fundamentalist religious groups, or because their rather restrained treatment of sexual reproduction was held to be lascivious. Such irrational pressures prejudice the acceptance of new curricula and can rob thousands of students of educational advantages that ought to be theirs. One part of the process of curriculum improvement is the development of an intellectual climate appropriate for its acceptance.

University Education

The Setting

Instruction in the life sciences occurs in a wide variety of settings. In some universities, the elements of biology are drawn together under a department bearing that name; in others, there are separate departments of botany and zoology; and, in perhaps the largest number, an even greater array of separate structures exist.

The traditional departmental fragmentation that prevails in the biological sciences at many American universities, and at the land-grant institutions in particular, is the consequence of a peculiar historical development. Zoology departments, charged with the responsibility of training premedical students, become incorporated into colleges of arts and sciences. Botany, as a rule developed independently of zoology, often derives a major part of its support from schools of agriculture. As various other subdisciplines achieved strength of their own, separate administrative units were erected to accommodate their interests.

This trend to fragmentation, once initiated, has been reversed only with difficulty; individual departments tend to persist unchanged, even when the disciplines they represent can no longer flourish in isolation. Biology and the training of biologists have suffered as a result. For example, many departments, even those dealing with the more specialized biological disciplines, offer undergraduate as well as graduate degrees. The requirements for the major are frequently an overdose of specialized courses, taken at the expense of more fundamental subjects of a broadly encompassing nature. The outlook of the student is restricted, and he may be ill prepared for subsequent graduate work. The tendency is to train disciples rather than pioneers.

Fortunately, there appears to be a growing realization that early training must be broadened. To require some advanced mathematics of a student in systematics is no longer considered unusual, nor is the idea that a biochemist may be expected to master evolutionary principles. Such recognition of the common needs of their students is forcing many departments to reconsider the validity of the boundaries that separate them.

Similarly, there is an overabundance of highly specialized undergraduate courses demanding replacement with broader substitutes. Comparative anatomy, as traditionally taught, is the evolutionary history of vertebrates; ignoring both invertebrates and plants. Embryology is usually almost strictly a zoological course; developmental botany, if taught at all, is rarely integrated with its animal counterpart. Traditional courses, moreover, are sometimes retained even after there is no longer a demand for their un-altered continuance: Few medical schools now require comparative anatomy or embryology for admission, yet these courses, once designed for the pre-medical students, persist unchanged at many institutions.

How basic biology may be reorganized, and how its relation to the "applied" life sciences may be most fruitfully redefined, is a matter of concern to institutions that are now reappraising their biology-department structures. The curriculum needs simplification rather than diversification to reflect the growing intellectual unity of biology. Courses are needed that bind the different subdisciplines, rather than additional courses that deal with subspecialties. Experimentation with curricula is necessary if only because, on campus, the life sciences have been so extraordinarily frag-mented.

Undergraduate Curricula

Despite this fragmentation, an increasing number of undergraduate majors in the life sciences, perhaps a majority, proceed through an undergraduate curriculum that embraces most major aspects of biology, or at least most of botany or zoology. Together with other pressures, this has had the laudable result of encouraging integrated curriculum planning for students in the life sciences, often drawing autonomous departments together.

There seems to be agreement that there exists, in the intellectual content of biology, a common core of material that should form the basis for an undergraduate major, appropriate regardless of subsequent fields of speciali-zation. Thus, the same set of courses can serve for the premedical student and for the student who intends a research career. But there remain unique problems in the training of prospective secondary school teachers and of "terminal" majors.

Several institutions have independently designed new core curricula after departmental reexamination of teaching objectives. An independent curriculum-study group, the Commission on Undergraduate Education in the Biological Sciences, has begun work on the problem of encouraging and assisting curriculum reform and other improvements in the teaching of biology to undergraduates. Financed almost entirely by the National Science Foundation, this Commission consists of 25 professional biologists who form a steering committee, a small executive staff, and a dozen panels drawn largely from outside the Commission.

One Commission panel has compared the new core curricula installed in different universities. Among four quite diverse institutions, the simi-larity in content and in distribution of time among major topics is remark-ably high, reinforcing the conclusion that the trend is toward uniformity in what is taught. The content of these curricula differs in several important ways from that of their more classical predecessors. In general, more cognate courses in the physical sciences and mathematics are required, often as prerequisites, often actually reducing the number of hours in biology required to complete the major. The core curriculum has a much heavier emphasis on biochemistry, genetics, and cell biology, largely at the expense of systematic and comparative morphology. At many institutions, the courses themselves are structured and labeled by "levels of organization" (i.e., molecular, cellular, organismic, population biology) rather than by taxonomic group or functional system. Thus they resemble somewhat the organization of the life sciences proposed and used in the present study (see Chapters 3 through 5), a trend we warmly endorse. It is not fair, however, to characterize the new curricula as being merely "more biochemical"; they simply reflect more accurately current biological understanding. Neurophysiology, endocrinology, and several other subjects are also better represented than they were before. These changes have enlivened the undergraduate major in biology in only a few institutions and are not yet well disseminated nationally. One of the major efforts of the Commission is to provide a medium through which information about curricular experimentation can be swiftly propagated and to supply competent help to institutions desiring to make changes.

The diversity of the life sciences is a critical consideration in the design of undergraduate curricula. While the trend has been toward unification of biology departments and standardization of major curricula, the quite different requirements of different life science subspecialties will always pose special problems. Thus, although some departments have installed successful programs that prepare undergraduates regardless of their professional intentions, many others feel it necessary to provide optional "tracks." The appropriateness of any one solution probably depends upon the inclination and taste of the faculty involved, and the resulting diversity is likely to be useful. Often, the options provided are concerned more with cognate courses than with the biology program *per se*. In the training of molecular biologists, for example, as much course work in chemistry as in biology may be desirable, and the departmental program should allow ample time for the appropriate courses. For evolutionary biologists, on the other hand, more work in biology—as well as in such outside areas as mathematics and statistics—may be desirable. To the extent that standardized "core" curricula provide broader exposure to all areas of biology for all biologists, then, they are desirable; but they should not create a lockstep in which the unique needs of particular groups or individuals cannot be fulfilled.

Even where the capabilities and temper of the faculty allow such changes in curricular structure, vexing problems hamper completion of the transition. It has long been assumed that first-hand laboratory experience is a critical part of undergraduate education in any scientific discipline. Typically, courses of the traditional biology curriculum included one to two afternoons a week of laboratory work—usually dissection, light microscopy, and some rather simple experiments. Even wealthier institutions now find themselves organized to teach laboratory work with rooms designed only for simple "sit-down" work. They possess large inventories of medium-quality compound microscopes and modest supplies of balances, kymographs, and perhaps such devices as electronic stimulators, but they usually lack the more elaborate equipment and facilities to conduct more sophisticated biochemical, physiological, or genetic experiments. The typical biology undergraduate uses instruments in the laboratory that he will never again encounter except in a museum! Even in the better institutions, he is unlikely to have the opportunity to work directly, in a formal laboratory course, with a cathode-ray oscilloscope or a polygraph, with counting equipment, a good centrifuge, an electron microscope, or even a phase-contrast microscope. Even if the latter were available for research purposes, few students could have useful access to them in most circumstances.

The high cost of research instruments raises an important practical question: Is it realistic to strive to make expensive instruments available to all students as part of their undergraduate education? For example, in 1900 the optical microscope was used at the very frontier of research; at a university of that day most biology students had ready access to a fairly good one. Today, work at some segments of the frontier requires an electron microscope. Shall we strive to provide access to an electron microscope for each student, at a cost increase of perhaps 100-fold?

Granting that, ideally, electron microscopes and a variety of other costly instruments should be available to all students, economic considerations force us to consider alternatives. Is it really essential that a student handle an expensive instrument in order to understand its uses and limitations? Or can he gain sufficient knowledge by other means? The economic necessity now being faced in science has previously been met in other fields of education. Consider, for example, the young man who wants to become a conductor of a symphony orchestra. Desirable though it might be to provide him with an orchestra for practice, the fact that rehearsal time for a full orchestra may cost in excess of $500 an hour compels the nature of the decision and, for many years, the young musician must be trained with various inexpensive surrogates. Without actual contact with the real "instrument" (the orchestra), he must somehow be trained in the principles of its control. We must consider the development of equivalent surrogates for expensive instruments in the training of scientists.

Meanwhile, lacking direct *or* surrogate experiences with such instruments, the interested student can observe the contrast between activities in his professor's research laboratory and his own student experience. All too often he performs repetitions of old experiments, because nothing else can be done to "introduce him to the laboratory," a process usually initiated with the explanation that the purpose is to establish verisimilitude. Such programs tend to defeat rather than nourish the scholarly urge.

This is first a problem in educational philosophy and only secondarily one of fiscal inadequacy. Laboratory work even with simple, inexpensive materials can be made exciting if it is really explorative and demands thoughtful initiative of the student rather than mere following of "recipes." Over-reliance on "recipes" produces students who are unable to attack *novel* problems with experimental tools and with the confidence that they can provide solutions.

Nevertheless, whole areas of significant modern biology will remain closed to undergraduate experience unless laboratories can be re-equipped. Even without those major instruments for which we may have to provide only indirect experience, the conversion will be costly; *apart* from the building costs of new and adequate laboratory space, a large institution that graduates about 100 majors each year would require at least $250,000 to convert to a modern laboratory program for the core courses in its major curriculum alone. By this estimate, the national equipment deficit for undergraduate instruction in 1,200 institutions of higher education is currently $50 million to $100 million. Federal sources for such funds are now totally inadequate; the undergraduate instructional equipment program at the National Science Foundation is woefully underbudgeted and cannot make grants of the size required. The total national bill may be reduced by "sharing" programs between smaller institutions and part-time use of research equipment. Even without such reduction, we consider the cost small indeed in terms of our national scientific effort and urge that a program of adequate scale be mounted either at the National Science Foundation or the Office of Education of the Department of Health, Education, and Welfare.

The Teaching of Biology

TEACHING AS AN ACTIVITY

Most of the nation's professional biologists are teachers of one sort or another, at least part of the time. Moreover, the nature of teaching today is one of the determinants of the direction the research enterprise will take in the next generation. It is disturbing, then, that college and university teaching has received relatively little "scientific" scrutiny and that professional scientists who are also teachers quickly discover that their efforts to improve teaching are less productive of prestige and professional advancement than their research efforts. These two problems contribute enormously to the difficulties in the way of improvement of education in the life sciences.

Evaluation of teaching is difficult. Although a number of universities have attempted to involve students and other faculty members in the evaluation process, no method seems free of the potential criticisms that student judgments are flawed by recency or uncritical enthusiasm and that colleagues' judgments can be obtained only by objectionable monitoring. The latter concern is both serious and perplexing. Many a university teacher regards his classroom as quasi-sacred and what he says to his students as privileged communication. Anything that suggests an evaluation of his teaching arouses intense, sometimes irrational, defense mechanisms. With corrective feedback thus prevented, the lack of progress in teaching is easily understood. Nor have we confidence that the organized rating systems of undergraduate bodies will suffice to upgrade the general quality of undergraduate teaching.

The same faculty member may run his research laboratory in a completely different manner. The conspicuously successful trainer of research students typically maintains an extremely open atmosphere in the laboratory; the hopes, plans, frustrations, failures, and successes are all visible and shared.

Unfortunately, this atmosphere is not to be found in teaching; hence the largely nonprogressive character of teaching. Teaching, good or bad, is typically unmonitored by knowledgeable individuals; unproductive of adequate feedback, it may fall far short of its potentialities. Significantly improved teaching could occur in teaching "laboratories" in which a sufficient number of experimenters interested in the process of teaching conduct their work in the open way that characterizes the best research laboratories. Only in an atmosphere in which monitoring is so much the rule that it is not recognized as such will satisfactory progress in teaching be made. An especially useful--and unobjectionable—form of mutual monitoring takes place in the increasing number of departments in which small groups of faculty members cooperate in the teaching of a course and attend one another's lectures. This usually results in improved performance, and in helpful cross-evaluation of materials and techniques.

REWARDS FOR TEACHING

Although some institutions and a few national foundations recognize and reward good teaching, these efforts have not yet had enough weight collectively to make educational activity at all comparable with research activity in generating recognition and reward. This is especially unfortunate in view of the fact that many teaching activities are essentially scholarly themselves: the writing of a really superior textbook, the design and execution of new means of instruction, or the preparation of a new and exciting course. Such activities, with evaluations of their success, deserve dissemination and reward much as do other kinds of scholarly activity.

NEW METHODS OF TEACHING

The rising demand for teachers—and widespread dissatisfaction with the effectiveness of current educational methods—has forced life scientists, as it has teachers in other disciplines, to consider new methods of instruction. Efforts in this direction are bearing fruit at such a pace that we now find ourselves amidst a rapidly expanding technology of new ways of teaching. Among the most prominent innovations are the use of "programmed" instructional material; the use of television in lecture and smaller group teaching; the use of "audiotutorial" laboratory teaching, in which the student is able to make use of stored audio and visual instructions for the conduct of laboratory work; the use of computers in assisting instruction; and the use of film loops and other audiovisual materials. Fewer than 20 biology courses in the country now make use of television, and only a dozen use audiotutorial laboratory methods. Among this small number the overwhelming majority use these methods for freshman courses; to our knowledge only two advanced course programs in the country employ them.

Adoption of new methods often requires heavy capital expenditures, which planners expect to be amortized by only slightly decreased costs of instruction. In many cases, however, the net result is actually an increase in cost, and a redistribution, rather than a saving, in staff time. But new instructional methods should be undertaken not as an economy but as improvements of the quality of instruction. As student enrollments stretch the system, only such new methods can effectively multiply the effectiveness of truly accomplished teachers. Misconceptions about the new methods have frequently hindered their adoption. Among the prevalent myths are the notion that televised teaching is necessarily impersonal, the idea that programmed materials are essentially boring, useful only for the establishment of the cut-and-dried factual base of a discipline, and the view that audiotutorial and similar methods are useful primarily for spoon-feeding slow students. Studies of situations in which these methods have been effectively used show that they can achieve results in student performance and attitudes that are at least comparable with those from traditional procedures. At a time when the shortage of really effective teachers grows steadily more critical, these ways of *multiplying* the especially good teacher deserve careful study and experimental development.

THE TRAINING AND RETRAINING OF TEACHERS

A variety of programs are available for improving the quality of training of teachers at secondary school and college levels and for providing those already in service with ways of keeping up to date or alleviating inadequacies in their own backgrounds. The National Science Foundation has sponsored academic-year programs as well as summer and in-service programs. Those for secondary school teachers have engaged a surprising fraction of the teacher population; summer institutes in the mid-1960's had in attendance in any given year about 20 percent of the nation's high school teachers of science. However, to a considerable degree, this was a "repeating" fraction: more than half of all teachers had no exposure to such experience; only 1 percent have had a full academic year. Among college teachers the record is substantially poorer. Only about 1 percent of this population has ever attended a summer institute, and only an infinitesimal percentage has taken a full year for the sole purpose of retraining. It might be thought that college teachers would not require training of this sort, but they do. The Commission on Undergraduate Education in the Biological Sciences regards 1,000 of our 2,400 institutions of higher education as being entirely inadequately staffed to teach a modern program in the life sciences. The 4,000–5,000 full-time life sciences faculty members in these institutions would all benefit from exposure to a program of retraining.

Thus one of the most vexing problems confronting education in biology is improvement of the existing situation at both secondary school and college levels. Several approaches to this problem may warrant consideration. One would be to expand the opportunity available for retraining by the provision of a massively supported plan for financing full-year sabbatical leaves for teachers who need retraining. Most of those who have taught in academic-year and summer institutes feel that the full-year program is much more desirable than a larger number of shorter periods. A second proposal is to enable college departments to achieve major curriculum revisions. This would involve supplying funds on a 3-to-5-year basis for particular departments, allowing them to release the time of one or more members to plan and organize new programs. A third approach would aim at alleviation of some of the major problems of relatively small institutions of higher education that lack research facilities and adequate research programs. Such a plan could include funds to facilitate exchanges between larger and smaller institutions in an area, to apprentice undergraduate students in small institutions to research enterprises in larger ones, to permit faculty in smaller institutions to spend summers in research activity in major universities or other laboratories, or to have postdoctoral fellows at large universities "extern" as teachers in relatively nearby smaller institutions.

Several general observations may be made in conclusion. The scientific community, in general, fails to project a positive attitude toward teaching at any level as a career, and, predictably, this prejudice rubs off on students. Students engaged in the intensive research-oriented training for the doctorate must be shown by example that teaching is a significant and creative aspect of their future careers. Federal and other funding arrangements for the support of graduate education should include proper provisions for structured teaching experience. We view this as an intrinsic part of graduate work and suggest that a well-thought-out program of this kind may well be a preferable alternative to the adoption of a special "teaching degree" as a means of training college teachers. Finally, to deal with present inadequacies in the teaching of the life sciences at secondary school and college levels, programs for retraining and for maintaining contact with recent advances are increasingly necessary.

BIOLOGY AND LIBERAL EDUCATION

The pattern of the introductory collegiate course in biology for the future professional biologist becomes increasingly clear: it must assume a considerable expertise in elementary chemistry, physics, and mathematics. As an incidental consequence, these prerequisites often require that the beginning biology course be postponed until the sophomore year. More importantly, the prerequisites for the beginning course for future professionals make this course increasingly unsuitable for students with other major interests. What should be done for them?

We reject the suggestion that biologists should abandon the attempt to educate the general college student in biology. Knowledge of the principles and facts of biology is required to make intelligent decisions in innumerable matters of social and political importance: air and water pollution, radiation hazards, biological warfare, agricultural policies, voluntary and compulsory quality control of food and drugs, and population control, to name only a few. Biologists cannot expect public understanding or acceptance of their advice on public issues unless the college-educated segment of the community is biologically literate.

What sort of a college course in biology should be given non-biology majors? Attempts to meet the need for such a course by "watering down" the major course have met with uniform failure for two generations. Because the non-major course cannot be strongly couched in chemical and mathematical language, it is more elementary than the major course. But the interests of the general student require that the course addressed to them emphasize certain advanced topics that the major student will not study until some years later: ecology, population genetics, and human genetics, for example. These topics must be treated thoroughly enough that students majoring in sociology, anthropology, psychology, political science, or the humanities see the overwhelming relevance of biology to their problems. Only a few courses that do this job well are being taught. There is a crying need for this type of biology course, and for its distinctness from the major course, if the profession is to build the broad base of educated understanding needed for the future support of biological research.

There is yet one more type of course to which biologists would be well advised to turn their attention—a course that might be thought of as "Elementary Biology from an Advanced Standpoint." The intended audience would be undergraduate majors in chemistry, physics, mathematics, or engineering. They would take this course as juniors, seniors, or even as graduate students. Since the level of their sophistication in the physical sciences would be very high, this select group of students would very rapidly be brought to a deep understanding of the fundamentals of biology.

Such a course would serve several purposes: to present biology as a cultural subject to physical scientists; to prepare physical scientists for collaborative research with biologists; and to proselytize from this group, which has furnished so many excellent investigators in biology in the recent past. Though enjoyable, it would be a difficult course to give, but it should be possible to present it at increasing numbers of institutions as the pioneer teachers develop the textbooks for such an advanced treatment of elementary biology.

Research Training: Graduate Education in the Life Sciences

THE INSTITUTIONAL SYSTEM: FUNCTIONS AND DIVERSITY

The system for educating life scientists at the graduate level is as complex and diversified as the roles biologists serve in society, but since this complexity largely reflects historical accident, it invites scrutiny.

In the main, appropriate graduate training is needed for: (1) school-teachers at elementary and secondary levels, (2) college and university teachers, and (3) research workers. For the latter two categories, specialization ranges through the applied fields (themselves internally very diverse) of medicine, agriculture, and forestry to the equally varied disciplines of academic biology. The training of a teacher or investigator in molecular biology, systematics, or embryology necessarily is very different.

The complexity of the institutional system serving the total enterprise is reflected in the fact that almost 40 subdisciplines in the biological sciences offer separate masters or doctoral programs and have sought traineeship support from the National Science Foundation. The departmental categories were found among 876 individual departments in colleges of arts and sciences, colleges of agriculture and forestry, and colleges of medicine. Of all graduate students, 53 percent are in Ph.D.-granting programs in the biological sciences in the professional schools (agriculture and medicine).

THE STUDENT POPULATION: SIZE, ATTRITION, LOCATION

Approximately 7,500 students enter the system each year, largely from the approximately 25,000 students who annually receive baccalaureate degrees in the biological sciences. Although the academic fields of biology are attracting an increasing number of students with bachelor's training in physics or chemistry, the absolute number of these is probably still very small. Immigration from the physical sciences is probably greatest at the postdoctoral level.

This annual input of students suffers substantial attrition; we can make a rough measurement from data reported in the questionnaire sent to department chairmen by this Survey Committee.

In the academic year 1966–1967, there were 23,287 graduate students in the responding life science departments; of these, 15,755 were Ph.D. candidates. These candidates were distributed among 876 departments, of which 560 awarded 2,332 Ph.D.'s in 1966–1967. These data allow some estimate of the "efficiency" of graduate education, or, at least, of Ph.D. production. Sixteen thousand doctoral candidates, under the ideal circumstances of a four-year Ph.D. without attrition, should produce 4,000 Ph.D.'s a year. In fact, the "pool" represented by our sample produced a little over half that number. This efficiency figure may actually be even lower than it appears, since the ratio should actually be determined by the number of candidates that *enter* during a period of time, rather than the number enrolled at a given instant. Also, some graduate programs define Ph.D. candidates as those who have completed a master's degree or passed a qualifying examination. This practice will reduce the apparent pool of active candidates, and thus inflate efficiency figures.

Another way of measuring efficiency (which is subject to the same reservations) is to use only those departments that awarded Ph.D. degrees during 1966–1967 and calculate the ratio of candidates to degrees awarded in those departments. Such calculations yield an average ratio of 6.0, which would translate to an efficiency figure of roughly 75 percent. There is some variation between types of departments; smaller ones, even though they may be less productive, tend to have lower attrition. The fates of the careers of students who fail to complete the doctoral program are largely unknown and merit future investigation, along with the causes of their "failure." Women are only half as likely as men to finish their doctoral degrees—an aspect of the general failure of our society to take advantage of the creative resources of its female population.

An entirely different measure of the effectiveness of a graduate program is the production of Ph.D.'s relative to the numbers of individual faculty members who train them. Fifty-five percent of the faculty members represented in our sample teach in Ph.D.-granting departments, and 73 percent are located in departments in which there are Ph.D. candidates. On the average, in those departments that produced Ph.D.'s in 1966–1967, there were four faculty members per Ph.D. produced. In departments with Ph.D. candidates, there was 0.8 faculty member per candidate. These figures varied from one kind of department to another. For example, departments of biochemistry averaged 0.5 faculty member per candidate and in 1967 produced a Ph.D. for every 2.5 faculty members. Anatomy departments, by contrast, had 1.3 faculty members per candidate and produced only one doctorate for every eight faculty members.

FINANCIAL SUPPORT OF THE GRADUATE STUDENT AND HIS EDUCATION

The cost of educating graduate students in biology, as in other sciences, is rarely met by the student directly; his training is subsidized in a variety of ways. Half of the graduate students in the life sciences are supported federally: 42 percent enjoy federal fellowships and traineeships, 38 percent have nonfederal (institutional) support, 13 percent are paid from faculty research grants (most of which are federal), and 8 percent are supported by other means, including their own resources.

The pattern of federal support differs strikingly by type of school and department. Sixty-eight percent of the graduate students in schools of medicine are supported by federal traineeships and fellowships, compared with 40 percent of students in graduate schools of arts and sciences and 22 percent of those in schools of agriculture and forestry. Federal support is provided for 66 percent of the students in departments of biophysics and 52 percent of those in departments of biochemistry, whereas only 20 percent of graduate students of botany receive such stipend support.

These discrepancies arise from the sources of these federal funds. Training grants support over half of the federally financed Ph.D. candidates. Only about 20 percent of the students hold national fellowships, and institutional grants take care of another 20 percent. Since the traineeship funds come very largely from the National Institutes of Health, the criterion of health-relatedness markedly affects their distribution. Accordingly, students in departments of botany and schools of agriculture and forestry must seek other means of support.

Competitive national fellowships, which support about 10 percent of the nation's doctoral candidates in the life sciences, establish stipend and prestige standards. Such fellowships are twice as important in private arts and sciences graduate schools as they are in the overall support picture because a relatively small number of high-prestige graduate schools attract a disproportionate share of the students able to compete successfully for these fellowships. Hence, competitive national fellowships loom larger in the support picture in the Northeast and on the Pacific Coast—concentration points for such institutions.

Of nonfederally supported students, most (75 percent) hold university teaching assistantships. Only a trifling number are supported by university fellowship funds.

The importance of graduate student support is reflected in department chairmen's statements of the relative priority of different types of funds needed to "improve the department's research endeavor." Increasing predoctoral stipend funds were repeatedly cited as a prime need in those departments that currently train graduate students; it was mentioned by a fourth of all the departments as their first priority and by over half as one of the three most important funding categories.

These data have led us to several conclusions:

Stipend Levels Stipend levels, previously adequate, are no longer so (1970–1971). For six years, national fellowship and traineeship stipends have been set at a 12-month standard of $2,400 for first-year students, $2,600 for intermediate-level students, and $2,800 for terminal Ph.D. candidates, with an allowance of $500 for each dependent. In the meantime, the cost-of-living index has been rising at a rate of over 4 percent per year. Especially in the more expensive areas of the country, graduate students— even those without dependents—are in serious economic straits. The availability of stipends for summer work unrelated to thesis research tends to prolong the total period of graduate work for many students.

Diversity of Sources of Support The diversity of sources of support is intrinsically desirable. The federal agencies responsible for science generally (National Science Foundation), health (National Institutes of Health), and education (U.S. Office of Education), all make contributions directed at different segments of the total enterprise. We note and deplore the absence of any significant contribution from the Department of Agriculture, the National Aeronautics and Space Administration, the Atomic Energy Commission, and the Department of Defense, because, until the National Science Foundation, the National Institutes of Health, and the Office of Education acquire broader mandates or more adequate appropriations, agencies that use substantial numbers of Ph.D.'s in basic sciences should contribute to the cost of their training. Diversity of sources of support is intrinsically desirable, and relieves the non-mission-oriented sources (National Science Foundation, Office of Education), making them freer to ensure adequate support for the least mission-oriented programs and institutions.

In this connection, we note that, until recently, the National Institutes of Health has, for the most part, been reasonably broad in its interpretation of the "health-related sciences" it is charged with supporting. It has correctly recognized that biology is "all of a piece," and that it is inherently impossible and historically fallacious to identify some aspects as related to health and others as unrelated. Indeed, it is difficult to imagine *any* biological problem of major importance that is without relevance to human welfare. Nonetheless, we ought to be cognizant of the danger inherent in the dependence of so large a fraction of the biological educational enterprise upon an agency that has an applied-science mission. Although we hope that mission will always be broadly viewed, it is clear that our judgment as scientists is not wholly shared by those entrusted with the establishment of federal policy. If the National Institutes of Health were to adopt a more restricted interpretation of its mission, a process currently in progress, the present pattern of dependence on the Institutes for research and training support could result in catastrophe for education in some areas of the life sciences. For this as well as other reasons, it would be unfortunate if the National Science Foundation and the Office of Education were to reduce their responsibilities to biology on the basis of past availability of funds from the National Institutes of Health. At the same time it is most strongly urged that the National Institutes of Health—particularly the National Institute for General Medical Sciences—continue to mount a vigorous, broad program of support for research training in the biological sciences.

Federal Support An analysis is required to ascertain the wisest combination of national competitive fellowships and traineeships available for allocation by departments.

National competitive fellowships enable the successful outstanding student to choose freely among the institutions that offer him admission and to choose his sponsor and course of study without the constraints that sometimes accompany local support, and their distribution affords some useful measure of the attractiveness and the quality of departments. The program of fellowships and traineeships as currently operated fails to provide adequate support to the host institutions; the $2,500 cost-of-education allowance that accompanies National Science Foundation and National Institutes of Health predoctoral fellowships (and National Science Foundation traineeships) barely covers the *nominal* tuition fee in many of the private institutions, and such fees cover less than half of the real costs of education.

The National Institutes of Health training-grant system is more wisely constructed in this important respect. Awarded to a department or multidepartmental group, training grants frequently provide an amount (determined by negotiation and the justification offered in the initial proposal) for equipment and other needs of the training program concerned. The annual cost to the agency of a National Science Foundation fellowship or traineeship or a National Institutes of Health fellowship is $5,000; on the other hand, the average cost to the National Institutes of Health of a student on a training grant is $8,000, which is much closer to the real costs of his education.

In our view, ultimately federal agencies should utilize the traineeship and training-grant system for all but a rather small fraction of graduate student support. Training grants offer virtually all the freedom of choice of institution and of research mentor available with the fellowship device and, by their geographical distribution, assure the continual strengthening of graduate education across the country. As long as the grants are awarded on the basis of periodic peer review, there is little chance of misleading prospective graduate students while matching the training capability of each departmental unit to the actual magnitude of its training support.

Cost of Education   The cost of education for graduate student biologists merits separate and special notice. This cost is continuously increasing as the space requirements increase and the equipment needed becomes more sophisticated; to this extent it imposes on the universities an increasing burden that they are less able to bear as time passes. The unrealistic provision for these costs in the $2,500 allowed per student has two consequences: it has not alleviated the generally inadequate state of training facilities across the country, and it permits the training process to remain too strongly dependent upon funds allocated primarily for faculty research. Few of the electron microscopes, spectrophotometers, and ultracentrifuges that are indispensable in the education of a graduate student today were purchased for that purpose. Thus, access to them is not easy for the student: the faculty member has a responsibility to his research and its sponsor for their maintenance in first-class research order. A conspicuous and urgent need in most graduate departments is adequate curricular provision for training in the great diversity of laboratory techniques required for effective research in modern biology. The lack of such formal training stems almost exclusively from the lack of funds for acquisition of equipment. The typical graduate student learns a sophisticated technique only if his dissertation work demands it—i.e., if it is an important element in his faculty sponsor's research. That very fact can serve as an inhibitor to the imaginative approach to research problems in his subsequent career.

THE GRADUATE PROGRAM

The M.A. The M.A., as a terminal degree, has been steadily declining in importance; a doctorate is an indispensable qualification for all university teaching and for the vast majority of college openings. Terminal M.A.'s find their principal academic opportunities in the secondary schools or junior colleges, but many school biology teachers take other routes, such as the Master of Arts in Teaching. To the extent that it survives as a distinct program, the M.A. usually consists primarily of the same formal course work that is a major element in the training of the "precandidacy" doctoral student.

Doctoral Training The Ph.D. level is, then, the goal of virtually all graduate education in biology, and that fact itself has important consequences. Ostensibly, the same degree is the required entry for teachers at all levels above the secondary school (junior colleges, colleges, and graduate schools), and for research scholars. In biology as in other disciplines this leads to national confusion concerning standards, curricula, and duration of study for the doctorate.

The overwhelming majority of Ph.D.'s are awarded in research-oriented departments supported by substantial federal funding. Nearly 90 percent of Ph.D. candidates are being trained in departments that have already produced doctorates. Graduate education has always been concentrated in a relatively small number of departments in "prestige" universities, and this feature shows little tendency to change. Seventy-seven percent of Ph.D. candidates are in departments that also have postdoctoral fellows.

As to type of institutions, 46 percent of the candidates are in graduate schools of arts and sciences, 29 percent are in schools of forestry and/or agriculture, and 25 percent are in medical schools. Three times as many candidates are found in public as in private institutions. This distribution contrasts with that of the *departments* that train graduate students: 28 percent of the departments that have awarded Ph.D.'s are in arts and sciences, 29 percent in agriculture and forestry, and 40 percent in medical schools.

In short, departments in arts and sciences graduate schools have larger numbers of students and accomplish more doctoral training than either of the other two kinds of institution; but the smaller departments found in schools of medicine have been expanding their capacity most rapidly.

It is perhaps more meaningful to ask about the kinds of departments in which doctoral candidates are trained. Although 29 percent are found in schools of agriculture and forestry, only 14 percent are receiving degrees in the agricultural sciences; the difference is accounted for by students in basic life sciences departments in these professional colleges. Eighty-six percent of the candidates are in biological sciences departments, 14 percent of them in biochemistry, 18 percent in biology–ecology, 20 percent in zoology–entomology, and 5 percent in physiology. The remainder are scattered among smaller, more specialized disciplinary units.

Geographically, the large land-grant institutions of the midwestern states are the largest producers of doctorates. Only in the Northeast does private surpass public education in Ph.D. production, and here medical schools play a prominent role. Graduate education in the life sciences is characterized by great regional asymmetries: over a third of the states produce fewer Ph.D.'s than a single medium-sized private institution, and both the South and the Rocky Mountain region are producing relatively few doctorates.

Finally, one may ask how the distribution of doctoral-degree candidates is correlated with the distribution of research funds. Not surprisingly, the correlation is high: If one eliminates the departments of clinical medicine, which are a special case in this regard, then 75 percent of the total federal allocation for research in the life sciences goes to those departments that have actually granted Ph.D. degrees.

THE FUTURE OF GRADUATE PROGRAMS

Increasing Ph.D. Production   In view of the frequently expressed need for increased numbers of research workers and teachers who have doctoral degrees, it is important to estimate the "elasticity" of the training system—i.e., how much more training it is capable of. Department chairmen queried by the Committee were surprisingly optimistic about their ability to expand enrollments even without additional space and faculty. Seventy percent of the responding departments claimed that they could grow without these usual prerequisites; in the aggregate, they assert that they can increase their graduate school enrollment by about 25 percent.

All departments combined predicted an expansion from 17,172 to 23,370 faculty members by 1970–1971. This projected increase is largest in the departments of basic medical sciences. Nearly 40 percent of this expansion was stated to represent the filling of already-budgeted positions. Those departments already producing Ph.D.'s anticipated an increase in space of approximately 40 percent, two thirds of which was already under construction.

In summary, the system seems to have a built-in expansion capacity that amounts to approximately 25 percent of its present production. By 1970–1971, it is predicted that it will expand by about 50 percent from 1966–1967 figures; this expansion will be accommodated by increasing space (by 40 percent) and faculty (by about 25 percent). These predictions do not appear realistic, even though half the projected expansion in space and faculty was thought to be underwritten at the time it was made. No more realistic is the prediction made by department chairmen of their own capacity to produce Ph.D.'s. In 1964–1965, the departments in our sample produced 2,000 Ph.D.'s; in 1966–1967, 2,330. For 1969–1970, the number of new Ph.D.'s projected by these departments was 4,300. If efficiency is maintained at its present level, Ph.D. production should lag behind increases in graduate enrollment by about two years. The maximum increase likely, even ignoring effects of the draft and reduced federal funding, is perhaps 10 percent a year, somewhat more than the 7 percent annual increment that has been characteristic of the sciences for many years.

Duration of the Doctoral Program Since the training system has a limited capacity for expansion, and since present funding limitations are curtailing its operation drastically, it becomes even more important to improve the quality and efficiency of the enterprise. In particular, both attrition and the time required to obtain the doctorate should be curtailed. The national average for the time elapsing between the B.A. and Ph.D. degrees in all fields is still an alarming 8.5 years, although in many leading institutions, where sufficient support and supervision are provided, four to five years is now established as adequate. The tendency has been, and will continue to be, a general curtailment of the duration of study, with four to five years becoming a universal norm. Its attainment demands a more prescribed and carefully organized graduate curriculum than has prevailed in many institutions; freedom from excessive burdens of teaching and from research assistantships that are not directly related to the thesis; and continuation of the trend, now more common in science than in the humanities, to prescribe appropriately modest research goals for the dissertation.

The trend is thus to establish as well-regulated a program for the doctorate as for the bachelor's degree, converting it from an ill-defined entry into the world of independent scholarship to a well-defined level of education. That level should be set realistically, at a point that will just satisfy the purposes of the doctorate. Students contemplating research careers now have the opportunity of postdoctoral education; the doctoral program is not the end of formal training.

The growth of postdoctoral education nationally has, in fact, developed hand in hand with this trend in the doctoral program. At the same time, there are plans to revitalize (or merely to rename, as M.Phil.) the M.A. degree to accommodate some of the functions now assigned to the doctorate. The prestige of the doctorate is such that even the smaller colleges are unlikely to be relieved of the pressure "to count their Ph.D.'s" and thus to render such subdoctoral degrees unacceptable. The regularization of doctoral programs to a four-year norm and the evolution of a carefully administered postdoctoral program seem, therefore, to constitute the more desirable policy. The postdoctoral student is not actually "postponing" his entry into the teaching system. He is, to be sure, free to benefit from a unique research experience unhampered by either a thesis deadline or the heavy burden of developing his own courses; but, in his daily contacts with graduate students in the same laboratory, he is discharging one of the most important and valuable teaching functions in the university while preparing for a life dedicated to research.

The Relation between Student and Faculty Sponsor The core of the doctoral program remains the dissertation, and the goal of a four-year Ph.D. makes some form of apprenticeship almost inevitable in performing thesis research. That apprenticeship can, of course, degenerate into menial research assistance for a prescribed fraction of the sponsor's own research program, but this abuse is rare. There is little doubt that the intimate personal relationship between a student and his thesis supervisor is the most important formative element in his training as a scientist. Misuse of the apprenticeship system may be fostered by the prevalent dependence of student research opportunities on the faculty's research funds. Departmental funds are needed for student equipment and facilities, for research as well as for formal course work in those not infrequent cases in which an able student's own research goals do not coincide fully with those of his sponsor. For the same reasons, continued growth of fellowships and traineeships is desirable because it relieves departments of the necessity of funding students via the research-assistantship route.

Formal Training and "Curriculum" Examination Decades ago, the Ph.D. general examination could pretend to test—or demand—detailed knowledge not only of a subdiscipline but even of the entire field of biology. That time, if ever it really existed, is past; re-evaluation of the function and nature of the general, or qualifying, examination has thus become a preoccupation of faculties, and it is rarely satisfactorily defined. The diversity of the biological sciences makes the problem more acute than in the physical sciences. At the other extreme, the inclination to be "realistic" about such examinations has too often resulted in an examination—and hence a whole curriculum orientation—that is unnecessarily specialized.

It is clear that the diversity inherent in the biological sciences, and the need for specialists ranging from economic entomologists to molecular biologists, will continue to demand a wide range of specialty-oriented graduate departments. But this hardly explains the extreme diversity of programs that have in fact evolved. The total biological enterprise in one of our leading graduate schools is fractionated into over 20 departments, each administering its own doctoral program. In fact, there is little excuse for persistence of even the traditional botany–zoology cleavage. Biology has undergone a century of maturation capped by two decades of extremely rapid advance. There is a central core of empirical generalization and theory whose existence every biological specialization must recognize, and doctoral programs should no longer be based on selection from the à la carte menu of courses that a department happens to offer.

We are impressed by the fact that major national efforts have been devoted to the development of curricula in biology at the high school and undergraduate levels, but insufficient thought has been given to the development of a core curriculum at the graduate level. Graduate students in biology, unlike those in physics, come to the university with widely disparate undergraduate backgrounds. Those who were biology majors often have an inadequate background in mathematics and the physical sciences, and even within biology present a preparation that is too unpredictable and disorganized to constitute a secure basis for advanced work. The slowly increasing number who enter from the physical sciences, on the other hand, have had virtually no exposure to the "classical" areas of biology. The majority of graduate departments fail to remedy this deficiency as they succumb to the twin pressures for a "realistic" four- to five-year program and adequate coverage of advanced work.

There is no reason why graduate departments cannot, with appropriate curriculum planning, overcome the problems posed by the welcome diversity of their entering students. These problems cannot be expected to disappear; increasing numbers of students with primary preparation in the physical sciences will continue to undertake graduate study in the life sciences, and the trend toward more uniform preparation within biology itself will move slowly. Because there is a fundamental unity in biology, it is desirable for the graduate program to provide basic competence in the fundamentals of genetics, evolutionary theory (and the major outlines of the history of life), physiology and biochemistry of cell and organism, developmental biology, and population biology. Our concern is, moreover, not just that the students have some adequate course preparation in these five areas, but that their interrelations in contributing to a truly general science of life be fully developed.

There is no doubt, in our view, that a student with an exclusively physical science background could be given this proper overview of the central core in biology within a year, yet we know of no single department that has seized the challenge of developing a general biology at the graduate level. Nor is its utility to be thought of as limited to those with physical science training. It is a compelling irony that the last time most biological scientists even attempted to see the subject in perspective was before they knew it—as freshman or sophomore undergraduates!

Beyond this point, the graduate experience will properly take a series of separate paths, depending upon the special requirements of the discipline for which the student is preparing. But it should not be supposed that, for most research workers in the life sciences, this special training can simply be laid over an appropriate background in physics and chemistry without an understanding that extends across the whole of biology. There is a genuine general biology today that is more than a shotgun marriage of subdisciplines; its focus is the organization of living things, and it recognizes that the analysis and understanding of cellular and organismic organization is the goal that characterizes or defines the enterprise of the biologist as against that of the purely physical scientist.

CHAPTER SEVEN

DIGITAL COMPUTERS
IN THE
LIFE SCIENCES

Throughout organized society we are engaged in a massive experiment to learn how to integrate the technology of information processing by computer into the working fabric of each area of endeavor. This experiment started in some areas two dozen years ago; in some it is just starting, and in others it is still largely in the planning stage. These experiences have begun to provide a picture of the process through which a given field assimilates the computer. No field is much beyond the beginnings of this experiment, as a retrospective study 20 years from now will undoubtedly demonstrate. To assess the role of the computer in a particular field—in this case, the life sciences—some general lessons must be understood and some questions examined. After presenting the general lessons rather briefly and dogmatically, we shall ask some appropriate questions concerning the life sciences. Together, these two approaches lead to a number of recommendations concerning management of computers for the life sciences in the immediate future.

GENERAL FACTS ABOUT COMPUTER USAGE

1. *The computer is a general processor of symbolic information.* Most importantly, the computer is a general device for processing symbolic in-

formation. It is not just a device for doing arithmetic fast, and information need not be in numerical form for the computer to "understand" it. Because all tasks employing symbolized information are candidates for performance by the computer, there is a great diversity of uses of computers.

2. *Limits on computer application in a given field.* Though *all* symbolic tasks are *candidates,* not all such tasks are appropriately performed by the computer at any moment. The most important determinants are:

The power of existing computers: speed, memory, file space, and reliability.

The ways existing computers can interact with the outside world: printers, visual displays, direct coupling to instruments, typewriters, and so on.

The cost of existing computers: dollars per million operations, or per thousand bits of memory.

Our ability to make the computer perform, viz., programming techniques; of all the multitude of things computers can potentially do, we have succeeded in getting them to do very few.

The financial resources available to the field.

The sophistication of practitioners in the field with respect to computers, especially their programming and operation.

3. *Cost effectiveness continuously and radically increases.* More than any other device man has ever built, the power of computers continuously increases from year to year, while costs per computation decrease. The yearly increment of these changes is huge, compared with such changes in other technological products (e.g., the airplane or metal products). Figure 39 shows the inverse nature of these changes. Estimates of costs and computer capability must be radically revised every five years. The interfaces between the computer and the world are also changing, but much more slowly and conventionally.

4. *Learning to accomplish a new type of task with a computer is analogous to "development" in other fields.* The term "development" is used in its standard industrial meaning, i.e., development as opposed to research, prototype construction, and production. When initiating research it is not certain the job is possible, for essential knowledge and techniques are missing and must be discovered or invented. At the development stage the job can be done, but since it has never been done before, a multitude of unknown difficulties are certain to emerge, each of which can generally be overcome by further application of manpower and time. When prototype construction is undertaken, all the elements of the process are understood, but the total process must still be organized into a smoothly working

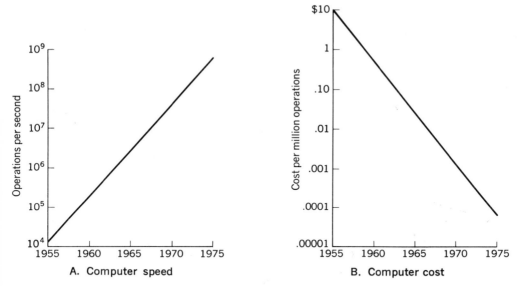

FIGURE 39   A. Computer speed. Storage speed is expressed in thousands of additions per second. B. Computer cost. Storage cost is expressed in dollars per million
additions. (Adapted from P. Armer, "Computer aspects of technological change,
automation, and economic progress," in *The Outlook for Technological Change and
Employment,* Appendix Volume 1, *Technology and the American Economy,* The
Report of the Commission, Studies Prepared for the National Commission on Technology, Automation, and Economic Progress, U.S. Government Printing Office,
Washington, D.C., February 1966.)

whole, until, at the production stage, only maintenance activity is required
to keep things working efficiently.

In many instances, the application of computers still requires research,
e.g., to achieve automatic processing of microscopic images for chromosome
analysis. But the bulk of computer applications begin at the development
stage; development is expensive and always takes longer than anticipated.
Once we understand what we require the computer to do, we cannot simply
tell it to do so, *viz.,* write down instructions. Each time, a new, relatively
large program has to be written. Hence, getting the computer to achieve
the new task successfully is an expensive, time-consuming development job.

5. *Assimilation of computers by a field must be accomplished by the
practitioners themselves.* It has often seemed that successful practice in the
use of computers in one field can be transferred to another (since they seem
to be "doing the same thing" regardless of the field) or that the computer
industry can develop the application systems and provide them to new
fields as a marketing service. Both assumptions are false, though both

considerations are important. *The computer becomes widely and success-fully used in a field only as the scientists in that field come to understand and use the computer.* Otherwise application remains isolated and peripheral. Despite superficial similarity of tasks among various fields, each field proves to have its own unique problems in processing information.

This is not to deny the importance of an autonomous computer science, or of attempts at interdisciplinary efforts (especially when they serve to seduce those with computer expertise ultimately to become life scientists!). It does, however, establish the essential condition of successful assimilation—and where the ultimate responsibility rests.

6. *Computer facilities must specialize.* The generality of the computer leads naturally to the view that a single computational facility can service all needs. The establishment of large university computation centers in the last decade reflects this philosophy in part. But so great is the diversity of computer use that, in fact, each computer facility, whether large or small, serves only a fraction of the whole range of needs. Every existing computer facility substantiates this notion. Time-sharing systems are not efficient for processing large numerical calculations; large systems for statistical calculations cannot accommodate laboratory monitoring, and so on. The generality of the computer is that it can be adapted to any symbolic task, not that it can be all things for all users simultaneously.

THE STATE OF COMPUTER APPLICATION IN THE LIFE SCIENCES

How quickly and thoroughly, and in what respects, computer application becomes effective in a given field will be determined by the interplay of the general facts cited above.

Extent of Use

Computing is widespread in the life sciences. Approximately one in three life scientists computes, and the total cost of their computing is in excess of $18 million a year. Unless otherwise specified, all data concerning the state of computing in the life sciences come from the census of individual life scientists conducted by the Committee and reflect use in academic year 1966–1967. See Appendix A, Individual Questionnaire, Questions 22–27. Information was gathered on the number of hours of usage by type of machine (A, B, C, D, using the classification of the

Rosser report*). Hours can be converted (for some purposes) to their equivalents on a type B machine (e.g., an IBM 7090): 1 type B machine is equal to 0.25 type A, 4 type C, and 20 type D machines. 1 B-hour provides approximately 300 million basic operations (in practice divided between input, compiling, computing, editing, output, etc.).

The total hours of computer time reported was equivalent to 90,000 hours on type B machines. This amounts to approximately $18 million a year in rental. Extrapolation to the total computing population increases this amount by a factor of 1.5 to 2. Large users who own computers that are used 24 hours a day partially offset the cost extrapolation as they can purchase machine time at less than the standard rental.

Unfortunately we do not have comparable figures for other fields. The Rosser report, which confined itself to academic institutions, estimated that for 1968 the physical sciences would require 90, engineering 20, and the life sciences 21 type B computers. For all biologists the estimate from our census approximates 32 type B computers, with 17 of them in universities. A guess must be made as to how many B-hours equal a type B machine, as defined in the Rosser report. It must range between 2,000 and 7,000 hours (one to four shift operations); we chose 4,000 hours. Actual usage in the other fields undoubtedly also exceeded earlier estimates. The two-decade head start in the physical sciences still remains. In any case, we can be sure that biologists are doing large amounts of computing, and the life sciences are no longer to be viewed as "computationally undeveloped country."

As would have been anticipated, actual computer use is very unevenly distributed. This is illustrated by dividing the users into "light users" (less than 10 B-hours per year), "medium users" (10 to 99 B-hours per year) and "heavy users" (more than 100 B-hours per year). As Figure 40 shows, 69 percent of life scientists do no computing at all; 21 percent are light users and do 5 percent of the computing in the aggregate; 8 percent are medium users, doing 25 percent of the computing; and the remaining 2 percent are heavy users, accounting for 70 percent of the computing. One of the consequences of this skewed distribution is to make average values somewhat meaningless, since there is no "central tendency."† The

---

* *Digital Computer Needs in Universities and Colleges,* A report of the Committee on Uses of Computers, NAS-NRC, Publ. 1233, National Academy of Sciences, Washington, D.C., 1966.

† More precisely, if we imagine the present sample to be drawn from some underlying continuous distribution that gives the probability of a person using x number of B-hours of computing, then it appears that this distribution does not have a finite mean value, i.e., that $\int xp(x)dx$ does not converge.

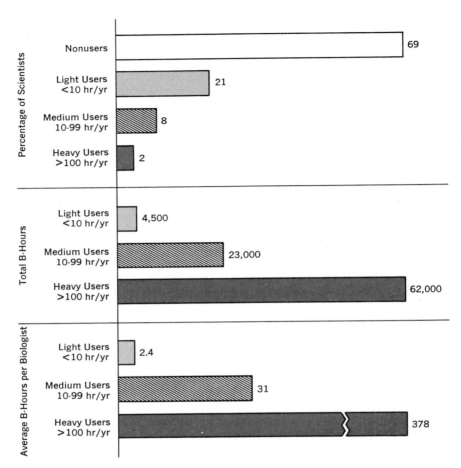

FIGURE 40   Distribution of computation use. (Source: Survey of Individual Life Scientists, National Academy of Sciences Committee on Research in the Life Sciences.)

total amount of computing is markedly dominated by the few heavy users.

   The diversity underlying this distribution is impressive. Seemingly equal amounts of computing power are not equivalent at the small and large ends of the scale. The composite figures summarized above include the use of vastly different computers, differing in power and facility up to a factor of 100. Our sample of almost 4,000 computer users claimed 14,000 hours of type A computer (like an IBM 360/65 or a CDC 6600), 23,000 hours of type B computer (like an IBM 7090), 19,000 hours of type C computer (like an IBM 360/40 or an SDS 940), and 100,000 hours of type D computer (like a DEC PDP-8 or an IBM 1130). This last category is especially noteworthy because it shows how widespread the use of small laboratory computers has become in the life sciences. Although accounting only for a small fraction of the total computing power (about 5,000 effective B-hours, or 5 percent of the total), it accounts for 60 percent of the "contact hours." Much of this form of computer use, as in on-line data acquisition, cannot meaningfully be exchanged for hours on larger machines.

Examining the data from our sample, diversity of arrangement meets one at every turn. The amount of actual type D computer use in effective time (equivalent to 5,000 B-hours) is of the order of the amount of computing time used by all light users (4,500 B-hours). However, light users utilized all types of machines; thus, approximately 40 percent of their computing was done on type A machines and 20 percent each on types B, C, and D. Conversely, approximately half of all time on the small computers (type D) was occupied by individuals who were heavy users.

Types of Use

Some information is available concerning the types of use of computers by broad functional categories. Almost all users performed some data analysis (93 percent), but half (47 percent) also did some other type of computing, and a substantial number (17 percent) engaged in several types. Table 57 shows the distribution of hours and number of scientists for each

TABLE 57   Tasks for Which Computers Were Used

APPLICATION	PERCENTAGE OF LIFE SCIENTISTS	PERCENTAGE OF COMPUTING (B-HOURS)
ALL USES	**100**	**100**
Data Analysis Only	54	21
Information Storage and Retrieval	16	7
Data Acquisition	3	3
On-line Experimental Control	1	1
Simulation	3	2
Theoretical Analysis	6	3
Multiple Uses	17	63

Source: Survey of Individual Life Scientists, National Academy of Sciences Committee on Research in the Life Sciences.

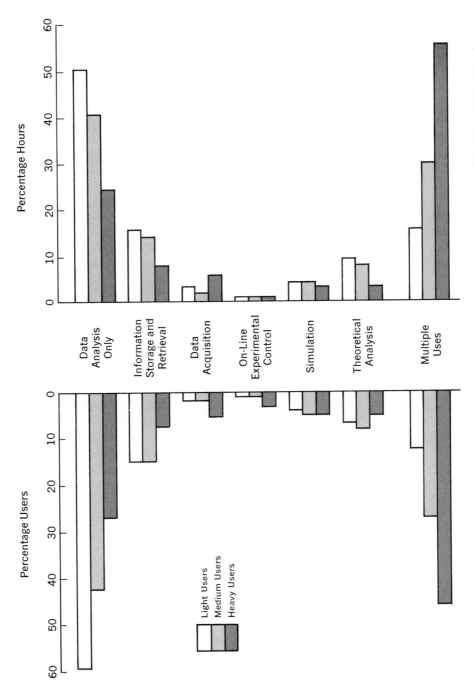

FIGURE 41 Tasks for which computers were used by questionnaire respondents. (Source: Survey of Individual Life Scientists, National Academy of Sciences Committee on Research in the Life Sciences.)

functional type. The indicated categories, other than "Data Analysis Only," imply that both data analysis and the specified activity were conducted, but nothing else. The "Multiple Uses" category indicates that data analysis and at least two other types of computation were reported.

The table shows that most of the computing hours are used by those in the "Multiple Uses" category, reflecting the fact that most heavy users use computers in several ways. Surprisingly, the light users were not much more likely to engage only in data analysis than was the population as a whole (59 to 52 percent). Other similarities between the three user categories are illustrated in Figure 41. Likewise, the machines themselves are not specialized for specific types of computing. As can be seen from Figure 42, only a few general aspects show through, e.g., type D computers are used relatively heavily for experimental control (which seems natural), and type A computers are used somewhat less than others for storage and retrieval. The reason for this is not so clear. However, it was unlikely that our gross categories could reveal the specialization of particular facilities to do particular jobs.

This description of functional types of computer use, based on answers to our questionnaire, is rather abstract and fails to describe the remarkable diversity of computer use in the life sciences. Only a few uses can be noted here. Included are very large numerical calculations, such as the processing of statistical data or deciphering the structure of an organic molecule from crystallographic data. Small data analysis may include relatively simple routine calculations from instrumental analysis, the striking of a nutritional balance, or the calculation of relatively simple reaction rates. The widest possible variety of functions is found within the category of "data acquisition." This may be an experiment in which electrical signals are obtained and converted to digital records for later processing, possibly with concurrent display to check whether a good record has been obtained. It could mean equally well the use of a currently existing system for gathering data about the feeding and milk-producing behavior of cows from all over the country, which (along with their genealogy) permits the evaluation of both feeding plans and the worth of bulls. Simulation could mean a study of enzyme kinetics or a study of the life cycle of a salmon. The relatively small amount of effort indicated for on-line experimental control is a reflection of the relative recency of the practicality of such exercises. The use of small laboratory computers to "run the show" is, qualitatively, a completely different use of computers than all the other categories.

The variation is such that no general pattern emerges, and the variety will surely expand in response to future demands for information-processing tasks.

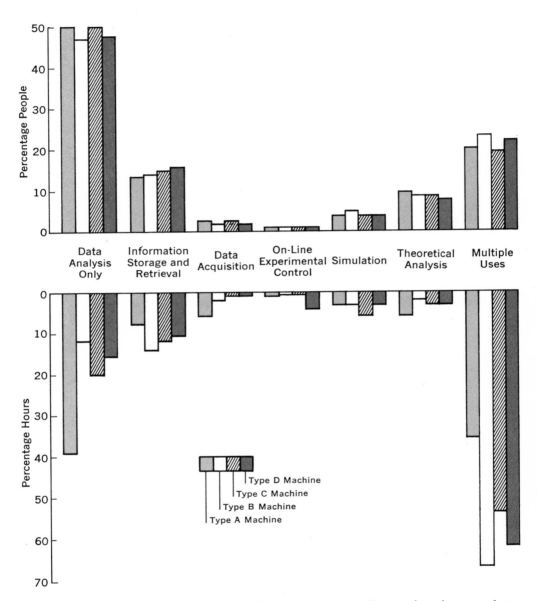

FIGURE 42   Tasks for which different sizes of computers were used by questionnaire respondents. (Source: Survey of Individual Life Scientists, National Academy of Sciences Committee on Research in the Life Sciences.)

Computer Use in Research Areas of the Life Sciences

Within each research area there occurs the same skewed distribution of computing, with many small users shading down to a very few heavy users. The more investigators active in a research area, the more computing they do; beyond that, no generalizations emerge. It is not useful to compute an average number of hours per scientist. Figure 43 displays this rather curious situation. This figure shows, for our sample, the number of computer scientists in a research area versus the amount of computing done by that area. Although it clearly rises linearly as the number of scientists who compute increases, the points become widely scattered. In terms of a computed mean value these "wild" points almost completely determine the slope, due to the small number of heavy users in each subfield.

Our survey shows that, for all research areas, the light users consume about 2.4 B-hours per year, the medium users about 30.6, and the heavy users about 295. Six people (0.2 percent of the population) indicated they used more than 1,000 B-hours of computing time per year. These "super-heavy" users averaged 2,390 B-hours per year apiece. Further, for any one research area, the percentage of light users is about 67, the percentage of medium users about 27, and the percentage of heavy users about 5.

As consumers of computing power, all subareas of the life sciences appear much the same. The deviations among areas do not appear to have any meaning.* Notwithstanding the uniformity of the distribution, there is a large variation in the amount of computing, depending on the exact behavior of the few heavy users. However, because these were truly very few, and subject to a large sampling bias, their exact values are not meaningful.

Hence the data suggest that *all research areas of the life sciences are engaged in computing.* There is no specialized subarea that is the "computing part" of the life sciences. Surely this reflects the considerable general advancement of the life sciences in making use of this major tool of modern research.

The Growth of Computer Usage

The movement into the use of the computer has been rapid and recent. Figure 44A shows the percentages of our sample that had been using the

---

* More precisely, the deviations appear to be due entirely to sampling variation. The deviations from the mean for each area are completely uncorrelated at the four levels of use. Furthermore, much of the seeming scatter results from the data for research areas involving relatively small numbers of scientists, so unusual behavior by a few scientists seems to create a large deviation.

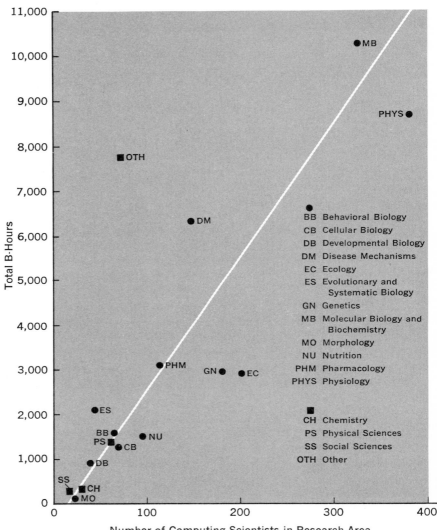

FIGURE 43   Number of computing scientists versus hours of computing, by field. (Source: Survey of Individual Life Scientists, National Academy of Sciences Committee on Research in the Life Sciences.)

computer for various lengths of time; five years or less, six to ten years, etc. Seventy-seven percent of current users had begun their use of computers within the last five years, and 19 percent have been computing six to ten years. Thus four times as many biologists began computing within the most recent five-year period as had begun during the previous five-year interval. When due account is taken of the steep curve, this is an entry rate of almost 20 percent per year. This certainly cannot continue. Eventually the rate must approach the growth rate of the life scientist population (currently about 8 percent). By now (early 1970), the number of life scientists computing is already between 40 and 50 percent, instead of the 30 percent reported by our sample in 1966–1967. At such time as the percentage approaches 50, it is likely that the rate of growth will have slowed markedly. Figure 44B shows the years-of-use curve for the two areas in which approximately 50 percent of the computing scientists commenced their computing within the last five years. These are genetics and nutrition, which have identical year-of-use distributions. Their curves have already begun to "bend over," and their entry rate is about 10 percent per year.

Table 58 illustrates that, at least for our sample of biologists, the assertion that "computing is a young man's game" is not so. An examination of the age distributions of all biologists whether they compute or not, and of those who compute shows that they are essentially the same. Also, all ages, whether light, medium, or heavy users, do their equivalent share of computing. Furthermore, as seen in Table 59, the distribution of computing effort (percentage B-hours) for biologists in each age group is proportionate to the number of individuals in that group. This proportionality holds for the three types of users—light, medium, and heavy. Table 60

TABLE 58 Age Distribution of All Biologists versus Computing Biologists

AGE GROUP (YEARS)	PERCENTAGE	
	All Biologists	Biologists Who Compute
ALL AGES	100	100
< 30	4	3
30–39	39	39
40–49	37	38
50–59	16	17
> 60	4	3

Source: Survey of Individual Life Scientists, National Academy of Sciences Committee on Research in the Life Sciences.

TABLE 59 Age Distribution of Computing Biologists versus Extent of Computing

AGE GROUP (YEARS)	PERCENTAGE DISTRIBUTION OF COMPUTING TIME IN B-HOURS			
	All Biologists Who Compute	Light Users (<10 hours)	Medium Users (10–99 hours)	Heavy Users (>100 hours)
ALL AGES	100	100	100	100
< 30	1	4	3	1
30–39	37	41	36	37
40–49	41	39	43	40
50–59	16	14	17	15
>60	5	2	1	7

Source: Survey of Individual Life Scientists, National Academy of Sciences Committee on Research in the Life Sciences.

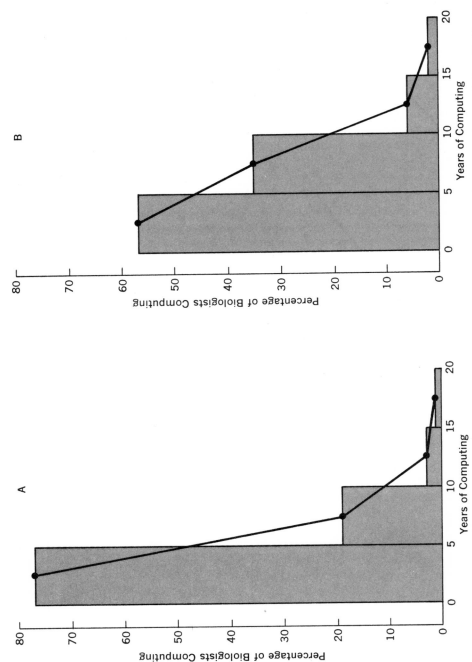

FIGURE 44 Distribution of computing scientists by years of computer use. A, Total biological computing population; B, Computing population of genetics and nutrition. (Source: Survey of Individual Life Scientists, National Academy of Sciences Committee on Research in the Life Sciences.)

TABLE 60   Shift of Percentage Distribution of Computing (Percent B-Hours) with Years of Computing Experience

YEARS OF COMPUTING EXPERIENCE	LIGHT USERS	MEDIUM USERS	HEAVY USERS
TOTAL	**100**	**100**	**100**
< 5	79	72	63
6–10	18	23	31
11–15	2	4	5
16–20	1	1	1

Source: Survey of Individual Life Scientists, National Academy of Sciences Committee on Research in the Life Sciences.

shows that as they become more experienced, computer users tend to shift into the higher use categories, a trend quite in keeping with expectations.*

However, research areas do differ in the percentage of their active scientists who compute (Figure 45). Here percentage participation for each field is plotted horizontally, and the percentage of these computing scientists who have used computers five years or less is plotted vertically. Thus genetics has the highest participation, with 49 percent of the field computing; and morphology has the lowest participation, with 18 percent. Hence, variation between different biological fields is considerable.† Furthermore, a field with very little participation should contain a high proportion of individuals just making the acquaintance of the computer. The regular decreasing sequence of Figure 45 clearly shows such a relationship. Extrapolation of the data to 100 percent participation indicates that about 30 percent of the users commenced computing within the last five years. This is equivalent to an approximate annual growth rate for computing participation of 6 percent. Such a growth rate is in tolerable agreement with the growth rate of biology as a whole. Hence it is plausible (though hardly conclusive from the evidence) to view all subareas as migrating down the curve of Figure 45, reinforcing the impression that all areas of biology are assimilating the computer. Their rate of assimilation differs only because of the point in time in which they commenced computing; their rate depends upon their position on the curve.

---

* However, the magnitude of this effect is not enough to help predict the amount of computing used in a research area by knowing its age distribution, even though the average number of B-hours per scientist is about 25 for a new user (less than five years) and about 50 for an old user (greater than five years).

† Again, there is no correlation with how much computing is done per scientist.

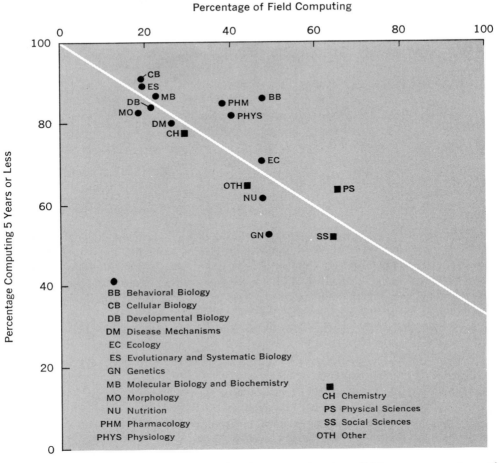

FIGURE 45   Percentage of biologists computing in a field versus percentage of recent entries to computing within that field. (Source: Survey of Individual Life Scientists, National Academy of Sciences Committee on Research in the Life Sciences.)

Institutional Arrangements for Computer Use

It remains only to look briefly at institutional arrangements. Here again, the main impression, as seen in Table 61, is diversity; all arrangements are used heavily for multiple purposes. The table shows the percentages of A, B, C, and D type computer hours used by researchers using three major types of computation facilities: (1) laboratories that own their own com-

TABLE 61   Percentage of Computing Done Using Different Types of Computer Facilities

SOURCE OF COMPUTER	COMPUTER SIZE					EXTENT OF USE		
	All Sizes	A	B	C	D	Light	Medium	Heavy
ALL SOURCES	100	100	100	100	100	100	100	100
Investigator's Laboratory	31	23	47	29	46	20	25	36
Life Sciences Computing Center	27	31	18	26	28	17	22	
Other (Including University Computing Center)	42	46	35	45	26	63	53	34

Source: Survey of Individual Life Scientists, National Academy of Sciences Committee on Research in the Life Sciences.

puters, (2) laboratories that use a life sciences computing center, and (3) laboratories that use some other computation center, usually a university center. Table 61 also shows corresponding percentages for light, medium, and heavy users. In this table it is possible to verify some facts that one might have expected. Thus, most type D computers belong to the scientists' own laboratories; but much computing by light users is done at university computation centers. The overall impression is one of multiple arrangements.

Funding of Computer Use

Table 62 shows the sources of funding for computing in the various research areas of biology. Overall, 42 percent of the support came from research grants to individual scientists; 29 percent from federal funds specifically allocated for life sciences computing, 9 percent from non-life-sciences funds (e.g., a university's own computing budget), 11 percent from other funds (e.g., state life sciences funds), and 9 percent from support whose source was unknown to the individual scientist. Given that about 75 percent of research grants are also provided by the federal government and that some fraction of the non-life-sciences funds and funds of unknown source undoubtedly is funded from National Science Foundation computer-facility grants to universities, it is clear that the federal government supports the great bulk of computation. Furthermore, Table 62 reveals that different areas of biology meet their computing costs in different ways. For

TABLE 62   Computing Costs Funded by Different Sources for Selected Research Areas

RESEARCH AREA	COMPUTING COSTS, BY SOURCE (PERCENTAGE OF B-HOURS)					
	All Sources	Own Research Grant	Federal Funds Life Sciences Computing Center	Non-Life-Sciences Funds	Other Funds	Funds of Unknown Source
AVERAGE ALL BIOLOGY	**100**	**42**	**29**	**9**	**11**	**9**
Behavioral Biology	100	49	39	5	6	1
Cellular Biology	100	72	13	2	11	2
Developmental Biology	100	28	13	1	57	1
Disease Mechanisms	100	31	58	2	6	3
Ecology	100	40	19	11	3	26
Evolutionary and Systematic Biology	100	34	4	42	6	14
Genetics	100	50	26	1	12	8
Molecular Biology and Biochemistry	100	38	30	7	14	8
Morphology	100	51	14	15	16	4
Nutrition	100	32	12	12	19	24
Pharmacology	100	68	11	1	11	9
Physiology	100	44	24	5	9	13

Source: Survey of Individual Life Scientists, National Academy of Sciences Committee on Research in the Life Sciences.

example, almost all the computing of the cellular biologists is supported from research grants, while a much smaller fraction is thus funded in the areas of ecology, and evolutionary and systematic biology. Biologists in the latter two areas reported a large percentage of unknown support, which probably implies the receipt of "free" computation at a university computation center.

CONCLUSIONS AND RECOMMENDATIONS

When this picture of computing in the life sciences in 1967 is combined with the basic lessons about the evolution of the computer industry and of computer use in any field, the following conclusions and recommendations emerge.

1. *Basic continuing support.* Computers are now an integral part of the life sciences, through all its subdomains. Computing costs must be

added to space, personnel, and instruments as basic continuing support items. As with these other resources, computing time should be awarded to the individual investigator or research project according to the merits of the research, through the regular funding channels. If funds permit, computing will continue to increase during the coming years as a fairly predictable increased percentage of life scientists (now probably of the order of 10–15 percent new scientists per year) come to use computers, and there is an unknown increase in the number of the already heavy users. The increased use by the new scientists will slow toward the base growth rate of the field in about five years, but the proportion of medium and heavy users will continue to increase for a long time. Offsetting the expense of this growth, while at the same time tending to increase its rate, is the decreasing cost of computation.

The emphasis on funding computation in relation to the quality and significance of the research to which it will contribute, is not meant to ignore the need for stability that large facilities require. Indeed the need for stability constitutes the main pressure for block funding of facilities. However, the experience with block funding of computer facilities (e.g., at the National Institutes of Health) makes it clear that one must move as rapidly as possible to associate with each research effort the cost of its computing and then assign to investigators both the freedom and the responsibility to obtain computing funds appropriate to their research.

2. *Multiplicity of facilities and decentralization of control.* The extreme diversity of uses of computers implies need for an equal diversity of computing facilities. As we have said, the generality of the computer is that it can be shaped (configured and programmed) for almost any type of information-processing task, but it cannot simultaneously be all things to all users. In fact, all computer facilities become highly specialized, the large university computing center being no exception. A part of the history of computing in the life sciences is written in the struggle of research groups to obtain computers of their own, which can be shaped to their own uses—as laboratory instruments or as data-retrieval and display systems, for example.

A second reason for decentralization of the selection, development, and control of computers is that only through the parallel attempt of many life scientists to adapt the computer to their needs will the computer play its appropriate role in life science research. If the technical development is isolated in a relatively few centralized centers, this assimilation will be substantially retarded.

3. *Development.* The life sciences can rely on the computer industry to continue to produce cheaper, more powerful technology. They cannot, however, look to anyone else for their development—that is, to assimilate

the computer into work on life science problems. Development is expensive and unsatisfactory, in that initial goals are seldom met. It is frustrating, in that frontier projects always have troubles that later projects (often exploiting better and cheaper technology) seem to avoid, making it appear that originally the wrong approach was taken. But this is characteristic of development, and is the price of getting the computer fully assimilated into particular fields.

There are no special "computer areas" in the life sciences. All subareas are assimilating the computer, though in somewhat different ways and, currently, at different rates. Every subarea has its unique forms of symbolic processing, which, as it is successfully developed, can make large differences in the progress of research in the subarea. For example: image processing, generally in microscopy, but also in ecology; large-scale simulations in ecology; laboratory computerized instruments in physiology and biochemistry; taxonomic retrieval systems in systematics. Almost all the heavy users in our census would reveal somewhat special developments.

The projects in the life sciences noted above have analogues in other fields, which may short-circuit the research effort but not the development effort. However, life scientists have some symbolic functions in common with all other scientists and all other technologically oriented professionals. The best example is small numerical calculations analogous to engineering calculations. The computer field can be relied on for development relating to these—for instance, in the multitude of time-consuming mathematical calculations necessary to untangle and to understand complex biochemical reactions. Another example is the development of time-sharing, which is of immense importance in getting the computer widely assimilated.

Massive development efforts repeatedly produce wedges that open up new technology. The life sciences must support such efforts for themselves. The amounts of money spent on such projects may often seem out of proportion to what the same amounts, distributed otherwise, could yield, but this is an illusion, for there is no other way to gain entry into new areas other than by paying the apparently "excessive" costs of large development efforts.

CHAPTER EIGHT

COMMUNICATION
IN THE
LIFE SCIENCES

In 1866 an Austrian monk named Gregor Mendel published, in an obscure periodical, a classic paper reporting his experiments in crossbreeding plants—experiments that delineated the basic laws of heredity. Mendel's paper was not just a suggestion; it was a rather convincing quantitative description of the operation of heredity, based on long experimentation and critical analysis. Because Mendel's concepts ran counter to those held by the knowledgeable biologists who were aware of his papers, and because the information channels were inadequate to direct Mendel's work to the attention of other scientists, it lay relatively unknown for almost four decades until its resurrection by Correns, DeVries, and Tschermak, an event that marked the beginning of the full-fledged study of genetics. The invaluable technique of absorption chromatography, by which rather similar compounds can be separated and purified, is another example of a scientific advance that lay dormant for years, only to be rediscovered later. Although there may be other such instances, the informal communication network now operating markedly decreases the likelihood of major losses of this kind. Such losses, however, could be brought to a minimum by further improvement of formal, appropriately designed information systems.

A modern information system must be designed to preclude the waste of time inherent in discovering the same thing twice, while managing the mushrooming volume of information published in journals both prominent and obscure. Investigators in all fields face the critical challenge of coping with the waves of information that threaten to swamp them, and they increasingly recognize their inability to scan all the reports directly related to their work, much less those of tangential interest. Yet only 15 years ago the situation was within bounds.

The number and variety of information services ostensibly designed to meet the needs of an increasingly diversified and compartmented clientele of biologists have grown dramatically in the last two decades. The total investment in dollars, trained manpower, and facilities required to implement these services now consumes an important fraction of the total investment in biological research and education.

Their rate of growth, their size, and their current level of investment make it desirable to study the cost effectiveness of biological information services. Such evaluation requires an in-depth review and appraisal of both the nature and the objective of these services in relation to the information requirements of today's biologists. Information-exchange organizations are confronted with an acute need to discern and adapt to the changing information requirements of a scientific community that presently appears to be in a state of flux. Alignment within this community is passing through a period of transition that will probably lead to a rearrangement and amalgamation of scientific disciplines and the structure of their organizational concomitants, leading to new information requirements better suited to the objectives of consolidated groupings as they evolve.

For example, one can now sense the beginnings of a spontaneous movement of the presently polarized factions of molecular biologists and "whole-animal" biologists toward cooperation and integration of their disciplines. The rapprochement of systems-oriented ecologists and ecologically oriented systematists is also much closer to realization. Meanwhile, new orientations to the use of biological understanding for dealing with problems of disease, toxicology, environmental health, pollution, and the general quality of the environment make demands on the information system quite different from those of only a few years ago.

Are the current organizational patterns of biological information systems such that they facilitate this reunion of disciplines so urgently required to break new ground? Or, by reason of investments in facilities and equipment and success in achieving their present modes of operation, have they generated a strong interest in preserving the status quo, thus constraining efforts to promote both interdisciplinary and multidisciplinary understanding and progress?

Federal agencies support both research and information services; however, this support occurs by independent mechanisms with no direct coupling and feedback. The present chapter will not offer specific recommendations for overhaul of the biological information services network, but will examine it from the standpoint of working biologists and offer some guiding philosophy.

SPECIAL PROBLEMS IN HANDLING BIOLOGICAL INFORMATION

While biologists, physicists, and chemists face similar problems in handling rapidly expanding information, and while all disciplines recognize the need for structured systems of information handling, storage, and retrieval, special needs arise within each discipline. For biologists, the overwhelming volume of published material is a special problem. Of the 26,000 distinct scientific and technical journals published annually, the life sciences claim no less than 50 percent (20 percent for agriculture, 13 percent for medical sciences, 4 percent for basic life sciences, and 10 percent for technology), or 13,000 serial publications. Not only are individual scientists obviously unable to deal with this plethora; libraries and abstracting services are inundated by it. It is important, therefore, to ask how much of this volume is critical. *Biological Abstracts,** in 1968, abstracted 7,400 periodicals, yet most of these are unlikely to publish truly significant new findings that will materially advance the progress of science. Indeed, it is possible to identify about 1,000 journals in which more than 90 percent of the truly significant original work in biology now appears.

Another special problem in biological information springs from the diversity of subject matter and of experimental approaches to understanding the living world. Thus, biology encompasses explorations of subcellular organization, of organisms from viruses and bacteria through the primates, of highly complex communities and ecosystems, from the kinetics of a chemical reaction to the behavior of populations. This diversity is reflected in the 20 major program categories the National Science Foundation finds necessary for biology, compared with four for chemistry and 10 for physics. The information needs of the individual working scientist, whose interests lie mainly within one of these 20 categories, are largely satisfied within 5 to 10 percent of the 1,000 journals mentioned above. For the rest, he is

* *Biological Abstracts.* BioSciences Information Service of Biological Abstracts. Philadelphia, Pennsylvania. 1926.

generally content with broad reviews and secondary information services to guide him within the remaining primary literature.

USERS OF BIOLOGICAL INFORMATION

Information system designers must found their blueprints on a clear understanding of the needs of their clients. Those who use biological or other types of scientific information can be placed, generally, in one of five categories:

1. *The biologist–scholar,* an investigator or university teacher who requires original articles or scholarly reviews and who should have access to such papers almost as soon as they are published.

2. *The practitioner,* who needs information pertinent to his particular ways of applying biological knowledge. For his purposes, digests of the results of research, rather than all the details of the process, usually suffice; although he needs these data with minimal delay, only in an unusual emergency does he demand rapid retrieval.

3. *The elementary or secondary school teacher,* who makes use of consolidated texts and teacher aids and can tolerate considerable delay.

4. *The policy-maker or administrator,* who requires generalized information about biology, rather than specific biological facts, from concise analyses such as those published in such general-essay magazines as *Science, Nature, Endeavour, Scientific American,* and the *New Scientist.*

5. *The citizen,* who needs information pertinent to major subjects, including conservation, pollution, population, and public health, and for such pragmatic matters as evaluation of advertising claims for new drugs. The citizen, like the teacher and policy-maker, seldom needs biological information urgently, though for him also extremely long delays are undesirable. The requirements of those who write for the citizen in the mass media should be assessed as carefully as those who write for the professional, and a high standard of accuracy in channels of mass communication should be assured.

INFORMAL INFORMATION TRANSFER

Those who would plan information systems for use by working scientists and practitioners should be aware of the myriad informal channels of information exchange.

Meetings

Biologists gather at meetings of many types and sizes, at which they discuss details of techniques, experiments that failed or should be tried, and hypotheses far too tenuous for print but nevertheless enormously useful to the working scientist.

In addition to providing opportunity for face-to-face conversations, meetings are structured to provide platforms for the presentation of papers —often hundreds, sometimes even thousands, at a single scientific congress—frequently supplemented by discussion from the floor. Most scientists have hardened their views against publication of the complete proceedings of such meetings, a process that may well take a year or two, during which time complete accounts of the best, most complete papers may be formally published in edited journals. Most of the rest might better be left out of print, and spontaneous comments from the floor seldom merit immortalization. It seems sensible to publish in full only those special symposia at which a few carefully chosen speakers review, as they see it, a delineated research area. For the rest, the best approach is succinct reviews, written promptly by leading participants and published in magazines such as *Science*.

To the investigator in the life sciences, attendance at scientific meetings has become an urgent necessity. These provide opportunities to learn of recent findings, to hear preliminary reports, to listen to summaries by distinguished experts of areas of research in which one is tangentially interested but whose literature there is not sufficient time to follow. Smaller meetings provide the settings for engaging in critical examination of much narrower research areas. Once relatively rare, small specialized meetings are now sponsored by a variety of organizations. When convened early in the history of a newly emerging research area, such meetings have frequently given direction for several years afterward to the entire development of the fields under discussion.

Exhibits of new instrumentation by commercial manufacturers have come to be an increasingly important aspect of large scientific meetings, affording the scientist an opportunity to compare the features of competing instruments, to query the instrument engineers and designers, and, frequently, to learn of the existence of new instruments.

The totality of such meetings constitutes a large enterprise. Of 10,325 research biologists, only 1,096 did not attend a meeting in 1966; 30 percent attended one meeting and half attended two or more (Table 63).

International meetings represent a special subset. They offer the scientist all the advantages of domestic meetings and also furnish the relatively rare opportunity to discuss matters of common interest (usually in English, the international scientific *"lingua franca"*) with otherwise rarely encountered foreign scientists. International meetings produce an additional meaningful bonus, the opportunity for the American scientist to serve as a goodwill ambassador to the citizens of foreign nations. As shown in Table 63, 16 percent of our respondents attended one such meeting in 1966 and a few attended two or more.

One other aspect of the face-to-face communications system warrants mention. Perhaps the most intensive such communications experience is the remarkable program of visiting seminar speakers that has grown up in the last two decades. In large measure supported by the National Institutes of Health training grants program, almost every qualified academic department brings to its seminar program 5, 10, or more visiting speakers annually. Each is invited because of interest in his own contributions, which are presented in detail under relatively informal circumstances. The entire local community—faculty, fellows, and students—participates in the ensuing, sometimes merciless, discussion. In turn, during his visit, the seminar speaker has opportunity to learn about recent developments in the laboratories of the host institution and brings this scientific "gossip" back to his colleagues at home. The sum of such activities has become an invaluable feature of current scientific life and is as vigorous in many federal and industrial laboratories as in academic departments. *Accordingly, it is urged that all organizations that support research remain sympathetic to reasonable requests for funds in support of travel that permits the applicant both to pursue his own research more effectively and to communicate its results to others.*

TABLE 63   Life Scientists Attending Meetings in 1966

	DOMESTIC MEETINGS		FOREIGN MEETINGS	
	No.	%	No.	%
TOTAL RESPONDENTS	**10,324**	*100*	**10,324**	*100*
Meetings Attended per Man				
0	1,096	*11*	8,330	*80*
1	3,062	*30*	1,631	*16*
2	3,082	*30*	279	*3*
3	1,647	*16*	59	*0.5*
4 or more	1,437	*14*	25	*0.2*

Source: Survey of Individual Life Scientists, National Academy of Sciences Committee on Research in the Life Sciences.

Informal Publications and Correspondence

Between the forum of a meeting and publication in primary journals lies a spectrum of informal routes of communication that includes correspondence and manuscript exchange among scientists, "invisible colleges," publication of newsletters, technical reports and book reviews, and listings of research in progress. Attempts to formalize manuscript exchange before editorial acceptance have met with considerable opposition, and unreviewed papers are probably best left out of formal information systems. The criterion recently established for primary publication by the Council of Biological Editors offers a realistic base line in deciding whether or not to include nonjournal materials in a storage and retrieval system. Citable publications are those in which sufficient details are given to enable peer scientists to assess the probable soundness of the results, to repeat the experiments if necessary, and to retrace the lines of reasoning that led to the conclusions.

Book reviews may be valuable if they contain new ideas or offer a critical analysis of the subject. An instance of how the content of book reviews can be extracted is provided by the annual *Mental Health Book Review Index,** which includes reviews in some relatively remote fields that are abstracted, however briefly, and stored in the information system.

Listings of research in progress serve the purpose of alerting scientists to research projects under way, but not yet published, that may overlap, duplicate, or even conflict with their own. Such listings are provided in special fields, for example, through a collaboration between the American Society of Plant Taxonomists and the International Organization of Plant Biosystematists, and in a more general way through the federally sponsored Scientific Information Exchange of the Smithsonian Institution. The latter records both federally and privately sponsored research projects actually in progress and covers much of primary life sciences research, distributing abstracts upon request to scientists, policy-makers, administrators, and reporters. However, while invaluable to federal science administrators, this service is rarely used by working scientists.

PRIMARY PUBLICATION

Advent of the computer age augurs a day when scientific publications will be recorded on magnetic tape; however, for the next several decades the

* *Mental Health Book Review Index: An Annual Bibliography of Books and Book Reviews in the Behavioral Sciences.* Council on Research in Bibliography, Inc., Flushing, New York. 1956.

printed journal will remain the most economical and common form of publication. Nevertheless, contemplated solutions to the current information deluge should take the future into consideration; a log jam in a computer network is no more tolerable than one in printed material.

The Journal

Not only has the number of journals in biological sciences (13,000) outpaced the ability of scientists and teachers to read them, but also the number of pages per journal has expanded. Most scientists are frustratingly aware that, however diligently they try to keep up with the literature, they can never read all that is potentially valuable or germane to their own interests. Although they may be concerned, few bench scientists are actually alarmed by this situation; most consider that they remain reasonably *au courant* with the leading edges of their own disciplines through a combination of regular reading, attendance at meetings, seminars in their own institutions, and the informal operation of "invisible colleges." This is particularly true of those whose research can be confined to a relatively narrow specialty, be it renal disease, virus structure, or resistance to smut infection. For the biological generalist seeking new insights into the forces directing evolution or behavioral adaptations to the environment, for example, the task is overwhelming, and no available information system meets his needs.

Some quantitative aspects of this enterprise may be illuminating. In 1966, of our sample of respondents, 8,801, or 72.4 percent, published one or more full-length articles describing original research in journals with national or international circulation. A total of 24,573 articles were reported, for a mean annual publication rate of 2.0 (including those who failed to publish), with 7 percent of the group publishing more than five articles a year. Eighty-eight percent of our respondents contributed at least one publication of the types listed in Table 64. Meanwhile, they also contributed 1,134 major reviews, 489 books and monographs, and 2,416 chapters that appeared in books edited by others! Figure 46 shows the proportion of biologists producing the various types of publications.

Improvement of the quality of primary journals, with the reader in mind, is a paramount need. Such improvement demands changes in the editorial process, a rigorous look at standards, considerations of cost, and some commitment to the education of scientists as authors as well as investigators. Many scientists are notoriously poor writers. A modest program of instruction in scientific writing during graduate training could lessen much of the future editorial burden. Some faculty time should be invested

TABLE 64 Publications Reported in 1966 by 12,364 Life Scientists

TYPE OF PUBLICATION	INVESTIGATORS REPORTING ONE OR MORE	NUMBER PUBLISHED	AVERAGE [a] NUMBER PER RESPONDENT
TOTAL, ALL TYPES	**10,727**	**50,858**	**4.1**
Full-Length Research Articles	8,801	24,573	2.0
In-House Publications	2,527	7,684	0.6
Books and Monographs	442	489	<0.1
Chapters in Books	1,710	2,416	0.2
Major Reviews	887	1,134	0.1
Abstracts of Original Research	4,668	9,674	0.8
Other Publications	2,028	4,888	0.4

[a] Average numbers based on 12,364 respondents to individual questionnaire.
Source: Survey of Individual Life Scientists, National Academy of Sciences Committee on Research in the Life Sciences.

in teaching a manner of data presentation and analysis that will be serviceable for a lifetime.

In selecting manuscripts for publication, editorial boards necessarily engage in value judgments. Publication priority generally goes to those manuscripts presenting significant advances in the understanding of nature. Methodology that could assist others in making important advances and work that is competent and fills gaps in understanding or knowledge but does not pave the way for new theoretical or practical advances are assigned lower priorities. The editor's task is to decline work that is duplicative, incompetent, incorrect, or totally pedestrian. *This set of editorial judgments is the backbone of the scientific information system; it protects the inexpert reader and those who provide research funds while assuring scientists in the field that published work has been performed with competence and that the findings are probably reliable.*

In almost every scientific subfield there is a hierarchy of journals that reflects the relative quality of published papers. Although it does not exist overtly, this hierarchy is known to all sophisticated scientists within the field. Occasionally, a paper is consecutively submitted to journals of diminishing quality until it finds acceptance.

Some of the deficiencies of journals can be attributed to the fact that editors usually serve on a voluntary basis, their editing competing for time with their own scientific work. Failure to provide editors with adequate assistance in handling the technical details of publishing a journal—usually because of the cost of such assistance—is a false economy. Such assistance

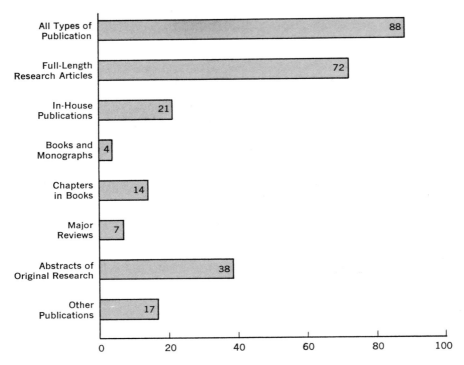

FIGURE 46   Percentage distribution of 12,364 respondents reporting various types of publication. (Source: Survey of Individual Life Scientists, National Academy of Sciences Committee on Research in the Life Sciences.)

could not only lead to improved quality but could also eliminate some of the time lag between submission of a manuscript and its publication.

Still unrealized is an appropriate mechanism for defraying the continually increasing costs of journal publication. These costs have risen drastically in the last two decades and, if built into subscription rates, would almost eliminate any but institutional subscribers, few of whom have the resources to purchase, bind, and store more than a small fraction of the 13,000 biological journals. The alternatives are to subsidize journals, either directly *en bloc,* or, as is frequently the case, by an assessment, per page, levied against authors. Since federal granting agencies have agreed that publication is intrinsic to research projects, these charges are legitimately defrayed from research grants, thus lessening the burden upon institutional and individual subscribers while serving as an incentive to investigator–authors to prepare tight, pithy manuscripts. In view of this policy, we find it absurdly contradictory that these agencies, as another matter of policy, consistently negotiate down requests for such page charges while awarding the research funds that support the research to be reported.

For many years, sale of reprints to authors has been a major source of revenue to journals. Again, rising costs, passed on to the authors, have become an inordinate drain on research funds. In the future, however, as photocopying equipment becomes more generally available, this practice should subside and the drain on authors' research funds should diminish, but the financial plight of the journals will become still more acute. *Federal research-supporting agencies should give serious consideration to these problems while there is yet time and before journals either become bankrupt or price themselves out of business.*

New Forms of Primary Publication

Current experiments in handling the overwhelming volume and costs of published material include the use of microcards, microfilm, and microfiche, selling individual journal articles to scientists who may have no use for entire volumes, and selling abstracts and indexes separately from journals. (The journal, *Wildlife Disease,** for example, which must print long names of species or geographic locations that may be of interest only to select readers, now publishes exclusively in microfiche.) As the number of papers in a single issue of a journal that are of special interest to a given reader diminishes, consideration is being given to the provision of only those few he wants, although complete journal issues would continue to be bound for libraries. An attractive alternative is publication of volumes containing only summary abstracts, as tried by the new publication, *Communications in Behavioral Biology*†; after scanning the abstracts, readers order complete texts of the articles of interest to them.

Future Forms of Primary Publication

Rapid development of computer technology and the prospect that scientists eventually may have computer consoles on their desks suggests that the days of printed journals are numbered. There is no question that electrical

---

* *Wildlife Disease.* Wildlife Disease Association, Ames, Iowa. 1965.
† *Communications in Behavioral Biology.* Academic Press, Inc., New York, New York. 1968.

transmission of information will be faster than the U.S. Mail, but the cost, at least for the foreseeable future, is high. Computer networks being planned at Project MAC (Machine-Aided Cognition) and EDUCOM (Interuniversity Communications Council) may take 10 years to construct in such a way as to place one or two terminals in each major research institution in the United States, another 10 to provide terminals for small groups of scientists, and another 10 to link with other continents. *But even when such a network is complete, the role of editors and reviewers will remain unchanged; indiscriminate release of unedited reports to a computer network could well be even more disastrous than indiscriminate publishing would be today.*

The International Literature

Science is an international venture. Each piece of information, regardless of where it is learned, is fitted into the total intellectual framework. The volume of the biological literature and its range of subject matter are such that, without some qualitative judgments, a simple count of the number of primary journals or the number of papers is an insufficient criterion of the contribution of the life scientists of a given nation to the development of the science. To make some assessment of the American contribution to the world literature, the Biological Sciences Communication Project of the George Washington University was commissioned to undertake a limited study, some of the results of which are summarized in Tables 65 through 68.

A group of 3,100 journals was identified, and the frequency with which publications are cited therein was used as a criterion of significance. Of these, 2,377 are published in 15 nations, as shown in Table 65. The United States publishes about one fourth of the world total of these journals.

TABLE 65   Number of Primary Journals in the Life Sciences Published in the 15 Leading Nations

United States	797	Netherlands	67
Japan	284	India	65
France	186	Spain	62
Italy	165	Norway	59
U.S.S.R.	150	Argentina	54
United Kingdom	143	Switzerland	53
West Germany	143	Brazil	52
Canada	97		

Source: Data from the Biological Sciences Communication Project of the George Washington University, Washington, D.C.

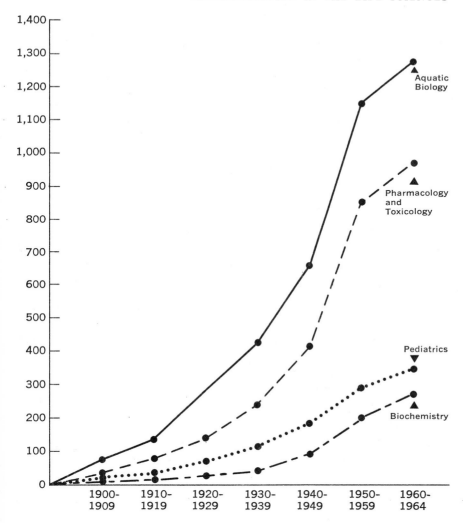

FIGURE 47 Comparative growth rates of serials in the fields of aquatic biology, biochemistry, pharmacology/toxicology, and pediatrics, ten-year periods, 1900–1964. (Source: Biological Sciences Communication Project, George Washington University, Washington, D.C.)

Spot checks were made in four disciplinary areas: aquatic biology, pediatrics, biochemistry, and pharmacology and toxicology. Figure 47 shows the comparative growth rates of the numbers of serial publications in these four areas from 1900 to 1964; this figure does not reflect the numbers of actual papers published, nor does it include any qualitative judgments. Some light on the forces behind the tremendous growth is provided by Table 66, which compares the sponsorship and/or ownership of journals in the four areas. Twenty-eight distinct patterns of sponsorship were found, of which only four are considered in this table. Whereas scientific societies predominate in sponsoring biochemical and pediatric journals, government agencies are the chief sponsors of journals devoted to aspects of aquatic biology, and industrial organizations contribute heavily to sponsorship of serials concerned with pharmacology and toxicology.

The survey committee selected three leading journals concerned with biochemistry and molecular biology, pediatrics, and ecology and population biology, published in each of the following countries: the United States, England, France, West Germany, Japan, and the U.S.S.R. The bibliographic citations reported in all issues of these journals published in 1966 were then examined. The bibliographies of American journals were inspected for references to journals published in the United States and in the other five countries (Table 67); the bibliographies cited in the foreign journals were examined for references to American work (Table 68).

In this set of selected foreign journals, citations of American references constituted one third of the bibliographies of all papers concerned with biochemistry and molecular biology and pediatrics, whereas American papers contributed only one sixth of the bibliographies of articles concerned with ecology and population biology. Figure 48 shows that the pattern of citation did not vary significantly from country to country.

American biochemists and pediatricians found it appropriate to cite publications from the other five nations decidedly less frequently than the

TABLE 66   Sources of Sponsorship of Journals in Selected Fields

SPONSOR	BIOCHEMISTRY		PEDIATRICS		AQUATIC BIOLOGY		PHARMACOLOGY	
	No.	%	No.	%	No.	%	No.	%
TOTALS	**177**	*100.0*	**194**	*100.0*	**696**	*100.0*	**829**	*100.0*
Societies	65	*36.7*	92	*47.4*	196	*28.2*	348	*42.0*
Government	18	*10.2*	18	*9.3*	450	*64.7*	53	*6.4*
Commercial Publisher	83	*46.9*	71	*36.6*	49	*7.0*	258	*31.1*
Relevant Industry	11	*6.2*	13	*6.7*	1	*0.1*	170	*20.5*

Source: Data from the Biological Sciences Communication Project of the George Washington University, Washington, D.C.

TABLE 67   Reference Citations in Selected U.S. Publications to Foreign Journals in 1966 in Selected Fields

	BIOCHEMISTRY AND MOLECULAR BIOLOGY		PEDIATRICS		ECOLOGY AND HYDROLOGY	
	No.	%	No.	%	No.	%
TOTAL	52,458	100.0	9,459	100.0	5,281	100.0
Total Five Selected						
Countries	6,455	12.3	1,262	13.3	808	15.3
United Kingdom	4,839	9.2	1,032	10.9	468	8.9
France	251	0.5	37	0.4	156	3.0
West Germany	909	1.7	174	1.8	97	1.8
Japan	372	0.7	13	0.1	49	0.9
U.S.S.R.	84	0.2	6	0.1	38	0.7

Source: Data from the Biological Sciences Communication Project of the George Washington University, Washington, D.C.

foreign scientists cited U.S. publications, whereas the two patterns in ecology were much the same (Figure 48). This figure must be interpreted with some major caveats. While truly reflecting the increasingly self-sustaining nature of American science, it also reflects (a) the painful fact of American incompetence in languages other than English and (b) an increasing tendency among scientists outside the United States to send some of their most significant papers to leading American journals for publication. The latter practice certainly accounts in part for the inordinately low level of citation of French, German, and Japanese journals. The former, however, does not account for the low level of citation of Russian journals, since the leading Russian journals are available in English translation.

In sum, this examination of a segment of the world biological literature confirms the impression gained from many other sources—that American biological science leads the world in the sheer magnitude of the endeavor and produces science of the first caliber, especially when considered in proportion to its quantity and scope.

REVIEW ARTICLES AND DATA COMPILATION

Because of the scale and diversity of the primary literature, there is continual need for comprehensive review articles. At their best, these contain new interpretations and ideas as well as systematic examinations of current

TABLE 68   Reference Citations in Selected Foreign Journals to U.S. Publications in 1966 in Selected Fields

SELECTED COUNTRY	BIOCHEMISTRY AND MOLECULAR BIOLOGY			PEDIATRICS			ECOLOGY AND HYDROLOGY		
	Total Citations	U.S. Citations		Total Citations	U.S. Citations		Total Citations	U.S. Citations	
		Number	% of Total		Number	% of Total		Number	% of Total
TOTAL	**39,280**	**14,164**	*36.1*	**22,109**	**7,108**	*32.2*	**12,777**	**2,246**	*17.6*
United Kingdom	5,832	37.4		3,845	1,424	37.0	3,352	632	18.9
France	6,638	2,304	34.7	4,354	1,575	36.2	925	161	17.4
West Germany	5,614	1,999	35.6	7,733	1,974	25.5	1,659	208	12.5
Japan	6,889	2,603	37.8	6,177	2,135	34.6	714	117	16.4
U.S.S.R.[a]	4,557	1,426	31.3	—	—	—	6,193	1,128	18.2

[a] It proved impossible to identify any Russian journal devoted exclusively to pediatrics.
Source: Data from the Biological Sciences Communication Project of the George Washington University, Washington, D.C.

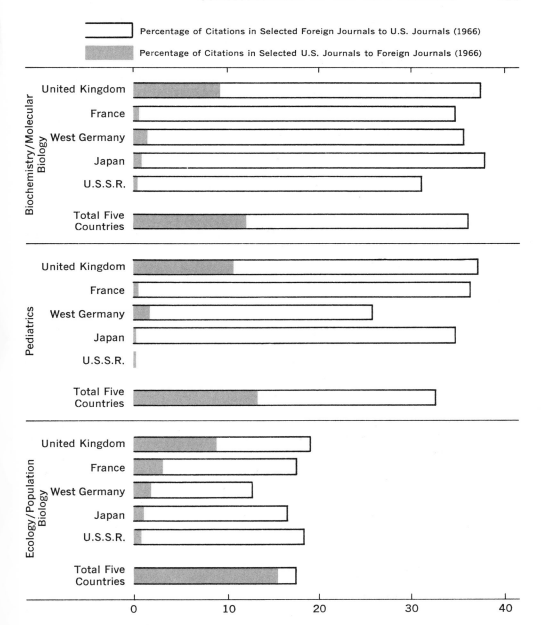

FIGURE 48   Comparison of the frequency of American citation of foreign literature with the frequency of foreign citation of American literature. (Source: Biological Sciences Communication Project, George Washington University, Washington, D.C.)

knowledge in given subjects. Such articles can serve not only active members of the scientific community but also teachers, policy-makers, mass-media writers, and other users of biological information.

The *Quarterly Review of Biology, Physiological Reviews,* the *Annual Review Series,** and the monograph literature fill this urgent need and should, of course, be stored in an information system. Expansion of such activity has been urged as essential. Modest financial reward has been an unsuccessful inducement to scientists at the forefront of research to leave their laboratories and undertake a task that seems repetitious or an interruption of their work. Provision of an abundant supply of reprints has proved more effective persuasion, presumably by satisfying a more fundamental urge. Other effective inducements might include provision of editorial assistants who would undertake the time-consuming chore of searching the literature for relevant papers and perform other mechanical tasks such as checking and arranging long lists of references.

However, it is not clear from the personal experience of our panels that expansion of the biological review literature is imperative. Our study revealed an annual output of monographs, book chapters, and reviews that, at least quantitatively, should be adequate to the task of surveillance of the primary literature and summary and correlation of cogent new developments. To be sure, the quality of reviews could benefit from more critical analyses, but it seems likely that coverage in many fields is adequate to the needs of the user community.

Compilations of biological data, based on careful critical evaluations of the relative accuracy and reliability of data reported by different investigators, are becoming another useful source of information for life scientists. Such data books have long been available to physicists and chemists. Because quantitative data in biology have been relatively few, biologists have taken little part in activities of this sort in the past, but the situation is rapidly changing. The *Handbook of Biochemistry,†* which first appeared in 1968 and which presents data on properties of a great number of biologically important molecules, has proved tremendously useful. The

---

* *Quarterly Review of Biology.* Stony Brook. Foundation, Inc., Stony Brook, New York. 1926.

Physiological Reviews. American Physiological Society, 9650 Rockville Pike, Bethesda, Maryland. 1921.

Annual Review Series. Annual Reviews, Inc., Palo Alto, California. Biological fields already under periodic review by June 1970 include biochemistry, entomology, genetics, medicine, microbiology, pharmacology, physiology, phytopathology, plant physiology, and psychology.

† *Handbook of Biochemistry: Selected Data for Molecular Biology.* H. A. Sober, Chemical Rubber Co., Cleveland, Ohio, 1968.

Atlas of Protein Sequence and Structure, 1967–1968* is likewise an important contribution in this area. The handbook series initiated under the auspices of the National Academy of Sciences and continued by the Federation of American Societies for Experimental Biology now represents an important source. And, traditionally, critical compilations have been a major segment of the taxonomic literature. Inducements, similar to those proposed for authors of review articles, should be offered to encourage scientists to undertake critical compilations.

SECONDARY INFORMATION SERVICES

Secondary services consist of all media, techniques, and activities by which scientists are made aware of and assisted in obtaining access to published information. Traditional abstracts, indexes, and bibliographies fall into this category. So do newer approaches, including programs to alert scientists to recent publications on the basis of "interest profiles" submitted to alerting services, custom-tailored reference services in specialized areas, computerized indexes, question-and-answer services, and preparation of summaries of groups of papers. Proprietary services produce some types of this secondary information.

Efficient operation of on-going programs and development of new ones demand a complex interaction and cooperation among authors, editors, staffs of indexing and abstracting services, and users of those services. For optimum effectiveness, each party must have as full an understanding as possible of the needs and activities of the others, with a well-organized interplay among services catering to various scientific subfields so that the user can be sure of obtaining, with minimum difficulty, as much information as he needs without drowning in a sea of seemingly relevant or closely related undigested references. In the future, as these criteria are met, such secondary information services should grow in importance; for the present, they are relatively little used by the scientific community.

Functions and Desirable Characteristics of a Secondary Service

The abstracting and indexing function—provision of abstracts that can, if necessary, substitute for original documents and classify them conveniently. If all journals required "heading abstracts" at the beginning of

---

* *Atlas of Protein Sequence and Structure*, 1965, 1966, 1967–1968. Editors, R. V. Eck and M. O. Dayhoff, National Biomedical Research Foundation, Silver Spring, Maryland.

each article—abstracts specifically designed to be informative when divorced from the text—readers and abstracting services would profit. While most leading U.S. journals now require such abstracts, only 30 percent of world biological journals demand them, and many are inadequate.

The storage and retrieval function—storage of references to publications in a form suitable for rapid retrieval yet permitting browsing in defined fields of interest. This service is now provided by the tapes of the National Library of Medicine and the National Agricultural Library, the former for some 7,000 journals. *Biological Abstracts* has adopted a computerized indexing service, based on extraction by its staff of relevant information obtained from publication titles. Speed is the most appealing feature of this procedure, though the subtlety and refinement of the conventional index is lost. The process would be much improved if editors or authors supplied amplifying key words, since it is often impossible to include all useful clues in titles of reasonable length.

The alerting or current-awareness function—to announce the appearance of a document of potential value to individuals or groups in specialized areas of research. At present, formidable costs preclude the distribution of selected abstracts on a routine basis to such groups, but the Institute for Scientific Information is developing working programs for routine distribution of selected references (titles plus bibliographic details). However, the personal element will remain overriding. As long as printed journals retain their primacy in scientific communication, the taste and judgment of the individual scientist will continue to determine which papers to read and which to discard. Every scientist has found useful gems of information in surprising places, and has been sorely disappointed by the content of papers with rather grandiosely promising titles. At any rate, structured information systems do not offer a substitute for either the utility or the pleasures of browsing.

SPECIALIZED INFORMATION CENTERS

For some fields of biology, particularly the taxonomic sciences, coverage of the major primary journals, which may yield 85 percent of the significant information, is not enough. In these cases, specialized services should be encouraged to provide access to literature in highly limited journals, with a deliberate attempt to include the more expensive last 10 percent of the information. Centers providing services of this type can flourish only if they are in contact with active research and are staffed in part by active scientists competent to grasp changing aspects of their fields and to provide to their peers critical, in-depth indexing and synthesis of information. Similar considerations apply to other specialized information centers, e.g., in toxicology, vascular disease, and human genetic pedigrees. A single national computer network is, as yet, rather remote, and the interim growth of specialized information centers to meet clearly defined and societally significant needs is strongly encouraged.

LIBRARIES

Libraries are overwhelmed by the abundance of scientific literature. Purchase, bibliographic, and maintenance costs and sheer physical shelf space make it almost impossible for any single facility to house all the available material. The increasing gap between production and acquisition of materials, bibliographic deficiencies, and the mechanical obstacles to sharing resources among libraries are also handicaps.

Improved technology for inexpensive reproduction and dissemination of literature could greatly facilitate the sharing of resources among libraries. The New York Public Library and the National Library of Medicine maintain their reference collections intact and allow no interlibrary loans, to ensure availability of documents. Other libraries have indulged so freely in interlibrary loans that their own shelves are seriously depleted of reference material. Inexpensive copying, coupled with clarification of copyright laws in relation to such reproduction, could solve many problems. Better dissemination of information concerning the availability of bibliographic materials would be quite helpful.

A most useful development would be agreements to assign specific responsibilities to particular libraries, each of which would acquire extensively in its designated areas, organize and publish bibliographies, and provide lending or photocopy services, patently an extension of the specialized information center concept. The national libraries of agriculture and medicine are successful examples of this approach. If major academic libraries concentrated individually on agreed-upon subdivisions of the life sciences (plant physiology, taxonomy, environmental biology, ecology, etc.) while maintaining their general collections, the needs of the scientific community could be much more adequately served than at present.

Most importantly, while new forms of communication are evolving and the nature of the science library is in transition, existing libraries are facing grave financial difficulties. All but a few urgently require funds for construction, for acquisitions, shelving, computer systems, desk consoles for use with microfiche, staff, and related necessities. *We recommend that the three primary federal agencies that must accept responsibility for the welfare of the life sciences—the National Science Foundation, the Department of Health, Education, and Welfare, and the Department of Agriculture—mount significant programs of financial assistance to this network of libraries.* We have not ascertained the appropriate level of such funding, but the needs are acute, and their solutions expensive; at least $25 million per year will be required if these libraries are to continue to serve their various clienteles adequately.

LOOKING FORWARD

A national information network with interlocking federal and private components is slowly evolving. The emerging patterns for physical and chemical sciences rely heavily on the existence of single coordinating organizations, *viz.*, the American Institute of Physics and the American Chemical Society. For the biological sciences, three generalized information services now function as central reference points for the three major subsets of the community: the National Agricultural Library, the National Library of Medicine, and the BioSciences Information Service of *Biological Abstracts*. This seems to be a viable pattern. The diversity of the biological sciences, both in organizational structure and in subject matter, is paralleled by that of the federal system. No single all-embracing information system exists; if it did, it probably could not serve the needs of the community.

Although a national plan for information-handling in biology must be conceived and developed, it is best founded on existing institutions; it must involve both public and private sectors and must be based upon the cooperation of several organizations—small specialized services as well as the three central institutions listed above. Those three institutions will play essential roles as intersects in the information network and as switching points within and between systems, and thus must accept responsibility for serving the scientist, practitioner, policy-maker, and citizen, because no specialized center could give the broad view these information users need. The overall system should be monitored and planned by a continuing group representing the "umbrella" societies and the three major federal foci for biological information.

If the system is to be developed adequately, the biological community must accept its responsibilities and coordinate its efforts. Each biologist must consider his needs, demand action of his society or institution, and, when called upon, contribute to the design of an ultimate functioning system.

CHAPTER NINE

BIOLOGY AND
THE FUTURE
OF MAN

THE NATURE OF MAN

The forces shaping the short-term future of man, perhaps to the turn of this century, are apparent, and the events are in train. The shape of the world in the year 2000 and man's place therein will be determined by the manner in which organized humanity confronts several major challenges. If sufficiently successful, and mankind escapes the dark abysses of its own making, then truly will the future belong to man, the only product of biological evolution capable of controlling its own further destiny.

Social organizations, through their political leaders, will determine on peace or war, on the use of conventional or nuclear weapons, on the encouragement or discouragement of measures to limit the growth of populations, on the degree of increase in food production, and on the conservation of a healthy environment or its continuing degradation.

These and lesser decisions will affect the composition of the human species. Some major population groups will grow in numbers, others will decline, relatively or absolutely, as they have in the past. Thus, in the seventeenth century, Europeans and their descendants on other continents made up approximately 20 percent of the world's population; in 1940 they represented nearly 40 percent of all people. A relative increase of Asian and African peoples has developed more recently. Each trend was the result of such complex circumstances as the opening of sparsely inhabited continents to immigration, the industrial revolution, the introduction of contraceptive devices, improvements in medical knowledge, and introduction of public health measures.

Man, a highly social being, is an animal as well. In form and function, development and growth, reproduction, aging, and death, he is a biological entity who shares the attributes of physical life with the millions of plant and animal species to which he is related. This relationship, known since prescientific times, became part of established science long before the theory of evolution was proposed. It is the reason why studies of fungi and mice, flies and rabbits, weeds, cats, and many other types of organisms have contributed to understanding man and to improving his health and biological well-being, and why future studies with experimental organisms will bear on man's own future.

Man's mental attributes form a superstructure that does not exist independent of his organismal construction. Human thought is based on the human brain; the brain—and, hence, the mind, the "self"—of each person is one of the derived, developed expressions of the genes that he inherited from his parents. Man's capacities are, thus, inextricably linked to his genes, whose molecular nature is now understood to a very high degree, but which presently lie outside his control. The social creations of man— language, knowledge, culture, philosophy, society—have an existence of their own and are transmitted by social inheritance from generation to generation. But they depend for both their persistence and change on the genetic endowments of the biological human beings who are subjected to them and, at the same time, make them possible.

In these considerations, "man" should be taken as inclusive of all human beings on our planet. The brotherhood of all men is not only an ethical imperative; it is based on our common descent and on the magnitude of the shared genetic heritage. Man has organized "nations"—geographic, political, economic, military, and cultural units—that insist on their sovereign status. The short-range future of man will differ among these sovereign groups, particularly between the "developed" and the "under-developed" nations. This discussion is particularly concerned with prospects in the developed countries.

THE GREAT HAZARDS

War

The considerations of the future of man that follow presuppose that mankind will not be subjected to a nuclear holocaust. If such an event were to occur, the problems of retaining or re-establishing social organization, the breakdown of health services, including the production and distribution of life-saving drugs, and the ensuing threat of worldwide epidemics would take precedence over all other aspects of human affairs. Modern technology is sufficiently powerful to make complete extermination of man a possibility. To accomplish such a deed would require overwhelming use of nuclear weapons over large areas of the globe and a deliberate effort to distribute lethal levels of fallout over all inhabited regions. Barring such extreme measures, some of mankind would probably survive a nuclear war. The acute dose of irradiation required to kill human beings in a single, brief exposure is relatively small. Accordingly, direct radiation from nuclear explosives would take an immense toll, but the survivors would probably be able to repopulate the earth.

Those who survived the immediate impact of nuclear explosions would be subjected to chronic irradiation from fallout, which would lead to a variety of deleterious effects. In addition to the damage to their own bodies, the survivors would produce egg or sperm cells that would contain many new mutations leading to abnormal offspring. Nevertheless, the radiation dose that the survivors would have received from the initial exposure and subsequent fallout might often be low enough to permit them also to produce normal-appearing and normal-functioning children, provided the survivors would still want to create a new generation. Ironically, it is precisely under these circumstances that social inhibitions against control of human genetics would dissolve most rapidly; post-nuclear-war man would almost certainly utilize available genetic understanding and biological technology to guide the evolution of his species.

Should *Homo sapiens,* as such, survive nuclear war, there can be no guarantee that he could reconstruct his civilization. Our technologically developed society rests on a complex web of production that could be rebuilt only with extraordinary difficulty. Meanwhile, this would probably occur in a world significantly altered. The ecological consequences of worldwide fallout and long-term rise in radioactivity are virtually impossible to predict. But plant and animal species vary remarkably in their radiosensitivity, and surely current food chains would be disrupted with such profound ecological consequences that it is not clear that man could continue to find sustenance, warmth, and shelter.

Similar considerations may well apply to the possibility of widespread use of biological warfare. The constructive understanding of life that biology provides can also be used for wholesale destruction of life. Once undertaken, war, in the future as in the past, is liable to grow beyond control, whether it be conducted with physical, chemical, or biological means. A future for man can be assured only when the ultimate danger of modern war is fully recognized and mankind abandons warfare.

Man and His Environment

For thousands of years, since first he became a farmer, man has changed his general environment. Some such actions were favorable; for example, during a previous era the rainfall in the American plains was limited, enabling the Indians annually to burn the prairies to drive the buffalo. The resulting debris created our great grasslands and helped generate the deepest, grandest soils on this continent. But, deforestation and primitive methods of agriculture have denuded vast areas and exposed their soil to erosion. It was just such practice that silted the Tigris and Euphrates rivers, thus contributing to the demise of the great Sumerian civilization. With continuing loss, precious soil is essentially irretrievable. Return of the dust bowl of the south central United States to productive agriculture will require years of expensive and intensive effort, which will be economically rational only when national requirements are desperate. Hundreds of thousands of acres of forest were despoiled without provision for reforestation. Large areas of South American forest have been cleared for agriculture, despite the fact that, within a year or two, the rich forest is replaced by a concrete-like laterite soil; until research provides the technology to prevent this, such forest should remain in the native state. But when such knowledge is in hand, vast areas could be opened for productive agriculture. Excessive hunting of some animals for food, and of the large predatory animals either for sport or in self-defense, has wiped out many species either wholly or in many regions. Witness the slaughter of the giant, flightless Moa by the Maori after they found New Zealand (about A.D. 1200) and the decreasing numbers of virtually all the great birds of our own country. When Europeans first came to this country, it harbored 5 billion passenger pigeons and 50 million bison. The former are gone and only 6,000 of the latter remain. Less than 3 percent of the original acreage of redwoods now stands, and there is no record of the disappearance of great numbers of species from our prairies, lakes, and forests. And mankind is the poorer. The prospect of a planet populated exclusively by man and the few animal species he has domesticated is bleak indeed. How grim to think of a world without tigers, whales, condors, or redwoods!

The rise in population density, the operations of modern industry, and the diverse products of modern technology have led to pollution of air, water, and soil by a wide variety of chemical compounds and even undesirable proliferation of certain living forms, notably algae, while defiling the landscape and minimizing the exposure of urbanites and suburbanites to natural surroundings. Large-scale use of pesticides can start a chain in which these substances concentrate in plant and animal tissues and, when ingested, accumulate in the adipose tissue of the human body. As an illustration of this process, consider the record of Clear Lake, California, where DDD (a breakdown product of DDT) entered the lake at 0.02 part per million (ppm). A year later, its concentration was 10 ppm in the plankton, 900 ppm in fish that eat the plankton, and 2,700 ppm in fish that eat fish that eat plankton. No data are available concerning people who ate such fish. Similarly, the routine addition of antibiotics to feed of domestic animals leads to their ingestion by man. Exposure to repeated low levels of these drugs may inhibit growth of sensitive organisms and may thereby foster growth of resistant strains and so may decrease the effectiveness of these drugs to fight infection.

The effects of these changes in the environment on man himself are not known. Although it is possible that some of the agents to which man is now inadvertently exposed will cause serious disease, shortening of the life-span, decreased fertility, or deleterious mutational changes in genes, none of these has as yet actually been demonstrated to have occurred. This, however, does not imply that dangers do not exist. Such possible effects may be numerous, yet difficult to discover. It has taken decades to establish the statistical relation between cigarette smoking and lung cancer; the same may also hold true for the relation between new factors in our environment and other diseases.

Unlike the effects of acute, heavy doses of deleterious substances that rapidly lead to severe illness, pollutants are taken up in small amounts over long periods. Their effects, therefore, may be delayed for years or decades. Moreover, different individuals probably react differently to the same low level of exposure to a foreign substance. Some may excrete more of a given compound than others, thereby avoiding accumulation. Some may decompose the agent in their tissues, while others leave it unchanged. Some may be more resistant to its effects. If, as an arbitrary figure, one in 1,000 individuals will suffer ill effects from a specific agent, causal relationships can be revealed only by very large-scale studies of whole population groups. Yet if one incident in 1,000 seems a small effect, consider that, in a population of 200 million, as many as 200 thousand individuals would experience damage. Conceivably, the incidence of heart attacks may have been increased to this extent by the carbon monoxide of automobile exhausts in regions where smog formation is heavy, but this has yet to be demonstrated. Appropriate studies would have to make use of large cohorts of people followed in their pattern of diseases, fertility, and life-span over very long periods—perhaps longer than the professional life-span of a single generation of investigators. There is precedent for such studies in the research on the relation of smoking to lung cancer, but the scale of such studies must be greatly expanded.

Until reliable evidence thus obtained becomes available, public health measures designed to minimize exposure to such pollutants are patently advisable. But surely a rule of reason should prevail. To only a few chemicals does man owe as great a debt as to DDT. It has contributed to the great increase in agricultural productivity, while sparing countless humanity from a host of diseases, most notably, perhaps, scrub typhus and malaria. Indeed, it is estimated that, in little more than two decades, DDT has prevented 500 million deaths due to malaria that would otherwise have been inevitable. Abandonment of this valuable insecticide should be undertaken only at such time and in such places as it is evident that the prospective gain to humanity exceeds the consequent losses. At this writing, all available substitutes for DDT are both more expensive per crop-year and decidedly more hazardous to those who manufacture and utilize them in crop treatment or for other, more general purposes.

The health problems engendered by undesirable contaminants of the environment may also be raised by substances that are intentionally ingested. Only large-scale, long-term epidemiological research will reveal whether the contraceptive pills, pain killers, sleeping pills, sweeteners, and tranquilizers, now consumed on so great a scale, have any untoward long-range effects on their consumers.* Man has always been exposed to the hazards of his environment and it may well be that he has never been more safe than he is today in the developed nations. Food contamination is probably minimal as compared with that in any previous era, communal water supplies are cleaner, and, despite the smog problem, air is probably less polluted than in the era of soft coal or before central heating systems were the norm. Witness the fact that jungle-dwelling natives of South America exhibit a considerably greater incidence of chromosomal aberrations in their somatic cells than does the American population. But modern man also increasingly exposes himself to the chemical products of his own technologies and has both the biological understanding to ascertain the extent of such hazards and the prospect of technological innovation to minimize them where they are demonstrated. To do less would be improvident and derelict.

The federal record systems, particularly those of the Veterans Administration and the Department of Defense, are already available for epidemiological follow-up studies among veterans. They could be made still more useful by utilizing record linkages, i.e., linking together the many inde-

---

* This sentence was written in June 1969. Revelations of the untoward effects of both steroid contraceptives and cyclamates were made public months later.

pendent records of births, illnesses, deaths, and other vital statistics in the defense and social security agencies and many others. While this entails the possibility of intrusion into the privacy of individuals, it should be possible to erect safeguards against misuse. Such safeguards will be effective if the prevailing climate of opinion welcomes the attainment of useful information and forbids authoritarian attempts at improper exploitation of linked data. Man's biological future depends on knowledge of his experiences, good and bad, and record linkage is an important means of acquiring such knowledge.

Even more subtle than the effects on man of pollutants and of specific agents may be the effects of changes in his general pattern of living. Urban aggregation has removed many men from natural surroundings. The increased level of environmental noise, caused by industrial procedures and automobile and airplane engines, has added a new dimension to sensory exposure. Crowding together in overpopulated regions has greatly changed the interrelations among people who, only a few thousand years ago, lived in small bands with minor contact with one another.

Development of the science of animal behavior is beginning to give some insight into the interrelations between genetically founded behavioral attributes, the effects of early training, and the effects of the immediate environment on overt behavior. We are prone to think of hostility, crime, and other antisocial behavior as conditioned by social circumstances. And there is, indeed, ample evidence to support this belief. We do not know, however, how much personal unhappiness and social distress is a consequence of man's basic biological nature in conflict with an unnatural, essentially nonhuman environment. The stereotyped movements of caged polar bears, *viz.,* the reaction of bears to a non-bearlike environment, may well have analogies in mentally ill patients. Undesirable modes of behavior as well as various psychosomatic illnesses may frequently be extreme expressions of maladjustments of the human animal. Hopefully, research in behavioral biology will furnish deeper insights into man's nature, and application of these insights may lead in the not-too-distant future to fundamentally new parameters for environmental engineering. These will endeavor to fit the environment to man instead of leaving man unfit for the environments he created. Important to such studies are the nonhuman primates—apes and monkeys; every effort should be made to assure the survival of as many such species as possible.

Scientific advancement has enabled man to triumph over his environment. With technological skills and machinery he is able to move, change, and control natural resources for his agriculture, forestry, fisheries, recreation, and urban and industrial development. The wasteful, injurious practices of yesterday have largely been abandoned. Forests are replanted as they are timbered; soils are fertilized so as to compensate for the minerals removed by the plant harvest; saline soils are restored to useful tillage by large-scale leaching; a beginning has been made at reversal of thoughtless practices that could result in the almost irreversible death of large lakes. Although Lake Erie is in serious trouble, Lake Washington is being recovered, and Lake Tahoe may yet be saved. Civilization depends and will continue to depend upon the renewable resources of the environment—land, water, air, and populations of plants and animals, both wild and cultivated. Fortunately, the public and its representatives are increasingly anxious about the status of these resources and the vital role they play in our survival and general well-being.

Environmental pollution becomes of increasing concern as the human population congregates in cities and occupies more of the landscape. The public is pressing for greater understanding of the function and interaction of the biological and physical elements of the environment and for application of this understanding to the management of the renewable resources that supply man's food, clothing, recreation, and shelter. This sense of urgency, arising originally from the desire simply to assure the viability of life on our planet, is heightened by growing public appreciation of the importance of beauty, natural and manmade, in our surroundings for the improvement of the quality of life.

Many reports have directed attention to the more obvious gross problems of managing the environment, problems that derive from a combination of population growth, advancing technology, and increased technological productivity. Concern has been expressed with respect to rising atmospheric CO_2; increasing particulate content of the atmosphere; buildup of radioactivity; accumulation of diverse chemicals in lakes, streams, rivers, coastal waters, and the ocean itself; soil erosion and destruction; replacement of fertile, green farm and woodland by highways and towns; rising noise levels; and "thermal pollution." Figure 49 displays the record and projection of various endeavors of our society, each of which must necessarily adversely affect the environment. If these projections are even approximately correct, and if each of these enterprises grows without appropriate monitoring for its ecological consequences, the totality could constitute the saddest, most brutal, and most disastrous act of vandalism in history. Economic growth, insofar as it leads to higher standards of living, better health, and national security, is clearly to be desired and fostered. But life should be worth living, and this generation should assure that it can transmit to posterity a land whose beauty and resources have been safeguarded, embellished, and protected. To do so demands a level of regional and national planning, with due regard for ecological understanding, without historic parallel. Moreover, we must assume the burden of transmitting

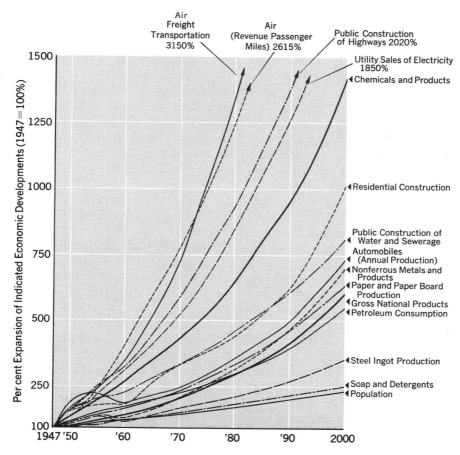

FIGURE 49   Projections of expansion of certain industries that have an influence on pollution. Medium-level projections, 1970–2000. (Data from H. H. Landsberg, L. L. Fischman, and J. L. Fisher, *Resources in America's Future,* published by The Johns Hopkins University Press for Resources for the Future, Inc., Washington, D.C., 1963.)

such understanding and planning capability to the developing nations, most of which still retain the native qualities of their environments and are in danger of galloping destruction of their resources as they race to develop their technological capabilities and to achieve economic independence and a reasonable standard of living.

Biologists and engineers should work jointly to design cities of quality and beauty. Dwellings and industrial structures should be surrounded by at least minimal lawns and plantings of ornamental trees, shrubs, and flowers. There can be no relaxation of efforts to assure "clean" air and water, although useful operational criteria for these remain to be established. The burden of responsibility must be made to lie with those whose activities introduce contaminants into the environment. Even now, there are tech-nological means for dealing with most pollutants at the source. A rational society will insist that these means be utilized and that new, cheap, and efficient means be continually sought. Where none exist, it becomes essential that the advantages afforded by polluting activities—nondegradable detergents and pesticides, heavy use of fertilizer, the exhaust of internal-combustion engines, the sonic boom, and the contrails of a supersonic transport—be weighed against their cumulative effects on mankind. It is doubtful that, at this writing, the evidence on which to rest such judgments exists. Patently, this evidence must be sought.

Through care, planning, and utilization of the sciences of agriculture and forestry, the landscape of the country can be conserved, returning it to its simple charm, with neither billboards nor automobile graveyards. More-over, a great and complex effort based on ecological understanding will be required to cope with the pressures that annually result in conversion of one million acres of farm and wildland to highways and building sites. Certain areas representative of nature—seashore, mountains, desert, forest—should be preserved forever wild for recreation and on a scale adequate to preserve the natural biota. Ecologists and environmental biologists, together with other scientists, should combine to develop a strategy for the wise use of our renewable resources and the preservation of an attractive environment. The attainment of these goals does not depend alone on the technical skills of biologists and other scientists and engineers; people generally must desire to live in harmonious, healthful environments. Without broad social motivation supporting their use, the knowledge and the skills of the specialists will lie fallow.

The Size of Human Populations

In the seventh century, according to the records of the Church of Mayo, two kings of Erin summoned the principal clergy and laity to a council at Temora, in conse-quence of a general dearth, the land not being sufficient to support the increasing population. The chiefs . . . decreed that a fast should be observed both by clergy and laity so that they might with one accord *solicit God to prayer to remove by some species of pestilence the burthensome multitudes of the inferior people.* . . . St. Gerald and his associates suggested that it would be more conformable to the Divine Nature and not more difficult to multiply the fruits of the earth than to destroy its inhabitants. An amendment was accordingly moved "to supplicate the Almighty not to reduce the number of men till it answered the quantity of corn usually produced, but to increase the produce of the land so that it might satisfy the wants of the people." However, the nobles and clergy, headed by St. Fechin, bore down the opposition and called for a pestilence on the lower orders of the people. According to the records a pestilence was given, which included in its ravages the authors of the petition, the two kings who had summoned the convention, with St. Fechin, the King of Ulster and Munster and a third of the nobles concerned. . . .

W. J. Simpson,
—A Treatise on Plague

The upsurge in the growth of human populations constitutes the major problem for the immediate future of man. Accordingly, it is difficult to exaggerate the urgency of deepening our biological knowledge of man and his environment. There is every reason to expect that, by the end of the century, a brief 30 years from now, the world will have twice its present population. Unless forestalled by a worldwide holocaust, in the year 2000 the world population will surely be not fewer than 6 billion people, and may well exceed 7 billion. (See Figure 50.) Since the means for improved control of disease are already at hand, if the food supply keeps pace, a

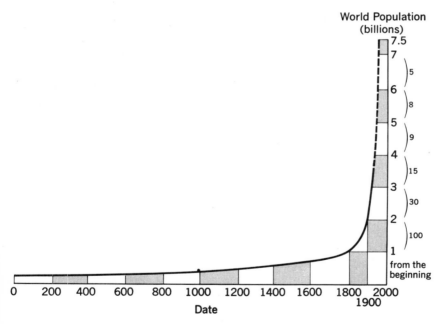

FIGURE 50 World population growth (projected with assumption of constant fertility levels and declining mortality). (Small numbers outside parentheses indicate the rapidly decreasing number of years required to increase world population by a billion people.) (From *World Population: A Challenge to the United Nations and Its System of Agencies*, UNA-USA National Policy Panel on World Population, May 1969.)

world population of 7 billion will be reached even if present birth rates are considerably reduced. Population growth is occurring most rapidly in the newly developing nations, where abject poverty is widespread, the mass of the population is uneducated, and the industrial sectors of their economies are poorly developed. (See Figure 51.)

Moreover, future demands on the biosphere are not to be measured by simple extrapolation from the present. Much of the present population is badly nourished, while, because of improved communications and appreciation of the living standards of developed nations, popular aspirations for improved living conditions are high in almost all nations. Even a doubling of per capita consumption of protein, clothing, and shelter would leave present aspirations unfulfilled. If the projected doubling of the world's population is realized, and if political order is to be maintained, it is not unreasonable to expect that the demands on the biosphere by the end of the century will be three or four times those of the present! The problem of achieving such increases becomes staggering when we realize that the end of the century is so close that this year's infants will then be in the middle of their childbearing period. Moreover, regard for the human heritage requires that these needs be met without despoiling either the quality of man or the material base from which he draws life.

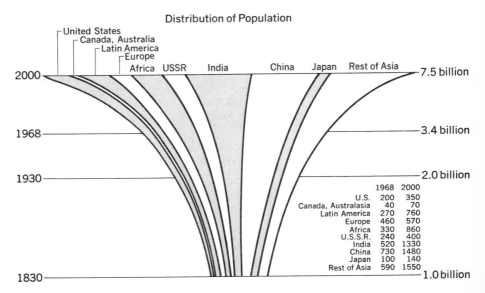

FIGURE 51 Projected population of developed and underdeveloped nations in the year 2000. (From *World Population: A Challenge to the United Nations and Its System of Agencies,* UNA-USA National Policy Panel on World Population, May 1969.)

Clearly a catalog of needed knowledge is a catalog of all natural and social science—the nature of man and his modes of change, the extra-human biosphere and its interaction with man and with the physical environment. This is just another way of saying that man's destiny turns on his knowledge of himself and of his total environment—inanimate, biological, and social. In the face of a desperate situation, almost nothing in the entire spectrum of basic and applied science is irrelevant.

It is clear that, in the long run (1) population growth cannot continue indefinitely, and (2) if the major populations can be brought to an adequate level of education and technical achievement, the ultimate constraints to population growth will not appear until populations become significantly larger than they are now. Such a world would probably find it desirable, on humane and esthetic grounds, to check its growth far short of the numbers that could be supported by a world economy that continues to develop and apply scientific knowledge. To such a world, knowledge of the human gene pool, and of its potentialities for change, should be of paramount importance because, in the last analysis, man's destiny lies in his nature.

But today's most urgent problems are not yet those of a world of highly educated and prosperous populations. They are the problems of moving from worldwide poverty, hunger, and illiteracy to worldwide education and prosperity. To do so, we must survive the coming crisis of population growth with sufficient political and social coherence to permit the sensible application of our developing science. Unfortunately, crises divert attention from matters of the future to those of immediate relevance. Patently, many of the immediate problems are social, economic, or political, but it is to the biological aspects of the emergency that attention is here directed.

It is necessary only to note that (1) much of the world is undernourished; (2) populations in the Americas south of the United States, in Asia, and in Africa are growing at between 2.3 and 4 percent per year, i.e., they are doubling in 17 to 30 years; and (3) agricultural production in some large areas is falling behind population growth. Production per acre has been increasing rapidly in the highly developed areas of the world, but, until recently, in most underdeveloped nations gains in production have come primarily from the extension of cultivated acreage rather than from increased production per acre. Moreover, the constraints to the extension of acreage are becoming all too visible in many of the most densely settled parts of the world.

FOOD PRODUCTION: THE SHORT-TERM PROBLEM

These facts suggest that (1) there may be mass starvation on a tragic scale within this century unless there is a prompt and major rise in production, and therefore (2) there is urgent need for the application of already existing knowledge. Indeed, effective practical application of present scientific understanding could certainly suffice to manage the problems of food supply during the present century. Basic theory does not result in increased production without a great deal of scientific work on local soil and water, crop management, development of seed strains, animal husbandry, pest control, and, most importantly, fertilizer usage. Basic science gives us the principles and tells us how to go about learning to apply them, but it still does not indicate in precise local terms how to do the multitude of things that must be done. The emergency need in agriculture is for great increases in local applied work and in training for such work. The recent successes of the Rice Institute in the Philippines and in wheat production in Mexico are noble demonstrations that such efforts are both feasible and rewarding. Introduction of the new strains of rice and wheat developed in those areas has resulted in a startling increase in production in areas of Pakistan, India, the Philippines, and Mexico, an increase that appears to have brought several years of surcease from the threat of famine in these countries. (With only a small fraction of the land planted in the new strains, recent successful harvests in India and Pakistan were due, in the main, to unusually favorable monsoons.) But this effort alone will not suffice without a concomitant endeavor to supply credit, manufacture fertilizer, ensure a water supply, build roads, and arrange for food and other commodity distribution. And the values of the latter cannot be realized unless the scientific basis for intensive local agriculture has been established.

Meanwhile it should not be thought that basic science has contributed all it can to food production. It will suffice to note that no new species of animal or plant has been adapted for human consumption as a major foodstuff in recorded history. We still have urgent need to provide fundamental designs for extremely intensive, very high-yield agriculture, to learn how to take advantage of offshore opportunities for intensive aquiculture of molluscs and, perhaps, of higher marine organisms, to breed wheat of more useful protein content, to find suitable alternatives to the dependence of man, globally, on just a few staple crops—rice, wheat, and corn. This dependence on only three cereal types offers the terrifying prospect of a worldwide epidemic caused by a virus to which no strain of one of these species might be resistant.

The fact that much of basic science is a product of a few nations and civilizations does not mean that the basic sciences cannot be learned, developed, and applied by other populations. Even though we have no means at present of comparing the intrinsic genetic endowments of different ethnic groups, it is clear that great reservoirs of trainable human genotypes exist in all of them. The shortage of brainpower, in the world at large, that can be applied to the immediate problems of agricultural production is not due to biological limitations of genetic endowments in different human groups, but to the limitations of their education. Every region is potentially able to produce the numerous trained persons needed to explore and solve the specific problems of the region. Obviously, these problems are not restricted to agriculture. Each region requires a corps of local scientific and engineering specialists to make available to its population the beneficial results of industrial and scientific technology, and American technical assistance should give high priority to assisting in the endeavor to train such specialists.

The immediate demands on the biosphere have been generated and exacerbated by rapid population growth, which, particularly in the developing nations, is the consequence of the almost abrupt inauguration of public health and sanitation measures, producing drastic reduction of the death rate while the birth rate remained unchecked. Because no imaginable program of population control could restrain population growth significantly in the next decade, the emergency problem is to attain, as rapidly as possible, adequate levels of education and significant local programs in agriculture and related science and technology for the vast numbers of the human race. We are not optimistic that this can be achieved in time to avert disaster in the 1970's, but the attempt must be made. It is all too evident that the surplus agricultural productivity of a few developed nations —Canada, Australia, the United States—even if used to the full, can have little impact in this worsening situation. Moreover, population growth in these nations will, in time, require domestic utilization of their own production.

If this estimate of the situation is correct, the acute problem is not that of population size itself but of the speed of modernization and the extent to which the gains so realized are offset by the speed of population growth. The speed of modernization turns on many factors, on national and international allocation of resources for space, war, schools, and factories, *inter alia*. But basically it also depends, even at quite local levels, on the extent to which the growth of agricultural production can be made to exceed that of population. Investment in development can be made only after the current costs of growth have been met.

POPULATION CONTROL: THE LONG-TERM PROBLEM

The long-term prospects for a truly human civilization depend in very large measure on whether humanity can, in time, succeed in moderating its fecundity. Only if this effort is successful, and early, can our progeny be offered the opportunity to relish the gift of life and to maximize their own human potential. The dimensions of the problem are dramatically evident in the remarkable diminution in the time required for doubling of the world population, as seen in Table 69.

This remarkably accelerated growth, largely the result of decreasing death rates due to simple public health and hygienic measures accompanied by commensurate increases in agricultural productivity, occurred first in Europe and the United States and is now operative also in many of the developing nations, with startling consequences. Witness Brazil with a population of 17.5 million in 1900, 52 million in 1950, 71 million in 1960, 83 million in 1966, and an estimated 240 million by 2000, or a 14-fold increase within the twentieth century!

Yet concern for population growth is not new. In *Politics,* Aristotle warned that ". . . neglect of an effective birth control policy is a never-failing source of poverty which, in turn, is the parent of revolution and crime," and he advocated that parents with too many children practice abortion. He went unheeded through the following centuries as the Romans encouraged large families to man their wide-flung armies, the Judaeo-Christian ethic considered children as gifts of God, and St. Augustine stated the purpose of Christian marriage to be procreation, a view unmodified by the Reformation. Much earlier, Tertullian noted that "the scourges of pestilence, famine, wars and earthquake have come to be regarded as a blessing to crowded nations since they served to prune away the luxuriant growth of the human race." From time to time, advocates of population control appeared, most particularly Malthus, who stated that, otherwise,

TABLE 69   Time Required to Double World Population

WORLD POPULATION	YEAR	TIME REQUIRED (YEARS)
250,000,000	1	
		1,649
500,000,000	1650	
		200
1,000,000,000	1850	
		80
2,000,000,000	1930	
		45[a]
4,000,000,000	1975[a]	
		30[a]
8,000,000,000	2005[a]	

[a] Estimate.

population would always rise to the limits of food-production capacity, so that necessarily there would always be hunger and poverty. (Ironically, Malthus rested his case on the history of eighteenth-century United States.) Unfortunately, his teaching was rejected both by the Christian ethic and by Marxism, which taught that overpopulation is merely a capitalist notion invented to justify the poverty of working-class peoples and is rectifiable by enhanced production and improved distribution rather than by birth control.

Malthusian predictions have largely been justified, although he foresaw neither the consequences of the introduction of agricultural technology nor the demographic consequences of simple hygienic measures. Today, these problems must be considered separately in global and local contexts. The food crisis of some developing nations, considered earlier, is patently urgent. Yet, worldwide, since about 1950 agricultural productivity has grown by about 3 percent annually, while population increase has averaged less than 2 percent. Indeed, it is estimated that if worldwide per capita food consumption had held constant at 1955 levels, despite the population increase by 1975 there would have been a world surplus of 40 million tons of wheat and 75 million tons of rice. This will not occur because of both rising per capita food consumption and the controlled productivity practiced, in varying degree and kind, in the United States, Australia, New Zealand, Canada, France, and the Argentine. Meanwhile, the developing nations, caught up in the worldwide revolution of rising expectations, find themselves short of food and of capital for development.

Income for development can be generated by increased production and by decreased reproduction. Clearly, both are needed. Some inherently undesirable means to decrease the rate of population growth, e.g., war, famine, pestilence, are all too evident. On the other hand, populations that have learned to reduce their fertility to the point where they enjoy good health and the longevity characteristic of the more developed nations, and whose growth amounts to only 1 percent per year should encounter no substantial difficulty in reaching a stationary position if that is clearly desirable. Only as that occurs can increased production of agriculture and the extractive and manufacturing industries be utilized for development and increase in the general standard of living. It may well be asked how many countries can be expected to do this soon enough. Markedly increased agricultural productivity accompanied by population growth inevitably leads to rapid urbanization, frequently at a rate in excess of any prospect of gainful employment of the translocated individuals. Yet this process, at a moderate rate, is imperative if a developing agrarian society is to acquire a sufficient urban population to sustain its growing industry, educational system, etc.

As we have indicated, if the oncoming food-shortage crisis can be averted, known technology, if put into practice, can readily so enhance food production as to defer the world food problem almost indefinitely. It is the combination of new strains and application of fertilizer that has so remarkably increased agricultural yields in Europe, Japan, and the United States. In a general way, application of a ton of fertilizer nitrogen yields an increment in crop production equal to the basic yield of a 14-acre plot. Stated differently, there are about 3.5 billion acres of land presently under cultivation; application of $10 worth of fertilizer per acre would increase production by about 50 percent, i.e., for $35 billion per year or $10 per capita, worldwide, world food production would rise by the equivalent of 1.7 billion acres of average land and a 50 percent increase in available food, per capita. Moreover, it has been calculated that if all land now in tillage were cultivated as in Holland, the world could support 60 billion people on a typical Dutch diet; if it were managed as in Japan, it could support about 90 billion people on a typical Japanese diet. And all this is possible apart from the realizable expectation of yet another agricultural revolution based on improved control of agriculture, growth of food yeast, bacteria and algae, or synthetic foodstuffs based on petrochemicals. Approximately one acre is required to feed one man by efficient current agriculture, yet a one-square-yard tank growing algae can produce all his caloric, protein, and vitamin needs! All of which is to say that measures to upgrade agricultural practice in the developing nations could forestall a Malthusian crisis for more than a half century, even at current rates of population growth. But with what consequences?

The problem of population growth involves much more than merely increasing agricultural production. The constraints to population size are all too visible even in the traditional self-sufficient agrarian society. Such societies have rarely been able to combine high population density with good health and relative freedom from poverty. But it is hard to specify the limits to the density of population that can be supported in health and prosperity by a highly educated population making sophisticated use of energy and raw materials and continuing to develop both its basic science and technology.

There is reason to believe that, given the time and effort required to increase all forms of production sufficiently, this planet can sustain in relative abundance a total population considerably larger than the present one. Although no data are available to establish what the maximum might be, it is patently very much larger than at present. One can argue, however, that the maximum possible is decidedly greater than the optimum. Even the present population suffices to populate the planet, at all times, with the diversified human talent required to contribute to progress on all human fronts—science, the arts, industry, government, etc. At some point, industry must forego population growth as the underlying basis for economic expansion. Meanwhile, many of the most tragic ills of human existence find their origin in population growth. Hunger, pollution, crime, despoliation of the natural beauty of the planet, irreversible extermination of countless species of plants and animals, overlarge, dirty, overcrowded cities with their paradoxical loneliness, continual erosion of limited natural resources, and the seething unrest that creates the political instability that leads to international conflict and war, all derive from the unbridled growth of human populations. The fortunate nations are those that have, spontaneously rather than as a matter of national policy, achieved a low rate of population growth or an exact equilibrium of the birth and death rates.

Accordingly, another set of important emergency problems of a biological nature are those relevant to a reduction of human fertility. In the long run, birth rates must come down if death rates are to stay low, and in the short run, lower fertility would speed the process of modernization by widening the difference between the growth rates of population and production. Reductions of the birth rate in the underdeveloped countries have an additional advantage. High birth rates produce high proportions of young people. In virtually every country with a birth rate of 40 or more per 1,000 population, more than 40 percent of the total population is under age 15. Under these circumstances it becomes almost impossible for such a society to increase its working capital—to generate enough wealth for school construction, higher education, improved housing, or industrial plants. In consequence, for example, the illiteracy rate must surely rise, despite national determination to lower it.

In advanced countries with low birth rates, between 25 and 30 percent of the total population is under age 15. A reduction in birth rates brings down rates of growth and reduces the proportion in the ages of childhood dependence. Correspondingly, it increases the proportion of the population in the productive years of life.

Indeed, very high birth rates speed population growth in two ways: (1) they swell the entering stream of life; and (2) by creating young populations they cut the rate of depletion through death. Today, the lowest crude death rates (i.e., annual deaths per 1,000 population uncorrected for age) are not found in the most highly developed countries. The world's lowest crude death rates are found in such places as Taiwan, Singapore, Puerto Rico, and Chile, where health protection has become relatively good and a history of high birth rates has left a young population. In the long run, reductions in birth rates reduce growth both directly and indirectly by increasing the average age and, other things being equal, the crude death rate. Clearly, the possibilities of modernization would be greatly enhanced, globally, if rates of growth could be cut in 15 years from 3 percent to, say, 1 percent by reductions in birth rates. This would mean that populations now growing at rates that double in 23 years would come to a rate that would give them 69 years in which to absorb the increase, *viz.*, the burden of natural increase in the newly developing countries would then be about that experienced by the United States in recent years.

Most societies and individuals desire to limit the size of families, but they are usually not content with a family size corresponding to zero growth of the population. Although, at the present stage of population growth, any reduction in family size is important, ultimately the mean family size will have to be limited to a replacement number. Family planning is not equivalent to population control. Family planning is the rational and deliberate spacing of children in the number desired by the parents. But that number is determined by cultural considerations, family income, and ego satisfaction in the developed nations and by the economic utility of children in the underdeveloped nations. Accordingly, large families are the norm among the affluent and among the ignorant poor. Population control demands that families be limited to the replacement rate.

Nothing can do more to help obtain reductions of fertility than the development of more efficient, cheap, safe, reversible, and acceptable methods of contraception. People will use even the best of methods only when they want to have fewer children. Historically, whereas strongly motivated couples have even utilized inadequate methods of contraception and resorted to abortion, weakly motivated populations must be enticed to use even the best possible methods. Today, readiness to accept contraception is widespread. More than half the population of the developing nations live under governments that have decided, as a matter of national policy, to foster the spread of family planning and limitation. The list includes most of the countries of Asia and a goodly number in Africa and Latin America. Most of these countries are developing educational programs to interest and inform their people and service programs to give them supplies. Careful surveys of attitudes toward reproduction have been made in more than 20 countries. Virtually everywhere, the majority of women desire to limit their childbearing. This does not mean that they want only two or three children, but that they want to stop before their families get truly large. Moreover, where services and supplies are made available, women are beginning to seek them in large numbers. Taiwan, South Korea, Hong Kong, and Singapore have clearly reduced their birth rates through their family planning programs.

Today, in the developed nations, contraception has changed rapidly from use of the older conventional methods to the combination steroid pill. In underdeveloped countries, new contraceptors are mainly using the plastic intrauterine device. Neither method is perfect, but both are spectacularly effective and successful compared with the conventional contraceptives, "rhythm" methods, etc. The availability of the cheap intrauterine device has encouraged governments of underdeveloped nations to build the organizations they require to spread family planning practices. With such organizations in being and operative, the next technological innovation can be introduced much more rapidly. It is because they have effective methods, hope for better ones, and the organizations to make use of them, that such countries as South Korea, Taiwan, India, and Pakistan now hope to halve their birth rates in 15 years. If they could do so, their long-run problems of modernization and economic development would be greatly simplified.

The world needs better methods than are now available. The pill and the intrauterine device represent major innovations because, temporally, they separate contraception from coitus. The intrauterine device is probably at a very early stage of development. An unacceptable proportion of all users spontaneously eject it, bleed, or suffer discomfort. On the other hand, apparently the majority of those who accept it wear it without awareness and with very high effectiveness. Two to three years after acceptance, 50 percent or more of women continue to wear their devices. It is likely that better procedures, better materials, and better shapes will lead to a greatly improved experience. Clearly, it is imperative that appropriate, vigorous investigation be undertaken to solve the riddle of the mode of action of these devices and to ameliorate their occasional side effects.

Similarly, work is needed to reduce the side effects of steroid pills, to minimize their effects on lactation and on thromboembolic phenomena, to reduce their costs, and to establish systematically and in sustained fashion the actual experience of those who take them, to reduce the frequency of subsequent multiple births and, most importantly, to establish the biological consequences of long-term use with complete certainty. Similarly, an effort to find means of replacing the oral route of administration with a depot injection, *viz.*, an injection allowing a steadier rate of absorption and, hence, smaller and longer-lasting doses, would be well repaid.

Meanwhile, as this research proceeds, we deplore statements decrying use of steroid pills on the ground of their manifest occasional untoward side effects. Since the death rate from such usage is well below the death rate from pregnancy itself, these pills not only afford millions of families the opportunity for a richer, fuller life while checking the demographic explosion; on balance they also spare the lives of a significant number of women who would otherwise die of the complications of pregnancy.

It would be highly desirable, of course, to have a method that is, in effect, permanent until positive measures are taken to counteract the contraceptive. Children are often conceived as a consequence of careless contraceptive practice. Doubtless, birth rates would drop faster if there were a method in which carelessness meant failure to counteract a contraceptive. Clearly, however, such a development could pose serious problems of personal freedom unless the counteracting agent were freely available.

Only a beginning has been made in relevant basic research. Among other things, we need to know a great deal more about tubal events, including gamete transport, fertilization, and zygote physiology. The fields of neuroendocrinology, immunological suppression of reproduction, blastocyst nidation, gonadotrophin chemistry, and the mechanisms of sex hormone action urgently need development. An intensive program of basic research might produce important results that could facilitate population control. The fundamental knowledge, the techniques, and the requisite base of professional skill for such an effort are now beginning to appear.

In the long run, all practical results depend on basic research. But in the long run, unless birth rates are lowered rapidly, populations will become multiples of what they are today unless death rates rise. In the past decade, applied research based on many preceding years of basic research has made possible the contraceptive pills, the intrauterine devices and, consequently, the beginnings of a birth rate decline in some developing areas as well as in developed nations. Further basic research is greatly needed to prepare for yet further advances, and it is the only pathway to completely new approaches in population control. Meanwhile, intensification and enlargement of applied research is an urgent necessity.

The United States and other developed nations find themselves in the embarrassing position of advocating that *other* nations increase their efforts at population control. Granted the validity of this position, *viz.*, that it really does address itself to the self-interest of the affected nations, such a posture is not readily acceptable when the advice comes from a nation that has not itself adopted comparable internal policies. Since this country is in the fortunate position of enjoying a high economic level and a relatively low population growth rate, and while our total population is not yet excessive for our natural resources but is on the way to becoming so, the moment is opportune to examine our internal policies and alter these as seems appropriate.

Clearly, our relatively low rate of population increase reflects the fact that American parents have not been behaving in the manner seemingly encouraged by the national mores and laws. Is it not appropriate to reconsider laws that discourage abortion, forbid or make difficult distribution of birth control information and devices, and encourage large families by income tax forgiveness and by other social measures? These derive from an earlier ethos when an expanding population was required to develop the national frontiers. They seem entirely inappropriate today.

Nor is the United States immune from the population explosion. Until recently our population growth has been dominated by the extremely low birth experience of the depression years and the subsequent war. But the children of the "baby boom" are just entering the child-bearing population. Thus, our female population in the age range 16–44 was 32 million in 1940, 34 million in 1950, and 36 million in 1960, but it will be 43 million in 1970 and 54 million in 1980. The potential for an extraordinary burst in population is evident in the very fact of the existence of this breeding population. The current rate of population growth, i.e., the excess of births over deaths, is about 2 million per year, yet it is estimated that about 500,000 babies per year are "unwanted." Surely, provision to the mothers of these unwanted babies of information, contraceptive materials, or legal abortions would make for a happier society while reducing the societal burdens of population growth.

Moreover, it should be understood that the penalties for population expansion are far greater in an affluent society than in a marginal economy. This is already painfully evident in the United States. Rising per capita real income places less and less tolerable burdens upon the environment: vastly increased solid waste, nondegradable detergents, pesticides, containers and trash, more automobiles, heavier traffic, increased CO and CO_2 production, more smog, rapid erosion of fields for airports, highways, parking spaces and suburbia, rapidly increased water usage for anything but drinking, viz., airconditioning, swimming pools, metal-fabrication and paper-production plants, etc., while the same processes accelerate the depletion of all our nonrenewable resources—oil, iron, and copper ores, etc. Indeed, this is the lesson of Figure 49. Further, consider the seemingly impossible burden of coping with the demand for college education: college enrollments, which were 6 million in 1965, will be 8 million in 1970, 10 million in 1975, and 12 million in 1980. The effort required to meet all the expectations of this burgeoning population will be enormous and will utilize more and more of our precious land and irreplaceable resources. This is in clear contrast to the burden upon the environment generated by adding to the population of a developing nation more individuals whose mean income is but a few hundred dollars a year. Clearly, the national interest and our individual interests would be well served by all measures that would damp the demographic explosion at home as well as abroad.

There is, however, one aspect in which population control and the health of the population are at odds. Population control would be furthered by encouraging late marriage, a principal factor in the low birth rate in Ireland. But the incidence of such congenital defects as Down's syndrome and cleft palate, as well as of twinning, rises with the age of the parents. Hence, the optimal situation would be that in which marriage occurs at a young age and a family of two or three children follows shortly thereafter. Success in such a program then requires the full cooperation of society and 20 to 25 years of uninterrupted, successful contraception. Without sterilization, statistically, this seems an unlikely prospect unless research can provide much simpler and more effective contraceptive methods than those presently available.

Guarding the Genetic Quality of Man

The human gene pool is the primary resource of mankind, today and tomorrow. The present gene pool is the culmination of 3 billion years of evolution and natural selection. The physical vigor, long life, and intellectual capacity of most humans reflect the fact that, historically, natural selection has minimized the incidence of genes that, when expressed in the homozygous phenotype (an individual with two identical genes for the trait in question), would result in serious physical or mental incapacity. However, advances in medicine in the last few decades have dramatically altered this situation. The "engineering" of human development so as to permit survival despite the handicap of such genetic endowment is called "euphenics" ("eu" = well, "phen" = appearance). By ensuring survival and, thus, permitting the reproduction of such homozygotes, medical practice has relaxed the selection against such genes.

For example, formerly the intellectual deficit of most phenylketonuric children was such that they were unable to reproduce; when raised from birth on a suitable phenylalanine-poor diet they will now, presumably, marry and have offspring. Instead of "extinction" of the genes responsible for the disease in the nonreproductive homozygote, this should lead to an increase in the frequency of these genes in future generations and consequently an increase in phenylketonurics. A similar situation obtains for all other genetic afflictions that can now be neutralized by various treatments. Consider pyloric stenosis, an abnormal constriction at the junction of the stomach and intestine; this is a relatively common hereditary disease of the newborn, occurring in about 5 out of 1,000 live male births and in 1 out of 1,000 live female births. Formerly, most infants so afflicted died in very early life, but 50 years ago a surgical procedure was instituted that permits survival and normal health. The survival and later reproduction of children treated with that procedure has resulted in the perpetuation of this genotype; among their offspring the frequency of infants with pyloric stenosis is about 50 times higher than in the general population. And these children, having been operated upon, will again later produce a surplus of their own affected kind. Thus, a continuous increase of the disease must be expected in successive generations. Similar considerations apply to galactosemia and fulminating juvenile diabetes, and the list must grow as clinical medicine learns to circumvent the consequences of many other genetic disorders. The extent of this problem is evident from the fact that, even now, 6 percent of all infants have detectable genetic defects of greater or lesser seriousness, and all humans must be heterozygous (possessing two nonidentical genes, one from each parent, for a given trait) for at least a dozen or more disadvantageous genes.

The speed of accumulation of unfavorable genes in the population depends on many factors. Generally, it is a very slow process, which, for centuries, will have no easily recognizable effects. Many a "bad" gene whose effects are overcome euphenically may be said to have lost its "badness," wholly or to a large degree, so that its accumulation no longer represents a serious biological load even though it may represent a considerable economic load.

Such accumulation may be contained by genetic counseling, which leads some carriers of such genes to limit their families or even to refrain from having children. Genetic counseling can often assure worried persons that their fears of defective offspring are unjustified or exaggerated, but in some instances the predicted likelihood of severely abnormal offspring is high. Knowledge of the inheritance and the variability in expression of the numerous kinds of human defects accumulates steadily, and the outlook for improved foundations for counseling is favorable. It will be enormously enhanced as procedures are developed that might make possible positive recognition of those who are asymptomatic heterozygotes for specified undesirable genes. As medical euphenics becomes increasingly successful, it will become increasingly important that genetic counseling be universally practiced. Otherwise, in a few generations, the ethic that guides medical practice will have seriously damaged the heritage of countless previous generations. Having thwarted the historical process of natural selection against such disadvantageous genes, civilization must provide an acceptable substitute.

The possibility has been discussed that the great insights of molecular biology may make it possible, in the future, to replace specific undesirable genes in a person's cells with desirable ones brought in from the outside. Several strategies are available, based largely upon understanding of the mechanisms of viral infection. However, many biologists think that the prospects for such "genetic surgery" are doubtful in the foreseeable future. Even if succesful, this would probably simply be a more sophisticated euphenic technique. While there is some possibility that appropriate, desirable genes might, one day, be introduced into body cells, it seems unlikely that the new genes could be so inserted into appreciable numbers of germ cells.

THE OPPORTUNITIES

Biology and Medicine

It is in improved medical practice that advances in biological understanding make their most immediate impact on most people. As we have seen, biological and medical research are intimately intertwined; the fundamental discoveries that have found useful clinical application have been of such general character that they are best termed "biological," regardless of the institutional setting in which they are made. In our society, the benefits of such studies are to be found in innumerable individual events in medical practice from prenatal care through infancy, childhood, and adulthood to old age.

The prolongation of life expectancy at birth is one of the impressive overall measures of the success of medicine. However, even in affluent societies this prolongation has been due primarily to a dramatic reduction, if not a near abolition, of infant mortality. Increases in life expectancy past the age of 45 have remained smaller, although by no means negligible. Even medically advanced countries still experience different life expectancies and infant mortalities, and it is likely that at least a major share of the differences is the result of social differentials. It is regrettable that infant mortality in the United States still exceeds that in more than a dozen other advanced nations. The largest single contribution to our infant mortality is to be found among the economically and socially disadvantaged segments of our society, although it is also unacceptably high even in middle-class families. The problem then is not lack of knowledge but to provide prenatal and early pediatric care of the quality available to the rest of the population. (See Table 70.)

MOLECULAR DISEASES

Future advances in the control of disease will come from better epidemiological knowledge, improved control of the environment, and deeper understanding of the regulation of life processes. Many of these advances will be based on applications of the fundamental information provided in recent years by molecular biology. With sufficient understanding will come a more powerful armamentarium for chemotherapy of endocrine disorders and cancer, treatment of autoimmune diseases, prevention of the degenerative disorders of the circulatory system, and therapy for metabolic disorders. There seems to be no reason why deposition of lipids in the great blood vessels—atherosclerosis—should be a necessary concomitant of

TABLE 70   Countries Reporting Twenty Lowest Infant-Mortality Rates per 1,000 Live Births, 1967

RANK	COUNTRY	INFANT MORTALITY	RANK	COUNTRY	INFANT MORTALITY
1	Gibraltar	11.8	11	Australia	18.2
2	Sweden	12.6	12	United Kingdom	18.8
3	Japan	13.3	13	Luxemburg	20.4
4	Iceland	13.7	14	East Germany	21.2
5	Netherlands	14.7	15	United States	22.1
6	Finland	15.0	16	Canada	23.1
7	Norway	16.8	17	West Germany	23.5
8	Denmark	16.9	18	Belgium	23.7
9	France	17.1	18	Czechoslovakia	23.7
10	New Zealand	17.7	20	Ireland	24.4
			21	Israel	25.3

Sources: Data from United Nations, *Monthly Bulletin of Statistics* (October 1968), *Population and Vital Statistics Report* (July 1, 1968), *Demographic Yearbook* (1967).

human life. Research directed at rational prevention of this process by simple means should markedly reduce or delay the incidence of coronary artery disease, aneurysms of the great vessels, and stroke. Death will still come to all, but the quality of adult life should be markedly enhanced as the onset of debilitating disease is delayed into the latter years of a prolonged life. Life should be not only longer but ever more enjoyable and free of the ravages of ill health in consequence of increasingly effective preventive and therapeutic measures.

It is yet too soon to engage in speculation concerning future progress in prevention or therapy for the major psychoses. Only the success of tranquilizers and mood-elevating drugs offers any basis for hope. Until the underlying basis for schizophrenia and the other psychoses is understood, it will not be clear whether there are fruitful chemical or surgical approaches to therapy. This should not be true of peripheral neurologic diseases. Even now, there are hopeful bits of progress in understanding the demyelinating disorders; and there has just appeared a drug that seems quite specific for relief of Parkinson's disease. Insofar as diseases have specific molecular etiology, there is hope for specific molecular therapy or prevention.

INFECTIOUS DISEASES

Yet another major problem, concerning which assessment remains difficult, is the panorama of virus diseases. Man lives in equilibrium with large numbers of each of what appears to be an ever-increasing number of different viruses. Our chief protection against them remains our own intrinsic biology, particularly our immune mechanisms. The strategy appropriate to assisting the defense against an established infection is unclear. A truly effective synthetic substitute for the naturally occurring "interferon" would probably be a most useful drug. Hope in this direction is afforded by the demonstration that synthetic, double-stranded RNA is an effective stimulus to interferon release and can protect mice against otherwise lethal inoculations of the virus that causes hoof-and-mouth disease. Most drugs that interfere with replication of viral DNA or RNA must also seriously affect our own duplicating cells, as do most chemicals that interfere with protein synthesis. However, this approach is not necessarily hopeless because synthesis of viral nucleic acids within the host cell is accomplished by enzymes synthesized by the host-cell machinery under the control of the infecting viral nucleic acid—not by the original host-cell enzymes. A few antibiotics have already been found to block one or another of these virally induced processes without damage to the host cell. Increased understanding of the genetic machinery and exploration of drugs that interfere with its operation or create a temporary diversion are our greatest hope of finding a rational basis for antiviral therapy, although there is, as yet, nothing to assure success.

The most promising avenues of approach to the problem of cancer derive from the fact that a variety of neoplastic lesions in experimental animals are definitely associated with the presence of specific viruses in affected tissues. Only suggestive evidence presently links most tumors of man with similar viruses, but the evidence is incontrovertible in several instances. It is thus apparent that there is more than one "cause" of cancer, but to the extent that human neoplasia may have a viral etiology, the route to successful cancer chemotherapy may prove to have much in common with the search for antiviral therapy generally. Indeed, it is conceivable that current procedures that enjoy some measure of success—e.g., radiation to solid tumors, antifolic acid compounds and cytosine arabinoside for treatment of leukemias—actually operate by their effects on viral reproduction rather than by limiting cell division itself. In any case, it is imperative that studies designed to establish whether human tumors are, in effect, manifestations of infection with the equivalent of an otherwise silent "temperate lysogenic virus" should be prosecuted with utmost vigor.

Morbidity and mortality from bacterial infections have been dramatically reduced by the availability of antibiotics. But this battle is never won. Resistant strains of almost every bacterial pathogen have repeatedly appeared. Each must be met with yet another antibiotic or combination thereof, and the search for new, more effective antibiotics, conducted largely by the pharmaceutical industry, must be unceasing. No episode in the

Vietnam war is more dramatic, and certainly none more gratifying, than the successful development of therapy for a virulent, highly lethal form of malaria carried by the *Aedes falciparum* mosquito, based on a combination of two drugs that affect two different aspects of the life cycle of the malarial plasmodium. There remain many infectious disorders, particularly in the tropics, such as schistosomiasis, for which no adequate drug is available. Appropriate agents must be sought, based on detailed understanding of the unique metabolism and life cycles of these organisms. If historical precedent is valid, there is good reason to be sanguine concerning the prospects in these regards, but the attack must be intensive and relentless.

TRANSPLANTATION AND ARTIFICIAL ORGANS

Important advances will surely come from application of technology to deficiencies in the function of whole organs. An artificial lung, external to the body, can be used to replace the physical action of the muscles necessary for breathing; an artificial kidney can serve as a chemical device for removing metabolic products from the blood when the original excretory organs cannot perform that function. Artificial blood vessels made of synthetic tubing can serve as substitutes for defective natural vessels, and external artificial hearts can, at least for some hours, take over the pumping function of the inborn organ. And there is every reason to believe that each of these devices can be improved markedly in the future. But it should be emphasized that, while each of these is a triumph of bioengineering and invaluable to those who require them, each also represents a failure of biological research to have found a solution to the underlying biological problem.

If artificial structures are less adequate than natural ones, transplantation of organs from one person, alive or after death, to another offers another avenue to saving of life. Successful transplantation requires mitigation or abolition of the usual incompatibility of the transplanted tissue of the donor with the immune mechanisms of the host, which otherwise results in rejection of the transplant as well as undesirable systemic reactions in the host. Experiments with chickens, mice, and cultured human blood cells, as well as the heroic measures used in successful human heart and kidney transplants, indicate that such antagonistic effects can be minimized. Although numerous tissue types exist, only a limited number serve as antigens to elicit production of antibodies, which then react with and damage the transplant. Although this number is large compared with classical blood types, it is small enough to permit the development of reliable typing procedures so that donor and recipient may be matched reasonably closely. In addition, a growing number of drugs is available to depress the production of antibodies. A combination of these procedures should significantly improve future management of transplant patients. It will be clear, however, that this approach, while dramatic in the extreme, is of limited value. It is inconceivable that hundreds of thousands of such operations can or should be performed annually. For coronary artery disease, prevention and early therapy of atherosclerosis must surely be the more fruitful long-term approach. Moreover, a useful mechanical heart, responsive to body needs, offers more promise as a replacement for a seriously defective heart than does homotransplantation for the medium term, while avoiding the serious ethical problems occasioned by the latter procedure. Nevertheless, ever-improving ability to perform and manage transplantation should be of continuing value in some endocrine diseases, in nephritis, and in an occasional heart patient.

These insights and techniques are capable of prolonging life beyond the normal span, but they create great new difficulties. Millions of individuals would profit from the transplantation of organs or from the use of "spare parts," but for years the supply of these and of the medical teams required for their installation will be inadequate to fill the demand. The ethical conflicts that the physician faces are crushing when he is forced to decide who is to benefit and who is to be denied vital help. Currently the occasions for such decisions are still rare since the numbers of available natural organs or substitute mechanical "organs" are so limited. In the future, yet other life-saving devices will be invented, and the problems of assigning them to specific patients while withholding them from others will increase in frequency. Even if such devices could be made available in large numbers, the cost of keeping a small fraction of the population alive by these means may be so high a fraction of the gross national product as to compete seriously with other needs for the well-being of the population.

THE ETHICS OF TERMINAL MEDICAL CARE

Another difficulty goes even deeper. Relatively little progress has been made in prolonging the adequate functioning of the human brain. The perpetuation of the physical workings of many parts of the body has not been accompanied by a perpetuation of its normal mental aspects. The death of individual neurons in the central nervous system is an attribute of the "normal aging process," and brain damage occurs not infrequently in consequence of trauma or illness. Here lies a great challenge for basic research and the beneficial application of the insights to be hoped for. Some of the best minds among biologists, psychologists, and physical scientists have recently turned their efforts to neurophysiology and brain function. Their studies should help to understand the riddle of the physical basis of the mind, while helping also to discover the basis for procedures for alleviating the tragic situation of keeping the body alive without the full mental attributes that characterize a normal person. One may hope for progress in prolonging the physical and mental health of the aged, but then the problem of disharmonious functioning will simply be displaced to the end period of a more extended life-span. Biology, the science of life, has to be complemented by new insights into the biology of death. The application of such insights will intensify concern with questions already demanding answers. Is society justified in keeping the aged alive when those mental functions that distinguish human beings from vegetating bodies have ceased? Where is the limit of anguish and material burden that the relatives of such aged persons and society at large can bear—a problem that is increasing in frequency and severity?

Research in aging involves the whole range of biological phenomena, from a study of the molecular changes of such substances as the collagen in our connective tissues and bones to the study of the most complex functioning of the central nervous system with its basis for consciousness, learning, reasoning, memory, and other psychological attributes. It is imperative that such research be prosecuted as vigorously as possible. It should be clear, however, that there are no indications, as yet, that the aging process can be delayed or mitigated, except in a most limited way. Support of the aged cannot be considered separately from economic and demographic facts. The success enjoyed by other aspects of medicine places in society ever larger numbers of aged, nonproductive individuals. Some continue to enjoy life, but to some life is a burden. The cost and effort to provide care for this group becomes an increasingly large fraction of total goods and services. A highly advanced civilization will have to find an appropriate solution to this problem. Meanwhile, the need for support of research that will benefit the health of the newborn and of young productive people in general competes for the personnel and material resources required for research on aging. The need for dedicating large-scale support to children and adolescents may limit the effort society can devote to keeping the aged alive beyond a reasonable state.

Moreover, similar considerations must also apply to the dedication of resources to the care of nonproductive individuals as compared with that of potentially productive persons. Viral diseases of childhood, congenital malformations, hereditary disease, leukemias of childhood, endocrine dyscrasias, demyelinating diseases that strike young mothers, trauma, accidents, etc., all appear more worthy of attention than do the afflictions of advanced age. But these are harsh decisions and should not be made in an absolute manner. Our nation has the resources and can afford attack on the entire front. Moreover, as we have already noted, research cannot be forced. Alert clinical investigators in sufficient numbers should be poised to apply new understanding flowing from basic research to alleviation of the human condition as opportunity affords, avoiding attempts to apply the inapplicable.

GENETIC DISEASES

The main illnesses of man have changed greatly in importance during the last hundred years. Infectious diseases have been combated effectively as their biological nature has become clarified. For example, malaria was recognized as being caused by mosquito-borne protozoa, tuberculosis by bacteria, and influenza by viruses. Sulfa drugs and the antibiotics were found to kill the infectious agents without damaging the host. Other diseases, like rickets, pellagra, and scurvy, were shown to be caused not by the presence of an abnormal agent but by the absence of normally required substances, the vitamins. Improved nutrition in the light of this knowledge has greatly reduced the incidence of such deficiency diseases in the developed nations. However, kwashiorkor, the consequence of protein deficiency, and xerophthalmia, due to deficiency of vitamin A, afflict tens of millions of the young in the tropical and subtropical countries around the globe. Resolution of these problems does not require further understanding of human biology; it requires a vast effort to upgrade education, agricultural practice, and the economies of these nations.

Increasingly, the diseases that plague man are "inborn errors" that, as the effects of abnormal genotypes, lead to gross congenital malformations or to subtle derangements of metabolism. Many of these inborn errors are now understood in biochemical terms. The example that we considered earlier, phenylketonuria, is a rare but serious inherited condition in which a specific enzyme, formed in the liver of normal persons, is not synthesized in the livers of affected individuals. A single gene, present in normals and absent or present in a defective mutant form in phenylketonurics, is responsible for the difference. The result of the absence of the enzyme is accumulation of the amino acid, phenylalanine, which normally is transformed by the enzyme into some other substance. This accumulation results in brain damage expressing itself in mental defect. The presence of this genetic defect can be discovered soon after birth by the presence of an abnormal derivative of phenylalanine in the urine or by an excess of phenylalanine in the blood. It took 20 years to unravel the biochemistry and genetics of the disease. Only then did it become possible to devise a treatment for it. Phenylalanine, a constituent of proteins, is essential in our diet. If, beginning in early infancy, a phenylketonuric individual is given a special diet very low in phenylalanine, development may proceed in improved fashion and the mental abilities of the child may approach normality.

It is hardly more than 10 years since phenylketonuria has been treated this way, and the last word has not yet been said about the real degree of success. Nevertheless, phenylketonuria is an example of the modern attack on genetic disease. Although the defective gene itself, which is responsible for the absence of the enzyme, cannot be "cured," its effect can be circumvented in greater or less degree. In diabetes, a disease with a complex genetic basis involving more than one gene, the defect is overcome to a considerable extent by furnishing the body with insulin from the outside. In galactosemia, the ingestion of milk sugar generates a problem, evident as stunted physical and mental growth and cataracts. Avoidance of milk and use of a synthetic formula containing cane sugar is all that is required. Or, in a congenital malformation such as cleft lip and palate, plastic surgery can not only save the life of severely affected infants but also produce an esthetically acceptable appearance. Clearly, a euphenic solution has to be discovered separately for each untoward genetic effect. Only a few such solutions have been found, and further search will be an important area of biological effort. There is reason to expect that future work will extend the range of euphenics to many errors that are presently beyond repair. In turn this will engender the serious problem, considered earlier, of protecting the gene pool.

One special class of genetic disorder warrants specific comment. A variety of serious disorders are the expression of a deranged chromosomal pattern, broken chromosomes, or an extra chromosome (trisomy), which can now be detected during uterine life or at birth. The survival of an infant so afflicted is an emotional and economic burden to its parents and a drain on the society, which must support and maintain it in an institution for its entire life.

REGENERATION

After the first few years of life, accidents are a leading cause of death. Even more frequent than fatal accidents are damaging accidents that maim or cripple. Whereas human amputees are doomed to empty sleeves or trouser legs, some lower forms—the lobster or newt—can perform prodigious feats of regeneration, replacing entire limbs. There are few leads to indicate the underlying basis for superiority in this ability in a variety of species, as contrasted with man's inability. But it would seem that a determined effort in this regard is certainly warranted. The boon to humanity would be huge, indeed, were it to find any success.

THE DELIVERY OF MEDICAL CARE

We cannot leave the subject of the future of medicine without noting that a substantial fraction of American society, the urban and rural disadvantaged, today lack access to medical care of the quality available to their more privileged fellow citizens. Many may be permanently limited by the experiences of very early life. Chronic illness, excess rates of infant and maternal mortality, and unnecessarily foreshortened life-spans are their lot. Humane considerations, loss of potential productivity, and the heavy burden upon the rest of society all argue for early amelioration of these circumstances. No agency of our society has either the knowledge or the means to do so. Hence, we urge the organization and implementation of a series of substantial field trials addressed to the design of an appropriate system of medical care for this segment of society.

Early Environmental Influences

Inadequate environments may lead to defects in genetically adequate persons, whereas an appropriate environment may minimize the consequences of genetic defectiveness. Recently, much attention has centered on the effects of early environmental influences.

As commonly used, "early influences" denotes the conditioning of behavior by all the experiences of very early life. Early experiences, however, do more than condition behavioral patterns; they also profoundly and lastingly affect many biological characteristics of the adult. Events during the prenatal and early postnatal period condition the initial growth rate, maximum adult size, efficiency in utilization of food, resistance to malnutrition, to infection, and to other forms of stress.

Early influences affect some of the most obvious characteristics of human populations. Throughout the past century, for example, there has been a constant trend toward greater size and earlier sexual maturity of children. This phenomenon was first detected in the United States, then in other western countries; it is now particularly striking in Japan and in other areas that have adopted western ways of life. Evidence for increased growth is provided by the greater heights and weights of children at each year of age, by the faster growth rates during adolescence for both boys and girls, and by earlier onset of the menses.

Early nutritional influences, exaggerated by infection, can also deleteriously affect later growth, mental ability, and general health. This is readily documented in the underprivileged areas of the world; very high infant mortality, slow growth during childhood and adolescence, with physical and mental lethargy continuing throughout life, are among the pathological manifestations commonly observed in all seriously deprived social groups. These disorders are not racially determined. For example, they are found alike among the deprived Indians of Central America and among the populations of European origin who share the Indians' ways of life. In contrast, these manifestations are rare among both groups when born and raised in social and economic environments similar to those now prevailing in the prosperous communities of the United States and Europe. And there is every reason to suspect that similar influences may be at work among the urban and rural disadvantaged sector of American society, creating individual handicaps that can never be overcome.

The most important effects of the environment are those experienced during very early childhood or the last trimester of fetal life. Most importantly, when environmental phenomena act adversely on the human organism in early life, their anatomical, physiological, and psychological effects are to a large extent irreversible; the fact that the various tissues and organs develop at different rates accounts for the existence of several critical periods in giving complete or partial irreversibility to responses that the developing organism makes to environmental forces. In the human species, the critical periods for the development of various mental capacities probably occur before 6 to 8 years of age and most critically during the first year, a phenomenon of great relevance to the determination of "intelligence" in different socioeconomic and ethnic groups. Effects of the environment are much more likely to be reversible when experienced after the end of differentiation and development.

Virtually all effects of prenatal and early postnatal influences so far recognized in human populations occur also in other animal species. A large variety of stimuli, acting on the pregnant animal during gestation or on the young shortly after birth, affect diverse phenotypic expressions throughout adult life. Exposure to toxic agents, malnutrition, undernutrition or overnutrition, overt or subclinical infections, emotional disturbances of the mother or of the young, crowding, isolation, and other forms of social deprivation are some of the variables that have been used to design experimental models for the study of early influences. These influences have been studied with regard to their effects on anatomical structures, physiological characteristics, metabolic activities, behavioral patterns, and learning ability in adult life. In all cases critical periods have been recognized, differing as to initiation and duration, depending upon the nature of the early influence and of the effect studied.

Nevertheless, the body of knowledge concerning the effects of early influences is superficial and episodic. Even the phenomena that have been most extensively studied—such as imprinting, the fixation during a critical period of a young animal's life of a stimulus that invariably elicits a particular response—are poorly understood, albeit highly reproducible in their details. In the absence of broad scientific generalizations, it is not possible to extrapolate from one animal species to another, let alone to man. Yet there is no doubt that very early influences are of great importance in human life, a fact recognized by the earliest psychiatrists and repeatedly documented since. Knowledge of these potentialities points to the safest and most effective way of affecting the mental as well as the physical development of man. At the same time, if improperly exploited by an authoritarian government, they could be of enormous danger to society.

Future studies of the effects of early influences should be conducted at several different levels:

1. Epidemiological observations in man, taking advantage of the fact that different human societies exhibit a wide range of customs with regard to gestation, parturition, lactation, and physical and behavioral management during the early postnatal period. These differences in social patterns can be considered as experiments on man, performed without awareness of their consequences, demanding careful description and analysis.

2. Development and refinement of experimental models in various animal species. To yield the greatest scientific rewards, these models should use laboratory animals of known genetic and experiential history, observed throughout their life-spans and, preferably, for several successive generations. Such longitudinal studies will require appropriate animal quarters, extensive facilities for recording and retrieving information, and, possibly, a new type of scientific organization.

3. Detailed analysis of the mechanisms through which early influences exert their lasting effect.

Controlled Sex Determination

More than a half century has passed since it was learned that a man produces two kinds of sperm cells in about equal numbers. In addition to 22 chromosomes that are visibly alike in all sperm, half the nuclei possess the relatively large X chromosome and half the small Y chromosome. In conception, X-bearing sperm are female-determining; Y-bearing sperm, male-determining. It may be possible to separate the two kinds of sperm by biological or purely physical methods such as differential centrifugation. Admittedly, no success has been attained yet, in spite of some promising leads. If success comes, insemination with the X or Y fraction of semen would then assure control of the sex of the offspring. Application to animal breeding could be of considerable economic importance, as in the production of dairy cattle. If applied to man, subtle psychological changes in the population might be expected. It is likely that no great deviation from a 1:1 sex ratio would result since most parents of more than one child seem to desire children of both sexes. The sequence of sexes in a family may, however, change considerably. Instead of the random sequence of boys and girls, the majority of firstborn might be boys and that of second-born girls. Since position in the birth order has an effect on both physical and personality traits of the developing offspring, the consequences of the firstborns being all boys and secondborns all girls might well be reflected in behavioral shifts of the population. If widely available, this could also serve as an adjunct to programs of population control by assuring offspring of the desired sexes to all families.

An alternative method for deliberate choice of the sex of offspring is technically feasible even now, albeit less attractive. Relatively simple surgical procedures permit determination of the sex of the young fetus. A family that desires to limit its size should be permitted the option of such inspection and, having one boy, for example, abort the next fetus if it is not a girl. Patently, this technical feasibility raises the same ethical and legal questions as does abortion of a genetically defective conceptus or of an unwanted child, but with less justification.

Differential Fertility

From time to time serious questions are raised about the long-range bio-logical and social effects of differential fertility* on the characteristics of populations. These questions arise because there is generally an inverse relation between fertility and socioeconomic status, measured in terms of occupational status, education, or income, and because there are differences in the fertility of the major races of man. There are also substantial differences in the fertility of the major religious groups. Maximum fertility is found among Moslems, with Hindus and Buddhists next. Christians as a group have lower fertility, and among Christians in the United States, Catholics have higher fertility than Protestants. In worldwide terms, however, Catholics run the gamut from very low fertility, e.g., in North Italy, to the highest fertility in the world in some parts of Latin America. In general, Jews exhibit the lowest fertility of any of the world's major religious groups.

---

* In this context, the word "fertility" is employed in its demographic usage, indicating the number of offspring produced per 1,000 of population, and not in the sense of the opposite to sterility.

A few sweeping propositions can be ventured:

1. Many data are available that are descriptive of the differences in the fertility of broad socioeconomic, racial, and regional groups, but little is known about the significance of these differences for either the social or the biological heritage.

2. In biological terms, it is probable that differences in reproductive performances among individuals of varying characteristics within all groups are much more important than differences in the average performances among the highly heterogeneous social groupings for which data are readily available.

3. In the developed world, as birth rates have declined since the mid-nineteenth century, the inverse relation between socioeconomic status and fertility first became stronger and then weakened. The small-family pattern has tended to occur first in the urban upper classes and only later to spread throughout society. As governments in the newly developing countries mount national programs to spread the practice of family planning, it is likely that the trends in the lower social strata of the populations will follow those of the upper strata more closely.

4. In the white population of the United States the "class" differences narrowed substantially with the postwar rise of the birth rate, which was more pronounced in the urban and upper-class groups than in the rural and lower economic strata. The inverse relation remains, however, partly because of an earlier age at marriage in the lower economic groups. It is probable that much of the higher fertility of the lower status and income groups would disappear if contraceptive information and services were made readily available.

5. In the United States, the fertility of Negroes exceeds that of the white population, mainly as a correlate of their lower educational, economic, and social status. When similar educational, income, and occupational groups are compared, the differences are greatly reduced and even reversed.

From a genetic viewpoint, differences in fertility among groups of people are important only when these groups differ in their genetic endowments. From the societal standpoint, only those genetic differences count that may bear on the intellectual, behavioral, and social attributes characteristic of man.

Selection and the Variability of Man

Even without scientific knowledge of genetics, man created a great variety of genetically different strains of domesticated animals and plants by selecting for desired types and breeding. The wild ancestors of cattle, dogs, chickens, wheat, and corn, for example, appeared rather uniform. Nevertheless, deliberate breeding practice has shown that a great amount of concealed genetic variation was present beneath the apparent uniformity—variation that enabled man to select for traits that appeared desirable. Such selection led to the establishment of cattle specialized for milk or for meat production, the astonishing manifoldness of races in dogs, and chickens high bred for egg-laying or for meat yield. At the same time, selection led to disease resistance, heat tolerance, and other physiological states. Plant strains were selected for adaptation to many climatic and soil conditions, as well as for yields that surpass by far those attained in the wild state.

But, in man, selection occurred without conscious direction. Different groups of mankind differ from one another in many ways; the significance of most of these differences is unknown or only incompletely understood. Why do the average body sizes of populations vary from the pygmies of Africa to the tall Watusi of the same continent, from the shorter Southern Mediterraneans to the taller Scots? Why do the facial features of Orientals differ from those of the Caucasians? It is possible that some of these differences arose by the chance sampling of genetic types in the distant past, followed by long periods of physical, hence genetic, isolation. Other differences have been presumed to be the result of natural selection. Dark pigmentation of the skin is an asset in the tropics, where it protects the tissues against excess ultraviolet radiation. Light pigmentation is an asset in northern regions, where enough ultraviolet light must penetrate the surface to transform dihydrocholesterol into vitamin D. Long limbs serve as radiators of heat in desert peoples, while short extremities conserve body heat in arctic climates. What is useful, natural selection preserves, and what is of negative value it rejects.

While it is obvious that the racial groups of mankind differ from one another in specific ways, the multitude of differences in facial features, body build, height, and other physical as well as mental traits readily indicate the great heterogeneity of each major population group. Indeed, the extensive overlap of these groups with respect to many genetically determined traits is as impressive as the differences that distinguish the groups. Moreover, the polymorphic nature of any human group—and that of every other species studied intensely, e.g., cattle, chickens, flies—has become dramatically clear in recent years. Every human group contains a great variety of genes for alternative blood substances, serum proteins, hemoglobin, and enzymes, and new polymorphisms are constantly added to our knowledge. Thus, in every population there are people belonging to blood group M, others to N, and still others to MN. Why should there be a variety of genes determining these properties instead of a single type best fitted to survival and therefore having become fixed by natural selection? What is the biological and sociological significance of polymorphism? The inability to answer these questions for most, if not all, human polymorphisms indicates fundamental gaps in our understanding of the genetics of human populations. Selective forces must exist that operate to retain variety of genes rather than to eliminate all but one of each kind. How these forces act specifically, so as to enhance survival of a gene under certain genetic or environmental circumstances and to decrease its survival under other circumstances, is unclear and will have to be established in each individual case. If we do not know how we became extensively polymorphic in the past or how we retain polymorphism at present, we cannot expect to predict the genetic future.

The complexity of selective forces is such that to gain the necessary understanding there are required longitudinal studies, from birth to death, of exceedingly large cohorts; analysis of the data will require the use of powerful computers. The biological insights to be so gained should elucidate the causes of the great load of biological losses in the form of spontaneous abortions, stillbirths, deaths before the end of the reproduction period, reduced fertilities, and infertility.

Notwithstanding the complexity of genetic population dynamics, gross interference with natural conditions is clearly possible. It would not take many generations to breed Caucasians whose average adult body size is four feet or average Japanese of six feet. We could breed for obesity or leanness, blue eyes or black, wavy or wiry hair, and any one of the obvious physical attributes in which human beings vary. Presumably, we could also breed for mental performance, for special properties like spatial perception or verbal capacity, perhaps even for cooperativeness or disruptive behavior. Most of these traits vary not only genetically but also under the influence of environmental factors, as, for example, size and weight with food, or mental scores with impressed social attitudes and educational opportunities. This, however, does not negate genetic components in the determination of the variety of traits. The heritability of a trait, which is a measure of the part genes play in the observed variability of the trait, may be large or small. Although more research with respect to heritability of human traits is needed, it is abundantly clear that selection could be effective even with traits of quite low heritability.

Although potentially able to select his own genetic constitution, man has not made use of this power. Selection is a harsh process. To make speedy progress, reproduction should be limited primarily to those who possess genotypes for the desired traits. But who will decide what is desirable? How much genotypic and phenotypic variability would be optimal in the human society? Who would dare to prohibit procreation to a majority of men and women, limiting this activity only to an elite group? And to whom would society entrust such decisions? May we expect changes in attitudes of whole societies so that they would accept the self-control of human evolution at the cost of foregoing the private decisions of most people to propagate themselves in their own children? It is extremely unlikely that such changes in attitudes will come soon. The future of man, however, may well extend over incomprehensibly long times, long enough not only to ponder these possibilities but also to explore them in actuality.

In order to overcome some of the objections to all-out self-selection by man, the late H. J. Muller advocated partial selection for the betterment of mankind. Muller proposed deep-freeze storage of the sperm of the most distinguished men. Such storage was to extend over a long period, perhaps decades after the death of the sperm donors, in order to give perspective to the judgment of their being unusually distinguished. The sperm of those who withstood the test of time were then to be made available to married couples. The wife and the donor would become biological parents while the husband would, like an adoptive parent, influence the child by his personal attributes. This scheme has a low genetic efficiency as compared to procedures in animal breeding. Its emotional appeal, too, is limited. Yet its control over man's genetic future, granting its limitations, would be accomplished by methods that leave room for free choice. Moreover, the procedure is already employed in numerous cases of infertility of a husband, without, however, using the opportunity to choose unusually distinguished sperm donors. Careful selection of the mothers by an appropriate agency and subsequent inbreeding might, however, accomplish the goal of a "superior" breed of man exhibiting the criteria chosen, but this was not essential to Muller's suggestion since the loss of free choice diminishes the social acceptability of the scheme.

A much more efficient and a most revolutionary way of selecting for specific human genotypes has been suggested on the basis of experiments with frogs and other amphibians—experiments whose original purpose had nothing to do with plans for genetic selection. It is possible to remove the haploid nucleus (half the adult number of chromosomes) of a frog egg before fertilization, and, instead of fertilizing it with sperm, implant the diploid nucleus (a complete nucleus with a double set of chromosomes, one from each parent) of a body cell from a frog embryo. Such an egg can develop into an adult frog with the same genetic constitution as the frog embryo whose body cell provided the transplanted nucleus. If the method of nuclear transplantation with subsequent full development should become successful, utilizing the nuclei of body cells of adult individuals, and if it could be applied to man, a most powerful means of controlling the genetic constitution of future generations would become available. Since the nucleus of a body cell retains the totality of one's genes, a child produced by an enucleated egg that had been supplied with the nucleus of an adult body cell would, genetically, be an identical twin of the donor of that body cell. Moreover, any desired number of genetic twins could be produced. It would require the collection of unfertilized eggs from the oviducts of many women, removal of the egg nuclei, and replacement by the nuclei of body cells of the chosen man or woman. This would be followed by return of the eggs to the uteri of women who then would undergo normal pregnancies. In this way one could produce multiple identical copies of any person judged admirable.

Technically, it is still a long way from the use of frog eggs to the use of human eggs, but what can be done in frogs today will surely be possible in man tomorrow. The biological problem now is primarily one of skill and development of detailed procedures. The next step would probably be the extension of the techniques from amphibians to laboratory mammals. Once successful in mice or rabbits, there could be practical applications to animal breeding. Prize bulls or cows could be perpetuated by identical "offspring" derived from their body cells. From thence, technically, the steps toward potential human use would not be difficult, and if there were a strong wish to make such potentialities a reality, it could probably be accomplished within a few decades.

At this time, there is need to ponder the personal and social implications of this biologically possible procedure. Powerful social forces would as surely resist adoption of such practice as they would a deliberately undertaken breeding program with selected human beings. At the present moment of extremely dangerous population growth, social pressures are best directed to lower reproduction, in general, without qualitative considerations. But one day, when populations are stable, world peace is the norm, and man's social and political institutions are sufficiently mature to assure that biological understanding will not be utilized to perpetuate injustice or strengthen dictatorship but, rather, to expand human potential, man will be free to guide his own evolutionary destiny.

There is no doubt that much of the seeming variability of mental, behavioral, and social traits can be accounted for by graded differences in nongenetic factors such as wealth and poverty, intellectual stimulation and its absence, or environmental encouragement and discouragement, as well as malnutrition in early life. A major task before mankind is to see that these nongenetic factors are adjusted so that each individual realizes his genetic potential to the fullest. At the same time, however, while the performance of unchanged genotypes may be improved in this way, intensive studies of the existing genetic variability should make possible the design of realistic blueprints for the control of man's biological makeup. These plans will rest on a future deeper understanding of the "gene pool," the genic content of populations quite apart from its actual existence in living individuals.

In abstract terms, control of the genetic future of man consists of manipulation of the gene pool. In concrete terms, such manipulation is accomplished by specified reproductive patterns of individuals. Although the hypothetical production of multiple identical copies, discussed above, may become technically feasible, there can be no certainty that a given genotype, successful under one set of conditions, would be equally successful under different circumstances. The future of man is more likely to be rich and exciting, to progress to greater possibilities, by exploring the variety of the gene pool than by standardizing on some uniform *Homo sapiens*. Although it might be feasible, we forcefully reject the abhorrent thought of breeding subsets of humans specifically adapted to the performance of various tasks, thereby creating a highly efficient but antlike society.

The brain of man has not increased significantly in size since his Cro-Magnon ancestor, perhaps not for many millennia before. When one day man accepts responsibility for his acknowledged power to control his own genetic destiny, the choice between various plans must be based on value judgments. When he begins to use the power to control his own evolution, man must clearly understand and define the values toward whose realization he is to strive.

Man's view of himself has undergone many changes. From a unique position in the universe, the Copernican revolution reduced him to an inhabitant of one of many planets. From a unique position among organisms, the Darwinian revolution assigned him a place among the millions of other species that evolved from one another. Yet *Homo sapiens* has overcome the limitations of his origin. He controls the vast energies of the atomic nucleus, moves across his planet at speeds barely below escape velocity, and can escape when he so wills. He communicates with his fellows at the speed of light, extends the powers of his brain with those of the digital computer, and influences the numbers and genetic constitution of virtually all other living species. Now he can guide his own evolution. In him, nature has reached beyond the hard regularities of physical phenomena. *Homo sapiens,* the creation of nature, has transcended her. From a product of circumstances, he has risen to responsibility. At last, he is Man. May he behave so!

METHODOLOGY: SURVEY OF INDIVIDUAL LIFE SCIENTISTS

POPULATION SELECTION

Detailed information from over 12,000 of some 24,000 biologists identified as actively engaged in research and working within the United States of America or its possessions was obtained from July 20 through November 30, 1967. The questionnaire "Survey of Life Scientists" forwarded to each individual is reprinted here as Exhibit A-1. All individually identifiable biologists meeting the following four criteria were surveyed:

Possessed a doctoral degree (Ph.D. or D.Sc.) or a health-professional degree, regardless of field of training or nature of degree. (The British MB.BS. was equated with the American M.D. degree.)

Was employed full-time.

Was a self-classified biologist, either by training or by research activity.

Devoted 20 percent or more of the work week to research during 1966 or expected to do so in 1967.

The National Science Foundation National Register of Scientific and Technical Personnel had the most complete listing of scientists, which included the desired population. Through cooperation of the Foundation and its National Register 23,388 individuals answering the 1966 Register Questionnaire were identified who clearly met the first two criteria and who were considered likely to meet the third and fourth criteria above for the following reasons:

The first or second work activity had been identified as one of the following:

    Basic research
    Clinical research and investigation
    Applied research
    Management or administration of research or development
    Clinical practice

One of the following categories (including all subcategories) of the Register's 1966 Specialties List had been identified as the area of greatest specialization:

GROUP A

Agronomy
Anatomy
Animal Husbandry
Biochemistry
Biophysics
Biochemical
    Oceanography
Botany
Ecology
Entomology

Fish and Wildlife
Forestry
Genetics
Horticulture
Immunology
Microbiology
Nutrition
Pathology
Pharmacology
Physiology

Plankton
Plant Pathology
Range Management
Virology
Zoology
Other Biomedical
    Specialties
Biology
All Other (Biology)

GROUP B *(Interdisciplinary)*

Agricultural and Food Chemistry
Biometrics and Statistics
Climatology
Clinical Psychology

Experimental, Comparative, and
    Physiological Psychology
Hydrology
Paleontology
Soil Specialties

TABLE A-1   Proportion of Scientists with Certain Interdisciplinary
Specialties Included in the Life Sciences Survey

AREAS OF PRIMARY SPECIALIZATION	NUMBER LISTED BY NATIONAL REGISTER	MEETING SURVEY CRITERIA	
		Number also Associated with AIBS or FASEB	Percentage of Total in Category
TOTAL	**4,601**	**841**	*18.3*
Agricultural and Food Chemistry	912	210	*23.0*
Biometrics and Biostatistics	153	19	*12.4*
Climatology	71	5	*7.0*
Clinical Psychology	459	13	*2.8*
Experimental, Comparative, and Physiological Psychology	1,892	105	*5.5*
Hydrobiology	114	8	*7.0*
Paleontology	381	45	*11.8*
Soil Specialties	619	436	*70.4*

For scientists listing one of the interdisciplinary fields in Group B, inclusion in the survey further required that they had listed a member society of either the American Institute of Biological Sciences or the Federation of American Societies for Experimental Biology as their major professional society. Approximately 2,500 scientists reported Group B interdisciplinary specialties, however, only 843 individuals fulfilled the additional requirement of society membership. Table A-1 shows the number of scientists reporting these subspecialties to the Register and the percentage meeting the additional criterion.

Approximately 87 percent of the individual names finally included were provided by the National Register; and for these biologists the Register also provided the following data:

Year of birth
Sex
Citizenship (U.S. or foreign only)
Professional identification
Type of principal employer
Professional location (state only)
Support of research from federal funds (yes, no, or unknown)

Additional biologists, if their names did not appear on the National Register list, were identified within the following groups and were included in the survey:

Consultants (155) comprising the 22 panels of the Life Sciences Study.

Department chairmen (preclinical and clinical) of the 87 functioning medical schools listed in the 1966–1967 Directory of the Association of American Medical Colleges.*

All chairmen identified by the American Institute of Biological Sciences in its listing of "Life Science Departments."

Chairmen of additional life science departments identified by the Office of Scientific Personnel of the National Research Council.

The unduplicated membership, active and emeritus, of four clinical medical research societies:

The American Society for Clinical Investigation
Society for Gynecological Investigation
Society for Pediatric Research
Society of University Surgeons

Inclusion of the membership of the four societies compensated for incomplete coverage of clinical investigators by the National Register. In all, approximately 2,500 additional individual society members were identified who potentially qualified as research biologists. No presurvey information was available concerning their work activity; survey returns subsequently revealed that a high proportion of members of these clinical medical societies were not actively engaged in research.

The selection criteria excluded most paleontologists identified by the National Register; of the 381 individuals listing paleontology as their major specialization, only 45 were included in this survey. Systematic biologists, though not having to list society identification as a secondary requirement, also appear to be under-represented. Anthropologists, including physical anthropologists (1,172 scientists), were excluded.

Table A-2 summarizes the number and source of names in the survey mailing, and Table A-3 summarizes the percentage return of completed questionnaires.

---

* *Directory of Administrative Staff, Department Chairmen and Individual Members in Medical Schools of the United States and Canada,* Association of American Medical Colleges, Evanston, Illinois, 1967.

TABLE A-2   Composition of Mailing List, Individual Life Scientists

SOURCES	NUMBER OF LIFE SCIENTISTS
TOTAL VALID NAMES[a]	**25,946**
National Register	22,490
Consultants	23
Departmental Chairmen (Medical)	721
Departmental Chairmen (Other)	283
Clinical Research Society Members	2,429

[a] Biologists not meeting all criteria for inclusion in the study were subtracted from the total mailing, as were all questionnaires returned because of incorrect addresses, etc.

Tabular data in the body of this report include only those individuals providing all necessary information; therefore, the grand total of any individual table is somewhat less than the total respondents. Awkward placement of Question 19, which dealt with individuals' research areas, caused approximately 25 percent of the respondents to omit this question. Internal correlation of several parameters shows the omission to be random. Thus the percentages of life scientists working for the various types of major employer were similar for the total survey population and for the 8,139 biologists who also reported their research areas (Table A-4). Because

TABLE A-3   Summary of Returns: Individual Life Scientists

CATEGORY	NUMBER OF LIFE SCIENTISTS
A. TOTAL VALID MAILING[a]	**25,964**
B. Maximum Potential Researchers[b]	23,967
C. TOTAL RETURNS[a]	**14,362** (55.4% of A)
D. Life Scientists Meeting Research Criteria	12,383 (51.7% of B)

[a] Biologists not meeting all criteria for inclusion in the study were subtracted from the total mailing, as were all questionnaires returned because of incorrect addresses, etc.
[b] This number is the total valid mailing minus the *known* nonresearch biologists and represents a maximal figure since it is impossible to estimate what proportion of nonrespondents qualified as researchers.

TABLE A-4   Similarity of Distribution by Major Employer between Respondents to Question 18 and All Respondents

PRINCIPAL EMPLOYER	TOTAL SURVEY POPULATION		LIFE SCIENTISTS REPORTING CURRENT RESEARCH AREA	
	Number	Percentage	Number	Percentage
TOTAL	**12,364**	*100.0*	**8,139**	*100.0*
Institution of Higher Education	8,288	*67.0*	5,476	*67.3*
Nonacademic Total	**4,076**	*33.0*	**2,663**	*32.7*
Private Industry or Business	1,186	*9.6*	761	*9.4*
Federal Government	1,713	*13.9*	1,149	*14.1*
Federal Contract Research Center	135	*1.1*	81	*1.0*
State and Local Government	309	*2.5*	200	*2.5*
Nonprofit Organization	462	*3.7*	315	*3.9*
Independent Hospital or Clinic	219	*1.8*	133	*1.6*
Self-employment	23	*0.2*	8	*0.1*
All Other	29	*0.2*	16	*0.2*

only 19 people failed to report an employer, a total approximating 8,139 must be considered the reference base, representing 100 percent response, for all tabulations having "research area" as one category. Similar comparison of the percentages trained in different fields (Table A-5), of geographic location, and of type of doctoral degree earned also revealed no significant differences between the two groups.

TABLE A-5   Similarity of Distribution by Field of Training between Respondents to Question 18 and All Respondents

FIELD OF TRAINING OF DOCTORAL DEGREE	TOTAL SURVEY POPULATION		LIFE SCIENTISTS REPORTING CURRENT RESEARCH AREA	
	Number	Percentage	Number	Percentage
TOTAL	**12,151**	*100.0*	**8,005**	*100.0*
Agriculture Subtotal	**855**	*7.0*	**576**	*7.2*
Agronomy	347	*2.9*	227	*2.8*
Animal Husbandry	132	*1.1*	93	*1.2*
Fish and Wildlife	50	*0.4*	36	*0.4*
Forestry	88	*0.7*	69	*0.9*
Horticulture	158	*1.3*	98	*1.2*
Agriculture, Other	80	*0.7*	53	*0.7*
Biological Sciences	**8,269**	*68.1*	**5,549**	*69.3*
Anatomy	196	*1.6*	130	*1.6*
Biochemistry	1,834	*15.1*	1,271	*15.9*
Biophysics	160	*1.3*	115	*1.4*
Cytology	109	*0.9*	80	*1.0*
Embryology	105	*0.9*	80	*1.0*
Microbiology	1,010	*8.3*	656	*8.2*
Pathology, Animal	77	*0.6*	52	*0.6*
Pharmacology	374	*3.1*	225	*2.8*
Physiology, Animal	805	*6.6*	535	*6.7*
Botany	365	*3.0*	230	*2.9*
Ecology and Hydrobiology	234	*1.9*	159	*2.0*
Entomology	415	*1.8*	288	*3.6*
Genetics	408	*0.6*	287	*3.6*
Nutrition	221	*3.4*	153	*1.9*
Paleontology and Systematic Biology	72	*3.4*	50	*0.6*
Pathology, Plant	345	*2.8*	227	*2.8*
Physiology, Plant	353	*2.9*	241	*3.0*
Zoology	773	*6.4*	512	*6.4*
Biosciences, All Other	413	*3.4*	258	*3.2*
Health-Professional Subtotal	**2,315**	*19.1*	**1,391**	*17.4*
M.D.	2,118	*17.4*	1,270	*15.9*
D.D.S.	65	*0.5*	35	*1.4*
D.V.M.	109	*0.9*	74	*1.0*
Other [a]	23	*0.2*	12	*0.1*
Related Areas Subtotal	**712**	*5.8*	**489**	*6.1*
Chemistry	442	*3.6*	305	*3.8*
Physical Sciences [b]	114	*0.9*	77	*1.0*
Psychology	105	*0.9*	73	*0.9*
All Other Fields [c]	51	*0.4*	34	*0.4*

[a] Includes D.O., D.P.H., D.Pharm., and other health-professional degrees not specified.
[b] Includes biometrics and biostatistics, computer science, earth sciences, engineering, mathematics, physics, and statistics.
[c] Includes anthropology, other social sciences, and other related fields of training.

DATA ANALYSIS: DEFINITIONS AND TABULATION CONSTRAINTS

Definitions

The questionnaire insert, Exhibit A-2, defines the following:

> Life scientists/Life science field
> Postdoctoral appointee
> Continuing or senior research associate
> Research dollars to be reported

Where applicable, definitions used in this study are essentially the same as those employed in the Academy's study of Postdoctoral Education.*

Constraints

Constraints on certain items tabulated for all respondents include the following:

Principal employer: State, but not city, tabulated.

Previous employer: Tabulated only if individual reported more than one employer since earning a doctorate.

Education information: Requested and tabulated only at the baccalaureate, doctoral, and postdoctoral levels. No information was obtained pertaining to the master's degree.

Field of training of doctoral degree—special note: Field of training presented no special tabulation difficulties for individuals reporting only a Ph.D. or D.Sc. degree(s). The area of specialization reported for the most recent doctorate was used. However, for biologists reporting *only* a health-professional degree, e.g., M.D., D.D.S., D.V.M., it was considered inappropriate to assign an area of specialization at the doctoral level. Therefore, all such respondents, whether they indicated such an area of specialization at this level or not, were classified by the name of their professional degree. For respondents who had both a health-professional degree and a Ph.D. or D.Sc. degree, the field of training of the latter type of degree was used irrespective of the order in which the degrees were earned.

* *The Invisible University: Postdoctoral Education in the United States,* Report of a Study Conducted under the Auspices of the National Research Council. National Academy of Sciences, Washington, D. C., 1969.

The following information was tabulated only for academic life scientists:

Name of principal employer: University names, types of school, and departmental titles were tabulated. Since approximately 650 distinct departmental names were reported, this parameter was unsuitable for further analysis.

Salary source: Requested and tabulated as a percentage of the total salary.

Number and type of personnel in their research groups were requested only of *principal* investigators.

Source and amount of research funds were requested only of *academic* principal investigators. They were reported as dollars (direct costs) available to principal investigators on June 1, 1967, 1966, and 1965.

SUBCATEGORY LISTINGS REQUIRED TO ANSWER QUESTIONS 18 AND 19

The questionnaire insert, Exhibit A-2, contains the predefined range of categories to be used by respondents in answering questions pertaining to the following factors:

Field of training
Research area

Research materials
Research organisms

VALIDITY OF THE RESPONDENT POPULATION

The professional location of the 23,388 biologists identified by the National Register was compared with the professional location of the 14,362 biologists returning questionnaires (Table A-6). The percentage of each group working within a given state was essentially the same for those receiving questionnaires and those responding. Similar comparison, based on the type of doctoral degree(s) earned by individual life scientists, showed only slight differences (Table A-7). Thus, no readily detectable bias was evident in the population responding to this survey.

TABLE A-6 Geographic Location of Biologists Identified by National Register and Life Scientists Answering Survey Questionnaire

NORTH

Census Region	State	Percentage of Biologists in State	
		NSF File [a]	Survey File [b]
1	**TOTAL**	**7.9**	**7.3**
11	Maine	0.4	0.4
12	New Hampshire	0.4	0.4
13	Vermont	0.3	0.4
14	Massachusetts	4.6	4.0
15	Rhode Island	0.4	0.4
16	Connecticut	1.8	1.7
2	**TOTAL**	**20.5**	**19.0**
21	New York	11.9	11.0
22	New Jersey	3.1	2.6
23	Pennsylvania	5.5	5.4
3	**TOTAL**	**16.9**	**17.3**
31	Ohio	3.4	3.6
32	Indiana	2.4	2.5
33	Illinois	5.4	5.4
34	Michigan	3.4	3.5

SOUTH

Census Region	State	Percentage of Biologists in State	
		NSF File [a]	Survey File [b]
5	**TOTAL**	**17.7**	**18.0**
51	Delaware	0.4	0.4
52	Maryland	6.8	6.5
53	District of Columbia	2.2	2.3
54	Virginia	1.5	1.6
55	West Virginia	0.4	0.4
56	North Carolina	2.4	2.6
57	South Carolina	0.5	0.5
58	Georgia	1.6	1.6
59	Florida	2.0	2.1
6	**TOTAL**	**4.2**	**4.4**
61	Kentucky	1.0	1.1
62	Tennessee	1.8	1.9
63	Alabama	0.9	0.9
64	Mississippi	0.6	0.5

WEST AND POSSESSIONS

Census Region	State	Percentage of Biologists in State	
		NSF File [a]	Survey File [b]
8	**TOTAL**	**4.3**	**5.1**
81	Montana	0.4	0.4
82	Idaho	0.3	0.4
83	Wyoming	0.2	0.2
84	Colorado	1.2	1.4
85	New Mexico	0.5	0.6
86	Arizona	0.7	0.8
87	Utah	0.8	1.0
88	Nevada	0.1	0.3
9	**TOTAL**	**13.1**	**12.9**
91	Washington	1.9	1.8
92	Oregon	1.4	1.5
93	California	9.8	9.6

41	Minnesota	2.2	2.2
42	Iowa	1.4	1.8
43	Missouri	2.0	2.0
44	North Dakota	0.4	0.4
45	South Dakota	0.4	0.4
46	Nebraska	0.6	0.7
47	Kansas	1.0	1.1
71	Arkansas	0.5	0.5
72	Louisiana	1.2	1.2
73	Oklahoma	0.9	0.9
74	Texas	3.1	3.3
94	Alaska	0.2	0.2
95	Hawaii	0.6	0.6
96	Virgin Islands	—	—
97	Panama Canal Zone (Guam)	<0.1	<0.1
98	Puerto Rico	0.1	0.1

a Percentages based on 23,388 respondents to the National Register of Scientific and Technical Personnel, 1966, of the National Science Foundation.
b Percentages based on 14,362 respondents to Individual Questionnaire.

TABLE A-7 Distribution of Life Scientists by Type of Degree

LIFE SCIENTISTS	TOTAL FOR CATEGORY	PERCENTAGE OF TOTAL WITH:		
		Ph.D. or D.Sc. Only	H.P.D.[a] Only	H.P.D. Plus Ph.D. or D.Sc.
Identified by 1966 National Register[b]	**23,388**	72.8	21.7	5.5
All Biologists in Survey	**12,383**	74.5	19.1	5.4
Biologists Reporting Research Area	**8,005**	77.1	17.4	5.5

a Health-professional degree, includes M.D., D.D.S., D.V.M., D.O., D.P.H., D.Pharm., and other health-professional degrees not specified.
b Life scientists who responded to the 1966 National Register Questionnaire of the National Science Foundation.

SURVEY OF LIFE SCIENTISTS

This form will be held in confidence; subsequent published information will be statistical in nature, unidentifiable with individual organizations or respondents.

In calendar 1966 did you spend **20% or more** of your time in research? Yes ☐ No ☐
In 1967 do you expect to spend **20% or more** of your time in research? Yes ☐ No ☐

If the answer to EITHER question is "YES," complete the entire questionnaire.
If the answer to BOTH questions is "NO," please complete ONLY ITEMS 1 through 13, SIGN and DATE your questionnaire on the last page, and return in enclosed envelope.

Please **type** or **print** answers.

GENERAL INFORMATION

NAME _____
Last   First   Middle Initial

1. Age in Years _____   2. Sex M ☐   F ☐

3. Place of Birth _____
State   Country

4. Citizenship (check one)
USA ☐   Non-USA ☐   (Specify Country) _____
☐ Temporary USA resident
☐ Permanent USA resident

CURRENT EMPLOYMENT

5. Check current status of employment
Full time ☐   Part time ☐   Unemployed ☐   Retired ☐

6. Name of **Principal** Employing Institution/Organization. _____

   a. Circle type of subdivision and specify its exact title:
   Academic Department, Institute, Center, Laboratory, Division, Other

   _____
   Title

   b. If you hold a joint appointment, identify subunit _____

7. Your professional location
   _____
   City   State

8. How many years have you been with your present **principal** employer (whether or not your position within the organization has changed)? _____ yrs.

9. Title of Current Position—Please check **only one** item per class

Academic
☐ Professor
☐ Adjunct Professor
☐ Associate Professor
☐ Assistant Professor
☐ Instructor
☐ Research Associate ("Senior" or continuing appointment)
☐ Postdoctoral Appointee (whether "fellow," "trainee," "research associate," etc.)

Academic/Administrative
☐ Department Chairman
☐ Dean
☐ Other (specify) _____

Non-Academic Title
☐ (Specify) _____

2

EXHIBIT A-1   Questionnaire used in individual survey.

10. Principal Employer—Type of Organization—Check **all** appropriate items

a. ☐ **Academic**

Private ☐   State ☐   Municipal ☐   Federal ☐

☐ Junior College
☐ Undergraduate College (4 yr. Liberal Arts College)
☐ University: School or College of
  ☐ Arts and Sciences   ☐ Nursing
  ☐ Agriculture   ☐ Pharmacy
  ☐ Dentistry   ☐ Public Health
  ☐ Engineering   ☐ Veterinary Medicine
  ☐ Forestry   ☐ Other (specify) _____
  ☐ Graduate Studies   _____
  ☐ Medicine

INSTRUCTION: If you have checked any items in 10a proceed directly to item 12.

b. ☐ **Private Industry/Business**

INSTRUCTION: If you check part 10b omit the rest of item 10 and complete item 11.

c. ☐ **Federal Contract Research Center** (Managed by Profit, Educational and Other Non-Profit Organizations)

d. ☐ **Federal Government**

☐ Dept. Agriculture   ☐ Dept. Health, Education and Welfare
☐ Dept. Interior   ☐ PHS except NIH
☐ AEC   ☐ NIH
☐ NASA   ☐ Other HEW
☐ NSF   ☐ Dept. Defense
☐ Smithsonian Institution   ☐ Civilian   ☐ Military
☐ VA   ☐ Air Force
☐ Other (specify) _____   ☐ Army
_____   ☐ Navy
                                                ☐ Other DOD

e. ☐ **State Government**

☐ Hospital   ☐ Wildlife/Forest
☐ Health Dept. and Laboratories ☐ Other (specify) _____
☐ Agriculture Service   _____

f. ☐ **Municipal Government**

☐ Hospital   ☐ Museum
☐ Health Dept. and Laboratories ☐ Other (specify) _____
                               _____

g. ☐ **Independent Hospital/Clinic**

☐ Self-Employed: Yes ☐ No ☐

h. ☐ **Non-Profit, Non-Government Research Institute/Foundation/Museum**

j. ☐ **Self-Employed Other Than in Independent Hospital/Clinic**

k. ☐ **Other (specify)** _____

INSTRUCTION: If you have checked any items in parts 10c through 10k, proceed directly to item 13.

3

EXHIBIT A-1   Questionnaire used in individual survey.

11. Private Industry/Business (See separate sheet for more detailed instructions.) Check in column 1 the industrial product or service category listed at the left that is **the most descriptive** of the activity in your plant or establishment. Check in column 2 the activity **most closely related** to the responsibility of your "research unit." Place **one** check only in **each** of columns 1 and 2. **Do not fill in column 3.**

Industrial Category by Major Product or Activity	1 Major Activity of Your Establishment	2 Activity that Corresponds Most Closely to Your Research Unit's Work	3 FOR OFFICE USE ONLY
Manufacturing			
Aerospace and ordnance _____	_____	_____	_____
Food and kindred products_____	_____	_____	_____
Chemicals			
Drugs and medicine _____	_____	_____	_____
Agricultural chemicals _____	_____	_____	_____
Industrial chemicals _____	_____	_____	_____
All other chemicals_____	_____	_____	_____
Instruments and related products_____	_____	_____	_____
Lumber, wood, paper, and allied products_____	_____	_____	_____
Petroleum refining and extraction _____	_____	_____	_____
Professional and scientific instruments _____	_____	_____	_____
Other manufacturing—if above categories do not apply, see separate sheet of instructions and write in appropriate category. _____	_____	_____	_____
Non-manufacturing			
Agriculture, forestry, and fisheries _____	_____	_____	_____
Mining (including petroleum and natural gas)_____	_____	_____	_____
Commercial laboratory services			
Medical and dental (analytic and diagnostic)_____	_____	_____	_____
Research, development, and testing _____	_____	_____	_____
Other non-manufacturing activities and services. Describe briefly in your own words the activity relevant to column 1 and/or 2 and check columns _____	_____	_____	_____

INSTRUCTION: After completing this item, omit question 12 and proceed directly to item 13.

12. **INSTRUCTION:** Complete item 12 **only** if your **principal employer in calendar 1966 was an academic institution.**

Give the approximate percentage of your **professional salary** that is derived from each source listed below. INCLUDE "summer salary," if any. EXCLUDE royalties, consulting fees, etc. If you are in doubt about the requested information, please consult the appropriate administrative officer.

____ % Institutional budget (EXCLUDE: Federal training grants and Federal institutional grants)
____ % Research grant/contract
               Check (√) all appropriate categories
      ☐ Federal   ☐ Private Foundation
      ☐ Industry   ☐ Other

° ° *Table Continued on Next Page* ° °

4

EXHIBIT A-1   Questionnaire used in individual survey.

____ % Federal training grants
____ % Federal institutional grants
____ % Fellowships (**Check** all appropriate categories)

☐ Federal   ☐ Private Foundation
☐ Industry   ☐ Other

____ % Other (INCLUDE private practice)
100 % Total

13. Estimated time distribution in Calendar 1966 on basis of your **average total** workweek hours for **all** professional activities.

____ % Research, including graduate research training

Check (√) each appropriate item

☐ Basic research
☐ Clinical research
☐ Applied research other than clinical research

____ % Administration
____ % Instruction, including teaching ward rounds, university extension work, etc.
____ % Patient Care
____ % Other (Consulting, editorial, professional society activities, production, testing, sales, promotion, etc.)
100 % Total

LAST PREVIOUS EMPLOYMENT

INSTRUCTION: Complete only if employment was **after** receiving a doctoral-level degree and if you were employed by an employer **different** from that specified in item 6.

14. Location of **previous principal** employer _____

                       State _____   Country _____

15. Total years with **previous principal** employer (whether or not your **position** within that organization changed): _____ yrs.

16. Type of Organization of **previous principal** employer. Check (√) one

☐ Academic   ☐ Municipal Government
☐ Private Industry/Business   ☐ Independent Hospital/Clinic
☐ Self Employed   ☐ Non-profit, Non-government Research Institute/
☐ Federal Contract Research Center   Foundation/Museums
☐ Federal Government   ☐ Other (specify) _____
☐ State Government

INSTRUCTION: Complete Question 17 only if you were employed **within the last 5 years** by an organization **other** than your present principal employer.

17. Responsibilities: Estimate time distribution from the average of your **total** workweek hours for **all** professional activities **during the last year** before you joined your **present principal employer.**

____ % Research, including graduate research training
____ % Administration
____ % Instruction, including teaching ward rounds, university extension work, etc.
____ % Patient care
____ % Other (Consulting, editorial, professional society activities, production, testing, sales, promotion, etc.)
100 % Total

EXHIBIT A-1   Questionnaire used in individual survey.

EDUCATION

INSTRUCTIONS for completion of Questions 18a, 18b, and 18c on facing page.

18a. BACCALAUREATE DEGREE: Check type of degree and specify the year awarded, country of institution and your field of training (See LIST I on separate CODE SHEET).

18b. DOCTORAL DEGREES: List **all** earned doctoral degrees, **starting with your most recent degree,** and specify the year awarded and country of institution. Use LISTS I, II and III from separate CODE SHEET to complete other columns.

18c. POSTDOCTORAL APPOINTMENTS (FOR RESEARCH TRAINING, NOT EMPLOYMENT): If you have had a postdoctoral appointment (see separate sheet for DEFINITION of POSTDOCTORAL APPOINT-MENT) complete the bottom portion of the table, listing **all** such appointments you have held starting with the most recent.

Use the separate CODE SHEET to specify the items requested in the last three columns of the table.

From LIST I, **"FIELD OF TRAINING":** Select **one number** for the **most appropriate field**

LIST II, **"RESEARCH AREA":** Select **one lettered category** for the **most appropriate area**

LIST III, **"RESEARCH MATERIALS/ORGANISMS":** Select **one or two lettered categories** and **one or two numbered categories**

HISTORY AND PROJECTION OF YOUR RESEARCH FIELD INTEREST

19. From the separate CODE SHEET specify the items that best describe your PREVIOUS (1962). CURRENT (1967) and PROJECTED (1970) **RESEARCH AREA** (LIST II) and **RESEARCH MATERIALS/ORGANISMS** (LIST III). Complete the information for "PROJECTED" **only** if you **seriously** are contemplating a change.

	LIST II	LIST III	
	RESEARCH AREA (Select **one** category) R10 - R39	**RESEARCH MATERIALS/ORGANISMS** (Select **one** or **two** lettered categories) M10 - M25	(Select **one** or **two** numbered categories) #10 - #67
PREVIOUS (1962)	R ____	M ____ M ____	# ____ # ____
CURRENT (1967)	R ____	M ____ M ____	# ____ # ____
PROJECTED (1970)	R ____	M ____ M ____	# ____ # ____

6

EXHIBIT A-1   Questionnaire used in individual survey.

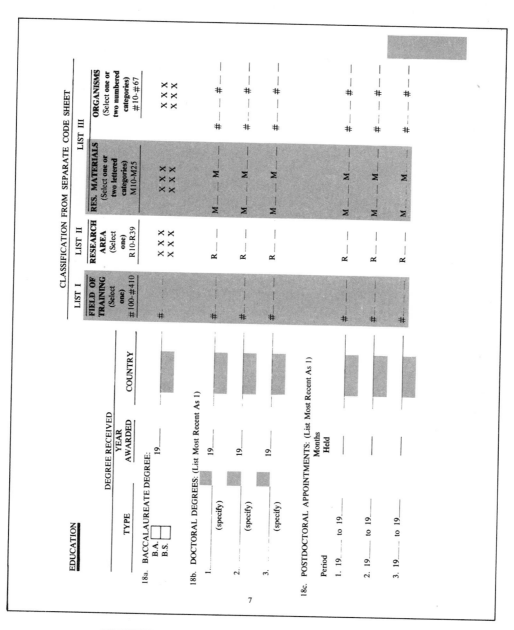

EXHIBIT A-1   Questionnaire used in individual survey.

SCIENTIFIC MEETINGS and PUBLICATIONS

20. Meetings Attended in Calendar 1966
    How many **formal meetings** of national professional societies and national-level symposia did you attend in Calendar 1966? ____
    How many **international** scientific meetings held in a foreign country did you attend in Calendar 1966? ____

21. Publications that Appeared in Calendar 1966
    List for each type of publication shown below the number that appeared in Calendar 1966 on which you were **either sole author** or a **co-author.**

    Number Appeared in
    Calendar 1966   Type of Publication
    ____   Full-length reports (articles) of **original research** that appear in scientific
            journals of greater than local interest
    ____   In-house publications/reports of original research
    ____   Whole books/monographs
    ____   Chapters in books
    ____   Major review articles
    ____   Abstracts of original research
    ____   Other (book reviews, editorials, general interest articles, etc.)

USE OF DIGITAL COMPUTERS IN RESEARCH

22. Are digital computers used in your personal research either by you or technical people directly responsible to you? Yes ☐   No ☐

    INSTRUCTION: If NO, proceed to item 28.

    If YES, please answer items 23 through 27.

23. Digital computers have been used in my research for ____ years.

24. For each of the types of computers listed below I used approximately the following amounts of computer time in 1966:

    Hours
    ____ of Type A: CDC 3600, 6400, 6600; IBM 7094II, 360/65, 67, 75; Univac 1108; GE 645; or equivalent.
    ____ of Type B: B 5000; CDC 3200, 3300; IBM 7044, 7094, 360/44, 50; PDP 6; Univac 1107; SDS Sigma 7; GE 625, 635; or equivalent.
    ____ of Type C: CDC G20, 3100; IBM 7040, 709, 360/40; SDS 930, Sigma 5; Univac 490's; GE 200's; or equivalent.
    ____ of Type D: CDC 160A, 1700; IBM 1400's, 1620, 1130, 1800, 360/30; PDP 1, 4, 5, 7, 8; LINC; SDS 910, 920, Sigma 2; DDP 24, 124, 116; or equivalent.

25. Was the computer you generally used in 1966 employed: (Check (√) **one**)
    ____ Primarily in **your own** research program?
    ____ Primarily by life scientists, including your own group?
    ____ By other than life science groups (e.g., university computation centers)?
    ____ Don't know?

26. I use computers for: (Check (√) **all** appropriate items)
    ☐ Data analysis   ☐ On-line experimental control
    ☐ Information storage and retrieval ☐ Simulation
    ☐ Data acquisition   ☐ Theoretical analysis
    ☐ Other (specify)____

8

EXHIBIT A-1   Questionnaire used in individual survey.

27. Give the approximate percent of your computer use that is paid for by:
    —— % Funds from **your own** research grants and contracts
    —— % Federal funds provided directly to your computation center in support of research in the life sciences
    —— % Non-life sciences funds
    <u>—— % Other</u>
    100 % Total

        Please check here if your source of computer funds is not known to you ☐

SPECIALIZED FACILITIES

28. Check (√) **all** appropriate items in Part A and complete Part B.

A. Item No.	SPECIALIZED FACILITY	Facility **Was** Available. I utilized it in last 12 months	Facility **Is** Available. I plan to utilize it in next 12 months	Facility **Is** Unavailable but I would utilize it if available
1	Field areas			
2	Zoo/aquarium			
3	Taxonomic research collection			
4	Organism-identification service			
5	Tropical terrestrial station			
6	Tropical marine station			
7	Marine station other than tropical			
8	High-altitude laboratory			
9	Low-pressure chambers			
10	High-pressure chambers			
11	Programmed climate-controlled rooms (phytotron, biotron, etc)			
12	Computer center			
13	Primate center			
14	Other specialized animal colony			
15	Germ-free facility			
16	Animal-surgery facility			
17	Animal-quarantine facility			
18	General animal-care facility			
19	Cell- and tissue culture-facility			
20	High-intensity radioactive sources			
21	Center for large-scale production of biological materials			
22	Clinical research ward			
23	Greenhouse			
24	Ships greater than 18 ft. (length: ft.)			
25	Electrically shielded room			
26	Instrument design and/or fabrication facility			

B. Select up to 3 items from those you have checked in the right-hand column (Facility **Is Unavailable** but I would utilize it if available) and rank in order of greatest priority.

Item No. _____   Item No. _____   Item No. _____
      1st Priority   2nd Priority   3rd Priority

9

EXHIBIT A-1   Questionnaire used in individual survey.

MAJOR INSTRUMENTS

29. Check ($\sqrt{}$) **all** appropriate items in Part A and complete Part B.

A. Item No.	MAJOR INSTRUMENTS	Instrument **Was** **Available.** I used it in the last 12 months	Instrument **Is** **Available** and I plan to use it in the next 12 months	Instrument **Is** **Available** but another is needed	Instrument **Is** **Unavailable** but would be used if available
	Acoustic				
1	Acoustic-analysis equipment				
2	Sonar				
3	Ultrasonic probes and sensoring systems				
	Centrifuges				
4	Analytical ultracentrifuge				
5	Preparative ultracentrifuge				
6	Refrigerated centrifuge				
	Chromatography				
7	Amino acid analyzer				
8	Gas chromatograph				
9	Programmed gradient pump				
	Counters				
10	Automatic particle counter				
11	Scintillation counter				
12	Whole-body counter				
	Microscopy				
13	Electron microscope				
14	Electron probe for microscopy				
15	Fluorescence microscope				
16	Metallograph				
17	Microtome-cryostat				
18	Phase-contrast microscope				
	Spectrometers				
19	Electron paramagnetic resonance spectrometer				
20	Mass spectrometer				
21	Nuclear magnetic resonance spectrometer				
	Spectrophotometers/-polari-meters/-fluorimeter				
22	Circular dichroism analyzer				
23	Infrared spectrophotometer				
24	Microspectrophotometer				
25	Spectrofluorimeter				
26	Spectropolarimeter				
27	Ultraviolet spectrophotometer				

10

EXHIBIT A-1   Questionnaire used in individual survey.

Item No.	MAJOR INSTRUMENTS	Instrument Was Available. I used it in the last 12 months	Instrument Is Available and I plan to use it in the next 12 months	Instrument Is Available but another is needed	Instrument Is Unavailable but would be used if available
	X-ray				
28	X-ray crystallographic analysis system				
29	X-ray diagnostic system				
30	X-ray source				
	Miscellaneous				
31	Apparatus for measuring fast chemical reactions				
32	Artificial kidney				
33	Cine and time-motion analysis equipment				
34	Closed-circuit TV				
35	Electrophoresis apparatus (various types)				
36	Intensive-care patient-monitoring system				
37	Infrared CO_2 analyzer				
38	Laser system				
39	Large-scale fermenter				
40	Light-scattering photometer				
41	Microcalorimeter				
42	Multi-channel oscilloscope				
43	Multi-channel recorder				
44	Osmometers				
45	Small specialized computer system (CAT/LINC, etc.)				
46	Stimulus programming and operant conditioning equipment				
47	Telemetering system				

B.

Select up to 3 items from those you have checked in the right-hand column (Instrument **Is Unavailable** but would be used if available) and rank in order of greatest priority.

Item No. _____   Item No. _____   Item No. _____
        1st Priority   2nd Priority   3rd Priority

EXHIBIT A-1   Questionnaire used in individual survey.

FACTORS LIMITING YOUR CURRENT RESEARCH PROGRAM

30. If **full development** of your current research effort is **very seriously** hindered by one or more of the following considerations, check (√) below **all particularly significant** factors.

YES   FACTOR

____   **Space:**
____   Inadequate for personal research
____   Inadequate Specialized Facilities (of type in Question 28)
____   Other (specify) _____

____   **Inadequate Budget** for:
____   Consumable supplies and minor equipment (items costing less than $2,000 apiece)
____   Specialized equipment (items costing more than $2,000 apiece)
____   Professional staff
____   Supporting technicians
____   Clerical/administrative personnel
____   Student fellowships
____   Postdoctoral/Investigator fellowships
____   Computer time
____   Travel
____   Other (specify) _____

____   **Time** limitations, because of:
____   Heavy teaching schedule
____   Service responsibilities (e.g., patient care)
____   Administrative duties

____   **Budgeted and Funded Positions** in the following categories were not filled for lack of available, qualified personnel:
____   Professional staff
____   Supporting technicians
____   Clerical/administrative personnel

____   **Insufficient Personal Training** in:
____   Chemistry
____   Statistics
____   Mathematics
____   Use of Computers
____   Electronics/Engineering
____   Physics
____   Supporting Biological Sciences

____   **Constraints Concerning Choice of Research Problem** and its direction arising from:
____   Conditions of your employment
____   Source of funds which support your research
____   Other (specify) _____

12

EXHIBIT A-1   Questionnaire used in individual survey.

YOUR PERSONAL RESEARCH PROGRAM: PERSONNEL, SPACE, AND LEVEL OF SUPPORT

INSTRUCTION: Questions 31 and 32 are to be completed **only if as of June 1, 1967** you were the **principal investigator** of a research program, regardless of source of your research support or type of your employing organization.

31. Personnel:
Please list below in Part A the **number** of individuals **(excluding yourself)** engaged in **your own personal research program** during either (check one) the academic year 1966-67☐ or the period July 1, 1966-June 30, 1967. ☐

A.

_____ Co-investigator(s)

_____ All other **professional staff** who give **all** their **research time** to **your** research program
These consist of

    _____ Visiting scientists
    _____ Other professional staff
    _____ Research/postdoctoral fellows or associates
        How many of these are:
        _____ Post-Ph.D./D.Sc.   _____ Post-M.D.
        _____ Post-D.D.S.   _____ Post-D.V.M.

_____ All **non-professional staff** (include **only** those who spend 50% or more of their time working on **your** research program)
    _____ Technicians/research assistants
    _____ Clerical/secretarial staff
    _____ Other

_____ All **students**
    _____ Pre-B.A./B.S.   _____ Pre-M.D.
    _____ Pre-Ph.D./D.Sc.   _____ Pre-D.V.M.
    _____ Terminal-M.S.   _____ Additional students of all types
    _____ Other   for the summer period

INSTRUCTION: Complete Part B if you directed the thesis research of Ph.D. candidates at any time between academic years 1964-65 and 1966-67.

B. How many students received Ph.D. degrees with you as major professor during the academic years:

    1964-65 _____   1965-66 _____   1966-67 _____

32. Space:

What is the approximate **indoor working laboratory space,** including space used by any graduate students and postdoctoral appointees working with you, which you actively use for your **own** research program? EXCLUDE **office space** and **common service areas** ("departmental" instrument or animal rooms, etc.).

☐ Less than 500 net sq. ft.   ☐ 1250 to 1499 net sq. ft.
☐ 500 to 749 net sq. ft.   ☐ 1500 to 1749 net sq. ft.
☐ 750 to 999 net sq. ft.   ☐ 1750 to 1999 net sq. ft.
☐ 1000 to 1249 net sq. ft.   ☐ 2,000 to 2,500 net sq. ft.
        ☐ Over 2,500 net sq. ft.

13

EXHIBIT A-1   Questionnaire used in individual survey.

THE LAST TWO QUESTIONS BELOW APPLY ONLY IF AS OF JUNE 1, 1967 YOU WERE PRINCIPALLY EMPLOYED BY AN ACADEMIC INSTITUTION. ALL OTHER RESPONDENTS PLEASE SIGN AND DATE QUESTIONNAIRE.

33. On June 1, 1967 were you the **principal investigator** of a research program?   Yes ☐ No ☐

    INSTRUCTION: If you answered YES to Question 33 above, complete item 34.

34. Level of Research Support and Supporting Organization:

    Complete **all** appropriate items in the table below in accordance with the instructions provided on the separate Definitions and Instructions Sheet.

SUPPORTING ORGANIZATION	Column A June 1, 1967 Number of Grants/Contracts	Column B June 1, 1967 Direct Costs (see instructions)	Column C June 1, 1966 Direct Costs (see instructions)	Column D June 1, 1965 Direct Costs (see instructions)
Your Present Institution _____	X X X	$ _____	$ _____	$ _____
Dept. Agriculture _____	_____	$ _____	$ _____	$ _____
AEC _____	_____	$ _____	$ _____	$ _____
Dept. Defense Air Force _____	_____	$ _____	$ _____	$ _____
Army _____	_____	$ _____	$ _____	$ _____
Navy _____	_____	$ _____	$ _____	$ _____
Other DOD Agencies _____	_____	$ _____	$ _____	$ _____
Dept. Health, Education & Welfare PHS (other than NIH) _____	_____	$ _____	$ _____	$ _____
NIH _____	_____	$ _____	$ _____	$ _____
Other HEW _____	_____	$ _____	$ _____	$ _____
Dept. Interior _____	_____	$ _____	$ _____	$ _____

LEVEL OF RESEARCH SUPPORT

14

EXHIBIT A-1   Questionnaire used in individual survey.

NASA _____   _____ $ _____ $ _____ $ _____

NSF _____   _____ $ _____ $ _____ $ _____

Other Federal (specify)_____   _____ $ _____ $ _____ $ _____

State and Municipal
   Agencies _____   _____ $ _____ $ _____ $ _____

Industry_____   _____ $ _____ $ _____ $ _____

Private
   Foundations _____   _____ $ _____ $ _____ $ _____

Voluntary Societies_____   _____ $ _____ $ _____ $ _____

Other Institutions _____   _____ $ _____ $ _____ $ _____

Other, if more than 5%
   of total (specify) _____   _____ $ _____ $ _____ $ _____

Return your completed questionnaire promptly, using
the enclosed self-addressed, stamped envelope:

MAILING
ADDRESS
  for
COMPLETED
QUESTIONNAIRE

Committee on Research in the Life Sciences
National Academy of Sciences
2101 Constitution Avenue, N.W.
Washington, D. C. 20418

_____   Date _____
     (Signature)

15

EXHIBIT A-1   Questionnaire used in individual survey.

DEFINITIONS, INSTRUCTIONS and CODE LISTS

DEFINITIONS

LIFE SCIENTIST/LIFE SCIENCE FIELD:

For the purpose of the present study by the Committee on Research in the Life Sciences and for this specific questionnaire, the term "LIFE SCIENTIST" is defined to mean any investigator who: (1) has been formally trained in a LIFE SCIENCE field (agricultural, botanical, zoological, biochemical, biophysical, and biomedical sciences), or (2) as the result of the nature of his research work, membership in professional societies, attendance at scientific meetings and self-identification on national surveys of professional manpower, considers himself to be a life scientist even though formally trained in a physical, behavioral, or social science (e.g., anthropology, psychology).

POSTDOCTORAL APPOINTEE:

Temporary (1-3 years) appointments that offer opportunity for continued education and experience in research usually, though not necessarily, under the supervision of a faculty member. INCLUDED are appointments to holders of professional (doctoral) degrees who are pursuing research toward second doctoral degrees, and appointments in government and industrial laboratories which resemble in their character and objectives postdoctoral appointments in universities. EXCLUDED are service or teaching appointments or residencies in which research training is not the primary purpose, and members of faculties of other institutions on sabbatical leave.

CONTINUING OR SENIOR RESEARCH ASSOCIATE:

Permanent members of the departmental research staff who possess a Ph.D. (or equivalent) degree but who are neither formally nor primarily involved in student instruction.

INSTRUCTIONS

QUESTION 11: Private Industry/Business

The industrial category you check in column 1 as "most descriptive of the activities in your plant or establishment" should be selected in terms of the product or service, or groups of products or services, produced or sold by your establishment. In large establishments performing a variety of activities, selection of the major category may be difficult. Make the best selection you can based on your knowledge of the establishment's production, shipments, sales, or research activities. Disregard your company's operations at other locations in making this decision (except for the special case of laboratory establishments which will be discussed subsequently).

The industries listed in the questionnaire are those believed to be employing appreciable numbers of life scientists. If your industrial category is not shown, please write in the appropriate broad category selected from the list that follows (or write in your own description), and check in column 1:

Tobacco manufactures   Fabricated metal products

Textiles, apparel, and leather products   Machinery, except electrical

Rubber products   Electrical equipment and supplies

Stone, clay, and glass products   Transportation equipment

Primary metal industries

Check in column 2 the industry category most closely related to your research unit's work. If necessary, write in a brief descriptive phrase indicative of this activity and check in column 2.

The "research unit" in which you are working may be a division, or department, or some other organizational component within the plant or establishment where you are employed. Or it may be the entire establishment, if you are working in an unattached laboratory.

EXHIBIT A-2   Explanatory insert for individual questionnaire.

If you are employed in an establishment devoted wholly to research, development, and testing or to provision of medical or dental diagnostic or other services, the following points should be considered:

1. If the major portion of the laboratory's services are sold commercially, the appropriate industrial category to be checked in column 1 should be one of the subcategories under the non-manufacturing heading. This holds whether the laboratory is a single unit enterprise or belongs to a multiunit company engaged primarily in manufacturing.

2. Some laboratory establishments will be non-commercial in the sense that they service primarily other units of the parent company or are engaged in research of general company interest. The industry category (column 1) should be one of the categories under the manufacturing heading, determined by the activity of the plant, division, or subsidiary the laboratory serves. If there is no specific tie-in of this type, base the classification on the major activity of the parent company.

QUESTION 33: Level of Research Support and Supporting Organization

In Column A please give for each listed source of support the number of research grants and contracts **active on June 1, 1967,** the funds of which **you had authority to draw upon and for which you had the principal responsibility.** In **no** case should the same funds be reported by more than one investigator—in those cases where equal responsibility for the research funds exists with a colleague, please select **one** person to report the amount.

Note: If part or all of your research also was supported by funds from a larger program grant, coherent area grant, program-project grant, sustaining grant, or other such type of grant, please include in your estimations the contribution(s) of that grant to your research.

In Column B for **each** designated Supporting Organization give **to the nearest thousand dollars** the **total direct costs expended and obligated** from these grants/contracts for the 12-month period **preceding June 1, 1967.** EXCLUDE the items noted below.

In Columns C and D give the same type of **direct cost** dollar information (**expenditures and obligations**) as for Column B, but use the 12 month periods **preceding** the dates of **June 1, 1966** (Column C) and **June 1, 1965** (Column D).

IMPORTANT—In estimating **direct cost** data for Columns B, C and D, please:

EXCLUDE: (1) All funds provided for indirect costs

(2) Funds provided for:
   training grants
   direct fellowships
   construction of buildings
   support of conferences/symposia

INCLUDE: (1) Salaries of **all professional personnel, including your own,** supported from these funds.

(2) Salaries of all research assistants, supporting technical and clerical/secretarial, and other personnel you **employed,** insofar as such salaries were charged against a **research account.**

(3) Stipends for graduate students and other funds for graduate student research training (exclusive of formal training grants).

(4) Funds you received for:
   purchase of equipment
   travel to meetings/field trips/etc.
   specialized and general research supplies
   renovation of your research area
   printing/publication of research results
   visiting scientist(s)
   library services/purchase of books and journals

(5) Other funds not specifically excluded in the list above

EXHIBIT A-2   Explanatory insert for individual questionnaire.

CODE LISTS

LIST I: FIELD OF TRAINING

Use the code number of the **single most appropriate** category.

Agricultural Sciences
100 Agronomy
101 Animal Husbandry
102 Fish and Wildlife
103 Forestry
104 Horticulture
105 Agriculture, Other
Biological Sciences
200 Anatomy
201 Cytology
202 Embryology
203 Physiology, Animal
204 Physiology, Plant
205 Pathology, Plant (see #311 for Animal)
206 Pharmacology
207 Biochemistry
208 Biophysics
209 Biometrics, Biostatistics
210 Botany
212 Ecology
213 Entomology
214 Genetics
215 Hydrobiology
216 Microbiology
217 Nutrition
218 Paleontology
219 Systematics
220 Zoology
221 Bio-Science, Other
Medical Sciences
300 Anesthesiology
301 Dermatology
302 Geriatrics
303 Internal Medicine
304 Endocrinology and Metabolism
305 Gastroenterology

306 Immunology
307 Infectious diseases
308 Obstetrics and Gynecology
309 Ophthalmology
310 Otolaryngology
311 Pathology, Animal and Human
312 Hematology
313 Pediatrics and pediatric specialties
314 Physical medicine and rehabilitation
315 Public health and preventive medicine
316 Psychiatry
317 Neurology
318 Radiology and nuclear medicine
319 General surgery
320 Cardiovascular surgery
321 Neurological surgery
322 Orthopedic surgery
323 Plastic surgery
324 Thoracic surgery
325 Urology
326 Tropical medicine
327 Dentistry and dental specialties
328 Osteopathy
329 Veterinary medicine
Related Areas
400 Anthropology
401 Chemistry
402 Earth Sciences
403 Engineering
404 Mathematics
405 Computer Science
406 Physics
407 Psychology
408 Social Sciences, Other
409 Statistics
410 Other

LIST II: RESEARCH AREA

Use the code number of the **single most appropriate** category.

R 10—Molecular Biology and Biochemistry
R 11—Cellular Biology
R 12—Developmental Biology
R 13—Genetics
R 14—Pharmacology
R 15—Physiology
R 16—Morphology
R 17—Behavioral Biology
R 18—Ecology
R 19—Evolutionary and Systematic Biology
R 20—Nutrition
R 21—Disease Mechanisms

Related Research Fields
R 30—Anthropology
R 31—Chemistry
R 32—Earth Sciences
R 33—Economics
R 34—Engineering
R 35—Mathematics
R 36—Physics·Astronomy
R 37—Psychology
R 38—Sociology
R 39—Other Related Field

EXHIBIT A-2   Explanatory insert for individual questionnaire.

LIST III: RESEARCH MATERIALS/ORGANISMS

Please select **up to two lettered categories** and up to **two numbers.**

Select the **lettered categories** which most appropriately identify your primary **research materials** of your research program.

Select **numbers** which most appropriately identify those **ORGANISMS** you study or employ in your research.

RESEARCH MATERIALS

M10 Mathematical models
M11 Atomic/molecular models
M12 Design/fabrication of apparatus
M13 Development of analytical procedures/ methodology
M14 Molecular systems
M15 Cell fractions/structural components of cells
M16 Disassociated animal or plant cells
M17 Cell cultures
M18 Tissue/tissue slice and organ/organ systems

M19 Artificial organs/limbs/devices
M20 Whole organisms
M21 Populations of organisms
M22 Ecosystem studies
M23 Comparative studies **within** a **single** phylum or plant division
M24 Comparative studies across **two or more** phyla or plant divisions
M25 None of above

ORGANISMS STUDIED or EMPLOYED

10—None
11—Occur in 3 or more phyla (If you select this item do **not** select a second number.)
12—Occur in 3 or more plant divisions (Do **not** select a second number.)
13—Virus
14— Bacteriophage
15— Animal
16— Plant
17—Bacteria
18—Actinomycetes, Mycoplasma and other Bacteria-like Organisms
19—Plankton
20—Protozoa
21—Algae
22—Fungi
23—Non-Vascular Green Plants other than Algae
24—Vascular Non-Flowering Plants

25—Seed Plants
26— Forest Species
27— Horticultural and field crops
28—Porifera
29—Coelenterata
30—Platyhelminthes
31—Nematoda
32—Rotifera
33—Bryozoa
34—Mollusca
35— Commercial
36— Other
37—Annelida
38—Arthropoda
39— Arachnida
40— Crustacea
41— Insecta
42— Other
43—Echinodermata
44—Tunicata

45—Vertebrata
46— Pisces
47— Commercial
48— Other
49— Amphibia
50— Reptilia
51— Aves
52—   Domestic
53—   Wild
54— Mammalia
55— Common Lab. Rodents
56—   Other Rodents
57— Carnivores
58—   Domestic
59—   Wild
60— Ungulates
61—   Domestic
62—   Wild
63—   Small primates (incl. rhesus monkeys)
64—   Large primates (exclude man)
65—   Man
66—   Other Mammals
67—Other Phylum/Division

EXHIBIT A-2   Explanatory insert for individual questionnaire.

APPENDIX B

METHODOLOGY: SURVEY OF ACADEMIC LIFE SCIENCE DEPARTMENTS

SOURCE OF DEPARTMENTAL MAILING LIST

Between July 1967 and November 1968, 2,277 identifiable departments, located in universities or health-professional schools received the question-naire entitled "Survey of the Life Sciences" (Exhibit B-1). Of these, 1,340 responded and 1,256 met the criteria for inclusion. Irrespective of whether the department had a doctoral program, it was included in the survey if its parent institution had granted one or more doctorates in any life science area.

Selection of departments was based on their titles as obtained from three sources:

The American Institute of Biological Sciences: listing of biology de-partments.

Office of Scientific Personnel, National Research Council.

Directory of American Association of Medical Colleges*: list of pre-clinical and clinical departments of the 87 functioning U.S. medical schools.

POPULATION SELECTION

Definition of a Department

Major divisions within clinical departments, or divisions of clinical departments in teaching hospitals, were not considered separately. Information pertaining to such divisions was requested from the chairmen of their parent departments. Similarly, agricultural field stations were not surveyed as separate departments. A summary of the departments surveyed is shown in Table B-1. By definition, a functioning department reported one or more full-time faculty members. Interdepartmental groups were rejected as both their faculty and their students were members of other departments.

Exclusion Criteria

To minimize overlap with the Academy's behavioral and social sciences survey,† responses from 35 psychiatry departments and 5 psychiatry and neurology departments were excluded. A total of 84 departmental returns were excluded for the various reasons shown in Table B-2. Valid responses were received from 1,256 departments.

DATA ANALYSIS

Two parallel series of correlations were carried out on the resultant list of valid responses. In the first series departments were grouped by their school

---

* Directory of Administrative Staff, Department Chairmen and Individual Members in Medical Schools of the United States and Canada, Association of American Medical Colleges, Evanston, Illinois, 1967.
† The Behavioral and Social Sciences: Outlook and Needs, A Report by The Behavioral and Social Sciences Survey Committee under the auspices of The Committee on Science and Public Policy, National Academy of Sciences; The Committee on Problems and Policy, Social Science Research Council. National Academy of Sciences, Washington, D.C., 1969.

TABLE B-1   Total Departments Surveyed in Each Type of School

| | NUMBER OF QUESTIONNAIRES | | |
SCHOOL AFFILIATION	Mailed	Returned Valid	% Returned Valid
TOTAL, ALL DEPARTMENTS[a]	**2,277**	**1,256**	*55.2*
Subtotal, All Departments Except Clinical Medical Departments	**1,450**	**924**	*63.7*
Departments in Schools of:			
Agriculture	384	252	*65.6*
Arts and Sciences	355	236	*66.5*
Dentistry	8	5	*62.5*
Engineering	4	4	*100.0*
Forestry	21	15	*71.4*
Graduate Studies	21	6	*28.6*
Pharmacy	17	7	*41.2*
Public Health	5	5	*100.0*
Veterinary Medicine	57	32	*56.1*
Medical Science Subtotal	**1,405**	**694**	*49.4*
Preclinical	578	362	*62.6*
Clinical	827	332	*40.1*

[a] Only departments from universities or professional schools that had granted a doctorate in a biological field were included in this survey.

affiliation, that is, medical schools, schools of arts and sciences, and so on. Where multiple schools were indicated, the single school of "highest precedence" was chosen. For all departments where the graduate school was one of two schools listed, the other type of school was given precedence, hence the number of graduate schools as represented in this survey appears low. If both the school of agriculture and the school of arts and sciences were checked, precedence was determined on an *ad hoc* basis in light of knowledge of the particular departments involved. Table B-3 gives precedence used.

Because of its large size, the Division of Biological Sciences, Cornell University, Ithaca, New York, deserves mention. A single questionnaire covering the activity of all seven of the major sections of this division was submitted. This aggregate return was listed as a private school of arts and

TABLE B-2   Departmental Responses Excluded from Data Analysis

NUMBER EXCLUDED	DEPARTMENTAL NAME	REASON FOR EXCLUSION
84	**TOTAL**	
	GROUP 1	
35	Psychiatry	Overlap with
5	Psychiatry and Neurology	Behavioral and Social
2	Psychology	Sciences Survey [a]
	GROUP 2	
6	Agricultural Chemistry and Soils	
1	Agricultural Engineering	
1	Agricultural Industries	
2	Agronomy and Soil Science	
1	Behavioral Sciences	
1	Biological and Agricultural Engineering	
1	Biomathematics	
1	Biometry	
1	Biostatistics	
1	Chemistry	
1	Environmental Sciences and Engineering	Not
5	Food Science	Considered
1	Food Science and Biochemistry	a Life Science
3	Food Science and Technology	Department
1	Food Technology	
1	Medical Statistics, Epidemiology, and Population Genetics	
3	Medicinal Chemistry	
1	Pharmaceutical Chemistry	
1	Plant, Soil, and Water Science	
1	Soil and Water Science	
2	Soil Science	
1	Soils	
1	Statistics	
	GROUP 3	
1	Division of Science and Mathematics	
1	Institute of Child Behavior and Development	Not an Academic
1	Water Resources Laboratory	Department
1	World Forestry Institute	

[a] *The Behavioral and Social Sciences: Outlook and Needs.* A Report by The Behavioral and Social Sciences Survey Committee, under the auspices of the Committee on Science and Public Policy, National Academy of Sciences; The Committee on Problems and Policy, Social Science Research Council. National Academy of Sciences, Washington, D.C., 1969.

TABLE B-3   Coding Precedence for Departments with Multiple School Affiliation

PRECEDENCE	NONAGRICULTURAL SCHOOLS	AGRICULTURAL SCHOOLS
1	Medical	Veterinary Medicine
2	Dental	Forestry
3	Arts and Sciences	Agriculture
4	Engineering	Graduate
5	Graduate	

TABLE B-4   Groupings of Departmental Names into "Class of Department" Categories

"CLASS OF DEPARTMENT" GROUPING	NUMBER OF DEPARTMENTS
TOTAL	**1,256**
Agricultural Sciences Subtotal	**186**
Animal Husbandry	90
Agronomy and Forestry	96
Biological Sciences Subtotal	**731**
Anatomy	65
Biochemistry and Nutrition	107
Biology and Ecology	111
Biophysics and Biomedical Engineering	16
Botany	52
Genetics	10
Microbiology	87
Pathology	65
Pharmacology	58
Physiology	75
Zoology and Entomology	85
Clinical Medical Sciences Subtotal	**339**

sciences, perhaps making the total response in this category somewhat high.

For final correlations, the following school types were analyzed together: arts and sciences, graduate studies, and engineering. A second category consisted of agriculture and forestry schools. The returns from schools of dentistry, pharmacy, public health, and veterinary medicine were listed under the title "Other Health-Professional Schools."

The second series grouped departments by discipline using departmental titles reported by their chairmen. Table B-4 shows the number of departments in each category. Table B-5 contains a complete listing of departments and the major category to which each was assigned. (All clinical medical departments were analyzed as a single group because they shared many common characteristics and they differed markedly from other life science departments.)

Definitions and Coding Restrictions

The questionnaire insert, Exhibit B-2, defines the following:

Full-time faculty
Instructor
Continuing research associate
Postdoctoral appointee
Graduate student
Potential Ph.D. candidate
Research space (Detailed instructions as to what constitutes research space as reported in Questions 20 and 21)

It should be noted that identical definitions for Continuing Senior Research Associate and Postdoctoral Appointee were employed in this questionnaire and the individual questionnaire. Where possible, all definitions were the same as those used in the National Research Council study on postdoctoral education.*

SPECIAL DEFINITIONS

For convenience and conciseness, two special departmental categories are utilized in certain analyses: "Performer Departments" and "Promiser Departments."

Performer department: Any department that reported one or more potential Ph.D. candidates and that had awarded at least one doctoral degree in either academic year 1964–1965 or 1966–1967.

Promiser department: Any department reporting at least one Ph.D. candidate, no degrees awarded during either academic year 1964–1965 or 1966–1967, and the expectation of awarding at least one degree during academic year 1969–1970.

* *The Invisible University: Postdoctoral Education in the United States,* Report of a Study Conducted under the Auspices of the National Research Council. National Academy of Sciences, Washington, D.C., 1969.

TABLE B-5 Departmental Disciplinary Categories Based on Departmental Titles

NUMBER OF DEPARTMENTS	CATEGORY
90	**TOTAL, ANIMAL HUSBANDRY**
1	Animal Husbandry
1	Vivarial Science and Research
1	Animal Diseases
2	Animal Industry
1	Large Animal Medicine
1	Small Animal Medicine and Surgery
2	Dairy and Food Industry
13	Dairy Science
1	Animal and Dairy Science
1	Poultry Husbandry
19	Poultry Science
1	Avian Diseases
9	Veterinary Science
3	Veterinary Anatomy
1	Veterinary and Animal Science
2	Veterinary Bacteriology
1	Veterinary Physiology and Pharmacology
1	Veterinary Public Health
1	Veterinary Medicine and Animal Physiology
3	Veterinary Pathology
2	Veterinary Clinical Medicine and Surgery
1	Veterinary Microbiology, Pathology, and Public Health
1	Veterinary Physiology
1	Veterinary Microbiology
2	Animal Pathology
17	Animal Science
1	Animal Range and Wildlife Science
96	**TOTAL, AGRONOMY AND FORESTRY**
1	Agriculture Science
2	Agriculture
23	Agronomy
1	Agronomy and Plant Genetics
1	Agronomy and Genetics
1	Farm Crops
2	Crop Science
1	Grain Science and Industry
1	Plant Breeding
1	Seed Investigation
1	Plant Industry
1	Conservation

TABLE B-5 Continued

NUMBER OF DEPARTMENTS	CATEGORY
1	Forest Chemistry
1	Forest and Wood Science(s)
14	Forestry and Conservation
1	Forestry and Range Management
1	Forest Resources
3	Forestry (School) and Forest Resources (School)
1	Harvard Forest
1	Resource Development
1	Silviculture
1	Watershed Management
1	Wood Technology and Forest Chemistry
22	Horticultural Science
1	Horticulture and Forestry
2	Floriculture and Ornamental Horticulture
2	Pomology
1	Park Administration, Horticulture, and Entomology
4	Vegetable Crops
1	Viticulture and Enology (Enology)
1	Soils and Plant Nutrition
65	**TOTAL, ANATOMY**
1	Biological Structure
61	Anatomy
1	Anatomy and Physiology
2	Anatomy and Cell Biology
107	**TOTAL, BIOCHEMISTRY AND NUTRITION**
1	Agricultural Biochemistry and Soils
3	Agricultural Biochemistry
1	Comparative Biochemistry and Physiology
3	Physiological Chemistry
1	Experiment Station Biochemistry
4	Biochemistry and Biophysics
1	Biochemistry and Microbiology
72	Biochemistry
3	Molecular Biology
1	Molecular and Cellular Biology
4	Nutrition and Metabolism
1	Nutritional Sciences
5	Nutrition and Food Science
1	Institute of Molecular Biology
1	Molecular and Genetic Biology
1	Pharmaceutical Biochemistry
4	Biological Chemistry

TABLE B-5 Continued

NUMBER OF DEPARTMENTS	CATEGORY
111	**TOTAL, BIOLOGY AND ECOLOGY**
26	Biological Science
1	Arctic Biology
1	Biological and Medical Science
1	Biological Science (Division)
1	Cellular Biology
1	Organismic Biology
1	Psychobiology
1	Marine Biology and Oceanography
1	Developmental Biology
1	Agricultural Biology
64	Biology
1	Division of Biology
1	Biological Research Center
1	Center for Theoretical Biology
1	Life Sciences
1	Natural Sciences
4	Wildlife Ecology
1	Paleontology
1	Museum of Paleontology
1	Oceanography
16	**TOTAL, BIOPHYSICS AND BIOMEDICAL ENGINEERING**
1	Radiation Biology
1	Radiation Biology and Biophysics
1	Radiological Science
1	Biophysics and Bioengineering
1	Biophysics and Physical Biochemistry
1	Biophysics and Microbiology
8	Biophysics
1	Physical Biology
1	Bioengineering
52	**TOTAL, BOTANY**
	(Excludes Plant Pathology and Plant Physiology)
1	Forest Botany and Pathology
1	Arboretum (Arnold)
1	Plant Biology
4	Plant Science
1	Plant Research Lab (MSU/AEC)
1	Herbarium
1	Botany and Biology
3	Botany and Microbiology (Bacteriology)
7	Botany and Plant Pathology
32	Botany

TABLE B-5 Continued

NUMBER OF DEPARTMENTS	CATEGORY
10	**TOTAL, GENETICS**
1	Animal Genetics
2	Human Genetics
7	Genetics
87	**TOTAL, MICROBIOLOGY**
7	Bacteriology
1	Bacteriology and Botany
5	Bacteriology and Immunology
1	Medical Microbiology and Immunology
5	Medical Microbiology
1	Microbiology, Medical Technology
1	Microbiology, Pathology, and Public Health
66	Microbiology
65	**TOTAL, PATHOLOGY**
14	Plant Pathology
1	Plant Pathology and Entomology
2	Plant Pathology and Genetics
1	Plant Pathology and Bacteriology
1	Pathobiology
1	Experimental and Anatomic Pathology
44	Pathology
1	Pathology, Parasitology, and Public Health
58	**TOTAL, PHARMACOLOGY**
56	Pharmacology
1	Biochemical Pharmacology
1	Pharmacy College
75	**TOTAL, PHYSIOLOGY**
1	Plant Physiology
1	Neurosciences
5	Physiology and Pharmacology
1	Experimental Endocrinology
67	Physiology
86	**TOTAL, ZOOLOGY AND ENTOMOLOGY**
1	Forest Zoology
26	Entomology
1	Entomology, Fisheries, and Wildlife
1	Entomology and Applied Ecology
1	Entomology and Limnology
1	Entomology and Parasitology
1	Entomology and Economic Zoology

TABLE B-5 Continued

NUMBER OF DEPARTMENTS	CATEGORY
3	Wildlife and Fisheries
2	Wildlife Management
2	Nematology
3	Zoology and Physiology
7	Zoology and Entomology
36	Zoology
339	**TOTAL, CLINICAL MEDICAL SCIENCES**
28	Medicine
11	Internal Medicine
1	Research Medicine
11	Dermatology
29	Obstetrics and Gynecology
21	Ophthalmology
1	Ophthalmology and Otolaryngology
16	Otolaryngology and Maxillo Facial Surgery
38	Pediatrics
8	Physical Medicine and Rehabilitation
3	Rehabilitation Medicine
2	Physical Medicine
3	Epidemiology and Preventive Medicine
12	Preventive Medicine
4	Preventive Medicine and Community (environmental) Health
2	Preventive Medicine and Rehabilitation
8	Community Health
1	Tropical Public Health
1	Environmental Health
1	Environmental Medicine
1	Occupational and Environmental Health
5	Preventive Medicine and Public Health
1	Medical Psychology
15	Neurology
32	Radiology
32	Surgery
21	Anesthesia
6	Neurosurgery
3	Neurological Surgery
2	Orthopedics
7	Orthopedic Surgery
8	Urology
1	Oral Biology
1	Dentistry and Dental Research
1	Experimental Medicine
1	Oncology
1	Clinical Pathology

SURVEY OF THE LIFE SCIENCES

FOR OFFICE USE ONLY
Control No.

Conducted by
The Committee on Research in The Life Sciences
under the Sponsorship of
The Committee on Science and Public Policy (COSPUP)
NATIONAL ACADEMY OF SCIENCES-NATIONAL RESEARCH COUNCIL

This questionnaire is to be completed **only** by the **Chairman** or **Head** (Acting, Rotating or Permanent) of a **College** or **University Academic Department** in the Agricultural, Biological or Medical Sciences.

All data requested herein refer to **Academic Year 1966-67** unless otherwise specified.

This form will be held in confidence; subsequent published information will be statistical in nature, unidentifiable with individual organizations or respondents.

Please **type** or **print** answers.

GENERAL INFORMATION

NAME ____
(Last) (First) (Middle Initial)

1. Name of University ____

2. School or College of:

- ☐ Arts and Sciences
- ☐ Agriculture
- ☐ Dentistry
- ☐ Engineering
- ☐ Forestry
- ☐ Graduate Studies
- ☐ Medicine
- ☐ Pharmacy
- ☐ Public Health
- ☐ Veterinary Medicine
- ☐ Other (specify) ______________________

3. Department of ____

4. Campus ______________________ ______________________
(City) (State)

5. Your University is? (Check (√) one)

Private ☐ State ☐ Municipal ☐ Federal ☐

EXHIBIT B-1 Questionnaire used in departmental survey.

STAFF

6. FULL-TIME FACULTY: (See separate sheet of DEFINITIONS)
   Complete the table below for Academic Year 1966-67.

FACULTY LEVEL	During Academic Year 1966-67		Probable Increase in **1970-71** Over Number Budgeted
	Number of Full-Time Faculty	Additional Positions Budgeted but Unfilled	
Professor			
Associate Professor ____	_____	_____	_____
Assistant Professor ____	_____	_____	_____
Instructor _____	_____	_____	_____
Other _____	_____	_____	_____
TOTAL	_____	_____	_____

7. FULL-TIME NON-FACULTY PERSONNEL: How many of each of the following were in your department during Academic Year 1966-67? (see separate sheet of DEFINITIONS)

   _____ Continuing or Senior Research Associates
   _____ Postdoctoral Appointees (whether "Fellow," "Trainee," "Research Associate," etc.)
   _____ Research Technicians/Professional Animal-Care Personnel
   _____ Business or Laboratory Manager
   _____ Other Supporting Personnel (Laboratory Assistants, Shop and Stockroom Personnel, etc.)
   _____ Clerical/Secretarial/Editorial Staff
          TOTAL

RESEARCH TRAINING ACTIVITIES

POSTDOCTORAL PROGRAM

8. How many **Postdoctoral Appointees** (see separate sheet for DEFINITION) were in your department during Academic Year 1966-67? _____ How many of these were supported from the following sources?

   Fellowships:
     Federal
       USPHS _____ ____
       NSF_____ ____
       Other Federal _____ ____
     State _____ ____
     Industrial _____ ____
     Other Non-Federal ___ ____
     Foreign Sources _____ ____

   Training Grants:
     Federal _____ ____
     Non-Federal _____ ____
   Institutional Grants:
     Federal _____ ____
     Non-Federal _____ ____
   Research Project Funds:
     Federal _____ ____
     Non-Federal _____ ____
   Other _____ ____

9. Of the total number of **Postdoctoral Appointees** who were in your department during Academic Year 1966-67 (See Item 8), how many were **foreign nationals?** _____

       Of these, how many received their **doctoral degrees outside** the United States? _____

10. How many MD's were in **research training** in your department during Academic Year 1966-67 as:

    a. Residents? _____ Typical Period: _____ years _____months

    b. Postdoctoral Appointees? _____ Typical Period: _____ years _____ months

2

EXHIBIT B-1   Questionnaire used in departmental survey.

11. How many **Postdoctoral Appointees** in your department during Academic Year 1966-67 earned their **doctoral** degrees in your own:

University? _____   School/College? _____   Department? _____

PREDOCTORAL PROGRAM: (See separate sheet for DEFINITIONS of **Graduate Students, Potential PhD Candidates,** etc.)

12a. How many **Graduate Students** were in your department during Academic Year 1966-67? _____

b. Of these, how many were **Potential PhD Candidates?** _____

13. How many of your **Potential PhD Candidates** (item 12b) received **stipend support** from the following sources? If the funds supporting any student(s) were from multiple sources, indicate source of **principal stipend** in Column I and **secondary stipend** source in Column II. Count each student **no more than once** in each column.

	Column I Principal Stipend	Column II Secondary Stipend
University-funded Teaching/Research Assistantships _____	_____	_____
Fellowships awarded to the **individual** from:		
Local sources (University, Alumni, etc.) _____	_____	_____
Federal competitive programs (NIH, NSF, etc.) _____	_____	_____
Other national competitive fellowship programs _____	_____	_____
NDEA Awards _____	_____	_____
Departmental Training Grants		
NIH _____	_____	_____
NSF _____	_____	_____
Other Federal _____	_____	_____
Other Non-Federal _____	_____	_____
Institutional-type Grants		
NIH _____	_____	_____
NSF _____	_____	_____
NASA _____	_____	_____
Voluntary Agencies/Foundations _____	_____	_____
Industry _____	_____	_____
Foreign Students supported by **home country** _____	_____	_____
Research Assistantships defrayed by funds appropriated or granted to support **faculty research** _____	_____	_____
Other Sources _____	_____	_____
TOTAL	should be same as 12b	

14. During Academic Year 1966-67 how many of your **Graduate Students** had support for?

11-12 months _____   8-10 months _____

15. How many of your **full-time Graduate Students** with 11-12 months support received **total stipend** support (**Excluding Family Allowances**) from **all** sources at a level:

a. **higher** than that given by NIH, NSF, etc. fellowships (first year, $2,400; intermediate years, $2,600; terminal year, $2,800)? _____

b. **lower** than that given by NIH, NSF, etc. fellowships (first year, $2,400; intermediate years, $2,600; terminal year $2,800)? _____

16a. How many **candidates** for the MD degree were engaged in **research** in your department during Academic Year 1966-67 who were:

(1) **also** candidates for the PhD degree? _____   (2) **not** candidates for the PhD degree? _____

b. How many **holders** of the MD degree **also** were candidates for the PhD degree? _____

3

EXHIBIT B-1   Questionnaire used in departmental survey.

17. With the faculty and space you had during Academic Year 1966-67 (or space available to you by January 1, 1968), could you accommodate **more Potential PhD Candidates,** given adequate funds to support them and their research?   Yes ☐   No ☐

    If YES, how many **additional Potential PhD Candidates?** _____

    If NO, is this due to lack of one or more of the following:

      ☐ Space?   ☐ Faculty?

      ☐ Other (specify)? _____

    _____

18a. How many **more Potential PhD Candidates** do you expect to have in **1970-71** than you had in Academic Year 1966-67? _____

    b. Will this require more **space?**   Yes ☐   No ☐

        If YES: How much research, office and instructional space?   _____ net sq. ft.

           Is such space **under construction** or in an **advanced planning stage?**   Yes ☐   No ☐

    c. Will this require more **full-time faculty?**   Yes ☐   No ☐   If YES, how many? _____

19a. How many **PhD degrees** were awarded in your department during the Academic Years:

    1964-65? _____   **1966-67?** _____

    b. Estimate how many will be awarded during the Academic Year **1969-70.** _____

RESEARCH SPACE

20. What was the approximate **total usable (net) research** space available in your department **as of July 1, 1967?** (see separate sheet for **Special Instructions**)   Answer: _____ net sq. ft.

21. If additional space of the type considered in Question 20 is **planned** for occupancy by **1970,** please cite the **expected increment** in space.   _____ net sq. ft.

DEPARTMENTAL RESEARCH BUDGET

22. How many research grants and research contracts were **active** in your department on **January 1, 1967?**

    Federal _____   Non-Federal _____

23. In **calendar year** 1966-67 (or your latest fiscal year: _____ , 196__ to _____ , 196__)
                                     (month)   (month)

what were the **total direct cost expenditures and obligations for research** projects carried on by the personnel of your department from funds **other** than those provided by the **university** budget? Include under 'Federal' all expenditures from Federal funds which have been provided as **institutional awards, provided** such reported expenditures were under the **control** of yourself or of members of your department. For inter-departmental research programs, include **only** those **expenditures and obligations** that relate to **your** department's contribution to the programs. **Count all funds only once.**

    Federal $ _____   Non-Federal $ _____

24a. How many of your **Full-time Faculty** listed in Question #6 are engaged in research **20% or more** of their time? _____

    b. Of these, what percentage of their **total academic** or **professional salaries** is derived from the **institution's current general funds?** _____ %

4

EXHIBIT B-1   Questionnaire used in departmental survey.

SPECIALIZED FACILITIES

25. For each SPECIALIZED FACILITY that is relevant to the research programs of your department check
(√) **the most appropriate** column in Part A and complete Part B.

Part A   CHECK ONLY ONE COLUMN PER ITEM

Item No. SPECIALIZED FACILITY	Dept. **Now Has or Has Contracted** to Obtain Facility	Dept. **Has Access** to Facility — Which is **Adequate**	But **Another is Needed** by Dept.	**Facility Unavailable** to Dept. But Needed
1 Field areas				
2 Zoo/Aquarium				
3 Taxonomic research collection				
4 Organism-identification service				
5 Tropical terrestrial station				
6 Tropical marine station				
7 Marine station other than tropical				
8 High-altitude laboratory				
9 Low-pressure chambers				
10 High-pressure chambers				
11 Programmed climate-controlled rooms (phytotron, biotron, etc.)				
12 Computer center				
13 Primate center				
14 Other specialized animal colony				
15 Germ-free animal facility				
16 Animal-surgery facility				
17 Animal-quarantine facility				
18 General animal-care facility				
19 Cell- and tissue-culture facility				
20 High-intensity radioactive sources				
21 Center for large-scale production of biological materials				
22 Clinical research ward				
23 Greenhouse				
24 Ships **greater** than 18 ft. (specify length: ____ft.)				
25 Electrically shielded room				
26 Instrument design and/or fabrication facility				

Part B

Select up to 3 items **from those you have checked in the right-hand column** (Facility Unavailable to Dept. But Needed) and rank in order of greatest priority.

Item No. _____   Item No. _____   Item No. _____
              1st Priority   2nd Priority   3rd Priority

5

EXHIBIT B-1   Questionnaire used in departmental survey.

MAJOR INSTRUMENTS

26. For each MAJOR INSTRUMENT that is relevant to the research programs of your department check (√) **the most appropriate** column in Part A and complete Part B.

Part A   CHECK ONLY ONE COLUMN PER ITEM

Item No. MAJOR INSTRUMENT	Dept. **Owns** or **Has Contracted to Obtain** Instrument		Dept. **Has Access** to Instrument		Instrument is Unavailable to Dept. But Needed
	Which is **Adequate**	But **Another is Needed** by Dept.	Which is **Adequate**	But **Another is Needed** by Dept.	
Acoustic					
1 Acoustic-analysis equipment					
2 Sonar					
3 Ultrasonic probes and sensing system					
Centrifuges					
4 Analytical ultracentrifuge					
5 Preparative ultracentrifuge					
6 Refrigerated centrifuge					
Chromatography					
7 Amino acid analyzer					
8 Gas chromatograph					
9 Programmed gradient pump					
Counters					
10 Automatic particle counter					
11 Scintillation counter					
12 Whole-body counter					
X-ray					
13 X-ray crystallographic analysis system					
14 X-ray diagnostic system					
15 X-ray source					
Microscopy					
16 Electron microscope					
17 Electron probe for microscopy					
18 Fluorescence microscope					
19 Metallograph					
20 Microtome-cryostat					
21 Phase-contrast microscope					
Spectrometers					
22 Electron paramagnetic resonance spectrometer					
23 Mass spectrometer					
24 Nuclear magnetic resonance spectrometer					

6

EXHIBIT B-1   Questionnaire used in departmental survey.

CHECK ONLY ONE COLUMN PER ITEM

Item No. MAJOR INSTRUMENT	Dept. **Owns** or **Has Contracted to Obtain** Instrument		Dept. **Has Access** to Instrument		**Instrument is Unavailable** to Dept. But Needed
	Which is **Adequate**	But **Another is Needed** by Dept.	Which is **Adequate**	But **Another is Needed** by Dept.	
Spectrophotometers/-polarimeters/-fluorimeter					
25 Circular dichroism analyzer ___	_____	_____	_____	_____	_____
26 Infrared spectrophotometer ___	_____	_____	_____	_____	_____
27 Microspectrophotometer _____	_____	_____	_____	_____	_____
28 Spectrofluorimeter _____	_____	_____	_____	_____	_____
29 Spectropolarimeter_____	_____	_____	_____	_____	_____
30 Ultraviolet spectrophotometer_	_____	_____	_____	_____	_____
Miscellaneous					
31 Apparatus for measuring fast chemical reactions _____	_____	_____	_____	_____	_____
32 Artificial kidney_____	_____	_____	_____	_____	_____
33 Cine and time-motion analysis equipment _____	_____	_____	_____	_____	_____
34 Closed-circuit TV_____	_____	_____	_____	_____	_____
35 Electrophoresis apparatus (various types) _____	_____	_____	_____	_____	_____
36 Intensive-care patient-monitoring system _____	_____	_____	_____	_____	_____
37 Infrared CO_2 analyzer_____	_____	_____	_____	_____	_____
38 Laser system _____	_____	_____	_____	_____	_____
39 Large-scale fermenter _____	_____	_____	_____	_____	_____
40 Light-scattering photometer _	_____	_____	_____	_____	_____
41 Microcalorimeter_____	_____	_____	_____	_____	_____
42 Multi-channel oscilloscope ___	_____	_____	_____	_____	_____
43 Multi-channel recorder _____	_____	_____	_____	_____	_____
44 Osmometers_____	_____	_____	_____	_____	_____
45 Small specialized computer system (CAT/LINC, etc.)__	_____	_____	_____	_____	_____
46 Stimulus programming and operant conditioning equipment _____	_____	_____	_____	_____	_____
47 Telemetering system _____	_____	_____	_____	_____	_____

Part B

Select up to 3 items **from those you have checked in the right-hand column** (Instrument is Unavailable to Dept. But Needed) and rank in order of greatest priority.

Item No. _____   Item No. _____   Item No. _____
1st Priority   2nd Priority   3rd Priority

7

EXHIBIT B-1   Questionnaire used in departmental survey.

DEPARTMENTAL NEEDS

27. Does your department **very seriously** require **additional funds** for the support of **research and/or research training:**

    (1) To **improve the research endeavor** of your department at the level of **Graduate Student enrollment** you had during Academic Year 1966-67?

    •Yes ☐   No ☐

    (2) To permit a **planned increase** in the **research and/or research training endeavors** of your department?

    Yes ☐   No ☐

    If YES is marked for **either** question, check (√) **all** appropriate items in Part A and complete Part B

Part A

Item No.	FUNDS REQUIRED FOR	Column I To Remedy Current Inadequacies	Column II for Expansion
1	Stipends and tuition for predoctoral students _____	_____	_____
2	Stipends for postdoctoral appointees_____	_____	_____
3	Salaries for additional faculty_____	_____	_____
4	Salaries for additional support personnel _____	_____	_____
5	Specialized research facilities (of the **types** listed in Question 25)_____	_____	_____
6	Major research instruments _____	_____	_____
7	Minor equipment and consumable supplies_____	_____	_____
8	Travel _____	_____	_____
9	Publication costs_____	_____	_____
10	Research funds for specific use of junior faculty_____	_____	_____

Part B

Select up to 3 items **from those you have checked** in Column I and up to 3 items from Column II and rank in order of greatest priority.

Column I

_____   _____   _____
1st Priority   2nd Priority   3rd Priority

Column II

_____   _____   _____
1st Priority   2nd Priority   3rd Priority

*   *   *

PLEASE SIGN, DATE AND RETURN
this questionnaire
to:

MAILING	Committee on Research in the Life Sciences
ADDRESS	National Academy of Sciences
for	2101 Constitution Avenue, N.W.
COMPLETED	Washington, D. C. 20418
QUESTIONNAIRE	

_____   _____ Date
(Signature)

8

EXHIBIT B-1   Questionnaire used in departmental survey.

DEFINITIONS

FULL-TIME FACULTY: Full-time (as defined by your institution) members of your department with an **academic rank** of Instructor or above and whose major responsibilities are concerned with the **academic** programs of the department. INCLUDE: **Yourself** and all members of the continuing faculty, including those who were on leave but who are expected to return. Include those faculty who held joint appointments and who received the **major** portion of their professional salary from your department. EXCLUDE: Professors Emeritus, Senior or Continuing Research Associates and Voluntary Unpaid Staff.

INSTRUCTOR: The lowest academic rank from which, following the usual promotional policies of your institution, tenured faculty status may be acquired.

CONTINUING OR SENIOR RESEARCH ASSOCIATE: Members of the continuing departmental research staff, on **indefinite appointment** who possess a doctoral degree, who are **neither** formally nor primarily involved in student instruction, and who are **not** on the academic promotion ladder in your institution.

POSTDOCTORAL APPOINTEE: Temporary (1-3 years) appointments that offer opportunity for continued education and experience in research usually, though not necessarily, under the supervision of a faculty member. INCLUDED are appointments to holders of professional (doctoral) degrees who are pursuing research toward a second doctoral degree. EXCLUDED are service of teaching appointments, internships or residencies in which research training is **not** the primary purpose, and members of faculties of other institutions on sabbatical leave.

GRADUATE STUDENTS: All potential MS degree or doctoral degree candidates who do **not** possess a professional doctoral degree and who spend **50 percent or more** of their time fulfilling graduate degree requirements.

POTENTIAL PhD CANDIDATES: all graduate students as defined above except those students specifically enrolled in or specifying interest in a **terminal** master degree program.

SPECIAL INSTRUCTIONS FOR QUESTION 20

INCLUDE

1) All laboratory space used on a year-round basis by the faculty, postdoctoral fellows and graduate students.
2) Common research facilities (cold rooms, instrument rooms, etc.)
3) Faculty and graduate student offices
4) **Research** museum space

EXCLUDE

1) Space associated with formal teaching (lecture rooms, course laboratories, etc.)
2) Libraries
3) Greenhouses
4) Corridor space and other non-research space
5) **Seasonal** field and marine stations
6) **Storage** space for museum collections

EXHIBIT B-2   Explanatory insert for departmental questionnaire.

PANELS
AND
CONTRIBUTORS

The Materials of Life and Their Transformations

HANS NEURATH, University of Washington School of Medicine, *Chairman*
KONRAD E. BLOCH, Harvard University
ERWIN CHARGAFF, Columbia University College of Physicians and Surgeons
EUGENE A. DAVIDSON, The Pennsylvania State University, Milton S. Hershey
    Medical Center
BERNARD L. HORECKER, Albert Einstein College of Medicine
DANIEL E. KOSHLAND, JR., University of California at Berkeley
HENRY LARDY, University of Wisconsin, Madison

Cellular and Subcellular Structure and Function

DAVID M. PRESCOTT, University of Colorado, Boulder, *Chairman*
BERNARD D. DAVIS, Harvard Medical School
HARRY EAGLE, Albert Einstein College of Medicine
MAURICE GREEN, St. Louis University School of Medicine
GEORGE E. PALADE, The Rockefeller University
HERBERT STERN, University of California at San Diego

Developmental Biology

JAMES D. EBERT, Carnegie Institution of Washington, *Chairman*
ALFRED J. COULOMBRE, National Eye Institute
MAC V. EDDS, JR., Brown University
PAUL B. GREEN, University of Pennsylvania
CLIFFORD GROBSTEIN, University of California at San Diego, Medical School
WILLIAM S. HILLMAN, Brookhaven National Laboratory
CLEMENT L. MARKERT, Yale University
HEINRICH URSPRUNG, Swiss Federal Institute of Technology

Function of Tissues and Organs

HORACE DAVENPORT, The University of Michigan Medical School, Ann Arbor,
    Chairman
ROY O. GREEP, Harvard Medical School
JOHN B. HANSON, University of Illinois, Urbana
TERU HAYASHI, Illinois Institute of Technology
ANTON LANG, Michigan State University
WILLIAM G. VAN der KLOOT, New York University School of Medicine

Anatomical Sciences

DON W. FAWCETT, Harvard Medical School, *Chairman*
DAVID BODIAN, The Johns Hopkins University School of Medicine
MILTON HILDEBRAND, University of California at Davis
ARNOLD LAZAROW, University of Minnesota Medical School, Minneapolis
RONALD SINGER, University of Chicago School of Medicine
W. GORDON WHALEY, The University of Texas at Austin

Structure and Function of the Nervous System

STEPHEN KUFFLER, Harvard Medical School, *Chairman*
ERIC R. KANDEL, New York University School of Medicine and The
    Public Health Research Institute of the City of New York
IRWIN J. KOPIN, National Institute of Mental Health
VERNON B. MOUNTCASTLE, The Johns Hopkins University School of Medicine
WALLE J. H. NAUTA, Massachusetts Institute of Technology
SANFORD L. PALAY, Harvard Medical School

Behavioral Biology

DANIEL S. LEHRMAN, Rutgers—The State University, Newark, *Chairman*
BENSON E. GINSBURG, University of Connecticut, Storrs
DONALD R. GRIFFIN, The Rockefeller University
PETER H. KLOPFER, Duke University
CARL PFAFFMANN, The Rockefeller University
KENNETH D. ROEDER, Tufts University
ELIOT STELLAR, University of Pennsylvania School of Medicine
HANS-LUKAS TEUBER, Massachusetts Institute of Technology

Ecology

ARTHUR D. HASLER, University of Wisconsin, Madison, *Chairman*
GEORGE A. BARTHOLOMEW, University of California at Los Angeles
JOHN E. CANTLON, Michigan State University
LAMONT C. COLE, Cornell University
EDWARD S. DEEVEY, JR., Dalhousie University
DAVID M. GATES, Missouri Botanical Garden
RICHARD S. MILLER, Yale University
FREDERICK E. SMITH, Harvard University
KENNETH E. F. WATT, University of California at Davis

Evolutionary Mechanisms and Population Biology

JAMES F. CROW, University of Wisconsin, Madison, *Chairman*
RICHARD C. LEWONTIN, The University of Chicago
G. LEDYARD STEBBINS, University of California at Davis

The Diversity of Life

ERNST MAYR, Harvard University, *Chairman*
RICHARD D. ALEXANDER, The University of Michigan, Ann Arbor
W. FRANK BLAIR, The University of Texas at Austin
PAUL ILLG, University of Washington

BOBB SCHAEFFER, The American Museum of Natural History
WILLIAM C. STEERE, The New York Botanical Garden

Biological Science and the Production of Food and Fiber

STERLING B. HENDRICKS, U.S. Department of Agriculture, *Chairman*
R. W. ALLARD, University of California at Davis
F. N. ANDREWS, Purdue University, Lafayette
NYLE C. BRADY, Cornell University
THEODORE C. BYERLY, U.S. Department of Agriculture
KARL MARAMOROSCH, Boyce Thompson Institute for Plant Research
JOHN L. MCHUGH, National Science Foundation
WILLIAM T. S. THORP, University of Minnesota, St. Paul

Biological Science and the Advancement of Medicine

† JOHN B. HICKAM, Indiana University, Indianapolis, *Chairman*
KARL H. BEYER, Merck Sharpe and Dohme Research Laboratories
RICHARD O. BURNS, Duke University School of Medicine
ALBERT H. COONS, Harvard Medical School
BEN EISEMAN, University of Colorado Medical School, Denver
MARTIN GOLDBERG, University of Pennsylvania School of Medicine
DAVID A. HAMBURG, Stanford University School of Medicine
NORMAN KRETCHMER, Stanford University School of Medicine
IRVING M. LONDON, Albert Einstein College of Medicine
KENNETH S. MCCARTY, Duke University School of Medicine
MACLYN MCCARTY, The Rockefeller University
CHARLES R. PARK, Vanderbilt University School of Medicine
OSCAR D. RATNOFF, Case Western Reserve University School of Medicine
MADISON S. SPACH, Duke University School of Medicine
DELFORD STICKEL, Duke University School of Medicine
LEWIS THOMAS, Yale University School of Medicine
MAXWELL M. WINTROBE, University of Utah College of Medicine
JAMES B. WYNGAARDEN, Duke University School of Medicine

Human Development and Changes With Time

ALBERT I. LANSING, University of Pittsburgh School of Medicine, *Chairman*
ROBERT A. ALDRICH, University of Washington School of Medicine
ALBERT DAMON, Harvard University
LEONARD HAYFLICK, Stanford University School of Medicine
HEINZ HERRMANN, University of Connecticut, Storrs
HOWARD C. TAYLOR, JR., Columbia University College of Physicians and Surgeons
† Deceased.

Biology and Industrial Technology

ERNEST H. VOLWILER, Abbott Laboratories, *Chairman*
ARNOLD O. BECKMAN, Beckman Instruments, Inc.
OTTO K. BEHRENS, Eli Lilly and Company
ROBERT D. COGHILL, Stanford Research Institute
KARL FOLKERS, The University of Texas at Austin
CARL H. KRIEGER, Campbell Institute for Food Research
EMIL M. MRAK, University of California at Davis
J. R. PORTER, University of Iowa College of Medicine
W. A. SKINNER, Stanford Research Institute
JOHN A. ZAPP, JR., E. I. du Pont de Nemours and Company, Inc.

Biology and Education

DONALD KENNEDY, Stanford University, *Chairman*
DONALD H. BUCKLIN, University of Wisconsin, Madison
THOMAS EISNER, Cornell University
GARRETT HARDIN, University of California at Santa Barbara
COLIN S. PITTENDRIGH, Stanford University
HOWARD SCHNEIDERMAN, University of California at Irvine

Biology and Renewable Resources

DAVID PIMENTEL, Cornell University, *Chairman*
JOHN L. BUCKLEY, Office of Science and Technology, Executive
    Office of the President
W. T. EDMONDSON, University of Washington
JUSTIN W. LEONARD, The University of Michigan, Ann Arbor
GEORGE SPRUGEL, JR., Illinois Natural History Survey

Biology and the Future of Man

CURT STERN, University of California at Berkeley, *Chairman*
THEODOSIUS DOBZHANSKY, The Rockefeller University
RENÉ J. DUBOS, The Rockefeller University
DAVID R. GODDARD, University of Pennsylvania
JOSHUA LEDERBERG, Stanford University School of Medicine
JAMES V. NEEL, The University of Michigan Medical School, Ann Arbor
FRANK W. NOTESTEIN, The Population Council
ROGER REVELLE, Harvard Center for Population Studies

The Role of Computers in the Life Sciences

ALLEN NEWELL, Carnegie-Mellon University, *Chairman*
G. OCTO BARNETT, Harvard Medical School
JEROME R. COX, JR., Washington University School of Medicine
MAX V. MATHEWS, Bell Telephone Laboratories
BRUCE D. WAXMAN, Public Health Service, U.S. Department of Health,
    Education, and Welfare

2

Environmental Health

NORTON NELSON, New York University School of Medicine, *Chairman*
ROY E. ALBERT, New York University School of Medicine
C. O. CHICHESTER, University of California at Davis
THEODORE F. HATCH, University of Pittsburgh
HAROLD C. HODGE, University of Rochester School of Medicine
J. CARRELL MORRIS, Harvard University
JAMES L. WHITTENBERGER, Harvard University
JOHN A. ZAPP, JR., E. I. du Pont de Nemours and Company, Inc.

Biology and National Defense

H. ORIN HALVORSON, University of Minnesota, St. Paul, *Chairman*
STEPHEN ALDRICH, Central Intelligence Agency
IRA L. BALDWIN, University of Wisconsin, Madison
JOHN H. DINGLE, Case Western Reserve University School of Medicine
DAVID E. GOLDMAN, The Medical College of Pennsylvania
RILEY D. HOUSEWRIGHT, United States Army

INDIVIDUAL CONTRIBUTORS TO LIFE SCIENCES STUDY

J. RALPH AUDY, University of California at San Francisco
JOHN E. BARDACH, The University of Michigan, Ann Arbor
CHARLES F. COOPER, The University of Michigan, Ann Arbor
JARED J. DAVIS, U.S. Atomic Energy Commission
MERRILL EISENBUD, New York University School of Medicine
EDWARD V. EVARTS, National Institute of Mental Health
STEPHEN GLICKMAN, University of California at Berkeley
LESTER GOODMAN, National Institutes of Health
JAMES DANIEL HARDY, Yale University Medical School
WALTER S. HOPKINS, U.S. Department of Agriculture
ROBERT F. INGER, Field Museum of Natural History
EDWARD F. KNIPLING, U.S. Department of Agriculture
SEYMOUR J. KRESHOVER, National Institute of Dental Research
ROBERT C. LASIEWSKI, University of California at Los Angeles
M. J. LAVOIPIERRE, University of California at Davis
RICHARD E. MARLAND, U.S. Department of Health, Education, and Welfare
JAMES MCCARROLL, University of Washington
RICHARD L. RILEY, The Johns Hopkins University
JOHN E. ROSS, University of Wisconsin, Madison
CURTIS W. SABROSKY, U.S. Department of Agriculture
MICHAEL B. SHIMKIN, University of California at San Diego
WILLIAM ALDEN SPENCER, New York University School of Medicine
JAMES H. STERNER, The University of Texas at Houston
EARL L. STONE, Cornell University
DANIEL Q. THOMPSON, Cornell University
JOHN WALDHAUSEN, The Pennsylvania State University College of Medicine
JAMES G. WILSON, Children's Hospital Research Foundation

† Deceased.